AF708991

WITHDRAWN

BIOREMEDIATION TECHNOLOGIES

HOW TO ORDER THIS BOOK

BY PHONE: 800-233-9936 or 717-291-5609, 8AM–5PM Eastern Time

BY FAX: 717-295-4538

BY MAIL: Order Department
Technomic Publishing Company, Inc.
851 New Holland Avenue, Box 3535
Lancaster, PA 17604, U.S.A.

BY CREDIT CARD: American Express, VISA, MasterCard

BY WWW SITE: http://www.techpub.com

PERMISSION TO PHOTOCOPY–POLICY STATEMENT

Authorization to photocopy items for internal or personal use, or the internal or personal use of specific clients, is granted by Technomic Publishing Co., Inc. provided that the base fee of US $3.00 per copy, plus US $ .25 per page is paid directly to Copyright Clearance Center, 222 Rosewood Drive, Danvers, MA 01923, USA. For those organizations that have been granted a photocopy license by CCC, a separate system of payment has been arranged. The fee code for users of the Transactional Reporting Service is 1-56676/98 $5.00 + $ .25.

BIOREMEDIATION: Principles and Practice

Volume III

Bioremediation Technologies

Edited by

SUBHAS K. SIKDAR
Cincinnati, Ohio

ROBERT L. IRVINE
Department of Civil Engineering and Geological Sciences
University of Notre Dame
Notre Dame, Indiana

TECHNOMIC
PUBLISHING CO., INC.
LANCASTER · BASEL

TD
192.5
.B557323
1998
v.3

Bioremediation Technologies, Volume III
a **TECHNOMIC**® publication

Published in the Western Hemisphere by
Technomic Publishing Company, Inc.
851 New Holland Avenue, Box 3535
Lancaster, Pennsylvania 17604 U.S.A.

Distributed in the Rest of the World by
Technomic Publishing AG
Missionsstrasse 44
CH-4055 Basel, Switzerland

Copyright ©1998 by Technomic Publishing Company, Inc.
All rights reserved

No part of this publication may be reproduced, stored in a retrieval system, or transmitted, in any form or by any means, electronic, mechanical, photocopying, recording, or otherwise, without the prior written permission of the publisher.

Printed in the United States of America
10 9 8 7 6 5 4 3 2 1

Main entry under title:
Bioremediation: Principles and Practice–Bioremediation Technologies, Volume III

A Technomic Publishing Company book
Bibliography: p.
Includes index p. 659

Library of Congress Catalog Card No. 97-61433
ISBN No. 1-56676-561-7

To the memory of Arthur Winston Busch, 1926–1994: Arthur W. Busch served as Professor Irvine's mentor from the moment that they met. His thoughts and ideas created an atmosphere which fostered creative research and development. The Sequencing Batch Reactor (SBR) is a technology that emerged from this environment and a technology which demanded fundamental understandings from science and creative applications from engineers. Art Busch's memory serves as a beacon lighting a path on which challenge will bring us closer to truth.

To Professor Alan D. Randolph, who despite severe physical difficulties imposed by multiple sclerosis still keeps his mind engaged in the pursuit of advancing the science and engineering analysis of particulate processes, in which field he is the most significant pioneer. Subhas K. Sikdar fondly remembers the graduate research experience under Professor Randolph, who still remains an inspiring figure in his (SKS) quest for innovative solutions to engineering problems.

Contents

Preface

This three-volume series, *Bioremediation: Principles and Practice,* will provide a state-of-the-art description of advances in pollution treatment and reduction using biological means; identify and address, at a fundamental level, broad scientific and technological areas that are unique to the subject or theme and that must be understood if advances are to be made; and provide a comprehensive overview of new developments at the regulatory, desk-top, bench-scale, pilot-scale, and full-scale levels. The series will cover all media—air, water, and soil/sediment—and will blend the talents, knowledge, and know-how of academic, industrial, governmental, and international contributors.

Developing this series has been a lengthy process because of the size of the project and because of the standard of excellence that has been set. We wish to thank the Technomic Publishing Company for its support throughout the process.

Glancing through the tables of contents, you will see that our objective for this series was to provide the theoretician and practitioner with an overview of bioremediation that will allow new research programs to be formulated and bioremediation technologies to be improved. This series is intended for the student and consulting engineer and the scientist and industrialist alike. Accordingly, we have addressed the removal of both hazardous and nonhazardous contaminants from the liquid, solid, and gas phase using biological processes. This includes the biological treatment of wastes of municipal and industrial origin; bioremediation of leachates, soils, and sediments; and biofiltration for contaminated gases.

As you read through many of the excellent chapters, you may wonder why some chapters were selected. In fact, one chapter that did not make the publisher's desk is one that dealt with incineration. We wanted a chapter on incineration so that we could better demonstrate, by contrast, how and when bioremediation is the appropriate choice. But as was the case with several other chapters, we simply could not convince the best authors to prepare chapters for what they perceived to be unrelated topics. Because of that, we would like to add a few words here about biological processes.

As we all know, microorganisms will grow if provided the proper conditions. In fact, it is often difficult to stop microorganisms from growing. Consider your soured milk, moldy week-old bread, or slippery shower room floors as examples. In fact, if you put a match to a substance, place your hands over the resulting flame, and keep your hands warm, that substance is very likely biodegradable, even if someone has labeled the substance as hazardous! The best that we can determine, unless the substance is drinking alcohol or rubbing alcohol, is that the compound is likely to be found on one hazardous substance list or another or that someone simply forgot to put it there.

Microorganisms need a carbon and energy source. As just discussed, both hazardous and nonhazardous contaminants satisfy either or both of these requirements. The design engineer must ensure that all of the necessary ingredients are supplied in sufficient quantities during any bioremediation project. Just like the grass on our public golf courses, microorganisms need nutrients such as nitrogen, phosphorus, and trace metals. Aerobic organisms need a source of oxygen. The temperature and pH must be controlled as needed, and sometimes, substances that are toxic to the organism (e.g., heavy metals) must be removed. In general, it is not difficult for the designer to engineer a system that will meet all of these requirements for the biological treatment of contaminated liquids, soil or sediments, and gases.

A basic messsage is that most hazardous and nonhazardous organics can be treated biologically, provided that the proper organism distribution can be established. This is how our authors fit in. Some will tell us how a substance that is hazardous for one group of organisms is a valuable food source for another group. It is only by viewing biological treatment systems as a dynamic population of microbes whose composition is subject to various selection pressures that proper systems can be designed, maintained, and corrected when problems arise. As a result, these chapters emphasize how fundamental principles can be used to develop commonsense approaches to establishing the desired microbial consortium, either in biological reactors designed for the treatment of heavily contaminated leachates or, in soils, using in situ and on-site decontamination techniques or, in gases, using biofilters or biotrickling filters.

Using 20 chapters, Volume I of the series, *Fundamentals and Applications,* begins with the law, the "fundamental force" that drives remediation; continues with some basic principles of mathematics that help us understand the present and predict the future; dwells on soils and the complexity that they add to the overall remediation process; fixes on the microbes and biochemical pathways that give bioremediation its uniqueness; and ends with an academic understanding of practice.

Volume II, *Bioremediation Technology Developments,* uses 20 chapters to lead us through a plethora of subjects ranging from the control of filaments, removal of phosphorus, and the treatment of chemical warfare agents in activated sludge processes to exotic biological systems, including those that use the white rot fungus, halophiles, and higher forms of plants, and fixed-film systems that treat chlorinated organics.

Volume III, *Bioremediation Technologies,* is divided into two sections: "International Bioremediation Agenda" and "Pilot-, Full-Scale, and Commercial Demonstration." In 22 chapters, this volume provides insight into how bioremediation is viewed throughout the world and gives full-scale examples of how it has been applied for the treatment of contaminated liquids, soils, and gases.

All of the chapters in this series have been peer-reviewed. To these reviewers, we owe our deepest gratitude. We would also like to thank the graduate students at the University of Notre Dame, who spent many hours helping select authors (that they would like to see in the book), editing, and organizing.

SUBHAS K. SIKDAR
ROBERT L. IRVINE

Editors

PROJECT EDITORS

Subhas K. Sikdar, Ph.D.
11043 Huntwicke Place
Cincinnati, OH 45241

Robert L. Irvine, Ph.D.
Department of Civil Engineering and Geological Sciences
University of Notre Dame
Notre Dame, IN 46556-0767

EDITORIAL BOARD

Motoyuki Suzuki
Institute of Industrial Science
University of Tokyo
7-22-1, Roppongi, Minato-Ku
Tokyo 106, Japan

Dennis D. Focht
Soils and Environmental Sciences Department
University of California, Riverside
Riverside, CA 92521

Jost O. L. Wendt
Department of Chemical Engineering
University of Arizona
Tucson, AZ 85721

Peter A. Wilderer
Technical University of Munich
Water Quality Management
D-85748 Garching, Germany

Howard Klee, Jr.
Amoco Corporation
200 E. Randolph Drive
Chicago, IL 60601

List of Reviewers

Carolyn Acheson, *USEPA*
Michael D. Aitken, *University of North Carolina*
Robert Arnold, *University of Arizona*
J. K. Bhattacharjee, *Miami University*
Dolloff F. Bishop, *USEPA*
Alec Breen, *University of Cincinnati*
Richard Brenner, *USEPA*
Daniel P. Cassidy, *University of Notre Dame*
Harlal Chaudhuri, *USEPA*
Stave Chiesa, *Santa Clara University*
Robert Chozick, *Foster-Wheeler Environmental Corporation*
Al Cunningham, *Montana State University*
Chamindra Y. Dassanayake, *University of Notre Dame*
Kate Devine, *Devoe Enterprises*
Larry E. Erickson, *Kansas State University*
Dennis Focht, *University of California, Riverside*
Sybren Gerbens, *University of Notre Dame*
John Glaser, *USEPA*
Rakesh Govind, *University of Cincinnati*
Janet Grimsley, *Texas A&M University*
James Heidman, *USEPA*

Michael A. Heitkamp, *Monsanto Company*
Ahmad Hilaly, *Archer Daniel Midlands*
Patrick Hirl, *University of Notre Dame*
Sa V. Ho, *Monsanto Company*
Robert L. Irvine, *University of Notre Dame*
C. Y. Jeng, *Amoco Corporation*
Douglas E. Jerger, *OHM Remediation*
David L. Kaplan, *U.S. Army Natick R&D Center*
Dhinakar Kompala, *University of Colorado*
Radha Krishnan, *IT Corporation*
Suzanne Lesage, *Environment Canada*
Ronald Lewis, *USEPA*
K. Y. Li, *Lamar University*
Bill Liu, *IGT*
Thomas Lupo, *Seyfarth, Shaw, Fairweather & Geraldson*
Ping-Ching Maness, *NREL*
Casey Miler, *University of Notre Dame*
William M. Moe, *University of Notre Dame*
Frank Mondello, *GE R&D Center*
Carlo Montemagno, *Cornell University*
Brij Moudgil, *University of Florida*
Debdas Mukerji, *USEPA*
Kimberly Ogden, *University of Arizona*
John Pellegrino, *NIST*
Jairaj Pothuluri, *FDA*
Vipin Rastogi, *Texas A&M University*
David Reisman, *USEPA*
Diane Saber, *IGT*
Steve Safferman, *University of Dayton*
Michael Shannon, *Envirogen*
Kanij Siddique, *City of Cincinnati*
Subhas Sikdar, *USEPA*
Rajib Sinha, *IT Corporation*
Makram Suidan, *University of Cincinnati*
David Szlag, *USEPA*
Henry Tabak, *USEPA*

Arthur Umble, *University of Notre Dame*
Ronald Unterman, *Envirogen*
Albert Venosa, *USEPA*
Walter Weber, *University of Michigan*
Daniel M. White, *University of Alaska, Fairbanks*

SECTION I

International Bioremediation Agenda

Bioremediation Risk Assessment: Merging Human Health and Ecological Methods

DAVID J. REISMAN, LARRY D. CLAXTON AND JAMES E. HARVEY
Office of Research and Development
U.S. Environmental Protection Agency

INTRODUCTION

During the past 20 years, the U.S. Environmental Protection Agency (EPA) has developed methodologies, guidelines, and documents for assessing the potential for adverse human and environmental effects from the use of chemical, physical, and biological agents. Almost every environmental statute passed by Congress in this era devoted some part to human health risk assessment. In order to comply with the statutes and accompanying regulations, the EPA established many types of guidance to produce integrated health and environmental assessments.

The previous volumes of this book provided information regarding the latest developments in the use of physical, chemical, and biological agents to remediate waste sites. Most types of remediation produce either intermediates or additional components that add to the complexity of evaluating the potential health effects of a contaminated site before, during, and after remediation efforts. The purpose of this chapter is to review the history of risk assessment as it has developed over the last 20 years, to provide brief descriptions of the major works and guidelines, and to review some of the bioremediation risk assessment research and plans. The primary intent of this chapter is to demonstrate how the emerging emphasis on health risk assessment influences research and, likewise, how research influences the health risk assessment process. Particular emphasis will be made to refer the reader to various sources in order that further study can be accomplished. Because risk assessment is dependent on many variables, the chapter will give specific examples of risk assessment research and application of these results in real world situations. Many of the references cited in this chapter form the basis from which risk assessments are cur-

This chapter is a government document and as such, is not subject to copyright.

rently being completed in all areas of the United States. It is beyond the scope of this chapter to describe or to illustrate actual risk assessments.

Biologically-based waste treatment methodologies have experienced remarkable growth in the last decade. Future uses are projected to increase as scientific advances fuel ever-broadening remediation applications (Dean and Kremer, 1992). Successful bioremediation trial experiments have the following components in common: characterization of the contaminated site, examination of the feasibility of enhancing *in situ* microbial populations, development of treatment technologies that target specific toxic chemicals or groups of chemicals, and development of treatment systems appropriate for the wastes, the site, and the treatment sequence. A trial is not complete without a determination of the level of risk reduction resulting from the treatment. The lack of validated scale-up information, the variability of contaminated field sites, and the potential for production of metabolites more toxic than parent chemicals complicate prediction of adverse ecological and human health effects. Studies examining fate and effects of degradative microorganisms, target chemicals, and metabolic by-products provide risk assessment information for environmental managers. Examination and discussion of risks associated with bioremediation technologies may help identify and avert potential adverse ecological and human health effects.

Probably the earliest reference to quantifying health risk assessment can be traced to the Clean Air Act (CAA), Clean Water Act (CWA), and the Toxic Substances Control Act (TSCA) in the 1970s. From these legislative mandates, government (EPA) health assessment documents were written on many different chemicals although there was no guidance on how to develop them. During this process, the EPA promoted both noncancer and cancer methods for evaluating health effects (USEPA, 1976, 1980); these methodologies were used to develop ambient air and water criteria for a variety of chemicals.

In 1983, the National Academy of Sciences (NAS) published *Risk Assessment in the Federal Government: Managing the Process* (National Research Council [NRC], 1983). One of the recommendations of that report was for federal agencies to establish "inference guidelines" for risk assessment. This publication produced one of the earliest approaches for defining and conducting health assessments and termed it the risk assessment paradigm. The "Red Book," as it became known, not only separated risk assessment from risk management, but it also divided the risk evaluation process into four interwoven phases: hazard identification, dose-response assessment, exposure assessment, and risk characterization. Figure 1 illustrates the interrelationship of these processes.

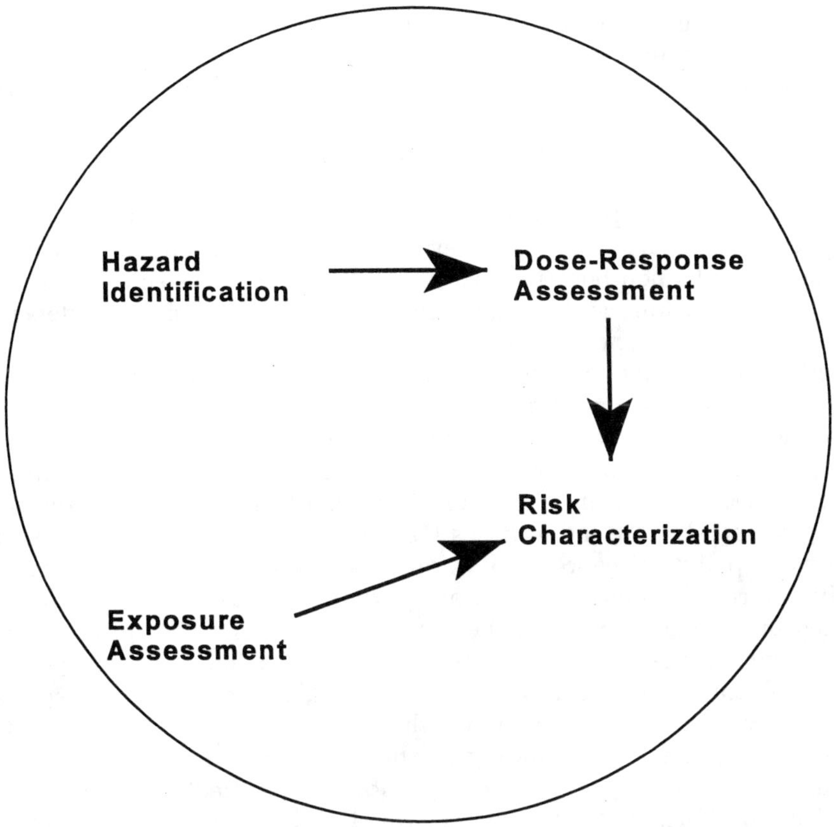

Figure 1. The four processes of the risk assessment paradigm and their interrelationship (NRC, 1983).

Originally, the EPA risk assessment guidelines following the NRC paradigm were developed for the protection of human health; however, in the late 1980s ecological risk assessment incorporated and expanded the use of the paradigm. Although this chapter explores both human and ecological risk assessment, we will place more emphasis on the ecological aspect as it relates to bioremediation, with the understanding that the principles have expanded in only the last few years.

Hazard identification is the first stage and involves the use of exposure and affects data from the laboratory and the field to determine whether the agent of concern can cause a particular adverse health effect (NRC, 1983). While in certain cases this may be a simple cause and effect relationship, more often than not, it is a very complex and detailed stage, especially in situations of multiple chemical or mixed chemical exposures. In this hazard identification stage, endpoint selection is critical to the ecological risk assessment process. Some chemicals that commonly cause severe effects to aquatic organisms, such as chlorine, ammonia, and aluminum, pose no risk or negligible risks to humans in drainage water (Suter, 1993). An ecological assessment endpoint is a formal expression of the environmental values to be protected (Suter, 1989). This step is often very difficult because endpoint selection is dependent upon both the ecosystem and the stress(es) of concern. How we determine and measure changes in the endpoint will depend upon if they can or cannot be quantified. For example, a reduction in population of a fish species subject to chemical X entering the stream from a point source such as a paper plant can be measured. However, the measurement may take 1–2 years in order to separate the chemical's effect from other variables (e.g., temperature, seasonal climate changes) that affect the measurements. An alternative in this example may be to calculate lethal doses (LD_{50}) and to show the relationship between pollutant concentration and species survival, but this acute (short-term) test has many limitations that restrict its use in the evaluation of chronic (long-term) health effects. Suter (1990) provides an excellent discussion of endpoints and their selection.

The hazard identification stage should also incorporate extensive information gathering, including the initial data collection effort to fill gaps in the understanding of the problem and any special site characteristics. Key data needs must be identified to ensure that collection efforts are dedicated to providing both useful and relevant information leading towards risk characterization and eventually feasible remedial alternatives in the risk management process. For example, a CERCLA RI/FS (remedial investigation/feasibility study) provides decision makers with a technical evaluation of the health threats posed at a site, a characterization of the potential routes of exposure, an assessment of remedial alternatives (including their relative advantages and disadvantages), and an analysis of the tradeoffs in selecting one alternative over another (USEPA, 1989a). Historically, the assessment of ecological risks was given less attention than the assessment of human health risks in the RI/FS process (Suter et al., 1995).

The second phase of the risk assessment paradigm pertains to dose-response assessment, which characterizes the relationship between the administered dose and the organism response to the effect of the chemical or stressor. Dose is simply the amount of material that is ingested, injected, or administered to the organism. Several variations of dose have been detailed in the literature with the most strict definition being delivered dose, and not absorbed dose. In ecological studies such as air pollution and its effect on plants, an alternate definition of dose is the product of concentration and time or, if concentration is not constant, the integral over time of concentration.

In the recent documentation of ecological risk assessment, EPA suggests using stress-response rather than dose-response in this stage since ecosystems can be adversely affected by many different types of anthropogenic stresses, not just toxic chemicals (USEPA, 1991a). This type of assessment can only be conducted once the appropriate endpoints have been selected. Ecological risk assessment must begin by defining a limited number of endpoints (Suter, 1990).

Dose-response takes the information from the previous phase to determine a level of daily acceptable dose that is likely to be without an appreciable risk of deleterious health effects during a lifetime. The U.S. Food and Drug Administration (FDA) and the World Health Organization (WHO) have used acceptable daily intake (ADI) terminology and quantification to develop acceptable exposure levels. However, EPA has reworked this approach and developed new, clarifying terminology in keeping with the paradigm, which calls for a clear demarcation between scientific and nonscientific factors in regulatory decisions. EPA now calculates a reference dose (RfD) or reference concentration (RfC) to estimate daily exposure to the human population, including sensitive subpopulations, that is likely to be without an appreciable risk of deleterious effects during a lifetime. Extrapolation and uncertainty are two important factors that affect the dose-response and its value in risk assessment. Many times we must extrapolate our data from laboratory animal data or from short-term to long-term. Because we can only do limited human testing, we must use animal data and extrapolate it to assess possible human effects. Uncertainty can come in several areas: the extrapolation from animal results to human possibilities, uncertainty that the effect in a shorter term will cause the same effect in a longer term, and many other types of uncertainties that must be factored into the risk assessment. Newer methods are unfolding for quantifying nonthreshold risk; however, those methods are not reviewed in this chapter.

The third phase of the NAS paradigm pertains to exposure assessment, which is conducted to measure or to estimate the magnitude, frequency, and/or duration of exposure and to characterize the human populations that are subject to exposure (NRC, 1983). In ecological terms, exposure more frequently refers to the concentration or magnitude of a contaminant or stress in the environment. A detailed exposure assessment requires that a "model" of the receiving environment be constructed that includes all media into which the pollutant is released or is likely to partition and to which the endpoint organisms are significantly exposed (Suter, 1993). While in a laboratory situation, many variables (e.g., temperature, light) can be controlled whereas in the environment, anthropogenic and natural stresses other than chemical inputs need to be reviewed. The ability of the system to recover from stress during and after bioremediation requires consideration. The ultimate effects of stress on an ecosystem depend not only upon the strength and duration of the initial stress, but also the ecosystem's ability to regenerate. Regeneration is influenced in turn by the scale of the stress-induced physical and biological alterations, the potential for habitat reestablishment, and sources of population for recolomigration (USEPA, 1991b).

The last stage of the exposure assessment is the description and quantification of this contact between organisms, their make-up and chemicals. In practical terms, this stage of risk assessment is the point of contact among the environmental chemists and engineers who estimate transport and fate and the biologists and toxicologists who estimate effects (Suter, 1993). Duration, temporal dynamics, and media concentration are three of the most important factors in the exposure scenario, as well as the health effects from exposure.

The final step of the NAS paradigm is risk characterization. The researcher now presents and summarizes the data collected in the earlier three phases and analyzes the data to estimate the incidence of adverse health effects under conditions defined in the exposure assessment. In the ecological risk assessment, the probabilities of adverse effects at estimated exposure levels are not usually determined (USEPA, 1991a). All uncertainties are considered, and assumptions and uncertainties are stated. EPA (1991a) provides a detailed discussion of these uncertainties.

Suter (1993) states that the risk characterization phase completes the integration of exposure and effects assessment, and these effects of exposure to a toxicant can be defined by a surface in a four-dimensional space. The dimensions are

- the concentration of the substance to which the organisms are exposed
- the duration of exposure
- the proportion of the organisms, populations, or communities responding
- the severity of the effect

All four dimensions are measurable in toxicity tests, but most times only one or two are used.

Characterization of risk is a difficult stage. For a variety of reasons, nonhuman organisms are often more highly exposed and more sensitive than humans to environmental contaminants (Suter, 1993). In the area of bioremediation where balancing health and ecological risks is not only part of the process but also characterizes the outcome, factors such as final land management and use may play a significant role in risk management, as well as risk assessment and characterization.

NAS specifically stated that risk assessment should follow the four steps (phases) and be totally scientific, thus separating it from risk management, which is the process of decision-making that attempts to reduce or minimize health risks without inflicting harm or hardship to other societal and economical values. While the risk analysis may be performed well, without proper communication to the risk manager and other interest groups, the decision may be made from the influence of other external factors such as social, legal, economical, or political considerations.

In order to improve on this two-tier process of risk assessment and risk management, the EPA administrator called for the establishment of a toxics integration task force in the summer of 1983 (USEPA, 1984). The recommendations emphasized the need for agency guideline development in six specific areas: carcinogenicity, mutagenicity, reproductive toxicity, substance effects, complex mixtures, and exposure assessment. The task force recommended that a separate entity in EPA address the health risk assessment issues and insure consistency, as well as technical accuracy and competence. The EPA administrator established the Risk Assessment Forum, a body of members from different offices within EPA involved in health risk assessment, to review major risk assessment issues.

The EPA risk assessment guidelines effort began in 1984 with technical working committees composed of scientists from each program office in EPA. The guidelines were developed and subsequently reviewed by scientific experts outside the agency. After revision, the draft guidelines were proposed and published in the *Federal Register* in November 1984 and January 1985 for the public to comment.

EPA RISK ASSESSMENT GUIDELINES OF 1986

On September 24, 1986, EPA published final risk assessment guidelines relating to five areas to promote technical quality and agency-wide consistency in the risk assessment process (51 FR 33992–34054). The guidelines were the product of a large, 2-year EPA effort that also included members of the scientific community outside EPA. Each guideline emphasizes the need to conduct risk assessments on a case-by-case basis, to gather the best available scientific information, and to use scientific judgment to interpret the data and to develop the most scientifically appropriate interpretation to assess risk. Furthermore, the guidelines stress the importance of discussing both the strengths and weaknesses of each assessment by describing uncertainties, assumptions, and limitations.

The Guidelines for Carcinogen Risk Assessment were started by an EPA work group in January 1984, were peer-reviewed, and opened to public comment (49 FR 46294).[1] It was the first revision since 1976 of the Interim Procedures and Guidelines for Health Risk Assessments of Suspected Carcinogens (41 FR 21402). As with the other guidelines, each follows the format proposed by the NAS but is unique in that each recognizes that chemicals that induce benign tumors frequently also induce malignant tumors and that benign tumors may progress to malignant tumors (51 FR 33992). The guidelines are consistent with those of the U.S. Office of Science and Technology, although in many cases, EPA guidelines are more specific.

In the cancer risk assessment guidelines, there are several major issues concerning qualitative and quantitative assessment. The qualitative or hazard identification stage begins with a thorough review of the chemical and any associated biological literature. Studies are evaluated both biologically (health effects) and statistically, and results and discussion are reached concerning the relevant portion to both human health and environmental risk assessment. In the cancer assessment, the studies are reviewed for whether or not they support the conclusion that a chemical or agent is a real or potential carcinogen. This well-researched and evaluated conclusion is termed "weight of evidence" and is stratified according to the amount of evidence for carcinogenicity.

The cancer risk guidelines originally were based upon a weight of evidence approach to determine if an agent is potentially carcinogenic in humans. A detailed discussion of these guidelines can be found in the *Federal Register* citations or in numerous other scientific documents devoted to health risk assessment and will not be presented in this chapter. Furthermore, as will be shown in the next section, these guide-

[1] At the time of this publication, EPA has released updated draft cancer assessment guidelines to be formalized in 1998.

lines are being updated continually by the EPA as new scientific studies further our knowledge in this area.

The second set of guidelines pertained to mutagenicity and were issued not for a detailed discussion of the mechanism, but to assess the risk associated with germ-cell mutations. In general, a mutation occurs when the genetic material in a cell is altered from its predecessor. When mutations occur in somatic cells, disease specific to that cell type may occur. When mutations occur in a germ cell, they may cause problems for future generations if the "mutation" is passed on.

In mutagenicity studies, adverse effects may be manifested at the biochemical, physiological, and cellular levels. In the guidelines, the mutagenicity of physical agents is not addressed. The endpoints of concern include point mutation, which is submicroscopic change in the base sequence of DNA, and structural or numerical chromosome observations. Some mutagens can directly induce alterations in the DNA, while others may interfere with normal DNA synthesis. The guidelines also refer to mutagenicity test systems that can be used to conduct these tests. The mutagenicity guidelines also have a weight of evidence statement with eight categories of decreasing strength of evidence for germ-cell mutations.

The third set of the 1986 guidelines concerns health risk assessment of chemical mixtures, which is probably the most difficult health risk assessment issue to address.[2] Currently, there is no single approach that can be recommended for risk assessments to multiple chemical exposures. Figure 2 shows the flowchart that was developed in the 1986 guidelines along with a 14-step procedure outlined in the guidelines (Table 1 in guidelines). It is important to note that dose-response models used for chemical evaluation are often based on biological mechanisms of toxicity for single compounds and may not be justified to use for complex mixtures. The well-known mixtures (gasoline, PCBs, and diesel exhaust), as well as lesser known (coke oven emissions), have been studied immensely. Another table in the guidelines provides a classification scheme for the quality of the risk assessment of the mixture. The subsections of this tabular classification are of three groups: (1) information on interactions, (2) health effects information, and (3) exposure information.

Other possibilities of examining risk associated with chemical mixtures depend upon whether there are data available on similar mixtures or whether data are available only for the mixture components. In the former instance, determination of "sufficient similarity" must be made

[2] In 1997, EPA had completed an update and revision to the chemical mixtures guidelines.

51 FR-34017

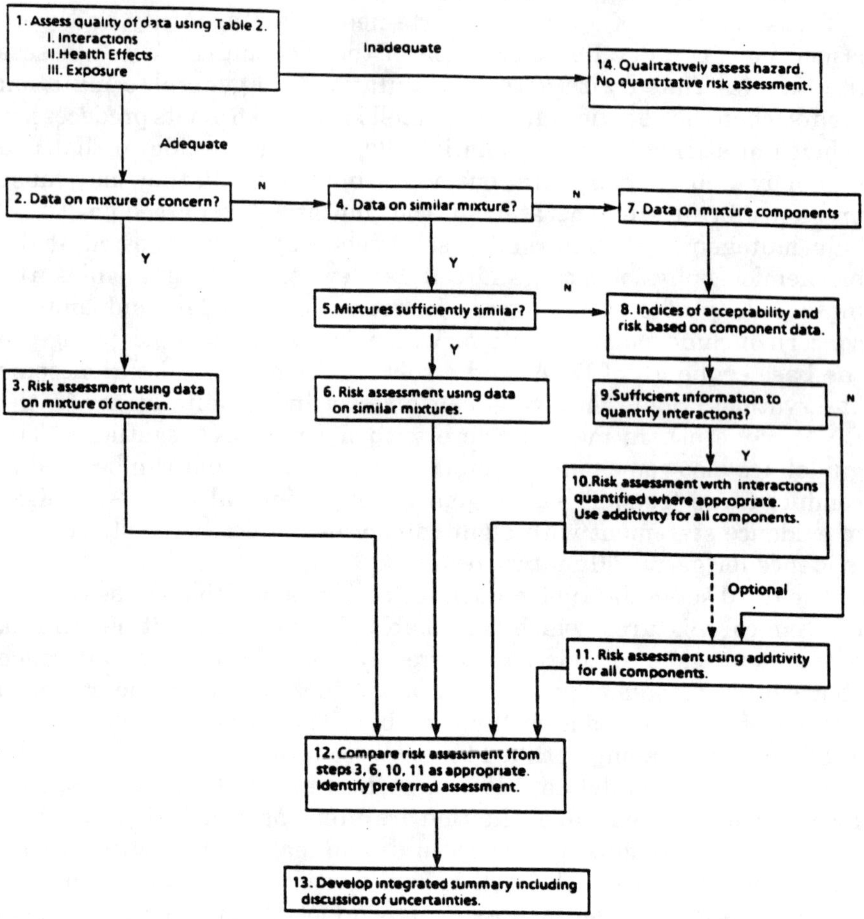

Figure 2. Flowchart of the chemical mixtures risk assessment approach. Note that it may be desirable to conduct all three assessments when possible (i.e., using data on the mixture, a similar mixture, or the components) in order to make the fullest use of the available data. See text for further discussion.

on a case-by-case basis. In the latter case, the risk assessment may be based on the toxic or carcinogenic properties of the components in the mixture. Other possibilities in mixtures risk assessment include synergism or antagonism, as well as other interactions including the mode and mechanism.

The fourth set of guidelines were for the health assessment of suspect developmental toxicants. These guidelines describe the procedures to be used in evaluating potential developmental toxicity from exposure to environmental toxicants. According to Wilson et al. (1977), roughly 20% of human developmental defects are from genetic transmission, 10% are attributable to known environmental factors, and the remainder are from unknown causes. There are many chemicals that have been shown to be developmental toxicants in both animal test systems and humans, with the most widely known being diethylstilbestrol (DES) and thalidomide.

The fifth and last set of the original EPA risk assessment guidelines are for estimating exposures, and these have also been updated (see next section). Exposure assessment is the estimation or determination of the magnitude frequency, duration, and route of exposure. The route of exposure impacts the extent of absorption, and the magnitude is determined by measuring or estimating the amount of the agent at the exchange boundaries (e.g., wings, skin). The guidelines are the first to include a decision path that provides a step-wise process to exposure assessment (Figure 3), as well as a suggested outline for an exposure assessment. A more detailed approach that is used daily in many risk assessments is presented in Part A of the Risk Assessment Guidelines for Superfund (RAGS) (USEPA, 1989a). These guidelines provide a

Table 1. Parameters for ecological effects of bioremediation.

Ecological Effects	Parameters
Community interactions	Population dynamics Population structure Competition Community structure Species abundance Species diversity Metabolic effects Physiological activity Alteration biomass compartment
Ecosystem function	Nutrient cycling Nutrient retention Thermodynamics Respiration/primary production Resiliency/homeostasis Food web complexity and detritus processing

general approach and framework for conducting human and animal exposure assessments for specified pollutants and include assessments based on both measured and validated modeled data. These guidelines stress the need for consistency in exposure assessments with respect to biological, chemical, and physical parameters. Exposure assessments often contain a number of assumptions with different amounts of uncertainty. It is essential that the exposure assesssment be done concurrently with the health effects assessment.

Besides the general introduction and information on the intended purpose of the exposure assessment, other information must be presented. This includes data on the chemical or toxic agents involved, sources that emit to the environment, and the rates of environmental releases, characterization of production and distribution, and especially exposure pathways and environmental fate. In the years since the guidelines were developed, more sophisticated measuring techniques and mathematical models have been developed to provide better data for exposure analyses and estimation of environmental concentrations. Other scientific products and techniques such as geographic information systems (GIS) have further developed integrated exposure analyses, a combination of the environmental concentrations with exposed population descriptions. Additional research into human dosimetry, biological measurements, and other areas of uncertainty in the assessment have furthered the exposure assessment process.

ADDITIONAL EPA RISK ASSESSMENT GUIDANCE

Since 1986, several of the guidelines, as well as recent environmental statutes, have led to research and additional guidelines to fit the specific requirements of a legislative act (Figure 4). One example of this type of effort was the Risk Assessment Guidelines for Superfund (RAGS) (USEPA, 1989). These particular guidelines are very specific in many ways, including exposure pathway development and health risk assessment, and apply to sites undergoing remedial investigation and cleanup under CERCLA (Superfund); however, the RAGS have been used in similar, non-Superfund site investigations, such as those under the Resource Conservation and Recovery Act (RCRA). Some of the research effort has led to other draft guidelines, and efforts by the EPA's Risk Assessment Forum (RAF) have had a profound influence on EPA risk assessment activities. Many of the RAF research activities pertain to specific assessment issues, and can be found in many separate publications available from the EPA or the National Technical Information Service (NTIS).

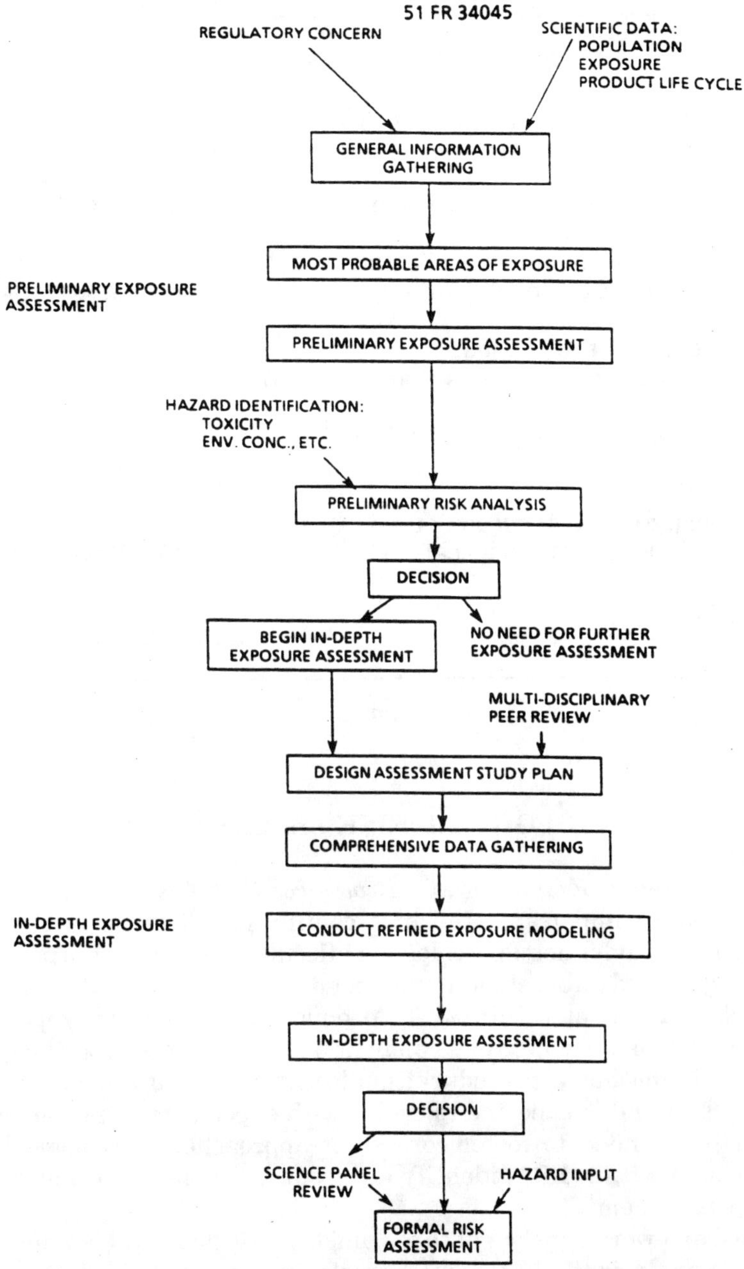

Figure 3. Decision path for exposure assessment.

Updated/New Risk Assessment Guidelines	
Guidelines for Developmental Toxicity Risk Assessment	FR 56 (234)-63798
Guidelines for Exposure Assessment	FR 57 (104)-22888
Guidelines for Reproductive Toxicity Risk Assessment (Draft)	EPA/630/R-96/009
Revisions to the Guidelines for Carcinogen Risk Assessment (Draft)	EPA/600/BP-92-003C
Neurotoxicity Risk Assessment Guidelines (Draft)	(In Press)
Dermal Exposure Assessment: Principles and Applications	EPA/600/8-91-011B
Proposed Guidelines for Ecological Risk Assessment	EPA/630/R-95/002B

Figure 4.

ECOLOGICAL RISK ASSESSMENT

An EPA report *Framework for Ecological Risk Assessment* (USEPA, 1992) was an important step in a continuing effort to develop risk assessment guidelines for ecological effects. Its primary purpose is to offer a simple, flexible structure for conducting and evaluating ecological risk assessment within EPA. Although the framework report will serve as a foundation for development of future subject-specific guidelines, it is neither a procedural guide nor a regulatory requirement within EPA and is expected to evolve with experience. The framework report is intended to foster consistent approaches to ecological risk assessment within EPA, identify key issues, and define terms used in these assessments.

The framework consists of three major phases: (1) problem formulation, (2) analysis, and (3) risk characterization. Problem formulation is a planning and scoping process that establishes the goals, breadth, and focus of the risk assessment based on regulatory policy and available information. The analysis phase is composed of hypothesis testing, experimentation, and information gathering. Risk characterization provides an end product that is a conceptual model that identifies the environmental values to be protected (the assessment endpoints), the data needed, and the analyses to be used for confidence levels for a predicted outcome.

To be effective, ecological risk assessment must also include strong components of data acquisition, verification, and monitoring studies. New data are frequently required to conduct analyses that are performed during the risk assessment. Data from laboratory and small-scale field trials can be used to validate the predictions of a specific risk assessment, as well as to evaluate the usefulness of the principles set forth in the framework. Ecological effects or exposure monitoring can aid in the validation process and suggest additional data, methods, or analyses that could improve future risk assessments (predictability).

The framework is conceptually similar to the approach used for human health risk assessment, but it is distinctive in its emphasis. First, ecological risk assessment can consider effects beyond those on individuals of a single species and may examine a population, community, or ecosystem. Second, there is no single set of ecological values to be protected that can be generally applied. Rather, these values are selected from a number of possibilities based on both scientific and policy considerations.

The framework does not specifically address accidentally or deliberately introduced species, genetically engineered organisms, or organisms used to control horticultural or agricultural pests, all of which are found in bioremediation efforts. While the general principles described in the framework may be helpful in addressing risks associated with these organisms, the capacity of such organisms for reproduction and biological interaction introduces additional considerations. Specific information on the fate, effects, and risks of introduced species, genetically engineered organisms, or pesticidal organisms has been provided by the EPA Office of Research and Development's Field Releases of Biotechnology Agents Research Issue (formerly the Biotechnology Risk Assessment Program) and the Bioremediation Research Issue. Like the framework, these research issues offer a basic structure and starting principles for EPA's bioremediation risk assessment. The process described by the framework provides wide latitude for planning and conducting individual risk assessments in many diverse situations, each based on the common principles discussed in the framework, and site information and treatability information provided by EPA's research issues. This process will help foster a consistent approach for conducting and evaluating risk by identifying key issues, data, and knowledge gaps and by providing operational definitions for bioremediation applications.

THE NEXT GENERATION OF RISK ASSESSMENT

When Congress enacted the Clean Air Act Amendments of 1990, the National Academy of Sciences (NAS) was to review the methods that the EPA uses to estimate toxicological risk. The NAS report (NAS, 1994) entitled *Science and Judgment in Risk Assessment* is a compilation of that review and is divided into three sections: (1) current approaches to risk assessment, (2) strategies for improving risk assessment, and (3) implementation of findings. Most laws, including the Clean Air Act, directed EPA to develop regulations to protect the public health with an ample margin of safety. EPA has previously based its regulatory decisions on quantitative risk assessment, but this act changed Section 112 and placed risk assessment secondary to technology-based regulation through the use of maximum achievable control technology (MACT).

The NAS review of risk assessment found several common issues to each type of assessment:

1. Default options—EPA guidelines used these options (values) in the absence of convincing scientific data and knowledge. These were guidelines to use when there was a paucity of data on which to set a value. NAS, as well as others, found that the EPA values were often too conservative. Thus, EPA's risk-based decisions were often overestimating the potential risk.
2. Data needs—In many cases, EPA used default options or extrapolated from other values when available information was lacking. Thus, risk-based decisions sometimes contained a number of the following factors in this list.
3. Validation—It is essential for scientists to test their models, and the degree of testing needs to be consistent with the latest scientific thinking. NAS reviewed this principal within the report.
4. Uncertainty—In risk assessment, there are always uncertainties. In many cases, EPA uses "uncertainty factors" to correct for these uncertainties. For example, in one calculation when only short-term data are available, EPA uses a factor of 10 to "correct" this value and make it cover long-term exposure. EPA, as well as other scientific organizations, has completed many research projects to determine if the use of this factor overestimates or underestimates the risk.
5. Variability—This issue concerns the extensive variation among individuals and their exposures to toxics. Additionally, genetic factors

may contribute to an individual's susceptibility to various adverse health effects. In the last few years, some studies have been completed on multiple chemical sensitivities whereby the combined exposure to several chemicals may influence different sectors of the population.

6. Aggregation—In combination with the above, health risk assessment needs to address the interactions among pollutants, multiple exposure pathways, and multiple adverse health effects. No longer can pollution be considered as simply as it was in the 1970s.

By addressing all of these factors in each phase of the risk assessment processes, NAS believed EPA could improve the accuracy, precision, comprehensibility, and utility of the risk assessment process. This 600+-page report provides an excellent overview of many of the issues associated with health risk assessment in 1990s, many of which will affect any future bioremediation risk assessment guidelines. The NAS committee did make strong suggestions to use an iterative approach to improve the current default-based approach. Another area cited by NAS as needing improvement is risk communication. While risk assessment is a set of tools, the utilization of them must be efficient enough to provide risk managers with options for reducing risk while improving the environment and protecting the public health.

RISK ASSESSMENT RESEARCH

The achievements in the 2 decades since the EPA's first cancer risk assessment methodology have been highlighted in the previous pages. That the data and information be useful in assessment activities and for decision making are an essential criterion for any health risk assessment research effort. The initial effort is identifying potential health and environmental hazards. Following this identification, the measuring of the extent of the risk should be characterized in a qualitative, comparative, and/or quantitative manner. For example, qualitatively, carcinogens may be deemed more threatening than skin allergens. Comparative assessments evaluate the risks of newly identified hazards against known well-characterized risks. Quantitative risk assessment tries to estimate directly actual numbers for increased mortality and morbidity. The NAS report (NRC, 1994) emphasized the need for an iterative risk assessment process in which research plays a key, formal role.

Although most researchers recognize that well-conducted epidemiologic studies provide the best basis for hazard identification and risk assessment, there are several features related to the design and analysis of epidemiologic studies that frequently limit their usefulness. Epidemiologic studies frequently lack accurate information on exposure (Smith and Rappaport, 1991; Anderson, 1988; Perera, 1987). Equally important is the lack of definitive causal relationships, including factors that limit the usefulness of epidemiologic studies such as statistical power, length of followup, and effect modification (Rothman, 1986; Stayner, 1992). Because of these inherent problems and an unwillingness to wait until human health problems occur to a recognizable degree, risk assessors have turned to nonhuman toxicology methods and studies to identify and quantitate risk.

Today, human health hazards are identified not only through human clinical and epidemiology studies but also through toxicological studies using a wide variety of organisms. Because these toxicological assays generally are designed to be predictive of specific human processes, no one test can identify all potential human toxicants. Also these tests may or may not give good quantitative indication of risk. Identified chemical hazards vary greatly in their potency; therefore, it is important to develop information on relative potency of pollutants and how it relates to human exposure. When using nonhuman systems, potency is generally estimated by generating dose-response (comparing responses of an organism at various doses of the substance) information and extrapolating that information to the human situation. A critical issue in toxicology, therefore, is understanding how to make appropriate extrapolations from the toxicological assay to thc human situation. When making extrapolations to estimate human toxicity, the toxicologist-risk assessor must evaluate items such as

1. The comparative dose in humans and test systems versus mechanisms that operate at different doses
2. Differences between test organisms and humans in the routes and timing of exposures
3. Differences in metabolism and pharmacokinetics
4. the comparative range of sensitivities found in toxicological tests when compared to the human population

In order to be able to use extrapolated effects appropriately, it is necessary to measure and/or estimate the numbers of humans exposed and their exposure levels. The risk assessor must combine presently available hazard information, toxicity dose-response data, and exposure information in order to provide the risk manager with estimation of the risk that a pollutant provides to the human population.

The risk assessor for environmental situations examines conditions with multiple pollutants and often with complex environmental mixtures such as crude oils. In many, if not most, cases, all of the individual toxicants will not be able to be identified. There will be little or no toxicological information on many of the individual substances, and there is the potential for synergistic or antagonistic actions when multiple pollutants exist. Assuming that only binary combinations could provide synergistic reactions, that only 15 compounds would be found at any one environmental site, and that only 0.1% of the combinations are synergistic, one can calculate that there are $N(N-1)/2$ or 105 binary combinations and that one would expect to find at least one synergistic combination. With 100 compounds present at a site, 4950 binary combinations and five synergistic combinations would be present. Over 2800 compounds have been identified in ambient air, and this probably accounts for less than 10% of the actual number (Gradel et al., 1986). If one assumes that human exposure consists of approximately 3000 compounds, there could be approximately 4,500,000 combinations and 4500 synergistic combinations. Substances such as crude oil may contain nearly 10,000 individual compounds, giving nearly 50 million combinations and 50,000 synergistic combinations. In addition to the pollutants that are emitted or placed in the environment, remediation methods (incineration and physical treatments as well as bioremediation) produce additional alterations. These alterations add complexity to the risk assessment process. New components may emerge. Some of these components are compounds that result from environmental transformation processes such as atmospheric oxidation processes or bacterial and fungal metabolism or are products of chemical interaction. Some of the components are the organisms used for the remediation. Any toxins or added organisms, such as engineered or nonindigenous ones, may alter the normal biota and processes, providing altered metabolism and a wider array of metabolites and interactive process. This, in turn, would alter which compounds and biological species would predominate in a particular environmental situation.

One must temper these calculations and observations with the view that, with increasing numbers of compounds, the concentration levels for any single compound or any binary combination should be substantially less, therefore diminishing the role of any one synergistic combination. Also, highly complex ecological systems have many mechanisms for recovery. There is, however, a simple lesson from this discussion; namely, the number of possible binary combinations and possible interactions is so large as to be beyond society's capabilities for testing.

Obviously, researchers in the field of toxicology will never have the resources to fully assess all of the toxic effects for all of the possible combinations of toxic substances to which humans are or could be exposed. Research strategies must be developed, therefore, to identify the more important (i.e., potent) toxicants emitted into the environment, to improve the quantification of the toxicity of environmental substances, to enhance our abilities to model and extrapolate results for the human situation, and to develop methods for examining actual complex environmental situations in a timely and useful manner. Researchers also cannot measure the exposure levels of each individual to all possible toxicants nor the relative sensitivity of all subpopulations to each type of toxicant. The goal of research toxicologists, therefore, is to develop the types of data and information that allow for reasonable approaches to estimating, on a population basis, the toxic effects of environmental substances.

As risk assessment research finds ways to evaluate the health hazards of environmental substances and situations, recognition that there is a very fuzzy line between basic and applied research quickly develops. Highly mechanistic research is needed in order to assess whether or not humans are likely to assimilate, metabolize, and respond to environmental substances in the same manner as the toxicity test being used to evaluate a pollutant. Applied research incorporating actual environmental samples must be done to identify unexpected outcomes and new issues and to assess the utility of more basic research to predict human hazard. Research that incorporates both basic and applied components emerges as the most effective, because research ideas for fundamental, basic research surface from questions that arise out of applied efforts. Likewise, basic research efforts provide the knowledge to understand basic principles needed to do the best possible applied research and to understand the results of applied research.

Figure 5 illustrates the complexity of environmental risk assessment research. Built upon a background of both basic and applied research, a large variety of assays determine the adverse effects of both chemical substances and biologicals. These types of efforts are made with single substances (e.g., a pesticide), artificial mixtures of substances (e.g., pentachlorophenol), complex mixtures (e.g., creosote), and environmental samples. Toxicity assays include subcellular systems, in vitro systems, whole animal systems, and human studies. Toxicological studies may examine potential carcinogenicity, neurotoxicity, immunotoxicity, and reproductive effects, as well as a variety of systemic effects. Within Figure 5, the chemical-biological hazards plane represents the hazards that result from the interaction of pollutants with microbes. Some of

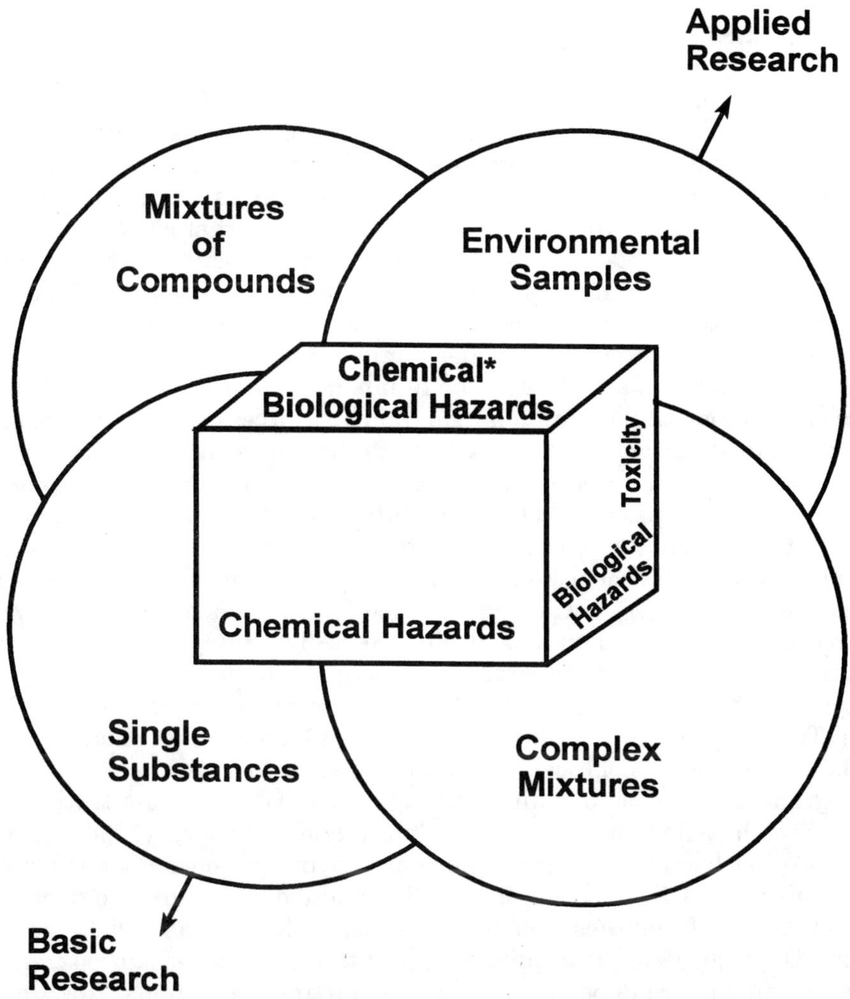

Figure 5. The complexity of environmental risk assessment (see text for further explanation). * Represents hazards that result from the interaction of pollutants with microbes or other biological organisms.

the more sophisticated efforts associated with bioremediation research examine the mechanisms of toxicity that results from the interaction of xenobiotic pollutants and remediation microorganisms.

ASSESSING TOXIC CHEMICALS IN THE ENVIRONMENT

The Committee on Environmental Epidemiology of the National Research Council (1991) recognized that many of the substances emitted into the environment and produced within the environment from anthropogenic materials are unidentified, as well as having unknown toxicity. The committee stated, "There is evidence that NCP's [nonconventional pollutants] are a potentially important source of hazardous exposure. Some preliminary toxicologic studies suggest that NCP's have important biologic properties, environmental persistence, and mobility. In the broadest sense, these unidentified substances represent risk of unknown magnitude." The numbers of compounds and sources of environmental pollutants is astounding. Sources of pollutants range from industrial waste streams to household cleaning chemicals put into the trash, from pesticides used in agriculture to gasoline spilled at service stations. Even a single pollutant placed in an environmental situation can create a complex situation for the risk assessor who must evaluate potential health effects. Pentachlorophenol (PCP) provides a good example. PCP, a widely used biocide, is used as a wood preservative, a fungicide, and an insecticide. Mikesell and Boyd (1988) showed that a consortium of anaerobic microorganisms found in municipal sewage sludge when mixed into PCP-contaminated soil converted PCP to TeCP (2,3,4,5-tetra chlorophenol), TCP (3,4,5-trichlorophenol), and 3,5-dichlorophenol. Mileski et al. (1988) showed that the main PCP degradation product of white rot fungus is TCHD (2,3,5,6-tetrachloro-2,5-cyclohexanediene-1,4-dione). Steiert and Crawford (1986) found that other bacteria generate chlorinated hydroquinones upon the metabolism of PCP. For PCP little is known about the toxicology of its metabolites. From these and other studies (Schenk et al., 1990; Lamar and Dietrich, 1990), it is quite apparent that products produced in the environment are dependent upon the starting compounds, the organisms present, and the environmental conditions. The public health researcher, therefore, must be concerned about the intermediates and by-products, as well as the originally identified toxic chemical.

Complex substances, such as crude oil and creosote, provide a more complicated situation (Claxton et al., 1995). Compounds within such mixtures differ in their volatility, water solubility, ability to diffuse within soils, and ability to be metabolized by indigenous organisms.

Because of these and other factors, chemicals within the mixtures migrate through soils and waters at differing rates; therefore, the concentration of individual compounds varies by location within a site and by time (Weinstein, 1993; Bonaventura et al., 1995a, 1995b; Gaver, 1995). With environmental transformation of the compounds occurring due to the interaction of compounds and to the metabolism of some compounds by microbial, plant, and animal systems, not only do the numbers of compounds increase, but the types (classes) of compounds also increase. It is this highly complex mixture of substances that the risk assessor must evaluate. As in the previous PCP illustration, these processes transform the substances present in the environment and alter the overall toxicity of the contaminated areas. These processes may either increase or decrease toxicity, and the effect may range from dramatic to undetectable.

When one considers cancer and heritable effects, the dimension of time also plays a major role. The time between exposure to the carcinogen that initiates the development of a cancerous cell and the diagnosis of the cancer may be as great as 20 years (Macieira-Coelho, 1986). The time between exposure to a mutagen and the recognition of adverse heritable effects is measured in generations. Although the health researcher is faced with these challenges, a variety of research tools exist for developing an understanding of the hazards and risks emanating from a hazardous compound, source, or site.

Researchers, therefore, must develop methods and strategies for evaluating these complex environmental situations. These newer methods must be scientifically sound and based on appropriate mechanisms, they must be more exhaustive in the numbers and types of toxic substances detected, they must be technically feasible, and they must be affordable. The researcher and risk assessor, therefore, cannot rely upon any singular approach but must use methods associated with epidemiology, analytical chemistry, toxicology testing, and biological analyses.

Intuitively, the most convincing evidence for hazard assessments comes from human studies—clinical studies and epidemiology. As mentioned previously, one of the greatest challenges to the environmental epidemiologist is the identification, understanding, and avoidance of confounding factors. A person diagnosed with cancer may have been exposed in the past to personal (e.g., cigarette smoke), occupational (e.g., coke oven fumes), and environmental (e.g., automotive emissions) pollutants that have genotoxic activity; therefore, one cannot distinguish which source initiated the observed disease and which sources contributed to the progression of the disease. Due to the lack of human exposure information, epidemiology also is not a suitable approach for a newly identified product or pollutant. In spite of these limitations, epidemiologists assimilated the evidence to identify asbestos, vinyl chloride, and cigarette smoking as cancer-causing agents (Sagan, 1984).

Because of the limitations of epidemiology, risk assessments must rely on information coming from nonhuman research and test systems. Toxicity testing typically examines one chemical at a time; therefore, the relationship between exposure dose and outcome (response) is relatively clear. However, the risk assessment power of toxicity test data is limited by our understanding of interspecies differences, extrapolations from the high test doses to the low environmental exposure doses, and a variety of other considerations.

The use of microorganisms introduces another level of complexity into the evaluation of potential health risks. Microbes can metabolize toxicants and other pollutants, thus changing the toxicity profile of an environmental mixture. Not only do microorganisms have the capacity to persist within the environment, but they may actually increase in number and be spread by a variety of mechanisms. Although pathogens are very unlikely to be used in bioremediation efforts, microorganisms can have adverse effects for humans by other mechanisms. The complexities introduced by the use of microorganisms, therefore, must be addressed by public health scientists.

Alternatively, nonhuman biological systems can be used to test actual environmental samples. While diminishing a number of concerns (the interactions of multiple compounds, the time between exposure and effect, and the need to identify and test the toxicity of all the pollutants within the environmental sample), other aspects remain. For example, how does one collect a truly representative environmental sample? There are also concerns about the cost, selecting appropriate assays, and relating dosage of selected environmental samples to human exposure. Biological assays have, however, played an important role in identifying complex mixtures that are potential human toxicants (Claxton and Huisingh, 1979; Claxton and Kohan, 1981; Claxton, 1981; Graham et al., 1984; Houk and Claxton, 1986; Driver et al., 1990; Houk, 1992a, 1992b).

The research scientist and risk assessor are not limited to the described approaches. The most useful information is generally gained from combinations of approaches, techniques, information gathering, and data analysis. For example, analytical chemistry methods enhance the understanding of exposure in epidemiology studies. Analytical chemistry methods can measure not only the level of pollutants in the environment, but also pollutants and their metabolites at critical target sites within human tissues. In prospective epidemiological studies, measuring the dose of a toxicant delivered to a target tissue or cellular component permits more accurate correlations between exposure and effect (Phillips et al., 1988, 1990; Hemminki et al., 1990). For a more thorough description of such biomarkers and their application, see Shugart (1995). Toxicological bioassays can now be used to identify the toxicants within a mixture, thus making the identification of environmental toxicants more efficient and more relevant (Claxton, 1982, 1983).

Hazardous waste sites are identified because of the presence of one or more toxic substances. However, other nontoxic pollutants are generally present at these same sites. A question that has not yet been addressed discernibly is the extent to which remediation efforts, including microbial methods, might convert some of these nontoxic substances into toxic components. A second consideration is that microbes designed for *in situ* remediation are chosen because they are expected to persist within the environment. Unlike chemicals within the environment, some microorganisms may actually survive and thrive. Other bacteria can be maintained in a small, nondetectable niche until the environment is more suitable for growth. From this, it can be concluded that (1) the environmental, bacterial concentrations are in a constant state of flux, (2) the concentration of bacteria to which individuals would be exposed is therefore variable, (3) changing numbers of bacteria influence the rate by which all pollutants are metabolized, and (4) because of the rate of bacterial growth and metabolism of pollutants, the type and levels of toxicants is in a dynamic state.

Even though organisms such as plants sometimes can be used, most bioremediation efforts use microorganisms. Bacteria and fungi produce a wide variety of enzymes that are capable of degrading organic compounds and mineralizing other substances. In some cases, remediation efforts take advantage of naturally occurring organisms. In other cases, indigenous microorganisms that can degrade known pollutants are isolated and enhanced through selection processes (customized) and applied in the remediation process. In still other cases, organisms can be genetically engineered to enhance their survival and/or improve their abilities in the remediation process. In most bioremediation efforts, a consortium of microorganisms, rather than a single isolated strain, is used as the degrader. Although pathogens are unlikely to be used in bioremediation efforts, microorganisms can have other types of adverse effects for humans.

The most obvious concern with microorganisms is whether or not they are pathogenic to humans or other biota. Very few bacterial and fungal species are recognized pathogens (Creager et al., 1990; Holt and Krieg, 1984; Sneath, 1986; Staley, 1986; Williams, 1989). A large number of microbes are found in a symbiotic association with humans, and some of these indigenous flora are pathogens such as *Staphylococcus aureus* (Creager et al., 1990). These indigenous pathogens seldom create problems because they cannot obtain the competitive advantage needed to cause disease. Other organisms, called opportunistic pathogens, must take advantage of special circumstances before they can cause disease. Homeostatic mechanisms that maintain beneficial indigenous microorganisms and prevent the harmful effects of indigenous pathogens and opportunistic pathogens are held in a delicate balance. Therefore, there are some human health reasons to be cautious about the application of large numbers of microorganisms in environmental situations. Also as Isenberg and D'Amato (1991) state, "No hard and fast rule divides the human microbiota into clear-cut categories of harmless, commensal organisms and pathogenic species."

POTENTIAL ADVERSE HUMAN HEALTH EFFECTS FROM BIOREMEDIATION EFFORTS

When protecting human health, researchers must consider several potential adverse effects. Under certain circumstances, nonpathogens may behave as pathogens. For example, the opportunistic pathogen *Legionella pneumophilia* may cause a lethal form of pneumonia when people are exposed to large numbers of aerosolized organisms (Creager et al., 1990). Such experiences engender caution when people are exposed to high concentrations of microorganisms. Because some pathogens are indigenous to humans and because humans are frequently exposed to opportunistic pathogens, exposure to large numbers of nonpathogenic organisms may disturb the normal balance of microorganisms to such an extent that indigenous pathogens are able to have an adverse effect (George et al., 1990; Deitch and Berg, 1987). Disturbance of normal flora may disturb normal gastrointestinal metabolism (George et al., 1989, 1991; Chadwick et al., 1992), which in turn may also cause adverse health effects.

Although not all potential health outcomes can be predicted, methods are needed to identify potentially adverse actions of microorganisms. The normal progression for determining potential human adverse health effects is to (1) survey known information, (2) use screening tests to identify problematic areas, and (3) use animal models and human studies to determine the magnitude of any potential problems. Excellent references and computerized data bases, coupled with modern identification methods, are available to assist in the identification of known pathogens (Holt and Krieg, 1984). All the bacterial strains that could play a role in human disease have not been necessarily associated with human disease. Legionnaire's disease (caused by *Legionella pneumophila*) is a reminder of this concept. Given the right circumstances, many different microorganisms can colonize, replicate, and cause disease. However, even before microbes are directly associated with human disease, it is sometimes possible to use other indicators of pathogenicity. Using either *in vitro* or *in vivo* systems, colonization, competitiveness, survivability, inflammatory potential, and the production of known pathogenicity factors can be monitored. Claxton et al. (1995) have proposed using a "comparative parallelogram approach" to estimate adverse human effects by comparing "human *in vitro* models," and animal *in vitro* and *in vivo* models. These comparative studies not only help to identify where animal models are inadequate or incorrect, but they also allow a means of ranking the potential severity of any identified adverse effects.

As bioremediation efforts progress, a very dynamic situation emerges in which the types and amounts of pollutants, metabolites of pollutants, nutrients, co-pollutants, and numbers of organisms are constantly changing. One can examine these dynamic situations in models that range from laboratory flask models to small microcosm studies that are used to study specific factors to large prototype field studies that mimic true hazardous situations. Field studies can also be accomplished during the actual bioremediation of existing polluted sites. In an example of a flask study, George et al. (1992) examined the genotoxicity of 2,4,5-T (2,4,5-trichlorophenoxyacetic acid), a component of Agent Orange®, as it was degraded. If degradation had produced 2,4,5-trichlorophenol in significant amounts, the genotoxicity would have increased. Instead, the mutation tests demonstrated that the genotoxicity decreased as the bacterial cell number increased. In a microcosm study, Donnelly et al. (1987a, 1987b) added wood-preserving waste to two different soils. Over a period of 1200 days, extracts of the contaminated soil were tested by the *Salmonella* mutagenicity assay. Mutagenicity decreased over time.

Claxton et al. (1991a, 1991b) examined the mutagenicity of a bioremediation site that was part of EPA's field study of the Prince William Sound Alaskan oil spill. Sites consisted of different types of beaches and the application of different types of fertilizers to enhance the growth of hydrocarbon-degrading microorganisms. Samples were collected over a 3-month period, extracted, and examined using the *Salmonella* mutagenicity assay. Results showed that the crude oil and its weathered counterpart were weakly mutagenic. The mutagenicity of the bioremediation samples diminished over time. Chemical analysis would not have been able to fully characterize the samples in any of these studies for the presence of mutagens since all the individual components in these studies have not been tested for genotoxic potential. Although not fully explained here, these studies also used chemical analysis to confirm that the biodegradation process was responsible for the loss of mutagenicity. By using a combination of bioassay testing with chemical analysis, a reliable and accurate picture of the usefulness of a bioremediation effort can be produced.

Bioremediation efforts present very complex environmental situations for health effects assessments. Initially, environmental situations requiring remediation typically present as mixtures of multiple toxicants and other pollutants. Bioremediation processes, like most types of remediation efforts, generate additional compounds, thus adding to the complexity of the environmental situation. This introduction and generation of chemicals limits the ability to perform risk characterizations on remediation sites when only analytical chemistry is used for exposure assessment studies. Coupled with analytical chemistry, bioassays can be used to identify otherwise unidentifiable hazards, to identify the components that contribute to the toxicity, and to do various levels of exposure and risk assessment.

ADVERSE ENVIRONMENTAL EFFECTS ASSESSMENT FROM USING BIOREMEDIATION AGENTS

Presently, most bioremediation applications rely on naturally occurring microorganisms, the majority of which are indigenous to the treatment site. A thousand different microbial species may be found in a square yard of land surface. Each of these species may have from 5000 to 20,000 genetically different strains or varieties, unique in their adaptive qualities and ecological requirements. Up to 90% of the soil microbes cannot presently be cultured in the laboratory (U.S. Congress, Office of Technology Assessment, 1988). A few firms manufacture and sell nonindigenous microorganisms or enzymes for environmental applications, and some companies have publicly announced research efforts to develop genetically engineered microbes (GEMs) for waste treatment. In most federal and academic research laboratories, nonengineered, nonpathogenic microorganisms are more commonly used for environmental remediation studies due to difficulties in transferring the genetic determinants that would enable a microorganism to persist in a habitat to which it is not already adapted (Lindow et al., 1989; Ramos and Timmis, 1987).

Nonengineered microorganisms targeted for bioremediation applications have not necessarily been characterized as to the genus and species of all participating organisms. However, uncharacterized microbial consortia can be used with appropriate risk safeguards (Levin and Strauss, 1991). Genetically engineered microorganisms created for in-the-field bioremediation applications will likely be composed of a host soil microbe and genetic information from one or more other microorganisms that produce catabolic enzymes or have other attributes pertaining to its success in degrading target chemicals. Inside the cellular fluid (cytoplasm), small circular pieces of DNA (plasmids), characterized to date possessing genes for xenobiotic catabolism, are from gram-negative bacteria, predominantly *Pseudomonas* species (LaJoie et al., 1992). This functional definition excludes microorganisms that have been engineered for tracking or monitoring purposes (noncatabolic engineering).

The existence of degradative microbes at a waste site does not necessarily result in the transformation of waste chemicals by natural *in situ* biodegradation processes. Because many degradable contaminants persist in the environment, the natural degradative processes may be prohibitively lengthy and/or biological, and environmental factors such as bioavailability, availability of growth substrates, pH, and temperature can act together to limit both the rate and extent of degradation. Chemicals that persist in the environment have a greater potential to elicit adverse effects (U.S. GAO, 1991).

Laboratory studies are often used to indicate the potential for transforming a waste chemical, but these studies usually target one chemical or class of chemicals. Mixtures, not individual compounds, are generated by specific industrial processes and represent the most likely form in which pollutants will be found. For example, wood-preserving wastes include creosote (aliphatic hydrocarbons and PAHs), pentachlorophenol, and metals (primarily copper, chromium, and arsenic); manufactured gas plant wastes containing PAHs, cyanides, and phenols; and refinery wastes containing PAHs, lead, chromium and benzene. Mixtures with compounds from different groups may not demonstrate purely additive effects. For example, ketones can mobilize sparingly soluble hydrophobic organic compounds such as PCBs, potentially making the latter pollutant more bioreactive. For that reason, additional levels of analysis would be required (USEPA, 1993a, 1993b).

A concern in the use of bioremediation agents is the potential for biotransformation that leads to the formation and accumulation of intermediate metabolites that are more hazardous than the parent compound. Biotransformation plays a pivotal role in both toxicity and fate of pollutants. For most organic chemicals, it appears likely that toxicity is a combination of hydrophobicity and reactivity. Toxic effects that are due to partitioning into the animal are the result of exposure to chemicals that cause physical changes. In such cases, toxicity is often related to hydrophobicity. In contrast, toxic effects that are due to reactivity are the result of exposure to chemicals that cause chemical changes. In the latter case, toxicity appears to be related to stereo-electronic interactions. Chemical structure, including the type and position of substituents, as well as physicochemical properties, influences relative toxicity and mechanism of action.

Risk must be assessed and ascribed to the parent chemicals about to undergo transformation, the formation of toxic metabolites, from that transformation, and the microorganisms facilitating the transformation. Unlike controlled laboratory situations, the prediction of the formation and quantities of metabolites and potential toxicity are especially difficult in chemically complex situations. Relevant, sensitive, inexpensive, and rapid toxicity screening systems for use in field/facility situations are needed. Once screened, appropriate toxicity tests may be selected, depending on chemicals and mixtures encountered. Discharging treated waste because it passes tests for organic chemical toxicity (based on extractable components) may be misleading because the waste may still contain nonextractable toxic inorganics. Comprehensive screening methods must be developed that include chronic toxicity estimators, as well as short-term incubation toxicity tests, and measurements of organism viability, fecundity, tumorigenicity, teratogenicity, and standard mortality tests. An empirical approach may be suggested that uses a case-by-case analysis in which routine toxicological tests of process effectiveness are run collaterally with chemical analyses (USEPA, 1993).

ECOSYSTEM EFFECTS

The buffering capacities of microbial communities are well documented. The nonproliferating portion of the microbial population has been suggested as the critical player in system-level responses to environmental changes (Lewis and Gattie, 1991). An ongoing flux exists between the proliferation zones where sufficient and appropriate energy and nutrient sources are available for cell division and the quiescent zones that harbor large portions of the ecosystem's previously growing microbial populations now in a resting stage. Within the proliferation zone, mutations and genetic exchanges allow the community to adapt to novel conditions, while organisms within the quiescent zones preserve capabilities addressing conditions historically encountered. These two zones enable an ecosystem to respond rapidly to stress.

ALTERATION OF BIOGEOCHEMICAL CYCLES

Organic components of the environment are recycled through a mineralization process mediated by microbial action (Krumbein, 1983). It has been estimated that over 90% of the conversion of organic matter to CO_2 is due to bacterial and fungal metabolism. Microbial community systems play a critical role in these biogeochemical cycles (Fenchel and Blackburn, 1979). The potential impact of adding microbial agents or of enhancing indigenous biological degradative capabilities on the carbon, nitrogen, sulfur, and other biogeochemical cycles may be significant (Levin and Strauss, 1991).

Although little agreement exists on a "normal" value for environment parameters, researchers have typically collected such measurements as oxygen uptake rates, microbial community structure, chlorophyll content, phosphate and nitrate levels, and thymidine incorporation rates. The finding that no effects could be attributed to the introduction of a microorganism in certain environments under certain conditions is promising, but inconclusive (Pipke et al., 1992). Other researchers have demonstrated environmental effects, again under certain circumstances (Short et al., 1991).

ALTERATION OF THE PREEXISTING ECOLOGICAL NICHES

Ecosystem stability, in which microorganisms utilized for bioremediation are introduced or stimulated, is an important consideration in assessing potential impacts or predicting effects. Researchers' opinions differ on long-term impacts that such introductions or stimulations might cause, but most agree that short-term fluctuations in the system are possible (Seidler, 1992).

Microorganisms that inhabit the rhizosphere or certain plant pathogens that adapt to living under conditions of a plentiful food supply place a high premium on competition. Conversely, microorganisms living in uncrowded, but possibly physically restrictive, environments such as pseudomonads are intermittently exposed to flushes of substrate. Rapid growth is most important to these organisms rather than competitiveness (Paul and Clark, 1989; LaJoie et al., 1992). Most microorganisms considered for bioremediation applications belong to this latter group.

Researchers have observed that, in most soils under the prevailing conditions, bacterial populations cannot be expected to divide more than a few times a year, unless there has been a substantial input of a readily available source of energy. Introduced bacteria that succeeded in the initial colonization of available soil sites might proliferate relatively rapidly up to the limits of substrate availability (van Elsas and Trevors, 1991). Other investigators suggest that changes in the soil microbial community are short-lived, with the long-term community equilibrium eventually reestablishing itself (Drake, 1990; Robinson and Kickerson, 1987; Vitousek, 1986).

Finally, ecological effects may be manifested at many levels, ranging from inter- and intra-species competition, symbiosis and parasitism, nutrient cycling and energy transformation, to system resiliency and homeostasis. While some of these effects are discernible, others are not. While it may be quite possible to quantify some effects at the microbial level, the ability to extrapolate these effects to some ecosystem-level perturbation is difficult. For instance, quantifying changes in community structure and diversity may succumb to modern molecular analysis in microbial ecology, but the relationship of these dynamics to areas of nutrient cycling and resiliency at the system level are weak. In part, this is due to a lack of fundamental operational knowledge of models such as climax microbial communities, resistance or susceptibility to species invasions, and resource competition.

FATE OF THE GENETIC MATERIALS

The issue of gene transfer in the environment and the relevance of genetically engineered microbes (GEMs) and mobile genetic elements to biotechnology risk assessment is well established (Hensche and Schmidt, 1990; Miller and Levy, 1989; Kokjohn, 1989). Simply because genes are transferred, an impact should not be implied. If a gene is transferred, it must still be expressed and elicit an effect. Genetic transfer represents only the first element of risk in comparing use of native organisms versus GEMs in bioremediation. The potential for effects from genes transferred among compatible (related or able to express the trait) species should drive the decision-making process on a case-by-case basis.

A quantitative evaluation of the significance of gene transfer in risk assessment for bioremediation should address a minimum of three factors:

- the intrinsic stability of genetic elements within the donor microorganism and the likely recipient host in the environment
- the probability and relative rate of gene transfer in the target environment
- environmental factors promoting or maintaining gene transfer and maintenance of the recipient host or genes

Parameters for evaluating the effect of bioremediation practices using native or engineered microbial process are summarized in Table 1. Ecological effects are basically anticipated as impacts on the structure, abundance, complexity, diversity, and dynamics of biological populations or functional processes mediated by these organisms. These impacts are conjectured to occur through interactions among microbial populations within the community and alterations of larger community structure and composition within the context of a functional ecosystem. For the most part, these effects and the parameter measurement and verification must be evaluated in complex experimental systems ranging from microcosms to mesocosms, small-scale field tests, and large-scale field tests and ecosystem surveys. Properties of native or engineered organisms and their assessment in exposure and effects testing requires a high level of dependence on methods employed in microbial ecology to enumerate and quantify microbial populations and activities. Historically, methodological limitations in this field have hampered quantitative assessment on microbial population dynamics and community structure in complex ecosystems. Often, it is quantitatively more accurate and efficient to measure activities of physiological processes in community or ecosystem function. Risk assessment must include the issue of exposure assessment for ecologically sensitive populations and processes. The parameters for ecological testing can be applied to organisms used in bioremediation, the complete bioremediation process, and potential chemical agents used or released during bioremediation.

Evaluating possible adverse effects of bioremediation to ecological systems and community interaction is difficult for a number of reasons, regardless of whether the potential impacts are due to the engineered remediation application or to natural or engineered organisms used in the remedial solution. The local ecology of the site has already been altered, perhaps adversely, by the acute or chronic contamination. The natural ecological baseline or background may be nonexistent as a reference for judging ecological effects. No fundamental reference or basis of comparison has been established to delineate an acceptable alteration of ecological properties or ecosystem function, assuming ecological effects are observed. Ecological relationships and community interactions at the microbial scale are difficult to observe, compounding the demonstration of ecosystem-level effects. How vigorously will selected bioremediation applications alter that impacted environment? Should one choose minimal site disruption or aggressive remediation?

The Alaskan Valdez oil spill presented many of these interesting dilemmas regarding relative ecosystem effect when comparing the quantifiable effects of the contamination to the effects of a proposed bioremediation solution. Similar concerns can and should be examined for other contaminated site scenarios such as coal coking and coal tar impacted sites, wood preservative works, and other large-scale, mixed organic waste disposal sites.

For sites that have been subjected to long-term pollution (such as industrial sites, Superfund sites, etc.), appropriate control or references for ecological interactions may not exist. Such sites, having experienced a high level of chronic contamination, are often observed to have highly altered and skewed microbial communities due to chemical selection and toxicity, as well as local land use (Sanseverino et al., 1993). Consequently, relevant changes in the microbial community suggestive of adverse ecosystem effects may be masked.

TIER TESTING SCHEMES AND INFORMATION LISTS FOR MICROORGANISMS

Only one fully developed tier testing scheme specifically for microorganisms exists, the subdivision M of EPA's microbial pesticides testing scheme (USEPA, 1989b). A second approach to addressing risks of recombinant microorganisms is offered by EPA's Office of Pollution Prevention and Toxics (OPPT), which has a listing of information items for microbials that would include bioremediation agents. The OPPT has also begun a draft testing scheme for ecological effects, although it is incomplete at this time.

EPA OFFICE OF PESTICIDE PROGRAMS SCHEME FOR MICROBIAL PESTICIDES

Subdivision M of the USEPA's pesticide testing guidelines (USEPA, 1989b) provides two-tier testing schemes, both with multiple tiers: one for health and domestic animal effects and one for effects on terrestrial wildlife, aquatic animals, plants, and beneficial insects (USEPA, 1989b).

The testing of microbial pest control agents (MPCAs) for possible effects on humans and domestic animals is performed in a sequence of three tiers. The tiers are triggered by use pattern and prior test results. The Tier I tests are designed to provide an evaluation of an MPCA with respect to these three endpoints. EPA recognizes that pathogenicity, infectivity, and toxicity comprise a complex set of microorganisms: host interactions that may not be easily resolved as independent endpoints in all studies. The following Tier I studies are appropriate for testing the MPCA preparation: acute oral toxicity/pathogenicity, acute pulmonary toxicity/pathogenicity, and acute intravenous toxicity/pathogenicity. The following studies are considered primarily to evaluate the toxicity of the chemical components of the MPCA (but are performed with the complete MPCA preparation): acute dermal toxicity, primary eye irritation, reporting of hypersensitivity incidents. There is also a Tier I study specific to virus effects on mammalian cell lines. Generally, Tier I tests administer a single dose of 10^7–10^8 cells per organism.

Tier II tests are designed to examine MPCAs whose Tier I testing indicates that the MPCA exhibits infectivity or toxicity without any evidence of pathogenicity. Data from acute toxicity/pathogenicity studies may indicate that the MPCA is able to infect test animals without eliciting definite signs of pathogenicity or toxicity, or the MPCA may be observed to persist in the test animals for periods longer than would normally be expected. EPA considers that these observations are sufficient to warrant subchronic testing under Tier II to determine whether repeated exposure to the MPCA preparation is sufficient to cause toxic or pathogenic effects. Subchronic tests may also be necessary if the preparation is unable to be freed of contaminating microorganisms, or contaminating microorganisms are insufficiently characterized, or there is sufficient reason to suspect that microbially produced toxins are present in the preparation (and these toxins would not exert effects during Tier I acute tests).

Toxicity of an MPCA preparation to test animals may be caused by substances produced by, or which are constituents of, the MPCA or any contaminating microorganism or by substances that are present as nonbiological components of the preparation. The Tier I toxicity/infectivity studies are designed to detect acute toxic effects on biological or nonbiological components of an MPCA preparation, in the absence of signs of pathogenicity or infectivity. The acute dermal toxicity study is designed primarily to evaluate toxic effects due to chemical components of the MPCA preparation. EPA believes that only rarely would a microorganism bypass the skin barrier or infect epithelial cells of a healthy host animal solely by virtue of being placed in contact with the skin. An acute intravenous toxicity/pathogenicity study has been placed in Tier I to evaluate possible adverse effects of an MPCA preparation once the skin barrier is deliberately bypassed.

If toxic effects are observed in the Tier I studies, in the absence of signs of infectivity or pathogenicity, then a Tier II acute toxicity study normally is required, with the toxic component of the MPCA preparation being used as the dosing material. Additional studies designed solely for evaluating toxicity effects of MPCA preparations are not included in the Tier II or Tier III testing scheme for MPCAs because appropriate protocols already exist for testing conventional chemical pesticides.

TIER TESTING SCHEMES FOR BIOREMEDIATION

There are very few testing schemes that address the risks of using bioremediation agents. Under development is the scheme for oil spill bioremediation agent evaluation. This scheme addresses both efficacy and risks of commercial products used for cleanup of oil spills. This draft four-tiered testing scheme has been developed to assess both the efficacy and safety of oil spill bioremediation products intended primarily for marine oil spill cleanup (USEPA, 1992). The scheme is specific to products for cleanup of marine spills, but portions of the scheme may be useful for development of a more general risk assessment scheme for bioremediation risk assessment. The scheme shows a meshing of efficacy and risk assessment tests: Tier III uses the effluents (containing oil, microorganisms, and metabolites) from microcosm efficacy tests to run short-term chronic toxicity assays. The safety testing is primarily oriented toward ecological toxicity; however, Tier I information requirements include identification of well-known mammalian pathogens.

Second, a draft general bioremediation risk assessment scheme has been developed, which may be useful as a starting point for a more detailed bioremediation scheme. Finally, there is a test battery that has been developed to screen aqueous effluents from remediation sites. Tests in this battery could be triggered in a tiered fashion, dependent on potential hazards due to the effluents.

SMALL-SCALE OR SINGLE-SITE ASSESSMENT VERSUS APPROVAL OF BIOREMEDIATION AGENT FOR USE AT MULTIPLE SITES

The testing schemes and information items presented above focus on approval of microorganisms for use at single sites (for intentional release to the environment). Fermentation approvals may be less dependent on site conditions. However, manufacturers of bioremediation microorganisms for commercial purposes to use *in situ* applications will probably only specify the pollutants the microbes are intended to degrade. The manufacturers will often not want to be restricted according to site characteristics (product only to be used at certain geographical locations, etc.). However, site characteristics may affect metabolites produced, survival and spread of bioremediation microorganism, etc. The microbial pesticide test guidelines address commercial release by hazard testing with maximum exposures—if no effects are observed at maximum exposures, testing need not continue.

DRAFT GENERAL BIOREMEDIATION RISK ASSESSMENT SCHEME

A draft general bioremediation risk assessment scheme was presented at the 1993 Duluth Bioremediation Workshop and is shown in Figure 6 (USEPA, 1993). In this scheme, there are several categories of information, grouped according to characteristics of the (1) waste, (2) treatment agent, (3) application system, and (4) process (which links the prior three items). For the waste, characteristics such as amounts, distribution, and physical chemical properties (volatility, aqueous solubility, etc.) should be considered in addition to the factors shown in the figure. Treatment agents may have mass balance information available from laboratory tests and characterizations of reaction products, stoichiometry, and intermediates. Engineered organisms should be described in terms of recipient (taxonomic identification) and introduced genes (function and source). For toxic risk of metabolites, estimation of metabolites from known pathways becomes more difficult with complex mixtures and variable site conditions. Toxicity screening of effluents may be needed if estimation techniques used to predict metabolites are uncertain.

RISK ASSESSMENT AT HAZARDOUS WASTE SITES

Before any treatment action may be undertaken at an identified hazardous waste site, a remedial investigation/feasibility study (RI/FS) must be conducted. Subsequently, based on the findings of the RI/FS, a proposed plan is developed. This plan is subject to public scrutiny and comment. Once comments are addressed from the public, potential responsible parties, EPA, and appropriate state regulatory authorities, a record of decision (ROD) is developed. The ROD is a summary of the

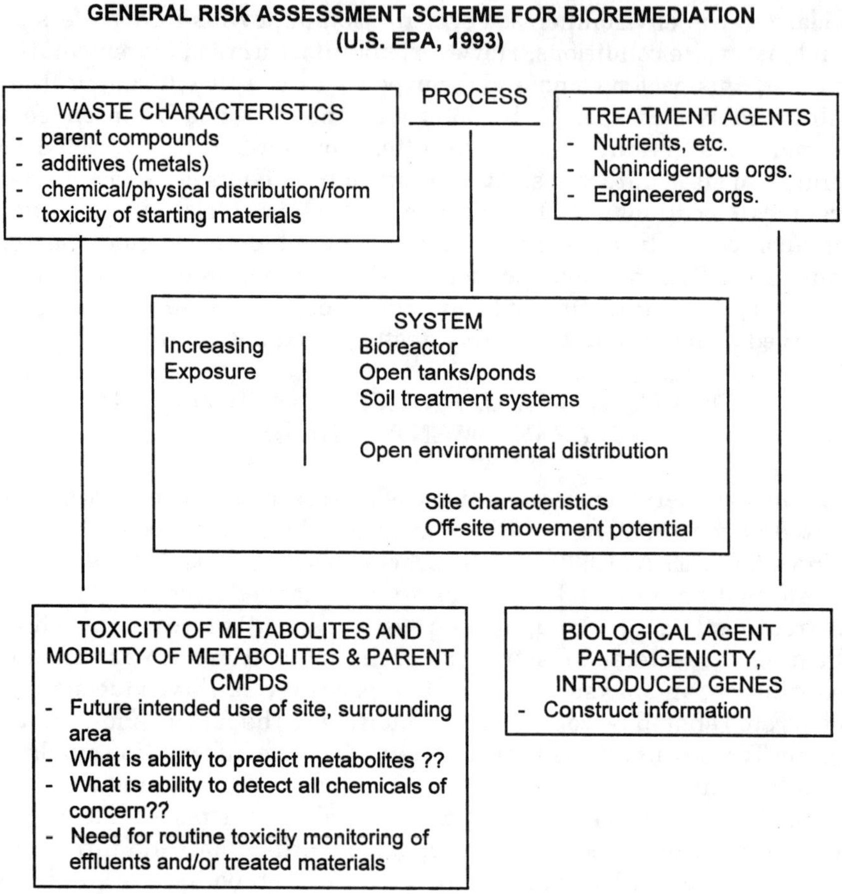

Figure 6. General risk assessment scheme for bioremediation.

RI/FS and presents actual and potential risks to human health and the environment and treatment for the site. The ROD generally certifies that the treatment selection process was conducted in accordance with the Comprehensive Environmental Response Compensation and Liability Act (CERCLA) and Superfund Amendments and Reauthorization Act (SARA). The ROD also outlines the engineering components and remediation goals of the selected treatment(s); provides the public with site history and risks posed by conditions at the site; and gives a summary of cleanup alternatives considered, their evaluation, and the rationale behind the selected remedy. An RI/FS generally has seven objectives:

- collection of information on the types, concentrations, extent and movement of contaminants found in subsurface soils, surface soils, surface water, sediment, and groundwater
- estimation of the volume of contaminated media and materials
- collection of information regarding site physical characteristics and site contaminants for use in the risk assessment, the feasibility study and the remedial design
- collection of data for use in treatability studies
- identification of geotechnical properties used in designing and locating any remediation structures during the remedial design
- identification of potential applicable or relevant and appropriate regulations (ARARS) for response actions
- identification and evaluation of remedial alternatives to address human health and environmental risks

Early in the RI/FS, a baseline risk assessment is conducted. This initial assessment examines the exposure pathways that need to be addressed by the remedial action and provides a "baseline," indicating what risks could exist if no remedial action were taken at the site. Typically, the baseline risk assessment identifies such exposure pathways as ingestion, inhalation, or direct contact. The assessment also identifies contaminants of concern. Potential receptors of these contaminants can include residents, workers, recreational users, and terrestrial and aquatic biota. Commercially valuable or sensitive terrestrial or aquatic species are ecological receptors of concern. Potential human health and ecological toxicities are then characterized. As with a human health toxicity characterization, ecological toxicity characterizations are composed of hazard identification followed by a dose response assessment.

Based on consideration of RI/FS, CERCLA requirements, the detailed analysis of remedial alternatives, and comment from involved parties and the public, an ROD is developed that selects a treatment(s) that represents the best balance of considerations for the intended use of the site and the safety of the public. The technology(s) selected must provide maximum source reduction and remediate contaminated media and materials to the greatest extent practicable permanently and cost-effectively. Since most remediations are expected to take several years, engineering controls such as fencing, posting, and restricting access must be implemented. Institutional controls such as zoning restrictions or conservation easements must also be in place. A long-term compliance monitoring program must also be developed. It is recognized that, while certain other alternatives may better satisfy certain individual selection criteria, the selected remedy best meets the entire range of the selection criteria and achieves, in the determination of regulatory officials, involved parties, and the public, the appropriate balance considering site-specific conditions and criteria identified in CERCLA and the National Contingency Plan (NCP).

FIELD TESTS

To date, there have been no field investigations using genetically engineered microorganisms (GEMs) to promote enhanced *in situ* bioremediation in the subsurface or any other environment (Fox, 1992). Bioaugmentation or field application of naturally occurring microorganisms for waste cleanup has been practiced for at least 30 years. In fact, when the Queen Mary was retired from service as a luxury ocean liner and converted into a floating hotel in Long Beach, California, oily bilge water was remediated utilizing commercial microorganisms and discharged under state permit into the bay.

Efficacy, not microbial distribution, survival, or safety, appears to be the recurrent theme in the non-GEM mediated field test literature. But a limited number of papers report greenhouse or field tests relying on naturally occurring microbes that address these questions. Researchers studied the survival of bacteria (including *Enterobacter, Pseudomonas*) during aerosolization (Marthi, 1990). Survival decreased rapidly only 10 meters from the point of application. Lower temperature and higher humidity seemed to contribute to the increased survival of this organism, while washed bacteria exhibited a markedly reduced survival. A second study by the same research team tested survival on plants and found that after 48 hours drift (at least 5 meters downwind from point of application), bacterial population *Pseudomonas syringae* was undetectable.

LABORATORY TO FIELD SCALEUP

The need to advance from laboratory research to bench-scale to field demonstrations with a knowledge of the limitations and extrapolations from increasing levels of ecological complexity has driven EPA bioremediation research. EPA's field demonstration bioremediation efforts have focused on waste types and site conditions; basic research and engineering efforts are directed to the degradation of wastes in matrices of greatest concern.

Five waste-site types were selected from nearly 1200 national priority list (NPL) sites that are candidates for bioremediation. These wastes include organic solvents, creosote and PAH-type wastes, halogenated aromatic hydrocarbons (HAHs), pesticides, and munitions wastes. The identification of these priority waste types is based on their prevalence at the NPL sites and the regulatory requirements governing their control or removal (CERCLA, RCRA, and UST). Although some of these wastes are more difficult to degrade than others, bioremediation is a potentially effective treatment technique. Current analysis of the NPL sites reveals that about 80% of the sites are contaminated with organic solvents, about 13% with wood-preserving chemicals, over 20% with halogenated aromatic hydrocarbons (HAHs), and approximately 12% with pesticides. Munitions wastes contaminate a number of military bases, some of which are listed on the NPL (USEPA, 1991b).

The site matrices in which the priority wastes are found include liquids/leachates, aquifers, the vadose zone (the unsaturated material below the surface soil and above an aquifer), surface soils, sediments, and air. Water that is removed from an aquifer and treated aboveground is considered to be part of liquids and leachates, whereas water in an aquifer treated *in situ* is considered part of the saturated (aquifer) zone. The saturated zone consists of both groundwater and saturated soil.

Bioremediation of the vadose zone has most of the constraints of aquifer remediation plus the added problems of a low-nutrient, low-moisture regime. Treatment of surface soils generally involves excavation followed by treatment as a slurry or in the solid phase. *In situ* soil treatment is possible if the soil is isolated from aquifer formations by natural or constructed barriers or if the soil and subsurface material are to be treated as a unit.

Sediments include contaminated natural sediments and sludges found at the bottoms of disposal ponds or lagoons. Current remediation scenarios for sediments include dredging and aboveground treatment as slurries or dewatered solids in storage silos or treatment *in situ*. Unlike soils that are generally aerobic, sediments are generally anaerobic.

This effort to prioritize wastes and site matrices echoes the emphasis that academic institutions and societies have placed on field research in bioremediation to address this widespread international problem. A recent report underscored the uncertainty of bioremediation site performance (Gibson and Sayler, 1992). Predictability is often limited by the lack of appropriate models, biological information, and accurate parameter estimation. To improve performance and predictability, representative field sites have been selected for demonstration purposes.

Bioremediation field studies should include appropriate microbiological, molecular, and ecological methods required for risk assessment and process monitoring. Risk assessment research is needed to monitor organism persistence and mobility, genetic stability, gene transfer, and effects within the soil microbial community. Sound ecological risk assessment for the release of a genetically engineered microorganism into the environment requires an estimate of the likelihood that it will (1) persist at a site with the potential to grow in population size and (2) cause effects to the ecosystem as a result of its interactions with other organisms or its metabolic functions (Tiedje et al., 1989). These factors correspond to the exposure and effects components of risk assessment. Given the high mobility of microorganisms in air and water and the potential for genetic transfer of plasmids, these are additional concerns in the release of GEMs. The ability of a GEM to invade successfully a diversity of habitats, soil types, or water bodies must be assessed. The likelihood of transfer of modified gene elements *and* the potential for expression of those modified elements should also be evaluated. The need to know under what conditions and for how long the GEM persists and what it does to its physical, chemical, and biological environments when it is growing should be the focus of ecological investigations in the field studies.

Expanded research on process monitoring methodologies is also critical to bioremediation field studies. Laboratory studies on environmental diagnostic techniques have reached a degree of sophistication and are now ready for field trials. Such microbial and molecular monitoring approaches include analysis to characterize the microbial community density and to determine the dynamics of the added microbes. These analytical approaches include measurements of aerobic and facultatively aerobic heterotrophic population densities, total direct cell counts, selective enrichment of genus, colony hybridization with gene probes, DNA extraction and blot hybridization [as well as attempting polymerase chain reaction (PCR) amplification], nucleic acid probes for specific genes, and mRNA analysis of in situ expression of "added" genes. Recovery of individual isolates and restriction fingerprinting (with possible PCR amplification) of DNA extracts could address stability of the GEMs.

FIELD DEMONSTRATIONS

Over 45 Superfund projects have selected bioremediation as a primary or ancillary cleanup technology. These sites demonstrate an ongoing commitment to employ state-of-the-art technology solutions. Monitored sites are distributed throughout all 10 EPA regions in the United States, with over 40% located in Regions 5 (Minnesota, Wisconsin,

Michigan, Ohio, Indiana, and Illinois) and 9 (California and Nevada). Analysis of the database reveals that petroleum-based hydrocarbons are most frequently treated, followed by wood-preserving wastes. Soil and groundwater are the media most frequently treated with bioremediation technologies. Identified sites are the subject of *in situ* and *ex situ* treatments, including land treatment, bioventing, bioreactor treatment, nutrient addition, and other techniques and combinations of techniques. Although many examples of field demonstrations serve to illustrate this effort, two will be presented here.

Libby, Montana, Site—Wood Treatment Wastes

The Libby groundwater Superfund site in Libby, Montana, illustrates application and comparison of several of these bioremediation technologies. Site soil and groundwater is contaminated with pentachlorophenol (PCP) and creosote (PAHs) from a wood-preserving facility formerly operated at the site. PAHs are the primary contaminants of concern associated with the soil phase. PAH-contaminated soils from three primary sources have been excavated and moved to a central waste pit (USEPA, 1993). EPA research scientists from the Robert S. Kerr Environmental Research Laboratory, in cooperation with Utah State University, are evaluating three biological treatment processes at the Libby site: (1) surface soil bioremediation in a lined, prepared-bed land treatment unit; (2) groundwater treatment in an aboveground fixed-film bioreactor; and (3) *in situ* bioremediation of the upper aquifer. Each process is being evaluated regarding design, operation, monitoring, and performance.

SURFACE SOIL BIOREMEDIATION

The land treatment unit consists of two adjacent 0.4-hectare cells lined with low-permeability synthetic liners to minimize leachate infiltration from the unit. Contaminated soil is applied to the cells in 20-cm layers and treated until target contaminant levels are achieved. Evaluation of the treatment's effectiveness includes soil sampling, studying field-scale treatment and toxicity reduction, analyzing the influence of moisture and soil structure, and calculating the mass balance of contaminants in terms of soil and leachate. Land treatment unit soil analysis data will be used to determine the statistical significance, confidence, and extent of biodegradation at this site. Degradation kinetics and toxicity reduction studies will generate data that can be used to help assess overall bioremediation effectiveness and to predict performance of this type of bioremediation process at similar sites. Current estimations based on the aforementioned criteria indicate that remediation of each soil layer takes 32 to 163 days. Based on these results, it is predicted that remediation of the 38,000 m^3 of contaminated soil will take 8 to 10 years.

ABOVEGROUND FIXED-FILM BIOREACTOR

Aboveground treatment of groundwater occurs in two fixed-film reactors, operating in series. Reactor effluents are amended with nutrients and reoxygenated prior to reinjection through an infiltration trench. Researchers are monitoring the performance of the bioreactors, including flow composited sampling, analysis of biofilm dynamics, calculation of mass balance of contaminants, and treatment optimization. Initial performance data on the fixed-film bioreactors indicate that PAH and PCP removal is taking place.

IN SITU BIOREMEDIATION OF THE AQUIFER

The *in situ* bioremediation system is based on the addition of hydrogen peroxide (to generate oxygen) and inorganic nutrients to stimulate growth of contaminant-specific microbes. Evaluation of this process includes determination of dissolved oxygen profiles across the site, sampling of aquifer material to identify contamination and correlate microbial content, distinguishing between abiotic and biotic effects, and correlating dissolved oxygen uptake with biodegradation and toxicity reduction. Results suggest that aquifer core samples have a chemically reduced condition, indicating that the site has an abiotic as well as a biological oxygen demand, complicating interpretation of remediative performance. However, contaminant reductions in monitoring wells have been demonstrated. Tests are being planned to differentiate between the abiotic and biotic oxygen demands.

Wurtsmith Air Force Base, Michigan—Chlorinated Solvents, PCBs, PAHs, and Heavy Metals

As an extension of the reasoning applied at the Libby site, federal agencies, including DoD, DOE, and EPA, EPA's Hazardous Substance Research Centers, regional and state authorities, and associated laboratories and corporations in the private sector have joined to establish a controlled field test facility for conducting the investigations required to develop, evaluate, and formulate a general basis for the design and engineering of integrated bioremediation systems. The project focuses on *in situ* remediation of surface soils, subsoils, surface waters, and groundwater contaminated by organic chemicals and heavy metals. The site, Wurtsmith Air Force Base, is located in Iosco County in northeastern Michigan. The closed airbase is approximately 18 km^2, containing over 40 specifically identified contamination sites and is somewhat representative of the over 1500 DoD installations listed in the installation restoration program.

Basic research, scaleup, and implementation of technologies will be conducted on-site. This well-characterized site will allow on-site and *in situ* technologies to be developed, tested, and refined by integrating biological, chemical, hydrological, geological and toxicological information with the appropriate engineering methods for cost-effective remediation with minimal environmental disruption. Advantages of experimentation at this site include

- good hydrogeology—sand, high conductivity, good recharge rate
- groundwater at 6 meters, clay aquitard at 18 meters
- aquifer connected to three different water systems—Lake Huron, Au Sable River, and a wetland
- 43 distinct contaminated sites containing chlorinated solvents, PCBs, PAHs, and heavy metals
- a variety of remediation conditions—aerobic/anaerobic, sorbed, dissolved, and aqueous phase
- geological and hydrological records established by 12 years of United States Geological Survey (USGS) surveys
- several conventional pump-and-treat systems are operating
- buildings are available on-site for administrative offices, laboratories, and test-cell installations

BIOREMEDIATION LIMITATIONS AND CONCLUSIONS

Most successful bioremediation demonstrations have depended on the abilities of native microbial communities or selected consortia to degrade targeted chemicals under environmental conditions, which managed to enhance their activity. Nonnative introduced organisms can effect some reductions in contaminants, but often hard data to link the disappearance of the chemical to the metabolic abilities of the introduced organism are lacking. Factors such as abiotic losses, volatilization, or stimulation of latent native degrading populations of microorganisms often can account for many of the chemical transformations. Nonnative microorganisms also die off quickly in the field, presumably due to competition with indigenous microflora or to predation in a nonadapted micro niche.

Land disposal regulations under RCRA dictate that waste concentrations must meet an established level, referred to as a *treatment standard.* Where a treatment standard has been established, any treatment method capable of achieving the specified contaminant concentration may be used. However, as a result of the 1984 RCRA amendments, treatment standards were implemented based on the best demonstrated available technology (BDAT), which favored conventional technologies. These technologies, such as incineration, set an artificially high decontamination level that alternative, biologically based technologies usually cannot achieve in a reasonable time frame. With some treatment standards, it is possible that degradation of the contaminant could approach 95% or more, yet be considered unsuccessful, depending on the concentrations and types of remaining chemical residues. The BDAT standards were originally established for process waste streams or for pure wastes, not for contaminated soils, sediments, or groundwater. To hasten the cleanup of hazardous waste sites, EPA is moving toward a more flexible site evaluation process and adoption of health- or risk-based treatment standards. Site-specific factors could be used to determine cleanup targets with an acceptable protective risk range, reflecting the potential use of the formerly contaminated site; for example, treated surface soils at an industrial site or under a parking lot would have less stringent cleanup levels than would similar soils in a residential or public park area.

Assuming that realistic, safe contaminant levels are targeted, reinoculation or other methods to stimulate or increase degradative biomass may not be successful. Two reasonable alternatives are possible: use an alternate physical or chemical remediation technique to clean up low levels of remaining chemicals or demonstrate that, despite the small amount of residual chemicals, environmental or human health risks have been reduced to acceptable levels. Developing risk-based cleanup standards is new to bioremediation but is gaining recognition as a cost- and time-effective alternative that is not based on unrealistically low contaminant levels. The key issue to accepting risk-based cleanup standards is to ensure human health and environmental stability for the projected use of the (formerly) contaminated area. Tests and measurements are providing basic information to assist regulators in determining realistic and safe site-specific cleanup levels. Measurements that help evaluate ecosystem health and resiliency include nutrient cycling, nutrient retention, respiration/primary production, resiliency/homeostasis, food web complexity, and detritus processing. Adverse effects disrupting community structure may be measured by examining population dynamics, population structure, competition, species abundance and diversity, metabolic effects, physiological activity, or alterations of biomass compartmentalization at appropriate spatial and temporal scales.

Human health risks from bioremediation are becoming better defined. Risks from microorganisms used for bioremediation or from metabolites produced from the bioremediation of waste chemicals include direct health effects such as toxicity, infectivity, pathogenicity, and allergenicity and indirect effects such as cytotoxicity and carcinogenicity. Microbes could also pose a threat as vectors for foreign genes. New bioassay techniques are continually being developed and improved to assess the extent and permanence of risk reduction. Bioassays are increasingly used to evaluate the potential for production of toxic metabolites. The *Salmonella* assay test has been the most commonly used assay for some time to evaluate mutagenicity and carcinogenic potential. Recently, additional bioassay tests have been developed that will allow monitoring of biodegradation processes to correlate decreases in target chemical concentrations with changes in toxicity (Mueller et al., 1991; Gradner et al., 1990). Thus far, these bioassays have proven to be very sensitive and represent one of the few assays that have undergone field testing.

Because of the limited ability to extrapolate laboratory biodegradation to field trials and the lack of knowledge regarding microbial community interactions (ecology), the potential for adverse ecological and human health effects is a significant concern. Additionally, contributing to uncertainty is the variability among field sites and differing environmental conditions and pollutants. It is clear that no single approach will be entirely adequate.

Parameters for evaluating the effects of bioremediation practices using native or engineered microbes are based on microbial ecological techniques and toxicology. It is anticipated that ecological effects would be manifested as impacts on the structure, abundance, complexity, diversity, and dynamics of biological populations or functional processes mediated by these organisms. These impacts are thought to occur through interactions among microbial populations within the community and through alterations of larger community structure and composition within the context of a functional ecosystem. For the most part, these effects and the parameter testing must be evaluated in complex experimental systems ranging from microcosms to mesocosms, small-scale field tests, and large-scale field tests and ecosystem surveys. Overall risk assessment must include the issue of exposure assessment for ecologically sensitive populations and processes. The parameters for ecological testing can be applied to organisms used in bioremediation, to the complete bioremediation process, and to potential chemical agents used or released during bioremediation.

Only one fully developed tier testing scheme specifically for microorganisms exists—subdivision M—EPA's microbial pesticides testing scheme legislated by the Federal Insecticide, Fungicide, and Rodenticide Act. There are very few testing schemes that address the risks (or efficacy) of releasing microorganisms for bioremediation or other purposes. Notably, under development by EPA's Office of Prevention, Pesticides and Toxic Substances and Office of Research and Development are draft risk assessment schemes for intentional field releases that may be useful for building detailed field release guidelines that incorporate human health and ecological concerns.

Genetically engineered microorganisms share many of the attributes and difficulties of other introduced nonindigenous microorganisms, and they bear some additional consideration in terms of the use of antibiotic resistance as a marker mechanism, the stability of their genetic construct, their potential to provide indigenous species with previously unavailable genes that have unknown potential for expression by indigenous species, etc. The potential for benefit from genetic technologies is real. It is perhaps one of the most powerful scientific techniques to be developed this century. As with any technological tool, however, the potential for good is balanced by the potential for harm from unintended consequences of the application of this technology. EPA and other federal agencies are examining advanced remediation technologies through cooperative research. Exciting strides are being made in such areas as environmental application vectors (reintroduction of enhanced environmental isolates acclimated to target chemicals), stimulation of degradation by addition of surfactants/detergents to increase bioavailability of chemicals, adaptation of degradative aerobic microorganisms to low oxygen environments, addition of nutrients to stimulate biomass, and development of screening methods to rapidly identify degradative microorganisms from the environment.

RECOMMENDATIONS

Researchers and risk assessors should consider the following recommendations:

1. Test the additivity hypothesis for mixtures of compounds that typically occur at hazardous waste sites.
2. Test the bioassay comparative approach method as a risk assessment approach for complex environmental sites.

3. Validate existing models for estimating exposure to both toxicants and environmentally applied microbes.
4. Compare presently used risk assessment methods for actual hazardous waste sites with newer alternative methods.
5. Reevaluate remediated sites using biological monitoring methods that would detect nonconventional pollutants.
6. Directly evaluate the cost, efficiency, toxicity levels, exposure evaluation methods, and available risk assessment methods by comparing cleanup options at single sites.
7. Determine whether or not nontoxic pollutants at hazardous waste sites can be converted into toxic substances by remediation efforts.
8. Test the parallelogram comparative potency approach for environmentally released microorganisms.

REFERENCES

Anderson, E.L. and A. Ehrlich. 1985. New Risk Assessment Initiatives in EPA. *Tox. and Ind. Health,* 1(4):64–22.

Anderson, M.E. 1988. Quantitative risk assessments and occupational carcinogens, *Appl. Ind. Hyg.,* 3(10):267–273.

Bonaventura, C., Bonaventura, J. and Bodishbaugh, D.F. 1995a. Environmental bioremediation: Approaches and processes. In: de Serres, F.J. and A. D. Bloom (eds.), *Ecotoxicity and Human Health: A Biological Approach to Environmental Remediation.* Lewis Publishers, Boca Raton, Florida, 183-210.

Bonaventura, C., Bonaventura, J. and Bodishbaugh, D.F. 1995b. Environmental bioremediation: Application and new horizons. In: de Serres, F.J. and A.D. Bloom (eds.), *Ecotoxicity and Human Health: A Biological Approach to Environmental Remediation.* Lewis Publishers, Boca Raton, Florida, 211-242.

Chadwick, R.W., George, S.E. and Claxton, L.D. 1992. Role of gastrointestinal mucosa and microflora in the bioactivation of dietary and environmental mutagens or carcinogens. *Drug Metabolism Reviews.,* 24:425–495.

Claxton, L.D. 1981. Mutagenic and carcinogenic potency of diesel and related environmental emissions: *Salmonella* bioassay. *Environment International,* 5:389–391.

Claxton, L.D. 1982. The utility of bacterial mutagenesis testing in the characterization of mobile source emissions: A review, in: Lewtas, J. (ed.) *Toxicological Effects of Emissions from Diesel Engines.* Elsevier Science, New York, pp. 69–82.

Claxton, L.D. 1983. The integration of bioassay and physio-chemical information for complex mixtures. In: Waters, M.D., Sandhu, S.S., Lewtas, J.,

Claxton, L.D., Chernoff, N. and Nesnow, S., *Short-Term Bioassays in the Analysis of Complex Environmental Mixtures III.* Plenum Press, New York, pp. 153–162.

Claxton, L.D. and Huisingh, J.L. 1979. Comparative mutagenic activity of organics from combustion sources, in: *Pulmonary Toxicology of Respirable Particles,* U.S. Department of Energy Symposium Series (CONF-7911002), Washington, D.C., 53:453–465.

Claxton, L. and Kohan, M. 1981. Bacterial mutagenesis and the evaluation of mobile-source emissions, in: Waters, M.D., Sandhu, S.S., Huisingh, J.L., Claxton, L.D. and Nesnow, S. (eds.), *Short-term Bioassays in the Analysis of Complex Environmental Mixtures II,* Plenum Press, New York, pp. 299–317.

Claxton, L.D., Houk, V.S. and George, E.S. 1995. Integration of Complex Mixture Toxicity and Microbiological Analyses for Environmental Remediation Research. In: de Serres, F.J. and A.D. Bloom (eds.), *Ecotoxicity and Human Health: A Biological Approach to Environmental Remediation.* Lewis Publishers, Boca Raton, Florida, 87-122.

Claxton, L.D., Houk, V.S., Williams, R. and Kremer, F. 1991a. Effect of bioremediation on the mutagenicity of oil spilled in Prince William Sound, Alaska, *Chemosphere,* 23, 643–650.

Claxton, L.D., Houk, V.S., Williams, R. and Kremer, F. 1991b. Oil spill cleanup, *Nature,* 353:24–25.

Creager, J.G., Black, J.G. and Davison, V.E. 1990. *Microbiology: Principles and Applications,* Prentice-Hall, Englewood Cliffs, New Jersey, p. 248.

Dean, N. and Kremer, F. 1992. Advancing Research for Bioremediation. *Environmental Protection,* pp. 19–25.

Deitch, E.A. and Berg, R. 1987. Bacterial translocation from the gut: A mechanism of infection, *J. Burn Care Rehab.,* 8:475–482.

Donnelly, K.C., Brown, K.W. and Kampbell, D. 1987a. Chemical and biological characterization of hazardous industrial waste. I. Prokaryotic bioassays and chemical analysis of a wood-preserving bottom-sediment waste, *Mutat. Res.,* 180:31–42.

Donnelly, K.C., Davol, P., Brown, K.W., Estiri, M. and Thomas, J.C. 1987b. Mutagenic activity of two soils amended with a wood-preserving waste, *Environmental Science and Technology,* 21:57–64.

Drake, J.A. 1990. The mechanics of community assembly and succession. *J. of Theoretical Biology,* 147:213–233.

Driver, J.H., Rogers, H.W. and Claxton, L.D. 1990. Mutagenicity of combustion emissions from a biomedical waste incinerator, *Waste Management,* 10:177–183.

Fenchel, T. and Blackburn, T.H. 1979. *Bacteria and mineral cycling.* Academic Press, New York.

Fox, J.L. 1992. Assessing the scientific foundations of bioremediation. *ASM News,* 58:483–485.

Fredrickson, J.K., Bolton H., Jr. and Stotzky, G. 1992. Methods for Evaluating the Effects of Microorganisms on Biogeochemical Cycling. *In:* Levin, Seidler and Rogul, eds. *Microbial Ecology: Principles, Methods and Applications.* McGraw-Hill, pp. 579–606.

Gaver, D.P. 1995. Chapter IX: Quantitative modeling and analysis in environmental studies. In: de Serres, F.J. and A.D. Bloom (eds.), *Ecotoxicity and Human Health: A Biological Approach to Environmental Remediation.* Lewis Publishers, Boca Raton, Florida, 243-270.

George, S.E., Chadwick, R.W., Chang, J.J., Kohan, M.H., Dekker, J.P. and Hayes, Y. 1992. 2,4,5-Trichlorophenoxyacetic acid influence on 2,6-dinitrotoluene-induced urine genotoxicity in the Fischer 344 rat: Effect on GI microflora and enzyme activity, *Fundamental and Applied Toxicology,* 18:240–246.

George, S.E., Kohan, M.J., Walsh, D.B., Stead, A.G. and Claxton, L.D. 1989. Polychlorinated biphenyl degrading pseudomonades: Survival in the mouse intestines and competition with the normal microbial flora, *Journal of Toxicology and Environmental Health,* 26:19–37.

George, S.E., Kohan, M.J., Whitehouse, D.A., Creason, J.P. and Claxton, L.D. 1990. Influence of antibiotics on intestinal tract survival and translocation of environmental *Pseudomonas* species, *Applied and Environmental Microbiology,* 56(6):1559–1564.

George, S.E., Kohan, M.J., Whitehouse, D.A., Creason, J.P., Kawanishi, C.Y., Sherwood, R.L. and Claxton, L.D. 1991. Distribution, clearance and mortality of environmental pseudomonades in mice upon intranasal exposure, *Applied and Environmental Microbiology,* 57:2420–2425.

Gibson, D.T. and Sayler, G.S. 1992. *Scientific Foundations of Bioremediation: Current Status and Future Needs.* American Academy of Microbiology, Washington, D.C.

Gradner, H., van der Schalie, W., Wolfe, M. and Finch, R. 1990. New methods for on-site biological monitoring of effluent water quality. In: S. Sandhu et al., eds., *In situ Evaluations of Biological Hazards of Environmental Pollutants.* Plenum Press, New York.

Graedel, T.E., Hawkins, D.T. and Claxton, L.D. 1986. *Atmospheric Chemical Compounds: Sources, Occurrences and Bioassay,* Academic Press, Orlando, Florida.

Graham, J.A., Claxton, L.D., O'Neil, J.J., Otto, D.A. and Goldstein, B.D. 1984. Automobile emissions: Primary Health Effects Concerns, *SAE Technical Paper Series,* 840908: 1–15.

Hemminki, K.E., Grzybowska, E., Chorazy, M., Twardowska-Saucha, K., Srocynski, J.W., Putman, K.L., Randerath, K., Phillips, D.H., Hewer, A., Santella, R.M., Young, T.L. and Perera, F.P. 1990. DNA adducts in humans environmentally exposed to aromatic compounds in an industrial area of Poland. *Carcinogenesis,* 11:1229–1231.

Hensche, R.B. and Schmidt, F.R.J. 1990. Plasmid mobilization from genetically engineered bacteria to members of the indigenous soil microflora in situ. *Current Microbiology,* 20:105–110.

Holt, J.G. and Krieg, N.R. 1984. *Bergey's Manual of Systematic Bacteriology, Volume 1.* Williams and Wilkins, Baltimore.

Houk, V.S. 1992. The genotoxicity of industrial wastes and effluents. A review, *Mutat. Res.,* 277:91–138.

Houk, V.S. 1997. The effect of waste treatment and remediation alternatives on the genotoxicity of industrial wastes, effluents, and waste sites, in preparation.

Houk, V.S. and Claxton, L.D. 1986. Screening complex hazardous wastes for mutagenic activity using a modified version of the TLC/*Salmonella* assay, *Mutation Research,* 169:81–92.

Isenberg, H.D. and D'Amato, R.F. 1991. Chapter 2: Indigenous and Pathogenic Microorganisms of Humans, in: Balows, A., W.J. Hausler, Jr., K.L. Herrmann, H.D. Isenberg and H.J. Shadomy (eds.) *Manual of Clinical Microbiology,* Fifth Edition, American Society for Microbiology, Washington, D.C. pp. 2–14.

Kokjohn, T.A. 1989. Transduction mechanism and potential for gene transfer in the environment. In: S.B. Levy and R.V. Miller (eds.), *Gene Transfer in the Environment,* McGraw-Hill Publishing Co., New York. pp. 73–88.

Krumbein, W.E. (ed.) 1983. *Microbial Geochemistry.* Blackwell Publications. Boston.

LaJoie, C.A., Chen, S.Y., Oh, K.C. and Strom, P.F. 1992. Development and use of field application vectors to express nonadaptive foreign genes in competitive environments. *Appl. Environ. Micro,* 58(2):27–35.

Lamar, R.T. and Dietrich, D.M. 1990. *In situ* depletion of pentachlorophenol from contaminated soil by *Phanerochaete* spp. *Appl. Environ. Microbiol.,* 56(10):3093–3100.

Leahy, J.G. and Colwell, R.R. 1990. Microbial degradation of hydrocarbons in the environment. *Microbiological Reviews,* 54(3):305–315.

Levin, M. and Strauss, H. 1991. An approach to risk assessment when uncharacterized microbial consortia are grown during bioremediation activities. Report prepared for the Environmental Health Directorate, Health and Welfare Canada, Ottawa, Ontario.

Lewis, D.L. and Gattie, D.K. 1991. The ecology of quiescent microbes. *ASM News,* 57(1):27–32.

Lindow, S.E., Panopoulos, N.J. and McFarland, B.J. 1989. Genetic engineering of bacteria from managed and natural habitats. *Science,* 244:1300.

Macieira-Coelho, A. 1986. Cancer and aging. *Exp. Gerontology,* 21:483–495.

Marthi, B., Fieland, V.P., Walter, M. and Sendler, R.J. 1990. Survival of bacteria during aerosolization. *Appl. Environ. Microbiol,* 56(11):3463–3467.

Mikesell, M.D. and Boyd, S.A. 1988. Enhancement of pentachlorophenol degradation in soil through induced anaerobiosis and bioaugmentation with anaerobic sewage sludge, *Environ. Sci. Technol.,* 22:1411–1414.

Mileski, G.J., Bumpus, J.A., Jurek, M.A. and Aust, S.D. 1988. Biodegradation of pentachlorophenol by the white rot fungus *Phanerochaete chrysosporium. Appl. Environ. Microbiol.,* 54(12):2885–2889.

Miller, R.V. and Levy, S.B. 1989. Horizontal gene transfer in relation to environmental release of genetically engineered microorganisms. In: S.B. Levy and R.V. Miller (eds.), *Gene Transfer in the Environment.* McGraw-Hill Publishing Co., New York, pp. 405–420.

Molholt, B. 1989. Chapter Six: Innovative biotechnologies for detection and disposal of hazardous materials and wastes, In: Majumdar, S.K., Miller, E.W. and Schmalz, R.F. (eds.), *Management of Hazardous Materials and Wastes: Treatment, Minimization and Environmental Impacts.* Pennsylvania Academic Press, pp. 58–70.

Mueller, J.G., Middaugh, D.P., Lantz, S.E. and Chapman, P.J. 1991. Biodegradation of creosote and pentachlorophenol in contaminated ground water: chemical and biological assessment. *Appl. Environ. Microbiol.,* 57(5):1277–1285.

National Research Council (NRC). 1983. *Risk Assessment in the Federal Government: Managing the Process.* National Academy Press, Washington, D.C.

National Research Council, Committee on Environmental Epidemiology. 1991. *Environmental Epidemiology: Public Health and Hazardous Waste. Volume 1.* National Academy Press, Washington, D.C.

National Research Council, Committee of Risk Assessment of Hazardous Air Pollutants. 1994. *Science and Judgment in Risk Assessment.* National Academy Press, Washington, D.C.

NETAC (National Environmental Technology Applications Corporation). 1992. *Oil Spill Bioremediation Products Testing Protocol Methods Manual.* Prepared for USEPA. University of Pittsburgh Applied Research Center, Pittsburgh, PA.

Paul, E.A. and Clark, F.E. 1989. *Soil Microbiology and Biochemistry.* Academic Press, San Diego, CA.

Perera, F.P. 1987. The potential usefulness of biological markers in risk assessment. *Environ. Health Perspect.,* 76:141–145.

Phillips, D.H., Hemminki, K., Alhonen, A., Hewer, A. and Grover, P.L. 1988. Monitoring occupational exposure to carcinogens: detection by ^{32}P-post-labeling of aromatic DNA adducts in white blood cells from iron foundry workers. *Mutat. Res.,* 204:531–541.

Phillips, D.H., Schoket, B., Hewer, A., Bailey, E., Kostic, S. and Vincze, I. 1990. Influence of cigarette smoking on the levels of DNA adducts of human bronchial epithelium and white blood cells. *Int. J. Cancer,* 46:569–575.

Pipke, R., Wagner-Dobler, I., Timmis, K.N. and Dwyer, D.F. 1992. Survival and function of a genetically engineered pseudomonad in aquatic sediment microcosms. *Appl. Environ. Microbiol.,* 58(4):1259–1265.

Ramos, J.L. and Timmis, K.N. 1987. *Microbial. Science.* 4:228.

Robinson, J.V. and Kickerson, J.E. 1987. Does invasion sequence affect community structure? *Ecology,* 68:587–595.

Rothman, K.J. 1986. *Modern Epidemiology.* Little, Brown and Company, Boston.

Sagan, L.A. 1984. Problems in health measurements for the risk assessor. In: Ricci, P.F., Sagan, L.A. and Whipple, C.G. (eds.), *Technology Risk Assessment.* The NATO Advanced Science Series (Series E), Martinus Nijhoff Publishers, The Hague, Netherlands, pp. 1–30.

Sanseverino, J., Werner, C., Fleming, J.T., Applegate, B.M., King, J.M.H. and Sayler, G.S. 1993. Molecular diagnostics of polycyclic aromatic hydrocarbon biodegradation in manufactured gas plant soils. *Biodegradation,* 4(4):303–321.

Schenk, T., Muller, R. and Lingens, F. 1990. Mechanism of enzymatic dehalogenation of pentachlorophenol by *Arthrobacter* sp. strain ATCC 33790. *J. Bacteriol.,* 172(12):7272–7274.

Seidler, R. 1992. Evaluation of methods for detecting ecological effects from genetically engineered microorganisms and microbial pest control agents in terrestrial systems. *Biotech. Advances,* 10:149–178.

Short, K.A., Doyle, J.D., King, R.J., Sendler, R.J., Stotzky, G. and Olsen, R.H. 1991. Effects of 2,4-dinitrophenol, a metabolite of a genetically engineered bacterium, and 2,4-dichlorophenoxyacetate on some microorganism-mediated ecological processes in soil. *Appl. Environ. Microbiol.,* 57:412–418.

Shugart, L.R. 1995. Biomarkers of DNA Damage. In: de Serres, F.J. and A.D. Bloom (eds.), *Ecotoxicity and Human Health: A Biological Approach to Environmental Remediation.* Lewis Publishers, Boca Raton, Florida, 123-142.

Smith, T. and Rappaport, S. 1991. *Exposure Assessment for Epidemiology and Hazard Control.* Lewis Publishers, Chelsea, MI.

Sneath, P. 1986. *Bergey's Manual of Systematic Bacteriology, Volume 2.* Williams and Wilkins, Baltimore.

Staley, J.T. 1986. *Bergey's Manual of Systematic Bacteriology, Volume 2.* Williams and Wilkins, Baltimore.

Stayner, L. 1992. Methodolgic issues in using epidemiologic studies for quantitative risk assessment, In: Clewell, H.J. (ed.) *Conference on Chemical Risk Assessment in the Department of Defense: Science, Policy, and Practice.* Am. Conf. of Governmental Industrial Hygienists, Cincinnati, OH.

Steiert, J.G. and Crawford, R.L. 1986. Catabolism of pentachlorophenol by a *Flavobacterium* sp. *Biochem. Biophys. Res. Commun.,* 141(2):825–830.

Suter, G.W. II. 1989. Ecological endpoints. In: W. Warren-Iticks, B.R. Parkhurst and S.S. Baker, Jr. (eds.) *Ecological Assessment of Hazardous Waste Sites: A Field and Laboratory Reference Document,* EPA 60013-89-013.

Suter, G.W. II. 1990. Endpoints for regional ecological risk assessments. *Environ. Manage,* 14:9–23.

Suter, G.W. II. 1993. *Ecological Risk Assessment.* Lewis Publishers, Chelsea, MI.

Suter, G.W. II, Combay, B.W., Hadden, C.T., et al. 1995. An approach for balancing health and ecological risks at hazardous waste sites. (In press)

Tiedje, J.M., Colwell, R.K., Grossman, Y.L., Hodson, R.E., Larski, R.E., Mack, R.N. and Regal, P.J. 1989. The release of genetically engineered organisms: A perspective from the ecological society of America. *Ecology,* 70(2):298–315.

Tilman, D. 1982. *Resource Competition and Community Structure.* Princeton University Press, Princeton, NJ.

U.S. Congress, Office of Technology Assessment. 1988. *New Developments in Biotechnology: Field Testing Engineered Organisms: Genetic and Ecological Issues.* OTA-BA-350.

U.S. General Accounting Office (GAO). 1991. *Hazardous Waste: Limited Progress Closing and Cleaning up Contaminated Facilities.* GAO/RCED 91-79.

USEPA. 1976. Interim procedures and guidelines for health risk and economic impact assessments of suspected carcingines. *Fed. Reg.,* 41:21402–21405.

USEPA. 1980. Water Quality Criteria Documents: Availability. *Fed. Reg.,* 45:79318–79379.

USEPA. 1984. *Risk Assessment and Management: Framework for Decision Making,* EPA/600/9-85/002.

USEPA. 1989a. *Risk Assessment Guidelines for Superfund Volume 1, Human Health Evaluation Manual (Part A).* Interim Final, EPA/540/1-89/002.

USEPA. 1989b. *Subdivision M of the Pesticide Testing Guidelines: Microbial and Biochemical Pest Control Agents.* Office of Pesticide Programs (H-7501C) U.S. EPA, Washington, D.C.

USEPA. 1990. *Methodology for Assessing Health Risks Associated with Indirect Exposure to Combustor Emissions.* Interim Final, EPA/600/6-90/003.

USEPA. 1991a. *Summary Report on Issues in Ecological Risk Assessment,* EPA/625/3-91/018.

USEPA. 1991b. *Innovative Treatment Technologies: Semi-annual Status Report.* U.S. EPA Office of Solid Waste and Emergency Response, Technology Innovation Office, EPA/540/2-91/001.

USEPA. 1992. *Framework for Ecological Risk Assessment,* EPA/630/R-92/001.

USEPA. 1993a. *Bioremediation Field Initiative Site Profile: Libby Ground Water Superfund Site,* EPA/540/F-93/510A.

USEPA. 1993b. Bioremediation Workshop, Duluth, Minnesota.

van Elsas, J.D. and Trevors, J.T. 1991. Environmental risks and fate of genetically engineered microorganisms in soil. *J. Environ. Sci. Health,* A26(6):981–1001.

Vitousek, P.M. 1986. Biological Invasions and ecosystem properties: Can species make a difference? In: Mooney, H.A. and J.A. Drake (eds.) *Ecology of Biological Invasions or North America and Hawaii.* Springer-Verlag, New York.

Weinstein, C. 1995. Ecotoxicology: Environmental Fate and ecosystem impact. In: de Serres, F.J. and A.D. Bloom (eds.), *Ecotoxicity and Human Health: A Biological Approach to Environmental Remediation.* Lewis Publishers, Boca Raton, Florida.

Williams, S.T. 1989. *Bergey's Manual of Systematic Bacteriology, Volume 2.* Williams and Wilkins, Baltimore.

Wilson, J.G., Ritter, E.J., Scott, W.J. and Fradkin, R. 1977. Comparative distribution and embryotoxicity of acetylsalicylic acid in pregnant rats and rhesus monkeys. *Toxicol. Appl. Pharmacol,* 41:67–78.

Bioremediation in Canada: Research Directions

SUZANNE LESAGE
Groundwater Remediation Project
National Water Research Institute
Environment Canada
867 Lakeshore Road
Burlington, Ontario, Canada L7R 4A6

INTRODUCTION

The Canadian research scene for the bioremediation of contaminated soils and groundwater has evolved over the past 5 years from the research efforts of a few selected individuals to being the major focus of remediation research. More than half of the proposals submitted for funding for the development of novel remedial technologies are based on biological processes, shifting from the physico-chemical methods, such as soil venting or pump-and-treat. In this respect, the Canadian research community seems to follow the worldwide trend towards natural, less invasive, and maybe less expensive technologies for remediation.

The research activities themselves range from studies of the fundamental processes at the genetic level to large field-scale comparisons of technologies. The rate of acceptance of bioremediation techniques, as they are developed, has been staggeringly rapid, such that some have moved from the laboratory to the field to commercial application within a few years. The willingness to move to full scale rapidly and being given the permission to do so by regulators have been the trademarks of many remediation successes in Canada. Biopiles are now accepted practice for the remediation of petroleum-contaminated shallow soils.

This chapter will give an overview of the research efforts in Canada. Most of the information is from reports and conference proceedings, which are not widely distributed outside the country. Hopefully this

Copyright: government of Canada; printed with permission.

account will give the reader enough information such that he will not hesitate to contact the individual researchers for more information.

FUNDAMENTAL PROCESSES

Research on the fundamental processes of bioremediation is being conducted mainly at the Universities of Waterloo and Guelph, Ontario, and at the École Polytechnique and the Biotechnology Research Institute (BRI) of the National Research Council of Canada, both located in Montreal, Quebec.

Genetics

One of the most innovative basic techniques that has been developed in recent years is the use of gene probes to find whether bacteria have the genetic ability to degrade certain compounds. A research group, under the direction of Dr. Charles Greer (Greer, Beaumier, et al., 1993) at BRI, worked on the application of catabolic gene probes to survey and monitor soils contaminated with petroleum hydrocarbons using the *alkB, xylE,* and *ndoB* gene probes, which are specific for aliphatic and aromatic degradation. The *alkB* gene codes for the enzyme alkane hydroxylase, whereas the *xylE* gene codes for catechol dioxygenase and the *ndoB* probe is one of the three genes encoding naphthalene dioxygenase. They found that the bacterial populations present in soils generally contained the appropriate genetic potential to degrade the hydrocarbons and that failure to do so must then be attributed to other environmental factors. The technique was also applied to PCB-contaminated soils, where it was found that the bacterial population could generally effect the initial step of the degradation from chlorobiphenyl to chlorocatechol, but that the occurrence of bacteria capable of complete PCB mineralization was very rare.

The gene probe technique can also be used as a tool to enrich bacteria with specific catabolic abilities (Gamati and Greer, 1993; Greer, Masson, et al., 1993). A bacterium that has shown a specific catabolic activity can be isolated, cultured, and reintroduced to enhance the overall degradation rate. The survival and activity of bacteria introduced into contaminated environments as bioremediation agents are essential if this technique is to be used successfully. Reporter genes were introduced into a 2,4-dichlorophenoxyacetic acid-degrading bacterium. The *lac*ZY and *lux*AB genes confer the ability to survive on lactose as a sole carbon source and bioluminescence, respectively. This allows monitoring the presence and viability of the bacteria amongst the much more prevalent native populations (Godbout et al., 1993).

A similar genetic probe system was used to characterize the bacterial population in an aboveground reactor, searching for chlorobenzoate degradability (Wyndham et al., 1993). It was found that bacteria capable of degrading the 2-chloro isomers were generally unable to degrade the 3- and 4-chloro isomers. The genetic ability was found to be encoded within a transposable element (Tn5271), which means that the degradative genes would be highly mobile within and between a variety of bacterial hosts. This group has also been involved in cloning of the transposable chlorobenzoate-3,4-dioxygenase gene (Nakatsu and Wyndham, 1993).

Metabolism

The number of potential contaminants in the environment is such that the bacterial metabolism for all of them is far from being known. For example, in the petroleum sector, most of the attention has been focused on polynuclear aromatic hydrocarbons, while petroleum is composed of hundreds of compounds, which are just as likely to be toxic, or whose metabolites have the potential to be toxic. One such group of compounds are the benzothiophenes. Researchers at the University of Alberta have been studying the fate of the benzothiophenes and the methylbenzothiophenes from petroleum and coal tar. They have synthesized metabolites, looked for their presence in the environment, and studied their toxicity (Fedorak et al., 1993). They found that the sulfur heterocycles are oxidized to hydroxyformylbenzothiophene and the corresponding sulfone and sulfoxides and that the metabolites identified in pure cultures can also be found in sediments contaminated with crude oil (Kropp et al., 1994).

Another important, yet seldom studied, phenomenon is the possibility of substrate interaction during bioremediation of complex mixtures. A very comprehensive study was done for a mixture of compounds present in creosote, using a factorial designed experiment (Millette et al., 1993) and a dynamic system of soil columns (Millette et al., 1994). The group studied the effect of phenanthrene, fluorene, and *p*-cresol on the aerobic biodegradation of carbazole in groundwater. The results obtained in the columns were different from those in the batch studies. *p*-Cresol caused an increase in the lag phase for the degradation of fluorene and carbazole but eventually allowed for complete mineralization, which did not occur in its absence. One important fact to note is the necessity to run dynamic experiments, not only batch studies, to describe in situ bioremediation and interaction between substrates accurately.

Other researchers are interested in manipulating bacterial metabolism, not to degrade specific compounds, but to produce compounds that may assist in bioremediation. The group under the direction of J. Trevors at the University of Guelph, has been isolating biosurfactants from bacteria isolated from oil-contaminated soils. The strain *Pseudomonas aeruginosa* UG2 was found to excrete a rhamnolipid biosurfactant that was found to enhance the bioavailablity of chlorinated biphenyls (Trevors et al., 1992) and of aliphatic hydrocarbons (Jain et al., 1992). The biosurfactants were found to be effective in enhancing the removal of selected hydrocarbons from soils and to be as effective as most synthetic surfactants (Scheibenbogen et al., 1994)

Microcosm Studies

In a microcosm study, the rhamnolipid surfactant from *Pseudomonas aeruginosa* UG2 was found to be more effective than sodium dodecyl sulfate for the solubilization of PAHs, but both surfactants seemed to be preferential substrates and hindered, rather than helped, the biodegradation of PAHs (Deschênes et al., 1994).

Many researchers report on the biodegradability of aromatic hydrocarbons, but few rigorous studies have been done on the kinetics of the transformation in soil and on the effect of the availability of nutrients on the rate. In their paper, Allen-King et al. (1994) suggest that, where oxygen is in ample supply, such as in the vadose zone, the observed variations in rates are due to the differing amount of biomass of the organisms responsible for BTEX degradation, to prevailing redox conditions, and to the availability of nutrients. Hawari et al. (1994) studied the correlation between biodegradation and sorptionl/desorption of nitrogen-containing PAHs. An initially rapid degradation rate was followed by a much slower phase. This was attributed to the degradation of contaminants present in the dissolved and sorbed states, respectively. The authors also identified different sorption mechanisms for the nitro- and amino-naphthalenes, naphthalene and quinoline.

Chlorofluorocarbons (CFCs) have been the target of much environmental research in recent years, mostly because of their adverse effects on the ozone layer. CFCs are also known as groundwater contaminants but have received less attention, probably because they are not on anyone's list of priority toxic chemicals. CFC-113 (1,1,2-trichloro-1,2,2-trifluoroethane) has been used extensively in the electronic industry and in the analysis of oil and grease content in the petroleum and waste sectors. CFCs are a problem in the stratosphere because of their recalcitrance to chemical and biological degradation. For years, these compounds were thought to be totally nonbiodegradable. Because of their similarity to the chlorocarbons, however, there was a distinct possibility for their biodegradation under anaerobic conditions. In microcosm experiments using landfill leachate as a biological inoculum, a half-life of 5 days for the reductive dechlorination of CFC-113 to HCFC-123a and 5.6 days for the removal of one more chlorine atom were measured (Lesage et al., 1992). The products HCFC-133 and HCFC-133b were identified by GC/MS.

Because landfill leachate is a major source of groundwater contamination, secure landfills are now required to have leachate collection and recirculation or treatment. Municipal landfills often contain chlorinated and aromatic organic solvents present in waste from either households or small industries. Treatment can be costly and exceed the budget of municipalities administering the facilities. The use of waste stabilization ponds for the leachate has been studied as a cost-effective treatment method. The fate of the organic solvents in such systems was studied both in microcosm and in pilot field scale (Lesage et al., 1993). It was found that, during storage in closed, but not airtight, containers, most of the chlorinated solvents would be biodegraded, within 2 to 3 months without the further addition of nutrients. Biodegradation was confirmed by the formation of reductive dechlorination products, notably from CFC-113, which was added as a positive control. The aromatic hydrocarbons were not significantly lost during the same period. The preferred disposal alternative would therefore be to store the leachate in closed containers for a period and subsequently aerate it, prior to discharge.

One of the major problems in the pollution of groundwater due to petroleum spills is that, while most of the components of gasoline are easily biodegraded under aerobic conditions and many aromatic compounds can also be degraded under anaerobic conditions, benzene, the most toxic component, has been recalcitrant in most anaerobic aquifers. The search, however, continues for the appropriate organism or growth conditions (Edwards, 1994). Some success was achieved in laboratory microcosms where it was shown that nitrate was not essential to this activity, but sulfate reduction seemed to be involved.

While biological processes have been shown to be of great advantage in the remediation of contaminated groundwater and soil, some difficulty in keeping the microbes alive in highly contaminated environments has hampered its widespread acceptance. One means to circumvent the problem is to use the isolated enzymes that are responsible for the transformation. This task is far from simple for many enzyme systems, but in the case of reductive dechlorination, it has been found that relatively simple molecules such as porphyrins (e.g., hematin) or corrinoids (e.g., vitamin B12) could effect the transformation, even in the absence of the protein that surrounds them in living organisms. This led to the possibility to apply vitamin B12, itself a product of microbial fermentation, to the dechlorination of tetrachloroethylene present as residual DNAPL (Lesage et al., 1993). The vitamin, with titanium citrate as a reducing agent, can be used either alone or in conjunction with methanogenic bacteria (Millar and Lesage, 1994).

BENCH AND PILOT SCALE

Bioreactors

The native biomass at a site may be insufficient to degrade the contaminant at the desired rate. In such a case, it may be preferable to isolate the bacteria and enrich them in a batch reactor under ideal growth conditions before reinserting them in the soil. One such experiment was conducted with a soil contaminated with a mixture of pentachlorophenol (PCP) and creosote (Otte et al., 1993). In the reactor, the rate of PCP degradation was increased 10-fold, and the rates of phenanthrene and pyrene were increased 30- and 80-fold, respectively. The use of bioaugmentation has the advantage of reducing the lag phase by reintroducing an acclimated bacterial population.

A similar technique was used to produce an inoculum acclimated to the presence of phthalate esters (Ward and Yu, 1993). This group first carried out a laboratory feasibility study followed by a pilot field application. Conditions were established that enabled soils contaminated with greater than 5000 μg/kg of plasticizer to be remediated to below 15 μg/kg in less than 80 days in plots with soil depths between 6 and 18″.

The performance of a combined anaerobic/aerobic reactor was found to be similar to that of the conventional upflow anaerobic sludge bed reactor for the degradation of tetrachloroethylene (Kuang et al., 1994), except that the coupled reactor demonstrated higher specific perchloroethylene degradation rates.

Bioreactors were designed to treat diesel invert mud residue (DIMR) in a high brine soil (Johnson et al., 1994). The experiments were aimed at optimizing reactor conditions and at comparing the rate of degradation of this petroleum waste with other residues. It was found that, while nitrogen was essential, it could be applied in any form and that a single application was needed. The pattern of microbial respiration of DIMR was similar to that of other petroleum wastes.

Fungi

The use of white rot fungi for the biodegradation of PAHs in soils contaminated with creosote is being investigated by a group in New Brunswick (Boyle and Richardson, 1993, 1994). Although field application is envisaged as part of the project, the only results reported so far are the bench microcosm experiments. One interesting feature of the work is the use of the ability of the fungi to decolorize a polymeric dye Poly-R as a visual mean of assessing the viability of the fungal population. Several fungi were isolated and their PAH degradation ability assessed. *Trametes versicolor* was found to give the best results for phenanthrene, fluoranthene, and benzo-*a*-pyrene, but none of the fungi were able to degrade chrysene. Fungi's growth in soil required the addition of lignin-rich amendments, and ground alfalfa gave the best results. The problem of population survival and competition with other species in soil remain to be addressed.

Coal Tar/Creosote

One of the major sources of soil contamination in Canada is coal tar/creosote, often with pentachlorophenol, widely used in the wood preservation industry. Coal tar and creosote have a very high content of PAHs, much higher than any petroleum product. In order to evaluate different commercially available technologies, a group of companies specialized in remediation was asked to do a bench-scale treatability study of the same soils, from an industrial area called Pacific Place, Vancouver, British Columbia (Whiting and Hayes, 1993). The groups used either solid-phase or slurry-phase bioreactors. All results were very disappointing, because most of the remediation criteria were not achieved. Where PAHs did disappear in the bioreactors, there was no difference with the control, thus indicating that bioremediation wasn't occurring. Most vendors agreed that the levels of contaminants were probably inhibitory to bacterial growth. The consultants found it difficult to assess the progress of treatment because the laboratory analytical data wasn't sufficiently reliable. This illustrates the fact that those who apply remediation technologies often underestimate the difficulty in analysing PAHs in highly contaminated soils.

Air/Biofilters

Soil and groundwater remediation technologies such as soil venting and air sparging result in off-gases that require treatment in most jurisdictions. The use of biofilters has been sought as an alternative to the current expensive technologies of carbon sorption, catalytic oxidation, and wet scrubbing (Lei et al., 1994; Rho et al., 1994; Parker et al., 1994). The contaminants to be treated here are benzene, toluene, and xylenes (BTEX). As with any bioreactors, the biofilter must be inoculated with appropriate strains and the suitable moisture and nutrient concentration maintained to achieve optimal performance.

In situ Treatment

A research facility to bridge the gap between laboratory and field scale was developed in 1993–1994 at the National Water Research Institute of Environment Canada in Burlington, Ontario. The controls necessary to produce scientifically defensible results are only possible when the experiments are conducted in an area when variables such as light, temperature, and precipitation can be regulated. The Aquatic Ecosystem Remediation Evaluation Facility (AQUEREF) will provide an area where pilot-scale experiments for the remediation of contaminated sediments, soils, and groundwater will be conducted such as to minimize impact to the environment. The groundwater component houses a stainless steel tank (8 ft × 6 ft × 20 ft) filled with aquifer solids and equipped with a very dense monitoring well system (more than 300 sampling points). The initial experiments will focus on the use of humic acids to enhance the solubility and rate of bioremediation of petroleum products.

FIELD

The number of full-scale applications that can still be considered research projects is somewhat limited. The research component of the field applications vary widely between projects. For soil bioremediation, the objective of many research projects revolve around improving the texture or nutrient distribution of the soil. For groundwater remediation, semipassive systems for the delivery of oxygen and nutrients are being applied to different contamination problems ranging from metals to nitrate to petroleum. Finally, the effect of cold climate on remediation is an obvious concern for most Canadian sites.

Soil Amendments

One such application involves the use of a soil treatment adjuvant registered under the trade name Daramend™. Its use on soils was found to increase the effectiveness of the biodegradation of chlorinated phenols, PAHs, and petroleum hydrocarbons (Seech et al., 1993, 1994). The soil amendment is a mixture of nutrients and bulking agents that help create the ideal conditions for bioremediation. They were able to treat soils contaminated with up to 700 mg/kg of pentachlorophenol to below CCME (Canadian Council of Ministers of the Environment) guideline levels. Total PAHs were reduced to below guidelines, whereas total petroleum hydrocarbons were reduced from 6000 to less than 50 $mg{\cdot}kg^{-1}$. The treatment was also used successfully at another site in British Columbia (Herbert et al., 1994). The addition of sawdust was investigated as a soil amendment for its biodegradation (Trudell, 1994) and found to reduce the half-life of PCP to 160 from 190 days for the unamended soil. Both cells received nutrients.

Amendments can also be used advantageously for the biodegradation of heavier fuels (Paquin, 1994). Both mechanical shearing and additives made of mixtures of biosurfactants and enzymes were used successfully to increase the accessibility of the microbial population and of nutrients to the oil and grease residues, thus greatly increasing the rate of biodegradation.

Atrazine, bromacil, tebuthiuron, and diuron have been used as soil sterilants in Alberta. When a formerly industrial site is to be rehabilitated or if the herbicides have moved off-site, it is desirable to neutralize the effect of these chemicals and allow revegetation to occur. Of a host of experimental adjuvants (activated charcoal, sawdust, manure, fly ash, commercial humates, commercial microbes, peat, and fertilizer), activated charcoal was the only truly effective compound (Cotton et al., 1994).

In situ Remediation of Groundwater

Some Canadian researchers conduct their field experiments outside the country. Cox and Major (1993) reported on an in situ biodegradation field experiment carried out in Merced, California. This site was unusual because trichloroethylene did not seem to be naturally biodegrading. The research team installed in situ microcosms (ISMs) (Gillham et al., 1992) and studied the microbial population present. The major product of transformation of [1,2-^{14}C]-TCE was not the reductive dechlorination products, *cis*- and *trans*-dichloroethene and vinyl chloride, but rather $^{14}CO_2$. The addition of methanol was successful in generating anaerobic conditions and the biotransformation was attributed to a consortium of both acetogenic and methanogenic bacteria.

The ISMs were also used at a site in North Toronto where tetrachloroethene degradation was observed to occur naturally (Major and Cox, 1994). The group did in situ injections of PCE, bromide, and methanol and found that the reductive dechlorination to ethylene and ethane was accompanied by the production of acetate and methane, thus providing evidence of the presence of an active methanogenic/acetogenic consortium.

Fixed-Bed Bioreactors for Nitrate Removal

The addition of nitrate in agricultural areas can lead to groundwater contamination in zones of high soil permeability. To prevent infiltration, in many areas, drainage tiles are installed in agricultural fields to channel the run-off water into the surrounding drainage channels and streams, then posing a threat to the surface water quality. Fixed-bed bioreactors were used on-site to treat this run-off water (Blowes et al., 1994b). Three reactors were made of 200-L barrels containing a mixture of sand, grow-bark, wood chips, and composted leaf mulch in varying proportions. The organic matter was added as a source of organic carbon to support the growth of denitrifying bacteria. After an acclimation period of 2 weeks, the effluents from the barrels were consistently below 0.02 mg/L, in spite of the fluctuation in water flow. The best results were obtained where a mixture of different types of organic carbon was used. These reactors are very simple in design and could be used widely in agricultural areas at a low cost.

In-Ground Reactive Walls for Metal Precipitation

Mine tailings impoundments generate leachates that are characterized by low pH and high concentration of sulfate, iron, and other reduced metals. A group at the University of Waterloo is involved in in situ bioremediation of the plumes generated downstream of these installations (Blowes et al., 1994a). A reactive wall was built across the zone to be remediated and filled with a variety of organic wastes easily accessible in a northern environment, such as leaf compost, sheep manure, sawdust, wood chips, and sewage sludge. Anaerobic, sulfate-reducing bacteria were from a nearby creek. The biological activity caused the reduction of sulfate to sulfide, which causes precipitation of most metals. Preliminary results from the field experiments show that neutralization and removal of Al, Ni, and Zn to below drinking water standard are possible.

Permeable Walls for Groundwater Remediation

The major hurdle in effectively implementing any in situ treatment is the effective delivery of the treatment to all affected areas. This is why it is now well established that there are ideal sites for in situ bioremediation: those sites where air and water can be delivered rapidly and efficiently, coarse sand and gravel aquifer. Unfortunately, it is not only the overall permeability that must be considered, but its distribution. Indeed, very permeable zones within less permeable ones cause channelling, which impedes the even distribution of nutrients and air. For this reason, a group at the University of Waterloo has been investigating alternate delivery systems. A permeable wall system was applied to a BTEX plume at the CFB Borden (Barker et al., 1993). The efficiency of the system was limited by the amount of oxygen that can be introduced in water. The funnel and gate scenario channels the water to be treated through a remedial zone (Smyth et al., 1994) in which an air sparging can be installed or in which an oxygen-releasing compound can be added (Bianchi-Mosquera et al., 1994). The compound, a peroxide salt, was added either as a powder or in a briquette of concrete. The latter form seemed to be the most effective, and oxygen was still being released 9 weeks after the installation (Weber and Barker, 1994).

Surface Application

Remediation of groundwater contaminated by gasoline is often accomplished by pumping and treating above ground, using either air stripping or adsorption on charcoal. However, in the vadose zone where an ample oxygen supply exists, biodegradation has been found to occur naturally. This suggested that surface application, often termed trickle irrigation, would be a potential cost-effective technique for groundwater treatment (O'Leary et al., 1993). In a controlled field study conducted at the Canadian Forces Base Borden, they found that, with nitrogen addition, a 25 m^2 plot of sandy soil could remediate 18 m^3 per day of water containing 11 mg/L of BTEX to below drinking water standards. Bioremediation was the major source of removal with volatilization accounting for only 6% of the losses.

Air Induction

At many sites where contamination is extensive and hydrogeologic conditions varied, no single system is likely to be able to remediate the site effectively. A combination of treatment, which they collectively termed "air induction" was devised by a consulting group in the Atlantic region (Elder et al., 1993). The system combines the sequential injection of air and/or water with nutrients, dewatering through the same wells, induction of air through the ground surface by extraction at the wells to enhance bioremediation. The operations are repeated on a cyclical basis where it appears that nutrients may have become rate limiting in the ground. This was applied successfully at bulk fuel distribution facility in Sydney, Nova Scotia, where the overall cost of \$20/m3 was half of the estimated cost for excavation and off-site treatment.

Biopiles in a Cold Climate

One of the Canadian realities is the nature of its climate. A full-scale demonstration of soil bioremediation using biopiles was done on the site of a former refinery in Montreal (Desrochers et al., 1993). In addition to the adverse climatic conditions, the soil to be treated had a large clay content. Therefore, the soils were subjected to a pretreatment consisting of sorting and mixing with a bulking agent such as sawdust or gypsum before being placed in the treatment cells. The group found that the major factor in successful remediation was the efficient aeration of the pile, more than moisture or the availability of nutrients. Intensive monitoring using gene probes showed that the bacteria specifically responsible for petroleum degradation (*xyl*E and *alk*B positive) were much more sensitive to cold conditions than the general heterotrophic population present in the soil (Samson et al., 1993). In less than a year, the criteria of less than 1000 ppm oil and grease was achieved. In spite of a slightly higher initial cost for set-up, this technology is considerably less expensive than the alternatives of excavation and landfilling, landfarming, or incineration.

A similar type of bioreactor was used for gasoline and diesel contaminated soils at the Canadian Forces Base Petawawa (McNicoll et al., 1993, 1994). The climate conditions were similar to those found in the Montreal area (winter −20 to −10°C, summer 20 to 30°C), but the soil was a fine to medium grained sand of deltaic origin. Attempts were made to control the summer solar heat by covering the soil pile with a white, instead of a black, membrane. One of the biopiles did not receive any added nutrients as a control. In spite of the more permeable soil type, there was a considerable difference in the distribution of contaminants between bioreactors. No significant difference in soil temperature was observed between the white- and the black-covered cells. The total petroleum hydrocarbon concentration was reduced by 97% after 6 months.

Two additional projects were carried out on poorly permeable soils, one targeting transformer oil, the other a soil highly contaminated with diesel (exceeding 50,000 mg/kg oil and grease) (Pouliot and Sansregret, 1993). In these cases, the operators periodically mixed the soils in the biopiles, while adding nutrients and amendments such as peat moss and sawdust. A 15-kW heater was also used in the winter to keep the temperature optimal. The addition of bacteria and surfactants was also assessed and found to be of no advantage. The treatment of the transformer oil was very successful, but the target criteria were not achieved for the highly contaminated soil. One of the reasons was that an excessive temperature (50°C) was reached, thereby destroying some of the hydrocarbon-degrading bacteria. The authors also comment that the older the spill, the more difficult it is to treat because of the accumulation of the more recalcitrant higher molecular weight hydrocarbons. The same group was able to apply this technology successfully on the east coast of James Bay (latitude 53° 12′) for diesel-contaminated soil (Faucher et al., 1994). Poorly permeable soils required special treatment in a site demonstration in the Toronto area (Itamunoala, 1994). The group studied the use of a high-flow/high-pressure pump to improve clay dissolution and a vortex agitator to reduce clay lump size. Bioventing was also found to compare advantageously to soil vapor extraction in a northern climate (Armstrong et al., 1994).

Passive and active biopiles were compared for effectiveness in treating contaminated soils at a polystyrene plant (Benazon et al., 1994). Both systems were capable of treating ethylbenzene, styrene, and semi-polymerized material, although the passive system required more time and space and was more difficult to monitor.

MONITORING, REGULATIONS, AND FUNDING

Monitoring

A systematic study was done at seven gasoline service stations to determine the most useful monitoring parameters for bioremediation

(Samson et al., 1994). The presence of bacteria capable of degrading hydrocarbons could not be used as a monitoring parameter because their quantity was not linked to the extent of contamination. While gene probes were interesting tools, the most useful biodegradation parameter was the mineralization rate of target compounds. Amongst the chemical tests, total petroleum hydrocarbons (TPH) and BTEX provided the most information. When more than one laboratory is used, it is important to standardize the methodology used for TPH.

Most of the assessments of bioremediation success are based on the chemical analysis of a selected group of target priority chemicals. However, the real goal of bioremediation is the return of a soil or water to its "before" state, that is to a soil typical of a forest or of an agricultural soil, able to sustain the growth of food suitable for human consumption. The preferable option, then, is clearly to use a bioassay that would measure the soil's toxicity (Renoux et al., 1993). One of the difficulties is that most soils, even when taken from pristine areas, do show a positive response in the bioassays that were mostly developed for water. Therefore, for every contaminated soil, a control soil must be obtained from a nearby source. The research group introduces the concept of acceptable ecotoxicological level for soils (AELS) to be used as the baseline for comparison. In assessing a petroleum-contaminated site, no correlation could be found between the reduced level of target chemicals and the toxicity of the soils. This shows that the ideal approach requires a battery of tests, probably combining ecotoxicological and chemical assays, and that some commonly used assays, such as Microtox, are not suited to soil testing.

Alternative testing for toxicity in soil contaminated with sour gas processing plants was proposed (Hamilton et al., 1994). The group studied the usefulness of seed germination and root development for lettuce, radishes, and barley, in addition to earthworm survival, as parameters that are more indicative of soil ecosystems rather than human health. For groundwater, bacterial luminescence, *Cerodaphnia dubia* survival and SOS-chromotest were added. The soil tests gave excellent results, but the tests were found to lack sensitivity for groundwater. A similar battery of tests was applied to a crude oil and diesel invert mud residues (DIMR) contaminated soils (Visser and Danielson, 1994) to assess the efficacy of bioremediation. The results did not correlate completely with the current TPH guidelines, and at the same TPH concentration, DIMR contaminated soil was more toxic than crude contaminated soils. The exact reason for this difference could not be confirmed, although the higher concentration of PAHs was identified as a possible cause.

Regulations

The Canadian Environmental Protection Act encompasses all the new products being introduced to the Canadian market. As such, the products of biotechnologies are also targeted by the Act. However, because of the complexity of the issues of biotechnology, the government has placed a tremendous effort in the development of regulations for the application of the Act, and at the time of writing, these regulations are not in force (McIntyre, 1993, 1994). The regulations are being developed, not only in consultation with stakeholders in industry, academia, governments, and environmental groups, but also in collaboration with the US EPA, in an effort to be scientifically relevant and to avoid duplication of efforts. One of the most interesting features of the proposed regulations is to regulate based on effects and not on the origin of the product. In that frame of reference, a genetically engineered bacterium would be assessed in the same way as a strain isolated from natural environments. The products of bacteria, such as enzymes or surfactants, are also included. The proposed regulations would also require notification of a new biotechnology product prior to importation and manufacture.

Mackay et al. (1993) and Cherry et al. (1992) have begun to reflect on the increasing number of new technologies being developed and on the realities of the goals to be achieved in groundwater remediation. Most of the full-scale applications are done without a complete bench- and pilot-scale study to back them up. Monitoring of field applications is driven by regulatory concerns, not scientific ones. This means that we may not learn as much from these expensive exercises as we would if they were conducted as large scale, but nonetheless experimental, treatments. Considering the mitigated successes of most remedial operations, it is useful to reassess the goals to be achieved. The likelihood of success is better if we can decide whether plume containment, plume removal, or full aquifer restoration is what we really want or what we can achieve given our current knowledge and the available resources.

Funding

In Canada, all university-based scientific research is generally funded through the Natural Sciences and Engineering Research Council of Canada (NSERC), which also administers research funds for industry and joint government-industry sponsored chairs. For example, a chair for research in bioremediation of contaminated sites was awarded at l'École Polytechnique in Montréal. There are very few research funds that are specific to site remediation. There are only two national programs of that nature: the Groundwater And Soil Remediation Program (GASReP) and Demonstration of Site Remediation Technologies (DESRT).

GASReP is a program funded almost entirely through the Panel on Energy Research and Development (PERD), a fund of the federal government, and primarily addresses the remediation problems of the petroleum sector. It is available to all researchers from industry, government, and universities, but the proponents must have cosponsors. Its purpose is to fund either fundamental or applied research at the bench or pilot scale. On the other hand, DSERT targets the demonstration of technologies in the field. It is approved and funded jointly by the federal government and the government of the province in which the project is conducted.

Some other research activities are funded regionally or provincially, but these are generally in response to specific problems or needs and are not for generic research. The Canadian government scientists also conduct some remediation research as part of the mandate of Environment Canada. Canadian researchers have also been able to access funds from either U.S. or internationally.

REFERENCES

Allen-King, R.M., J.F. Barker, R.W. Gillham and B. Jensen. 1994. "Substrate- and Nutrient-Limited Toluene Biotransformation in Sandy Soil." *Environ. Toxicol. and Chemistry,* 13(5):693–705.

Armstrong, J.E., B.J. Moore, P.E. Hardisty, V. Kühnel and D. Stepan. 1994. "A Field Study of Soil Vapour Extraction and Bioventing," in *Proceedings of the 4th Annual Symposium on Groundwater and Soil Remediation,* September 21–23, 1994, Calgary, Alberta, pp. 621–629.

Barker, J.F., J. Weber, J.S. Fyfe, J.F. Devlin, D.M. MacKay and J.A. Cherry. 1993. "Efficient Addition Technology for in situ Bioremediation." *Proceedings of the 3rd Annual Symposium on Groundwater and Soil Remediation,* September 8–10, 1993, Quebec City, QC, Canada, pp. 285–296.

Benazon, N., D. Bélanger, D. Scheurlen and M. Lesky. 1994. "Bioremediation of Ethylbenzene and Styrene Contaminated Soil Using Biopiles," in *Proceedings of the 4th Annual Symposium on Groundwater and Soil Remediation,* September 21–23, 1994, Calgary, Alberta, pp. 683–696.

Bianchi-Mosquera, G.C., R.M. Allen-King and D.M. Mackay. 1994. "Enhanced Degradation of Dissolved Benzene and Toluene Using a Solid Oxygen-Releasing Compound." *Ground Water Monitoring and Remediation.* Winter, 120–128.

Blowes, D.W., C.J. Ptacek, K.R. Waybrant, J.G. Bain and W.D. Robertson. 1994a. "In situ Treatment of Mine Drainage Water Using Porous Reactive Walls." in *Proceedings of the Environment and Energy Conference of Ontario,* November 15–16, Toronto, Ontario.

Blowes, D.W., W.D. Robertson, C.J. Ptacek and C. Merkey. 1994b. "Removal of agricultural nitrate from tile-drainage effluent water using in-line bioreactors." *J. Contam. Hydrol.,* 15:207–221.

Boyle, D. and A. Richardson. 1993. "Bioremediation of Polyaromatic Hydrocarbon Contaminated Soils Using White Rot Fungi." *Proceedings of the 3rd Annual Symposium on Groundwater and Soil Remediation,* September 8–10, 1993, Quebec City, QC, Canada, pp. 103–118.

Boyle, D. and A. Richardson. 1994. "Development of a Technology Using White Rot Fungi to Remediate Soils Contaminated with PAHs." In *Proceedings of the 4th Annual Symposium on Groundwater and Soil Remediation,* September 21–23, 1994, Calgary, Alberta, pp. 169–183.

Cherry, J.A., S. Feenstra and D.M. Mackay. 1992. "Developing a Conceptual Framework and Rational Goals for Groundwater Remediation at DNAPL Sites," in H. Ward, editor, *Proceedings of Subsurface Restoration Conference, Third International Conference on Ground Water Quality Research,* Dallas, Texas, June 21–24, 1992, pp. 14–17.

Cotton, M.M., M.P. Sharma and W.H. Vanden Born. 1994. "Reclamation of Soils Treated with Non-Selective Residual Herbicides (Soil-Sterilants)." In *Proceedings of the 4th Annual Symposium on Groundwater and Soil Remediation,* September 21–23, 1994, Calgary, Alberta, pp. 61–82.

Cox, E.E. and D.W. Major. 1993. "A Laboratory and Field Assessment of the Potential to Biotransform Trichloroethene in a Deep Aquifer in California." *Proceedings of the 3rd Annual Symposium on Groundwater and Soil Remediation,* September 8–10, 1993, Quebec City, QC, Canada, pp. 85–100.

Deschênes, L., P. Lafrance, J.-P. Villeneuve and R. Samson. 1994. "The Impact of a Biological and a Chemical Anionic Surfactants on the Biodegradation and Solubilization of PAHs in a Creosote Contaminated Soil," in *Proceedings of the 4th Annual Symposium on Groundwater and Soil Remediation,* September 21–23, 1994, Calgary, Alberta, pp. 217–224.

Desrochers, R., T. Hawkes, C.H. Nelson, R. Samson, M. St-Cyr and J.-R. Michaud. 1993. "Biodégradation aérobie en conditions nordiques dans un sol argileux contaminé aux hydrocarbures." *Proceedings of the 3rd Annual Symposium on Groundwater and Soil Remediation,* September 8–10, 1993, Quebec City, QC, Canada, pp. 539–563.

Edwards, E.A. 1994. "Laboratory Search for Microorganisms Capable of Degrading Benzene under Anaerobic Conditions," *Proceedings of the 4th Annual Symposium on Groundwater and Soil Remediation,* September 21–23, 1994, Calgary, Alberta, pp. 435–443.

Elder, D.M., S.A. Yuill, C.D. Barclay and B. Barreyre. 1993. "Air Induction: Efficient and Cost-Effective In Situ Remediation." *Proceedings of the 3rd*

Annual Symposium on Groundwater and Soil Remediation, September 8–10, 1993, Quebec City, QC, Canada, pp. 315–320.

Faucher, C., G. Lajoie, D. Bender, J.-L. Sansregret and Y. Pouliot. 1994. "Environmental Remediation and Field Scale Biological Treatment of Fuel Contaminated Material in Wemindji, James Bay," in *Proceedings of the 4th Annual Symposium on Groundwater and Soil Remediation,* September 21–23, 1994, Calgary, Alberta, pp. 457–466.

Fedorak, P.M., K.G. Kropp, J.A. Gonçalves and J.T. Anderson. 1993. "Bacterial Metabolism of Benzothiophene and Methylbenzothiophenes." *Proceedings of the 3rd Annual Symposium on Groundwater and Soil Remediation,* September 8–10, 1993, Quebec City, QC, Canada, pp. 331–352.

Gamati, S. and C.W. Greer. 1993. "Enhanced Bioremediation of Naphthalene-Contaminated Soil Using Gene Probe-Enriched Indigenous Bacteria." *Proceedings of the 3rd Annual Symposium on Groundwater and Soil Remediation,* September 8–10, 1993, Quebec City, QC, Canada, pp. 395–408.

Gillham, R.W. and S. O'Hannesin. 1992. "An In-Situ Device for Examining Biotransformation Processes." In *In-Situ Bioremediation Symposium '92,* S. Lesage, editor, Proceeding of the In-Situ Bioremediation Symposium held in Niagara-on-the-Lake, Ontario, Canada, Sept. 20–24th, pp. 83–94.

Godbout, J., A. Mihoc, Y. Comeau and C.W. Greer. 1993. "Survival and Activity of a Reporter Gene Marked Bacterium Introduced into Different Soil Environments," in *Proceedings of the 3rd Annual Symposium on Groundwater and Soil Remediation,* September 8–10, 1993, Quebec City, QC, Canada, p. 623.

Greer, C.W., D. Beaumier and R. Samson. 1993. "The Application of Catabolic Gene Probes to Survey and Monitor Soils Contaminated with Chlorinated and Non-chlorinated Organic Pollutants." *Proceedings of the 3rd Annual Symposium on Groundwater and Soil Remediation,* September 8–10, 1993, Quebec City, QC, Canada. pp. 419–428.

Greer, C.W., L. Masson, Y. Comeau, R. Brousseau and Samson. 1993. "Application of Molecular Biology Techniques for Isolating and Monitoring Pollutant Degrading Bacteria." *Water Pollution Research J. of Canada,* 28(2):275–288.

Hamilton, H., D. Kerr, S. Goudey, D. Chollak and J.G. Ward. 1994. "Development of a Protocol for Using Bioassays to Generate Site-specific Remediation Criteria for Sour Gas Processing Plants," in *Proceedings of the 4th Annual Symposium on Groundwater and Soil Remediation,* September 21–23, 1994, Calgary, Alberta, pp. 327–346.

Hawari, J., B. Al-Bashir, R. Leduc and R. Samson. 1994. "Sorption/Desorption and Biodegradation of N-PAHs in Flooded Soil: Substituent Effect," *Proceedings of the 4th Annual Symposium on Groundwater and Soil Remediation,* September 21–23, 1994, Calgary, Alberta, pp. 253–267.

Herbert, D.J., L.W. Rerup, D. Cheng, L. Dikson and R. DeWit. 1994. "Organic Amended Bioremediation and Near-Real-Time Monitoring of Contaminated Sites," in *Proceedings of the 4th Annual Symposium on Groundwater and Soil Remediation,* September 21–23, 1994, Calgary, Alberta, pp. 305–320.

Itamunoala, G.F. 1994. "Bioremediation of MTRCA Site Borough of East York, Ontario." *Proceedings of the 4th Annual Symposium on Groundwater and Soil Remediation,* September 21–23, 1994, Calgary, Alberta, pp. 497–510.

Jain, D.K., H. Lee and J.T. Trevors. 1992. "Effect of Addition of *Pseudomonas aeruginosa* UG2 Inocula or Biosurfactants on Biodegradation of Selected Hydrocarbons in Soil." *J. of Industrial Microbiology,* 10:87–93.

Johnson, R.L., I. Guo, H. Bertram, J. Xu, X. Li, P. Yeung and R.M. Danielson. 1994. "Treating Diesel Invert Mud Residue in a Bio-reactor: a Mass Balance Approach," In *Proceedings of the 4th Annual Symposium on Groundwater and Soil Remediation,* September 21–23, 1994, Calgary, Alberta. pp. 571–582.

Kropp, K.G., J.A. Gonçalves, J.T. Anderson and P.M. Fedorak. 1994. "Microbially Mediated Formation of Benzonaphthothiophenes from Benzo[b]thiophenes." *Appl. Environ. Microbiol.,* 60(10):3624–3631.

Kuang, X., R. Stephenson, C. Beaulieu, J. Hawari and S. Guiot. 1994. "Development of Anaerobic and Anaerobic/Aerobic Treatment for Tetrachloroethylene (PCE)," in *Proceedings of the 4th Annual Symposium on Groundwater and Soil Remediation,* September 21–23, 1994, Calgary, Alberta, pp. 103–115.

Lei, J., D. Lord, R. Arneberg, D. Rho, P. Mercier and B. Cyr. 1994. "Design, Field Operation and Performance of a Biological System to Remove Volatile Hydrocarbons from Waste Gases," in *Proceedings of the 4th Annual Symposium on Groundwater and Soil Remediation,* September 21–23, 1994, Calgary, Alberta, pp. 691–696.

Lesage, S., S. Brown and K.R. Hosler. 1992. "Degradation of CFC-113 under Anaerobic Conditions." *Chemosphere,* 24(9):1225–1243.

Lesage, S. and S. Brown. 1993. "*In-Situ* Biochemical Degradation of Perchloroethylene Present as Residual DNAPL," Presented at the *Emerging Technologies for Hazardous Waste Management, ACS I&EC Symposium,* Atlanta, Sept. 27–29. Extended abstract Volume 2, pp. 539–544.

Lesage, S., R.A. McBride, P.M. Cureton and S. Brown. 1993. "Fate of Organic Solvents in Landfill Leachates under Simulated Field Conditions and in Anaerobic Microcosms." *Waste Management and Research,* 11:215–226.

MacKay, D.M., S. Feenstra and J.A. Cherry. 1993. "Alternative goals and approaches for groundwater remediation," in *Proceedings of the Workshop on "Contaminated Soils—Risks and Remedies."* Stockholm, October 26–28, 1993, pp. 35–47.

Major, D.W. and E.E. Cox. 1994. "A Field and Laboratory Study to Assess and Demonstrate the in situ Anaerobic Biotransformation of Tetrachloroethene," in *Proceedings of the 4th Annual Symposium on Groundwater and Soil Remediation,* September 21–23, 1994, Calgary, Alberta, pp. 121–147.

McIntyre, T. 1993. "An Overview and Update on Development of New Substances Regulations for Biotechnology Products Under the Canadian Environmental Protection Act," in *Proceedings of the 3rd Annual Symposium on Groundwater and Soil Remediation,* September 8–10, 1993, Quebec City, QC, Canada, pp. 353–374.

McIntyre, T. 1994. Personal communication on November 3.

McNicoll, D.M., A.S. Baweja, M.J.L. Robin, F. D'Addario and P. Fernandez. 1993. "Enhanced Bioremediation of Petroleum Contaminated Soils Using an Engineered Bioreactor Design." *Proceedings of the 3rd Annual Symposium on Groundwater and Soil Remediation,* September 8–10, 1993, Quebec City, QC, Canada, pp. 583–592.

McNicoll, D.M., A.S. Baweja, M.J.L. Robin, C.W. Greer and F. D'Addario. 1994. "Operation, Monitoring and Performance of a Bioreactor Engineered to Treat Soils Containing Petroleum Hydrocarbons," *Proceedings of the 4th Annual Symposium on Groundwater and Soil Remediation,* September 21–23, 1994, Calgary, Alberta, pp. 545–556.

Millar, K. and S. Lesage. 1994. "Interactions between vitamin b12 and anaerobic bacterial cultures for the reductive dechlorination of PERC." Poster presented at the *24th International Symposium on Environmental Analytical Chemistry,* Ottawa, Ontario, May 16–19.

Millette, D., J.F. Barker, B. Butler, E.O. Frind, A.Y. Renoux, R. Samson and Y. Comeau. 1994. "Column Experiments to Study Substrate Interaction During Aerobic Biodegradation of Creosote-related Compounds in Groundwater." *Proceedings of the 4th Annual Symposium on Groundwater and Soil Remediation,* September 21–23, 1994, Calgary, Alberta, pp. 7–24.

Millette, D., J.F. Barker, Y. Comeau, B. Butler, E.O. Frind and R. Samson. 1993. "Substrate Interaction During Aerobic Biodegradation of Creosote-related Compounds: A Factorial Experiment." *Proceedings of the 3rd Annual Symposium on Groundwater and Soil Remediation,* September 8–10, 1993, Quebec City, QC, Canada, pp. 123–148.

Nakatsu, C. and R.C. Wyndham. 1993. "Cloning and Expression of the Transposable Chlorobenzoate-3,4-Dioxygenase Genes of *Alcaligenes* sp. Strain BR60." *Appl. Environ. Microbiol.,* 59(11):3625–3633.

O'Leary, K., J.F. Barker and R.W. Gillham. 1993. "Surface Application of Dissolved BTEX: A Prototype Field Experiment," *Proceedings of the Petroleum Hydrocarbons and Organic Chemicals in Ground Water,* Nov. 10–12, 1993, Houston, Texas, pp. 149–160.

Otte, M.P., J. Gagnon, Y. Comeau, N. Matte, C.W. Greer and R. Samson. 1993. "Activation of an Indigenous Bacterial Consortium for the Degradation of Pentachlorophenol in Contaminated Soils." *Proceedings of the 3rd Annual Symposium on Groundwater and Soil Remediation,* September 8–10, 1993, Quebec City, QC, Canada, pp. 69–84.

Paquin, J. 1994. "Demonstration of Dredging and an Additive/Aeration System to Promote the Biodegradation of Petrochemicals in Wastewater Sediments," in *Proceedings of the 4th Annual Symposium on Groundwater and Soil Remediation,* September 21–23, 1994, Calgary, Alberta, pp. 513–527.

Parker, W., L. Seed and R. Corsi. 1994. "Biofiltration of Offgases from Soil Venting," in *Proceedings of the 4th Annual Symposium on Groundwater and Soil Remediation,* September 21–23, 1994, Calgary, Alberta, pp. 699–713.

Pouliot, Y. and J.-L. Sansregret. 1993. "Biotraitement de sols contaminés par des huiles de transformateurs et du diesel: Deux projets de démonstration." *Proceedings of the 3rd Annual Symposium on Groundwater and Soil Remediation,* September 8–10, 1993, Quebec City, QC, Canada, pp. 637–650.

Renoux, A.Y., Y. Roy, R.D. Tyagi and R. Samson. 1993. "Ecotoxicological Evaluation of Soil Biotreatment: Comparison with the Inherited Toxicity of Non-Contaminated Soils." *Proceedings of the 3rd Annual Symposium on Groundwater and Soil Remediation,* September 8–10, 1993, Quebec City, QC, Canada, pp. 377–393.

Rho, D., P. Mercier, J.F. Jetté, R. Samson, J. Lei and B. Cyr. 1994. "Performances of a 30-L Biofilter for the Treatment of Contaminated Air with Toluene and Gasoline Vapors: Filter Media Selection, Microbial Activity and Substrate Interactions," in *Proceedings of the 4th Annual Symposium on Groundwater and Soil Remediation,* September 21–23, 1994, Calgary, Alberta, pp. 671–682.

Samson, R., C.W. Greer, J. Hawari, N. Matte, D. Beaumier, C. Beaulieu, D. Millette and A. Pilon. 1994. "Biotreatability Assessment for In-Situ Treatment of Sites Impacted with Light Hydrocarbons." *Proceedings of the 4th Annual Symposium on Groundwater and Soil Remediation,* September 21–23, 1994, Calgary, Alberta, pp. 225–252.

Samson, R., J. Hawari and C.W. Greer. 1993. "Demonstration of a New Biotreatability Protocol to Monitor a 1500 m^3 Soil Bioreactor." *Proceedings of the 3rd Annual Symposium on Groundwater and Soil Remediation,* September 8–10, 1993, Quebec City, QC, Canada, pp. 565–575.

Scheibenbogen, K., R.G. Zytner, H. Lee and J.T. Trevors. 1994. "Enhanced Removal of Selected Hydrocarbons from Soil by *Pseudomonas aeruginosa UG2* Biosurfactants and Some Chemical Surfactants." *J. Chem. Tech. Biotechnol.,* 59:53–59.

Seech, A., L.-A. Cornacchio and I. Marvan. 1993. "Bioremediation of Soils and Sediments Containing Chlorinated Phenols, PAHs and Petroleum Hydrocarbons." *Proceedings of the 3rd Annual Symposium on Groundwater and Soil Remediation,* September 8–10, 1993, Quebec City, QC, Canada, pp. 53–65.

Seech, A., I. Marvan and J.T. Trevors. 1994. "On-site/ex situ Bioremediation of Industrial Soils Containing Chlorinated Phenols and Polycyclic Aromatic Hydrocarbons," in *Bioremediation of Chlorinated and Polycyclic Aromatic*

Hydrocarbon Compounds. R.E. Hinchee, A. Leeson, L. Semprini and S.K. Ong, editors, CRC Press Inc., pp. 451–455.

Seed, L., B. Harper, W. Parker and R.L. Corsi. 1993. "Biofiltration of BTEX in Soil Remediation Off-Gas." *Proceedings of the 3rd Annual Symposium on Groundwater and Soil Remediation,* September 8–10, 1993, Quebec City, QC, Canada, pp. 297–314.

Smyth, D.J.A., J.A. Cherry and R.J. Jowett. 1994. "Funnel-and-Gate for in situ Groundwater Plume Containment." Presented at *Superfund XV,* Washington, D.C. Nov. 28–Dec. 1, 1994.

Taraoui, K., R. Samson and D. Rho. 1994. "Biofiltration of Waste Gases: Influence of the Microbial Activity on Filter Bed Temperature and Water Content," in *Proceedings of the 4th Annual Symposium on Groundwater and Soil Remediation,* September 21–23, 1994, Calgary, Alberta, pp. 731–742.

Trevors, J.T., H. Lee and M.I. Van Dyke. 1992. *Proceedings of in situ Bioremediation Symposium '92,* September 20–24, 1992. Niagara-on-the-Lake, ON, Canada, p. 65.

Trudell, M. 1994. "A Field Pilot Experiment for Bioremediation of Pentachlorophenol at a Wood-Preserving Site in southern Alberta." *Proceedings of the 4th Annual Symposium on Groundwater and Soil Remediation,* September 21–23, 1994, Calgary, Alberta, pp. 25–43.

Visser, S. and R.M. Danielson. 1994. "Ecotoxicological Testing of Hydrocarbon and Salt-Contaminated Wastes Following Treatment in a Bio-Reactor," in *Proceedings of the 4th Annual Symposium on Groundwater and Soil Remediation,* September 21–23, 1994, Calgary, Alberta, pp. 373–386.

Ward, O.P. and J. Yu. 1993. "Bioremediation of Phthalate Esters in Contaminated Soils," in *Proceedings of the 3rd Annual Symposium on Groundwater and Soil Remediation,* September 8–10, 1993, Quebec City, QC, Canada. pp. 119–122.

Weber, J.K. and J.F. Barker. 1994. "Field-Scale Remediation of Hydrocarbon Contaminated Soils and Water Using an in situ Funnel and Gate Strategy." *Proceedings of the 4th Annual Symposium on Groundwater and Soil Remediation,* September 21–23, 1994, Calgary, Alberta, pp. 45–51.

Whiting, S. and W. Hayes. 1993. "Demonstration of Bioremediation and Thermal Extraction Technologies for Treatment of Soils from the Pacific Place Site in Vancouver, B.C." *Proceedings of the 3rd Annual Symposium on Groundwater and Soil Remediation,* September 8–10, 1993, Quebec City, QC, Canada, pp. 513–536.

Wyndham, R.C., M. Peel, C. Nakatsu, J. Ng and F. Szilagyi. 1993. "Distribution and Structure of a Bacterial Transposon, Tn5271, in a Groundwater Bioremediation System." *Proceedings of the 3rd Annual Symposium on Groundwater and Soil Remediation,* September 8–10, 1993, Quebec City, QC, Canada, pp. 409–417.

Bioremediation for Hazardous Waste Management: The Indian Scenario

RISHI SHANKER, HEMANT J. PUROHIT AND PURUSHOTTAM KHANNA
National Environment Engineering Research Institute (NEERI)
Nehru Marg
Nagpur 440020, India

INTRODUCTION

India has a well established and diversified industrial structure with a sizable capacity in basic and heavy industry. Since the inception of planned development in the early 1950s, the number of chemical, petrochemical, fertilizer, and engineering industries has increased geometrically in India. The protection hereto provided to the industry and the emphasis on decentralization of industrial development have resulted in serious environmental pollution problems related to the release of a variety of toxic and hazardous substances. The industrial pollution problems now transcend the limits of industrial estates with the advent of pulp and paper, sugar, distilleries, leather tanning, and chemical units in the small-scale manufacturing sector.

Flyash, phospho-gypsum, iron, and steel sludges are the major industrial solid wastes generated in India. An estimated 35–40 million tonnes of flyash are generated annually by the thermal power plants, of which only 2–3% is utilized. The growing capacity of fertilizer, iron, and steel industries has also progressively enhanced the output of phospho-gypsum and slag wastes. The storage, treatment, and disposal of these wastes have emerged as a major environmental threat in the country.

Relatively few of the known hazardous waste management tools such as physico-chemical processes, thermal processes, and biotechnological processes are presently in use in India for the treatment and disposal of hazardous wastes. Incineration is used in large industries for destroying highly toxic wastes, while open-pit burning is commonly practiced for the disposal of sludges, along with factory garbage in small- and medium-scale industries. Secure landfill of wastes is practiced in a few cases as an exception to the general rule of uncontrolled dumping. Co-disposal, along with municipal solid waste, is generally in vogue in highly industrialized urban agglomerations.

In recent years, there is an increasing awareness of the quantity and diversity of hazardous wastes generated by the industry in India and the risks posed to human health and environmental quality, due to unscientific disposal methods for such wastes.

GOVERNMENT LEGISLATION AND POLICIES

The industrial activities involving hazardous chemicals have the potential to cause serious accidents and spills. Even the most commonly used chemicals, due to their indiscriminate and negligent use over the decades, have accumulated in water and soil environment creating serious pollution and human health problems. The need exists, now more than ever before in India, for caution during the manufacture, processing, storage, transport, use, and disposal of such chemicals, as well as for scientific risk assessment of such activities.

Environment Protection Act

The need to protect the environment brought in the Environment Protection Act (EPA) in 1986, which places on the central government the responsibility of laying down procedures and safeguards for handling of hazardous substances and prevention of accidents (Bhattacharjya, 1992). The following sets of rules have since been established under the EPA:

- Manufacture, Storage and Import of Hazardous Chemicals Rules, 1989
- Hazardous Waste (Management and Handling) Rules, 1989
- Manufacture, Use, Import, Export and Storage of Hazardous Microorganisms/Genetically Engineered Organisms or Cells Rules, 1989
- Regulation over transporation of hazardous waste/chemicals, Central Motor Vehicles Rules, 1989
- Public Liability Insurance Act, 1991

A central crisis group (CCG) has been constituted by the Ministry of Environment and Forests (MEF) with senior officials and technical experts for assessing the problems due to chemical accident(s) and providing expert guidance for handling potential disasters.

HAZARDOUS CHEMICALS

Large-scale production of a variety of chemicals in India and concomitant exposure of human population warranted the Ministry of Environment and Forests to classify and develop a directory of 434 chemicals. This exhaustive list also includes toxicological data to facilitate the implementation of various regulations (Mahadevan, 1993).

HAZARDOUS WASTE MANAGEMENT

The MEF formulated guidelines for management of hazardous wastes in 1991 for the effective management of hazardous wastes. In the same year, the Ministry issued directives to the Central Pollution Control Board (CPCB) for identification and assessment of abandoned hazardous waste sites. The consumer-oriented guidelines consider the labelling, packaging, transport, storage, treatment, and disposal of hazardous and toxic wastes (Rajamani, 1991). Two major growing industrial states, Gujrat and Maharashtra, have been provided with financial assistance for conducting environmental impact and risk assessment studies for identification of sites appropriate for the disposal of hazardous wastes.

An MEF-sponsored brainstorming session (National Hazardous Wastes Management Programme held at NEERI in 1991) discussed the current status of hazardous waste management practices and the need for adoption of cleaner technologies, waste minimization, and proper disposal of hazardous wastes. The session suggested an agenda for action at policy, plan, and program level implementation of component activities in hazardous waste management. The main features of the action plan include development of protocols for hazardous waste measurement and monitoring, promotion of environmentally balanced industrial complexing, extension of biotechnological routes for treatment, and design of databases for prevention, treatment, and disposal technologies (Approach paper NEERI, 1991). NEERI was retained by the MEF in 1993 to explore the application of clayey soils and indigenously available synthetic membranes as liners in secure landfills for containment of hazardous industrial wastes.

Environmental Audit

The environmental audit of industrial projects has been made mandatory by the MEF since 1993. The audit reports are submitted to the concerned State Pollution Control Board every financial year. The MEF also initiated a scheme in 1993 for conducting environmental audits in select units belonging to 17 heavily polluting industrial sectors.

Training programs have also been initiated for the personnel in statutory bodies for effective evaluation of pollution control systems.

In an endeavor to generate greater awareness, a National Register for Potentially Toxic Chemicals (NRPTC) is in the offing in consultation with the United Nations Environment Program (UNEP). The register will follow the London Guidelines for Exchange of Information in International Trade to facilitate the dissemination of information and monitoring of potentially toxic chemicals.

Public Awareness

An enlightened public awareness is an essential prerequisite to effective environmental management. A program for enhancing public awareness on prevention, abatement, and control of pollution has been initiated by the MEF in 1993. The initial thrust is on creation of awareness on water and air pollution and the impacts on the day-to-day life of people, possible preventive measures, and securing compensation for those affected by the pollution.

CURRENT PRACTICES OF HAZARDOUS WASTE MANAGEMENT IN INDIA

The geometrically rising levels of air, water, and land pollution in India and the consequences on aesthetic, human, and environmental risks call for an effective strategy for pollution control and a shift in the paradigm of developmental planning (Khanna et al., 1991).

Hazardous Waste Generation, Treatment, and Disposal in India: Situation Analysis

Various hazardous industrial wastes are currently managed with physico-chemical options transferring the contaminants from one environmental medium to another in addition to incurring high energy costs in their operation. The high treatment costs, coupled with ineffective implementation of statutory regulations, have led to indiscriminate dumping of hazardous wastes in industrial areas in the country.

In a typical case, the sludge generated during manufacture of H-acid, a dye intermediate, was dumped on land inside the plant premises. Due to high porosity of soil around the plant area, leachate contaminated the soil and water environment around the village Bichhari, near Udaipur. Water samples collected within 2 km downstream of the plant have shown a dark brown color (1600 pt-co scale), sodium (200–1232 mg/L), and sulphate (140–3500 mg/L). The soils also show high concentration of sodium (2500 meq/L), sulphate (2000 meq/L), and organic carbon (2.7%). Geographical and hydrological aspects of the affected area were assessed using electrical resistivity methods. A bioremediation program has been designed by the Institute with a view to restoring the environmental quality of the affected area in this project sponsored by the Rajasthan Pollution Control Board since July 1992.

A rayon plant manufacturing 150 tonnes of rayon grade pulp from hardwood through sulfite process generates 48,000–50,000 m^3/d of highly colored wastewater with high COD (2200–2400 mg/L) and a pH of 2.5. NEERI recommended reverse osmosis using cellulosic membranes that can remove over 90% of color and 70% COD from the wastewaters (NEERI, 1993).

The state of Himachal Pradesh represents a typical case of current hazardous waste generation and disposal regimes in India. The total process solid waste generated from industries is 1600 MT/year, the main contributors being metallurgical industry (55%) and oil wells (34%). In addition, 435 MT/year of sludge is generated from effluent treatment plants, which is classified as hazardous. The leachates from tannery, metallurgical, and lead battery units contain chromium (9.93 mg/L), lead (183 mg/L), and iron (36.3 mg/L) beyond the threshold limits. The tannery sludge exhibits a sulphide reactivity exceeding a threshold limit of 500 mg/L. The option recommended by NEERI for disposal of these sludges and solid wastes is secure landfill with clayey soil as liner.

The inventorization of hazardous waste generation in Maharashtra indicated 18 categories of hazardous wastes. The total waste generation in 362 industries was computed at a million tonnes per year. The industries manufacturing inorganic chemicals contributing 22%, textile and leather 15.5%, dyes 8.4%, organic and petrochemical 4.3%, engineering 15%, respectively, and the remainder was attributed to other industries. Nearly 86% of the waste was generated from industries located in the Thane district and greater Bombay.

BIOREMEDIATION AS A TOOL FOR HAZARDOUS WASTE MANAGEMENT

Bioremediation is emerging as a potential tool for the conversion of toxic organic chemicals and wastes into carbon dioxide, water, inorganic salts, and biomass. It is the process of stimulating degradative capabilities of microorganisms. Bioremediation requires integration of several scientific disciplines in soil, water, and subsurface environment. Full-scale application of this technology is not yet reported from anywhere. We present an overview of some of the concepts currently under investigation at NEERI for application in remediation of hazardous waste dump sites.

Microorganisms and Transformation of Xenobiotics

The degradation of natural organic constituents in the environment is primarily mediated by bacteria and fungi (Shanker et al., 1985, 1990, 1991; Shanker and Bollag, 1990; Shanker and Robinson, 1991; Jothikumar et al., 1986; Muller, 1992; Snyder, 1993). There is sufficient knowledge on the existence of central intermediary metabolites in many degradative pathways and the role of phenomenon like co-metabolism in degradation of xenobiotics (Omenn, 1988; Lynch and Hobbie, 1991; Muller, 1992). The identification of plasmids involved in hydrocarbon degradation in different geographical locations in the world further substantiates the observations on adaptability and metabolic capabilities of bacteria in nature (Harayama and Don, 1985; Assinder and Williams, 1988). Bioremediation demands an exacting monitoring approach (Macdonald and Rittman, 1993) due to several questions on the microbial infallibility and their potential use in controlled and quantitative bioremediation (Nicholson and Safaya, 1993; Omenn, 1993; Shannon and Unterman, 1993). The feasibility of bioremediation option is usually determined with recourse to available data on biodegradability of target molecule(s), confirmation of biotreatability with site samples on bench-scale bioreactor or microcosms, and final demonstration of bioremediation efficacy through pilot-scale experimentation or full-scale trials (Rubin and Setzer, 1993).

Forced and Programmed Bioremediation

FORCED BIOREMEDIATION

Forced bioremediation relates to the treatment of preexisting contaminated site(s) by enhancing desired microbial activity. Bioremediation in such a situation must be engineered to be an operational and effective process warranting optimization of growth conditions with additional nutrient supply for available and desirable microorganisms. The latter may be necessary to degrade specific recalcitrant molecules present in the hazardous waste.

An approach currently under investigation at NEERI is to design remediation protocol, such as in crude oil decontamination, on the inducer-modulator concept of catabolic operons of the degradative plasmids. The induction of TOL and NAH operons of degradative plasmids of the genus *Pseudomonas* are regulated by inducers such as *m*-toluate, benzoate, or salicylate (Sayler, 1991). The activity of organisms at the site possessing such catabolic operons can be enhanced by amending the soil with an appropriate inducer for complete degradation or mineralization of toxic hydrocarbons. The selective carbon source or inducer modulator strategies for increasing population size and biochemical potential can be exploited for both natural and genetically engineered microorganisms.

PROGRAMMED BIOREMEDIATION

Programmed bioremediation considers the establishment of an appropriately designed site for the disposal of specific waste. Such a system will only work when the naturally occurring rate of degradation exceeds the rate of contamination. This, in turn, can be controlled by maintaining an efficient catabolic pool. The degradation capability is dependent on maintenance of distribution of heterogeneous mixed populations acting as consortia and can be monitored by molecular techniques such as gene probes and polymerase chain reaction (Byrd et al., 1991; Sayler, 1991; More et al., 1994). The utilization of molecular techniques for determining the functional gene distribution and variations in ecosystem is intrinsic to successful bioremediation technology.

Construction of Genetically Engineered Microorganisms for Bioremediation

Genetic engineering of microorganisms can lead to stabilization and increased rate of desired activity, increased spectrum of activity, and development of safe microbial systems (Sussman et al., 1988). The versatility of bacterial communities towards metabolism of various contaminants can be explained by the availability of catabolic plasmids like OCT, TOL, NAH, and others that can be transferred from one bacterium to another (Wallace and Sayler, 1992). Certain catabolic gene sequences are located on transposable elements in plasmids, further increasing versatility (Tsuda and Iino, 1988; Byrne and Lessie, 1994). It is apparent that in ecological niches these plasmids seem to recombine to form new degradative pathways as and when necessary for survival. The concept that catabolic plasmids can be transferred from one bacterium to another is the genesis of construction of strains with desired catabolic abilities. However, genetically engineered microorganisms have not been used in remediation process either because they have not reached a stage of perfection or due to regulatory norms (Muller, 1992).

STRESS RESPONSIVE MICROORGANISMS

In the natural environment, microorganisms are subject to different stress factors and variations, such as nutrient, oxygen or water limitation, temperature, and pH, that influence their physiological state. Nutrient limitation and fluctuation is perhaps the major stress factor. Starvation survival in nature, especially in oligotrophic environments, is based on the adaptability to changing conditions and shift between growth and nongrowth states (Morita, 1988; Kolter et al., 1993). Bacteria respond to starvation by activating complex regulatory mechanisms. The expression of starvation genes allows the control of degradation of macromolecules as energy source, scavenging of scarce nutrients, and development of higher resistance to various stresses (Matin, 1991; Siegele and Kolter, 1992). Several starvation inducible genes are expressed via induction of starvation promoters in such a fashion that their basal level during growth increases as growth rate decreases. Hence, these regulatory proteins can serve as models to provide sustained high expression of biochemical activities in bacteria that can be utilized in in situ bioremediation (Kjelleberg, 1993). However, our current understanding of starvation survival is limited to organisms like *Escherichia coli* or *Vibrio* species that have little or no role in bioremediation technology (Kolter et al. 1993; Kjelleberg et al., 1993) and should be extended to *Pseudomonas,* a metabolically more versatile organism for bioremediation (Givskov et al., 1994). Recent observations indicate that controlling the catabolic genes by promoters that are starvation-specific or those that are not growth phase-dependent could be the approach to exploit the capability of *Pseudomonas putida* (Shanker et al., 1994).

An immediate application of stress tolerance and survival is in the cleanup of crude oil spills in hypersaline environment, which demands osmotolerance even in a metabolically versatile organism like *Pseudomonas putida.* An approach, undertaken at NEERI, involved subcloning the *pro U* operon of *E. coli,* encoding the osmoprotective function into a broad host range vector under the control of tac-lac promoter that was shuttled into *Pseudomonas putida.* The organism was able to degrade a wide variety of hydrocarbons from a model petroleum mix and constituted an active member of a bacterial consortium designed to degrade crude oil under hypersaline marine conditions (patent pending). Similar solutions for bioremediation of specific hazardous chemicals and wastes are currently being investigated at the Institute for cleanup of industrial waste dump sites in the country.

TRACKING OF GENETICALLY ENGINEERED MICROORGANISMS

The determination of viable microorganisms in a natural environment has considerable significance in ecological risk assessment following the release of genetically engineered microorganisms for bioremediation applications. Such organisms may be undetectable in traditional enumeration techniques, and this demands sensitive and reliable detection methodology (Duncan et al., 1994). Molecular techniques such as nucleic acid probes and polymerase chain reaction provide adequate sensitivity and selectivity but suffer from the absence of direct real-time information on the activity of cells and gene transfer due to their inability to distinguish amongst culturable, viable, and nonviable cells (Sayler, 1991). Evidence is now available regarding the physiological change in viable but nonculturable cells that affect the ability to amplify DNA by PCR (Byrd et al., 1991). A direct monitoring system currently under exploration is the integration of bioluminescence gene(s)—the lux operon of *Vibrio harveyi* into a single site in the chromosome of the target organism, providing a genetically stable, nontransferable marker (Shaw et al., 1992). The target strain can be differentiated from indigenous bacteria and detected microscopically by on-line detection at great sensitivity for use in bioreactor systems (King et al., 1990; Fox, 1994). At NEERI, the use of genetically engineered bacteria for application in biotreatment of halogenated hydrocarbon waste is being explored. The luciferase encoded by *Lux* AB genes of *Vibrio harveyi* have been transposed on to the chromosome of *Pseudomonas putida* possessing the cytochrome P450 cam monoxygenase on a broad host range plasmid vector for dehalogenation of halogenated hydrocarbons. The luciferase cassette possesses the aromatic hydrocarbon inducible TOL plasmid promoter *Pm,* which not only helps in tracking the organism, but also serves as a biosensor for aromatic pollutants (Shanker et al., 1994).

BIOLOGICAL CONTAINMENT

Bioremediation applications demand microorganisms designed and manipulated for limited survival in open environment. Hence, the use of such microorganisms is dependent on containment limiting distribution. A few studies have been made using the *hok* (host killer) *rel* F and *gef* (gene expression fatal) genes from *E. coli* that cause breakdown of transmembrane potential and concurrent drop in oxygen consumption followed by the death of cells (Bej et al., 1992). Construction of an efficient suicide system, with tightly controlled promoters linked to physiological markers on the host chromosome, will reduce the chances of survival of the target organism, limiting distribution of engineered organisms for application in contained bioreactors.

BIOREMEDIATION RESEARCH AND DEVELOPMENT IN INDIA

In the last decade, research efforts were directed towards microbial transformation of xenobiotics and industrial wastes in research programs funded by the Department of Science and Technology, Department of Biotechnology, and Ministry of Environment and Forests (Sundaresan, 1982). A Department of Biotechnology, Government of India-sponsored brainstorming session on environmental biotechnology identified NEERI for developing biotechnologies for pollution control in the following areas:

- environmental monitoring
- restoration of environmental quality
- resource/residue/waste: recovery/utilization/treatment
- substitution of nonrenewable resource base with renewable resources

The activities related to bioremediation are discussed in this chapter.

Cleaning up of Crude Oil Spills in Freshwater and Hypersaline Situations

The technology currently under investigation involves removal of 95% of the spilled crude oil by alkali-treated sawdust, followed by degradation of remaining oil using bacterial consortia. Bacterial cultures isolated from crude oil acclimated sludge were screened to design bacterial consortia that emulsified and degraded 70–80% crude oil in 48 hours. The consortia degraded Bombay high and Gulf high crude oil. An important aspect in the survival of organisms in a marine environment is the expression of osmoregulatory function. The *E. coli pro* U operon, which allows the organism to tolerate hypersaline conditions, was subcloned and expressed in *Pseudomonas putida*. The developed *pro* U expression system was also imparted to the select members of bacterial consortia, attacking crude oil to extend their survival in a marine environment (Chattre et al., 1994).

Treatment of Oily Sludges and Emulsions for Recovery of Crude Oil and Disposal of Sludge

A bioprocess has been developed with potential applications in the petrochemical industry for remediation of water-oil (w/o), oil-water (o/w), and sludge-in-water emulsions. The recovery of oil from such a matrix was associated with secondary metabolite of glucose metabolism from *Bacillus subtilis.* The bacterium has also shown the release of free oil from oily sludge. Over 90% recovery of oil from sludges by the glucose grown cells of *Bacillus* has been recorded. The pretreatment of sludges with surfactants is not a prerequisite in the process (Janiyani et al., 1994).

Biotreatment of Petrochemical Wastewaters

Pseudomonas pseudoalcaligenes KJBI, isolated from trickling filter media of a petrochemical waste treatment unit, efficiently utilized a wide array of aromatic substances bearing amino, carboxyl, chlorohydroxyl, methyl, and nitro groups as sole carbon source. Strain KJBI efficiently utilized benzoate and nitrobenzene as mixed substrate and effected rapid reduction of these substrates when immobilized on calcium alginate. Degradation of *p*-cresol in *P. pseudoalcaligenes* was plasmid mediated and via the protocatechuic acid. Plasmid pKJBI from *P. pseudoalcaligenes* has been efficiently transferred to *Pseudomonas* sp. and to the indigenous flora of the wastewater. Temperature, total suspended solids, and aeration greatly influence the gene transfer in the wastewater treatment. The removal of a wide range of toxic compounds by *P. pseudoalcaligenes* is useful for improvement in the performance of the wastewater treatment system (NEERI, 1992).

Decontamination of Hazardous Waste Disposal Sites

The handling of accidental chemical spills and abandoned hazardous waste dump sites is an important aspect of the bioremediation program at NEERI. A library of 180 bacterial strains has been developed with a wide spectrum of substrate specificity from natural niches such as dump sites, waste from industry, and oil drilling platform sites. The biodegradable characteristics cover the substrate specificity for various aliphatics, aromatics, substituted aromatics, or halogenated compounds. Studies have been initiated for characterization of genes involved in degradation and regulation of substituted phenols and halogenated molecules. A better understanding of molecular mechanisms of transformation of such molecules along with microcosm studies will help in understanding degradation of xenobiotics in the environment (Purohit et al., 1994).

Desulphurization of Fossil Fuels

The Institute is engaged in a project on preemptive and cost-effective desulphurization of fossil fuels, viz., natural gas, oil, and coal. A two-stage process for microbial desulphurization of gaseous fuels containing high concentrations of hydrogen sulphide, incorporating ferric sulphate as an oxidant for sulphur recovery, has been developed. In the first stage of the process, ferric sulphate under appropriate pH conditions chemically reacts with hydrogen sulphide where sulphur is precipitated and ferric sulphate is reduced to ferrous sulphate. The ferrous sulphate so formed is biologically regenerated to ferric sulphate in a rotating biological rope contactor (RBRC) for reuse. The results indicate that 0.98 kg elemental sulphur is obtained per kilogram of hydrogen sulphide reacted and that 92% conversion of Fe^{2+} to Fe^{3+} is achieved in rotating biological rope contactor with a hydraulic retention time of 1.9 h and influent loading of 1.25 kg ferrous sulphate/m^2/day.

Two reactor systems, viz., fixed-bed and coal slurry reactors (CSR), have been evaluated for desulphurization of coal. The results indicate that 72% removal of pyritic sulphur is possible in the CSR within 4 days using mixed cultures of *Thiobacillus thiooxidans* and *Thiobacillus ferrooxidans* as against 15 days in a fixed-bed system with initial biomass concentration of 306 mg/L. The calorific value of coal improved from 4436 to 6888 k·cal/kg with concomitant ash reduction by 76%. Also 60% removal of organic sulphur was possible with the use of a wetting agent, viz., methanol in the CSR within 96 h of using the mixed cultures. Studies are now underway on genetically engineered strains of *Thiobacillus* to reduce the cost in desulphurization of coal (NEERI, 1993).

The research efforts at NEERI are also directed towards development of technologies for scientific handling and treatment of hazardous wastes and conducting environmental impact–risk assessment studies for identification of sites for disposal of hazardous substances. This requires a comprehensive analysis of the information on types of industrial wastes, capabilities of microorganisms, and regulation requirements. The evaluation of the efficacy of bioremediation for treatment of soil, sediment, groundwater, or industrial wastes and waste streams indicates that there are innumerable possible applications (Gabriel, 1991). The relative critical steps from the flask to microcosm, bioprocessing scaleup, and pilot plant demonstrations have been made for certain wastes. It is likely that innovative processes for actual controlled and directed field applications will evolve to make bioremediation an economical "off the shelf" technology in the near future for developing nations that are confronted with the past and current environmental problems in the face of limited fiscal resources.

REFERENCES

Approach paper, National Hazardous Waste Management Program, NEERI, 1991.

Assinder, S.J. and Williams, P.A. 1988. "Comparison of meta pathway operons on NAH plasmid pWW 60-22 and TOL plasmid pWW 53-4 and its evolutionary significance," *J. Gen. Microbiol.,* 134:2769–2778.

Bej, A.K., Molin, S., Perlin, M. and Atlas, R.M. 1992. "Maintenance and killing efficiency of conditional lethal constructs in *Pseudomonas putida,*" *J. Indus. Microbiol.,* 10:79–85.

Bhattacharjya, A. 1992. *A guide to manufacture, storage and import of hazardous chemicals rules, 1989.* Hazardous Substances Management Division, Ministry of Environment and Forests, Government of India, New Delhi.

Boronin, A.M. 1992. "Diversity of *Pseudomonas* plasmids: To what extent?" *FEMS Microbiol. Letts,* 100:461–468.

Byrd, J.J., Xu, H.H. and Colwell, R.R. 1991. "Viable but non-culturable bacteria in drinking water," *Appl. Environ. Microbiol.,* 57:875–878.

Byrne, A.M. and Lessie, T.G. 1994. "Characteristics of IS 401, a new member of IS 3 family implicated in plasmid rearrangements in *Pseudomonas cepacia,*" *Plasmid,* 31:138–147.

Chattre, S., Das, S., Purohit, H.J., Shanker, R., Chakraborti, T. and Khanna, P. 1994. Biodegradation of various fractions of crude oil using a bacterial consortium. Presented at *36th Annual Meeting of Association of Microbiologists of India,* February 9–11, Ludhiana, India.

Duncan, S., Glover, G.A., Killham, K. and Prosser, J.I. 1994. "Luminescence based detection of activity of starved and viable non-culturable bacteria," *Appl. Environ. Microbiol.,* 60:1308–1316.

Eglin, R. 1994. "GATT and environment," *Encology,* 9:34–38.

Fox, J.L. 1994. "Field testing for bioremediation called the next logical step," *Amer. Soc. Microbiol. News,* 47:319–322.

Gabriel, P.F. 1991. "Innovative technologies for contaminated site remediation: focus on bioremediation," *J. Air Waste Manag. Assoc.,* 41:1657–1660.

Givskov, M., Eberl, L., Moller, S., Poulsen, L.K. and Molin, S. 1994. "Responses to nutrient starvation in *Pseudomonas putida* KT 2442: Analysis of general cross protection, cell shape and macromolecular content," *J. Bacteriol.*, 176:7–14.

Harayama, S. and Don, R.H. 1985. "Catabolic plasmids: their analysis and utilization in manipulation of bacterial metabolic activities," In *Genetic Engineering: Principles and Methods, vol. 7,* pp. 283–307, (eds.) Setlow, J.K. and Hollaender, A., Plenum Publishing Corporation, New York.

Janiyani, K.L., Purohit, H.J., Shanker, R. and Khanna, P. 1994. "De-emulsification of oil-in-water emulsions by *Bacillus subtilis,*" *World J. Microbiol. Biotechnol.,* 10:452–456.

Jensen, B.G., Ramos, J.L., Kaneva, Z. and Molin, S. 1993. "A substrate dependent biological containment system for *Pseudomonas putida,* based on *E. coli gef* gene," *Appl. Environ. Microbiol.,* 59:3713–3717.

Josephson, J. 1993. "Hazardous waste management: seeking least-cost approaches," *Environ. Sci. Tech.,* 27:2298–2304.

Jothikumar, N., Krishnamoorthi, K.P. and Rao, D.M.R. 1986. "Toxicity and biochemical responses carp to dinitrobenzene plant effluent" *J. of Air, Water and Soil Pollution,* 28:117–126.

King, J.M.H., Digrazia, P.M., Applegate, B., Burlage, R., Sanseverino, J., Dunbar, P., Larin, F. and Sayler, G.S. 1990. "Rapid, sensitive bioluminescent reporter technology for naphthalene exposure and biodegradation," *Science,* 249:778–781.

Kjelleberg, S. (ed.) 1993. *Starvation in bacteria,* Plenum Press, New York.

Kjelleberg, S., Albertson, N., Flardh, K., Holmqusit, L., Jouper-Jaan, A., Marouga, R., Ostling, J., Svenblad, B. and Weichart, D. 1993. "How do non-differentiating bacteria adapt to starvation," *Antonie van Leeuwenhoek,* 63:333–341.

Khanna, P., Kulkarni, V.S., Dutt, P.S., Joshi, V. and Aggarwal, A.L. 1991. "An approach to carrying capacity based developmental planning process," *Journal IAEM,* 18:12–19.

Knudsen, S.M. and Karlstrom, O.H. 1991. "Development of efficient suicide mechanisms for biological containment of bacteria," *Appl. Environ. Microbiol.,* 57:85–92.

Kolter, R., Siegele, D.A. and Tormo, A. 1993. "The stationary phase of bacterial life cycle," *Ann. Rev. Microbiol.,* 47:855–874.

Lynch, J.M. and Hobbie, J.E. (Eds.). 1991. *Microorganism in action: concept and applications in microbial ecology,* Blackwell Scientific Publications, Oxford.

Macdonald, J.A. and Rittman, B.F. 1993. "Performance standards for in situ bioremediation," *Environm. Sci. Technol.,* 27:1974–11976.

Mahadevan, T.N. 1993. "Air pollution from hazardous chemicals and liquid effluents," *Encology,* 8:34–42.

Matin, A. 1991. "Molecular basis of carbon-starvation-induced general resistance in *Escherichia coli,*" *Mol. Microbiol.,* 5:3–10.

More, M.I., Herrick, J.B., Silva, M.C., Ghiorse, W.C. and Madsen, E.L. 1994. "Quantitative cell lysis of indigenous microorganisms and rapid extraction of microbial DNA from sediment," *Appl. Environ. Microbiol.,* 60:1572–1580.

Morita, R. 1988. "Bioavailability of energy and its relationship to growth and starvation survival," *Can. J. Mirobiol.,* 34:436–441.

Muller, R. 1992. "Bacterial degradation of xenobiotics," In *Microbial Control of Pollution* (Eds.) Fry, J.C., Gadd, G.M. Herbert, R.A., Jones, C.W. and Watson-Craik, I.A., pp. 35–57, Cambridge University Press, Cambridge.

NEERI, Annual Report 1992, 44.

NEERI, Annual Report 1993, 48–54.

Nicholson, S.A. and Safaya, N.M. 1993. "Restoring hazardous waste sites," *Environ. Sci. Technol.,* 27:1022–1024.

Omenn, G.S. (Ed.) 1988. *Environmental biotechnology: reducing risks from environmental chemicals through biotechnology,* Plenum Press, Inc., New York.

Omenn, G.S. 1993. "Environmental biotechnology: biotechnology solutions for a global environmental problems, hazardous chemicals wastes," *Asia Pac. J. Public Health,* 6:40–45.

Purohit, H.J., Shanker, R. and Khanna, P. 1994. Unpublished observations.

Rajamani, R. 1991. *Guidelines for management and handling of hazardous wastes,* Hazardous Substances Management Division, Ministry of Environment and Forests, Govt. of India.

Rubin, D.K. and Setzer, S.W. 1993. "Bioremediation picking up," *ENR,* 231: 15.

Sayler, G.S. 1991. "Contributions of molecular biology to bioremediation," *J. Hazardous Materials,* 28:13–27.

Shanker, R., Ramakrishna, C. and Seth, P.K. 1985. "Degradation of some phthalic acid ester in soil," *Environ. Pollut. (Series A),* 39:1–7.

Shanker, R., Ramakrishna, C. and Seth, P.K. 1990. "Microbial degradation of acrylamide monomer," *Arch. Microbiol.,* 154:192–198.

Shanker, R. and Bollag, J.M. 1990. "Transformation of indole by methanogenic and sulfate-reducing microorganisms isolated from digested sludge," *Microbial Ecology,* 20:171–183.

Shanker, R. and Robinson, J.P. 1991. "Anaerobic degradation of dalapon in water logged soil and river sediment," *Letts. Appl. Microbiol.,* 12:8–10.

Shanker, R., Kaiser, J.P. and Bollag, J.M. 1991. "Microbial transformation of heterocyclic molecules in deep subsurface sediments," *Microbial Ecology,* 22:305–316.

Shanker, R., Purohit, H.J. and Atkins, W.M. 1994. Unpublished observations.

Shannon, M.J. and Unterman, R. 1993. "Evaluating bioremediation distinguishing fact from fiction," *Ann. Rev. Microbiol.,* 47:715–738.

Shaw, J.J., Dane, F., Geiger, D. and Kloepper, J.W. 1992. "Use of bioluminescence for detection of genetically engineered microorganisms released into environment," *Appl. Environ. Microbiol.,* 58:267–273.

Siegele, D.A. and Kolter, R. 1992. "Life after log," *J. Bacteriol.,* 174:345–348.

Sundaresan, B.B. 1982. *Proceedings: National workshop on microbial degradation of industrial wastes.* NEERI, Nagpur.

Sussman, M., Collins, C.H., Skinner, F.A. and Stewart-Tull, D.E. (Eds.) 1988. *The Release of Genetically Engineered Microorganisms,* Academic Press, London, UK.

Synder, J.D. 1993. "Off-the-shelf bugs hungrily gobble our nastiest pollutants," *Smithsonia,* 24:66–68.

Tsuda, M. and Iino, T. 1988. "Identification and characterization of Tn 4653, a transposon covering toluene transposon Tn 4651 on T01 plasmid pWWO," *Mol. Gen. Genetics,* 213:72–77.

Wallace, W.H. and Sayler, G.S. 1992. "Catabolic plasmid in environment," In *Encyclopedia of Microbiology,* Vol. 1, pp. 417–430. (Ed.) Lederberg J., Academic Press Inc., San Diego.

Bioremediation Research and Development in Israel

Y. Shevah
Tahal Consulting Engineers Ltd.
54 Ibn Gvirol St
Tel Aviv 61111, Israel

M. Waldman
Ministry of Science and Technology
Building 3, POB 18195
Jerusalem 91181, Israel

A. Adin
Environmental Science Div.
The Hebrew University of Jerusalem
Jerusalem 91904, Israel

INTRODUCTION

Contamination of aquifers has become a great concern in many countries where the population relies on groundwater resources for its drinking water supply. In Israel, where groundwater resources provide 60–70% of the water supply, about 1 billion cubic meter (cu m) of the average annual renewable potential of 1.6 billion cu m, the exploited aquifers are vulnerable to pollution from human waste and agricultural chemicals. A combination of a shallow phreatic water table, sandy soils, dense population, and intensive agricultural, industrial and urban activities are highly unfavorable for groundwater quality (Shevah and Waldman, 1989). Thus, nitrate concentration in nearly half of the operating wells in the coastal aquifer already exceeds the current drinking water standard of 45 mg/L (Wolman and Ronen, 1992). Furthermore, the occurrence of traces of pesticides in groundwater has been observed (Muszkat et al., 1993).

The deterioration of aquifers due to contamination has been enhanced by the reuse of sewage effluents for irrigation (Amiel et al.,

1990) and further aggravated by accidental spills of hydrocarbons, which draw much attention due to the abrupt release of large quantities of fuel (Kanfi, 1991). In Israel, the fuel pipes network extends over 1400 km, of which 780 km are laid above active aquifers. In addition, there are fuel depots and a large number of underground storage tanks in petrol stations, army camps, industrial plants, and residential areas. Within a period of 20 years, about 70 incidents of spillage were reported, with seepage of tens to thousands of cubic meters per case (Kanfi, 1991). Of particular concern are volatile organic compounds (VOCs), some of which are toxic and have been classified as carcinogens. Benzene, toluene, and xylene (BTX) are some of these volatile organic compounds, which comprise up to 20% of the weight of gasoline. Following a spill or leak of gasoline, VOCs dissolve in groundwater either by direct contact or by leaching from contaminated soil.

The magnitude of the hazardous contamination problem in Israel is not yet certain, but the slow deterioration of soil, subsoil, and the active aquifers below has created enormous demand for soil and groundwater bioremediation, as evident in other countries. In the United States, approximately 69,000 hazardous waste sites are being considered for remediation (*Science,* 1992). In Germany, there are more than 250,000 contaminated sites. In the UK, there may be at least 50,000–100,000 contaminated sites, with an estimated cleaning cost of 10–30 billion pounds. In the Netherlands, more than 6000 contaminated sites have been cleaned up since 1982, at a cost of 1.5 billion dollars (Steiber et al., 1994).

The huge scale of the problem is the driving force behind the extensive research and development and field studies that are conducted in Israel in relation to soil and groundwater bioremediation. Some of these studies on soil and groundwater bioremediation are reviewed in this chapter.

EVOLUTION OF ON-SITE AND IN SITU REMEDIATION

Soil and aquifer remediation is implemented in order to achieve the following objectives: statutory requirements, resource management considerations based on local use of aquifers, and achievement of acceptable levels for the protection of health and environment. Restoration of background levels of off-site contaminants is attempted by remediation attenuation through induced or natural processes, such as biodegradation, volatilization, adsorption, and dispersion (Yaron, 1989).

The released contaminants percolate downwards through the vadose zone to groundwater, where immiscible compounds will either float or sink, depending upon their specific gravity. Most hydrocarbons float, while halogenated compounds sink, forming a pool or dense nonaqueous phase liquid (DNAPL) above the bottom of the aquifer. The various contaminants are known to undergo a variety of both chemical and biological transformation, leading to various products, some of which may be more hazardous than the original compounds. On the soil surface or underground, the contaminants are subject to a variety of abiotic processes that influence the transport, distribution, and fate of the compounds, including immiscible transport, as a separate liquid phase, dispersion—mixing and spreading along the flow path, advection—solute transport, sorption—on the solid phase, abiotic transformation, or biotic conversion into products or intermediates (Yaron, 1989).

The mechanisms of transformation of the parent compounds, as well as derivative products, are receiving increased attention. Mathematical modeling, as well as laboratory and field studies, are being conducted in order to predict the impact of the increasing number of pollutants and the expanding list of hazardous chemicals to which environment and humans are exposed.

Modeling Studies

Transport of solutes in natural and heterogeneous soil formations is an intensively researched topic in Israel. Various modeling approaches have been developed for the simulation of contaminants' migration in soils and subsurface water. A single cell model was used by Mercado (1976) to get a conservative estimate of the rate at which solubilized fractions of an immobilized immiscible contaminant in the unsaturated zone is leached down. Moreover, since aquifer properties vary greatly and are uncertain, it is common to regard them as random Monte Carlo simulation modeling, generally producing data of nonpractical use. Recently, more elaborate modeling techniques referring to the whole flow or to different parts of the flow were developed, in which laboratory tests and theoretical equations are used to determine the flow characteristics (Kessler and Rubin, 1987). Determinants of simultaneous flow of water and contaminants with regard to the immiscible phase migration were also studied (Pistiner et al., 1990). Three phase movement of nonaqueous phase liquid (NAPL), water, and air in the unsaturated zone, accompanied by the process of volatilization, dissolution, partitioning, sorption, decay and advective-diffusive transport of the organic phase, was attempted by Zaidel and Russo (1993), demonstrating the importance of the kinetically controlled, nonequilibrium mass transfer between NAPL and other phases. Subsequently, one- and two-dimensional, steady-state organic vapor transport models pertinent to NAPL spills were suggested (Zaidel and Russo, 1993). Similarly, the contaminant upscaling transport equations, applicable to advanced numerical simulations on coarse grid with elements of a given size and shape, are being developed (Indelman and Dagan, 1993). This is in an attempt to generalize the procedure for three-dimensional structures. Due to the complex nature of the aquifers and despite the intensive modeling work, quantitative simulation based on improved models is still in the evolutionary development stages.

Soil Processes

SOIL FRACTIONS/INORGANIC INTERACTIONS

The transport of toxic metals in the unsaturated/saturated zones is strongly affected by dissolved and solid organic substances. The solubility and sorption behaviors of these metals depend on the presence of complex-forming organic substances. Irreversible, specific adsorption to solid organic substances removes relatively high percentage of dissolved metals from the soil pore solutions (Klamberg et al., 1989). The transport and deposition of solids in porous media depend on soil particle characteristics and soil particle size distribution. The latter physical parameter may be expressed for secondary effluents by a power-law function (Adin and Alon, 1993). The complexation of contaminants with organic substances and the role of clay-humic complexes in the binding of solutes to organic contaminants were further demonstrated in a controlled system of clay and humic acids (Rebhun et al., 1987).

Similarly, under different conditions, the mineral fraction may dominate the fate of toxic organic chemicals in the unsaturated zone. Low organic matter content below the root zone allows the mineral surface to dominate the abiotic interaction of the organic toxicants; in particular, the fate of cationic organic pollutants may be controlled by this fraction (Saltzman and Mingelgrin, 1984; Mingelgrin and Prost, 1989).

MICROORGANISMS

Microorganisms that have the genetic potential to bind, emulsify, transport, and degrade hydrocarbons are widely distributed in nature (Rosenberg et al., 1992). Furthermore, many compounds of a significant group of pollutants such as halogenated alkanes, which are extensively used in agriculture and industry, were found to be biodegradable, partially or fully, as single substances or as a mixture of pollutants by microorganisms (Belkin, 1992; Shochat et al., 1993). However, due to the low solubility of hydrocarbons in water, the degradation involves a membrane-bound oxygenase enzyme system, which makes it essential for the microbiota to come into direct contact with the hydrocarbon substrate. Furthermore, petroleum is a mixture of different hydrocarbon molecules, and any particular microorganism is only able to use part of the petroleum (Rosenberg et al., 1992).

The use of live organisms to degrade, detoxify, or transform hazardous compounds in the environment is practiced for in situ hazardous waste remediation of soil and oil spill cleanup of beaches and the open ocean. The process involves biostimulation, i.e., stimulating the indigenous microbiota with appropriate nutrients and co-metabolites or bioaugmentation, i.e., introducing nonindigenous microbiota to degrade the organic contaminants in place. However, bioaugmentation is a difficult process to put into practice because of competition from the natural microbial population and the inability of the cultured organisms to handle the stress of the natural environment. Improved genetically engineered microorganisms that can be tailored to degrade specific waste are essential for further development of bioaugmentation. The supplemental microbiota must be able to (1) survive and compete for nutrients with indigenous microorganisms and (2) be mobile from the point of injection to the location of contaminant and to retain selectivity for metabolizing the compound(s) for which they were selected. Degradative enzymes mass-produced by genetically altered microbiota were suggested in order to avoid the need to introduce foreign organisms into the environment. In addition, the enzymes might be more readily dispersed within a soil porous medium (Bowman, 1989).

NUTRIENTS

The limiting factor in biodegradation is not only the lack of appropriate microorganisms, but also moisture, oxygen limitation, and available sources of nitrogen and phosphate. Of these factors, the major limitation in biodegradability in soil and water is nutrient availability (Rosenberg et al., 1992). For hydrocarbon degradation, oxygen, nitrogen, and phosphate not sufficiently available in crude oil have to be added. Available commercial sources of nutrients are not suitable because of a high water solubility, which reduces their effectiveness in open systems, and because of rapid dilution. Controlled release hydrophobic fertilizers are being developed that are more effective in supplying the nutrients in a controlled manner. A modified urea-formaldehyde polymer (F1), containing 18% nitrogen and 10% phosphate as P_2O_5, was used together with an enriched culture of bacteria to clean a sandy beach on which approximately 100 tons of heavy crude oil was accidentally spilled 3 km north of Haifa in August 1991. Applying 5 g of F1/kg sand, the hydrocarbon content decreased within 28 days after treatment from 5.1 mg per gram to 0.6 mg (86%), as compared to 15% in the control, as measured by pentane extractable (Rosenberg et al., 1992).

PLANTS

A process of biofiltration of heavy metals by the water fern *Azolla* has been developed for the detoxification of effluents containing Cd, Cr, Ni, Cu, Zn, and Pb (Sela et al., 1988). This process was patented worldwide and the development of the biofiltration process and its product(s) is now under development by the industry, Makhteshim Ltd., Israel. Recently, the infrastructure for a large-scale *Azolla* farm, based on the experience with 100 m^2 outdoor *Azolla* ponds, was set up by Makhteshim. The routine operation involving cultivation, harvest, and quality control, with respect to the metal binding capacity, is under investigation.

In the laboratory, the development of the metal-binding process of *Azolla* is continuing, focusing on the binding of Pb and its removal from solutions containing 1–1000 ppm. The process for treatment of Pb-polluted solutions was optimized by Adin and Tel-Or (Tel-Or, 1995) with lab-scale biofilters. Flow rates, pH, and degree of purification were studied. This *Azolla* biofiltration process is currently being tested on other metal ions.

Groundwater Remediation Processes

The most practical method for remediation of groundwater-contaminated aquifers is a combination of pumping and treatment followed by recharge to the aquifer or off-site disposal. The first step is to pump out any floating oil that can be recovered and then begin groundwater extraction and surface treatment (pump and treat). A pump-and-treat system involves flushing clean water through contaminated zones, recovering the contaminated water, treating it in aboveground reactors and reinfiltrating the treated water to enhance the removal of the more soluble contaminants. Recently, on-site and in situ remediation has become an attractive alternative treatment technology to pump-and-treat systems due to its relative cost-effectiveness and in keeping with new trends that recognize the advantage of in situ and on-site treatment technology. In situ remediation is often the method of choice because it destroys contaminants in place, rather than merely transferring them to another environmental medium, as occurs with a pump-and-treat remediation procedure.

Soils and the vadose zone soil layers can be treated by a combination of various engineering approaches. Vertical and/or horizontal injection and extraction wells, strategically placed within the contaminated zone to control flow and direction in the vadose zone, are typically used to introduce the required electron acceptors, electron donors, and essential nutrients into the groundwater to enhance biodegradation. The approaches include the use of liquid and vapor extraction wells, with or without downhill pumps and infiltration galleries that can be used to percolate water; nutrients; and oxygen through the contaminated unsaturated zone. In situ bioremediation can be further supported by an on-site water treatment (flushing circuit) to remove hydrocarbons, Fe, and Mn and to increase temperature. The on-site treatment could include stripping, flocculation, sedimentation, and sand filtration. The stripped hydrocarbons are absorbed on granular active carbon, which is steam regenerated.

Technical groundwater remediation systems include

- dual vacuum extraction (DVE) wells—liquid and vapor extraction
- those with or without downhole pumps
- air stripping
- in situ vapor stripping method
- air sparging wells (in situ)
- dual vacuum strippers (DVS) wells (with nutrients, oxygen, and or microbial injection)
- infiltration galleries and trenches

VAPOR EXTRACTION

Vacuum-enhanced soil venting is one of the most efficient and frequently used tools to address fuel hydrocarbons in the subsurface. The technique takes advantage of the generally highly volatile nature of petroleum hydrocarbons. The system produces a vacuum that is transmitted to the subsurface soils through manifold piping and extraction wells. Vapors flow through the piping to an air liquid separation and collection vessel that contains the extracted groundwater. As the contaminant-laden vapors are removed, other phases of the VOCs will vaporize in place, which in turn will continually be driven out by the vacuum extraction process. Extracted vapor can be treated using catalytic oxidation or dispersed into the air at a rate in which air quality is maintained at a safe level. Air quality limitations may require the exhaust stream to be treated, typically by activated carbon adsorption.

EXTRACTION OF THE LIQUID PHASE (FREE OIL)

Liquid-phase hydrocarbons penetrate the soil layers to the more permeable sandy clay layers. Consequently, dissolved-phase hydrocarbons are prevailing in the groundwater. The free oil can be extracted by the application of vacuum, which enhances the hydraulic yield and is essential in low hydrocarbon yielding clay layers. Alternatively, downhole pumps and free product pumps can also be employed. These pumps work best for removal of liquid-phase hydrocarbon in the sands.

RECOVERY OF THE DISSOLVED HYDROCARBONS

Dissolved hydrocarbons are recovered as part of the liquid hydrocarbon recovery operation when a dual vacuum pump system is used. The bulk of the groundwater containing dissolved hydrocarbon can, however, be treated with a variety of engineered systems, depending on the flow rate, the contaminant concentration, and groundwater chemistry. Alternative treatment and disposal solutions include discharge to a water treatment system; discharge to a surface water outfall, if permitted; treatment and recharge to the aquifer; and establishment of a recirculation cell followed by treatment and reuse.

BIOSTIMULATION

Oxygen can be added by sparging the gas directly, achieving a concentration of 40 mg O_2/L as compared to 8–10 mg O_2/L if air is used for sparging (Piotrowski, 1992). However, because of oxygen-limited solubility, it is difficult to get sufficient amounts into the groundwater to produce significant degradation. Oxygen may also cause biofouling and/or the precipitation of dissolved metals (Johnson, 1994). The ability to perfuse the subsurface with nutrients and oxygen is in many cases limited by the specific site lithology; clay-type soils are poor candidates due to low hydraulic conductivity, while sandy-type soils are better suited. The treatment may therefore be limited to areas at which hydrogeological data are well established. Various geophysical methods, including a vertical electrical sounding method (VES), buried electrode method, and ground penetrating radar system, have also been suggested for mapping contaminated zones of aquifer (Mazac et al., 1990). These methods, if successful, can substitute expensive drilling and soil/water chemical analysis. Water chemistry also may limit its application as delivery of oxygen via H_2O_2 is precluded if levels of iron and/or manganese are excessive due to the catalytic nature of these chemicals.

Aquifer formations of sand, silt, and clay could trap the gas liberated as bubbles during the remediation process, resulting in clogging and preventing the flow of water (Abeliovich, 1992). High microbial growth rates can also occur in the near-field of the injection well, reducing permeability and well capacity. Ultimately, biofouling may decrease the injection capacity to the point where the well is no longer usable. The addition of surfactants has been suggested in order to release absorbed petroleum by increasing solubility, biodegradation, and reduced interfacial tension between the water phase and the soil phases (Zoller, 1992).

Groundwater Remediation Engineering

In situ groundwater treatment has been reported for removing particular contaminants from groundwater. Werner (1985) utilized the subsoil as a reactor for removal of mineral oil pollution from groundwater. Kruithof (1985) carried out in situ denitrification experiments at the J. H. Van Heek Water Treatment Plant in the Netherlands. The performance of a pilot plant for in situ nitrate removal in Czechoslovakia was reported by Chalupa (1985) and in the United States by Skinner et al. (1992). Alternative in situ treatment schemes were also defined in Israel by Mercado et al. (1988). However, these experiments and many other bioremediation projects completed to date have consisted of either laboratory- or demonstration-scale studies or of an empirical, site-specific approach applied to the cleanup of actual vadose zone contamination (Bowman, 1989).

IN SITU DENITRIFICATION

In Israel, in situ denitrification treatment was conducted by Mercado et al. (1988) using an aquifer with a high nitrate content. The testing well pumps water from a calcareous sandstone aquifer of an extremely heterogeneous formation, intersected by clay and loam layers. The water static level was at about 50 m below the land surface. Three small diameter injection wells, 90–100 m deep, were drilled at a distance of 15–25 m from the Shivat Zion, Rishon Lezion production well. Aqueous solution of sucrose, at a concentration of 50 mg/L, mixed with a small quantity of phosphoric acid, was injected in order to induce denitrification in the aquifer. At an injection rate of 2–4 m^3/h and a pumping rate of 5–60 m^3/h, a 10% reduction in the nitrate content was achieved. The use of sucrose as a carbon source induced microbial growth, and in order to minimize clogging, several injection wells were required, enhancing hydrodynamic dispersion and a continuous supply of treated water.

The observed clogging was further investigated in the laboratory in sand columns (Soares et al., 1989). The results indicated that gas bubbles, rather than biomass accumulation, were the main clogging factor. Of four biodegradable carbon sources investigated, sucrose, formate, acetate, and ethanol, formate brought about the lowest accumulation of gas and the highest permeability levels in the test column.

IN SITU VAPOR STRIPPING METHOD

Of the various remediation techniques, a combination of air-lift pumping and in situ vapor stripping technique was investigated by Gvirtzman and Gorelick (1992) for the removal of VOCs. The process includes the injection of air into a well, causing a lift of the contaminated water in the well, and at the same time the extraction of the contaminants as air bubbles. The VOCs are then collected at the top of the well by vapor extraction. The partially treated water is recirculated back into the unsaturated zone where it reinfiltrates without being brought to the surface. Aboveground installations are not required as remediation occurs while water is circulated within the well. The air injection creates groundwater circulation towards the well and in situ volatilization of VOCs, gradually reducing the concentrations of the VOCs. The lifted water bubble mixture is diverted through a series of drains installed within the subsurface back by infiltration into the aquifer. The organic rich bubbles are collected using vapor extraction techniques and treated by sorption into activated carbon.

TESTING FACILITIES

The concept of in situ vapor stripping, under various aquifer conditions, together with enforced ventilation and biostimulation, are being tested in a room size aquifer model facility ($2 \times 0.7 \times 1$m). This precedes a field study in which VOCs are to be degraded using air-lift to pump the contaminated water, followed by recharge of the treated water. In addition, the chemical, biological, and physical aspects of the remediation process, namely the biochemical activity, gas emission, colloidal production, transport, and hydraulic conductivity, will be further studied in a large-scale artificial aquifer facility under construction in the Weizman Institute, Rehovot (Ronen et al., 1994). This facility will allow the simulation of a three-dimensional gradient flow condition similar to those encountered in natural aquifers, as well as forced pumping flow conditions, thus allowing a better understanding of basic field processes and phenomena in a near-field but controlled conditions processes. Advanced water sampling techniques will be employed using a multilevel sampler most suited for the study of the processes occurring at the interface between the unsaturated and saturated zones (Ronen et al., 1986).

Site Remediation—The Case of Or-Akiva Site

A jet fuel spill of about 400 m^3 originated from a 10-inch pipeline on October 29th, 1990. About 70,000 m^3 of soil were affected, extending for an area of 100 × 30 m. Site investigation revealed that a fuel layer with a maximum thickness of 90 cm covers an area of 1000 m^2, amounting to about 235 m^3 of liquid-phase hydrocarbon and approximately 90 m^3 of dissolved-phase hydrocarbon in concentrations of 0.5 to 100 ppp, over an area of about 3600 m^2. Approximately 54 m^3 were estimated to remain adsorbed in the unsaturated zone. Twelve pumping wells that collectively extract 5 million m^3 per year of groundwater are located within the area of concern. Soil and groundwater samples taken from the site indicated that the soil and bedrock at the site consist of three identifiable layers; an uppermost layer of sand dunes 3 m thick, followed by a black heavy clay layer 6 m thick and finally a sandy clay (hamra) layer 6 to 12 m thick. The water table level is 7–8 m below the surface.

The remediation process includes a containment stage, with prevention of further migration of mobile hydrocarbon phases by controlling hydrocarbon plume movement (liquid and dissolved phases), and a cleanup stage, with removal of as much liquid hydrocarbons as possible to minimize spreading of the residual hydrocarbons in the saturated and unsaturated zones. A series of remediation processes that represent a combination of various technologies was proposed, including

- vacuum extraction (VE) for the unsaturated sand
- dual vacuum extraction (DVE) for the saturated clay and sandy clays
- pump and treat (P&T) for the dissolved-phase plume

The DVE procedure will treat the soils as well as the contaminated groundwater in the area of spill. Air sparging was not considered because the presence of the clay layer would cause the air bubbles to move laterally outside of the active zone. Groundwater will be treated by an oil/water separator, an air stripper, and a liquid-phase carbon adsorption system for polishing. The treated water will be reinjected into the aquifer. The reinjection of treated groundwater could include the injection of heated air to enhance volatilization and/or the addition of oxygen, nutrients, or microbes to enhance biodegradation. Reinjection may be carried out through horizontal wells placed at the water table perpendicular to the hydraulic gradient, on the entire upgradient side of the contaminated area, or through infiltration trenches. Modeling is to be performed to generate a preliminary capture zone as a starting point for the design and placement of DVE, VE, and P&T wells.

COST ESTIMATES

For estimation of costs, the plan assumed that, for full-scale treatment, approximately 20 VE and DVE wells and 7 pump-and-treat wells (50 m^3/h^{-1}) will be required, amounting to about US$2.5 million. Based on a quantity of 235 m^3 of free oil, 90 m^3 of dissolved oil, and 54 m^3 of absorbed oil, the cost per cubic meter will amount to about US$6650.

DISCUSSION AND CONCLUSIONS

The information on soil and groundwater remediation presented in this chapter indicates that contamination of groundwater resources by various contaminants will remain a problem for many years, and effective measures will be required to confine and alleviate the pollution hazard. Groundwater supplies contaminated by traces of oil products are likely to be difficult and expensive to treat. As a realistic alternative, in situ and on-site treatment processes are increasingly being considered. The few operating systems that are available have shown that properly designed extraction systems can reduce the concentration of dissolved contaminants in groundwater, although complete cleanup is often not possible using the available techniques. In Israel, an attempt to apply enhanced in situ microbial denitrification to treat the contaminated coastal aquifer proved to be impractical, both in the laboratory and in field studies (Soares et al., 1991). Biological processes, though capable of destroying more than 99% of the waste concentrations, usually decrease rapidly after initiation of pumping and then are not sufficient to meet the regulations. In addition, biological processes are temperature-dependent. Therefore, it is doubtful whether the maximum concentration level of 5 ppb mandated by the EPA could ever be achieved with the current technology (*Science,* 1992).

Results of field experience using in situ bioremediation are positive in the case of aromatic, aliphatic, and phenolic compounds, although in situ bioremediation of TCE-containing water is apparently limited to fluids containing less than about 100 ppm of TCE, as higher concentrations seem to be toxic. Contaminant concentrations usually decrease rapidly after initiation of pumping and then tend to level off at some asymptotic concentration (Swelt, 1992). For mixed contaminants, different requirements and a combination of different methods are necessary to overcome mass balance limiting factors, including bioavailability of the contaminants (spatial separation, solubility), lack of electron acceptors, and decrease of the cleanup target values followed by risk assessment of metabolites and dead-end products. However, bioremediation is considered as potentially advantageous in treating hazardous wastes, but the limited experience and the lack of reliable information on application and performance and cost are major impediments to its use. Most of the remediation work that was reported to date is still at the laboratory and field demonstration level although real progress can be observed since the early work of Werner (1985) in the early 1980s and the current results of the EPA work at the demonstration site in Traverse City, Michigan (Skinner et al., 1992).

The current limited experience with bioremediation is due to a lack of data on performance and costs of bioremediation technology. The dearth of cost data is particularly evident for biology-based waste treatment methodologies that were applied on a recycling and disposal of chemical waste site in South Carolina. Based on a remediation of 45,000 cubic yards of soil, cost estimates for various remedial technologies were given as follow: in situ soil vacuum extraction—US$1.1 million, on-site incineration with stabilization of treated soils—US$28.3 million, on-site thermal desorption with stabilization of treated soils—US$18.3 million, off-site disposal of contaminated soils—US$20.7 million, and off-site thermal treatment of contaminated soils—US$100 million (USEPA, 1990). According to recent estimates for the Superfund Program (U.S. Dept. of Energy, 1988), the average per site cost of characterization, assessment, and remedial action design is about US$1.7 million, very close to the cost estimate of US$2.5 million made for the cleaning of about 400 m^3 of jet fuel from the Or-Akiva site. These estimates are significantly lower, compared to the average cost of the conventional remedial action, usually a pump-and-treat system, which is about US$12.5 million.

Based on these cost estimates, the total cost of managing and remediating contaminated sites could be enormous. Current estimates for the cleaning of U.S. Department of Energy waste sites are estimated at about US$350 billion, as reflected by the new 1993 DOE budgeted expenditure of US$5.3 billion on waste remediation, an increase of 24% over 1992 (*Science,* 1992). In the European Union, soil rehabilitation efforts are expected to reach US$10 billion by the year 2000 and are expected to grow to US$30 billion in the next decade (Porta, 1992). These figures are indicative of a fast developing technology and a growing market for innovative cleanup processes. The magnitude of the polluted sites and the huge amounts of expenditure that will be required justifies the wide interest of the various concerns in groundwater remediation technology.

Concerted efforts are still required to develop techniques for cleaning of groundwater by a collaborative research approach, with scientists and engineers working together on integrated molecular genetics and physiology with environmental needs. Technologies related to the application of microorganisms to the soil, to the release of nutrients into the soil, to the enhancement of microbial decontamination through various additives and restoration of enzymatic activity for xenobiotic decontamination have yet to be defined. The promise of a feasible combination of in situ and on-site treatment technologies, favorable economics, reduced residuals, and little environmental impacts give reason for optimism. Yet further basic research and field demonstrations in bioremediation are needed before the potential of groundwater treatment can be realized. Specifically, more credible field experience is needed in order to build confidence in bioremediation among decision makers and the general public. It is to be noted, however, that attempts to recapture, isolate, and treat pollution is costly, and prevention of pollution by strict control of pollutants is far more desirable. A vigorous development control and enforcement policy for land close to groundwater sources is imperative to prevent the pollution of public supplies derived from aquifers.

REFERENCES

Abeliovich, A. 1992. Transformations of Ammonia and the Environmental Impact of Nitrifying Bacteria. *Biodegradation,* 3:255–264.

Adin, A. and G. Alon. 1993. The Role of Particle Characterization in Advanced Wastewater Treatment. *Wat. Sci. Tech.,* 27:131–139.

Amiel, A.J., Magaritz, M., Ronen, D. and O. Lindstrand. 1990. On the Mobility of Dissolved Organic Carbon in the Unsaturated Zone under Irrigated Land with Sewage Effluents. *J. Water Pollution Control Fed.,* 62:861–866.

Belkin, S. 1992. Bidegradation of Haloalkenes. *Biodegradation* 3:299–313.

Bowman, R.S. 1989. Manipulation of the Vadose Zone to Enhance Toxic Organic Chemical Removal. In Gerstl, Z., Chen, Y. and U. Mingelgrin (eds.). *Toxic Organic Chemicals in Porous Media,* Springer-Verlag, Berlin, Heidelberg, New York, pp. 275–286.

Chalupa, M. 1985. La Denitrification Artificielle des Eaux de Nappe. *Proceeding of the Workshop on "Nitrates Dans les Eaux."* October 22–24. 1985, Paris, France.

Gvirtzman, H. and S.M. Gorelick. 1992. The Concept of In-situ Vapor Stripping for Removing VOCs from Groundwater. *Transport in Porous Media,* 8:71–92.

Indelman, P. and G. Dagan. 1993. Upscaling of Permeability of Anisotropic Heterogeneous Formations. Part 1: The General Framework, *Water Resources Res.,* 29 (4):917–923.

Johnson, R.L. 1994. Enhancing Biodegradation with In-situ Air Sparging, In Hinchee, R.E. (ed). *Air Sparging for Site Remediation,* Lewis Publishers, CRC Press Inc., 14–22.

Kanfi, Y. 1991. *Fuel Spill on the Coastal Plain Aquifer, Or-Akiva Site.* Israel Water Commission Report, Tel Aviv.

Kessler, A. and H. Rubin. 1987. Relationship Between Water Infiltration and Oil Spill Migration in Sandy Soils. *J. Hydrol.,* 91:187–204.

Klamberg, H., Matthess, G. and E. Pekdeger. 1989. Organo-Metal Complexes as Mobility-Determining Factors of Inorganic Toxic Elements in Porous Media. In Bar-Yosef, B., Barrow, N.J. and J. Goldshmid, (eds.). *Inorganic Contaminants in the Vadose Zone.* Springer-Verlag, Berlin, Heidelberg, pp. 4–17.

Kruithof, J.C., Paassen, J.A.M., Hijnen, W.A.M., Dlerx, H.A.L. and C.A. Bennekom. 1985. Experiences with Nitrates Removal in the Eastern Netherlands. *Proceedings of the Workshop on "Nitrates dans les Eaux,"* October 22–24, 1985, Paris, France.

Mazac, O., Benes, L., Landa, I. and A. Maskova. 1990. Determination of the Extent of Oil Contamination in Groundwater by Geoelectrical Methods. *Geotech. and Environ. Geophysics,* 2:107–112.

Mercado, A. 1976. Nitrate and Chloride Pollution in Aquifers. A Regional Study with the Aid of a Single Cell Model. *Water Resources Res.,* 12:731–747.

Mercado, A., Libhaber, M. and M.I.M. Soares. 1988. In-situ Biological Groundwater Denitrification: Concepts and Preliminary Field Tests. *Water Sci. Technol.,* 20 (3):197–209.

Mingelgrin, U. and R. Prost. 1989. Surface Interactions of Toxic Organic Chemicals with Minerals. In Gerstl, Z., Chen, Y. and U. Mingelgrin (eds.). *Toxic Organic Chemicals in Porous Media,* Springer-Verlag, Berlin, Heidelberg, New York, pp. 91–132.

Muszkat, L., Ronen, D. and M. Magaritz. 1993. Penetration of Pesticides and Industrial Organics into the Soil Profile and Groundwater. *Archives of Insect Biochem. and Physiol.,* 22:487–499.

Piotrowski, R.M. 1992. In-situ Designs for Aquifer Clean-up. *Environmental Protection,* May, 34–49.

Pistiner, A., Sapiro, M. and H. Rubin. 1990. Similarity Solutions for Immiscible Phase Migration in Porous Media and Analysis of Free Boundaries. *Transport in Porous Media,* 5:491–515.

Porta, A. 1992. A Review of European Bioreclamation Practice. In Hinchee, R.E. & R.E. Offenbuttel (eds.). *In-situ Bioreclamation,* Butterworth-Heineman, II, 1–13.

Rebhun, M., Bracha, C. and E. Arvin. 1987. *Behavior and Fate of Organic Micropollutants in the Subsurface,* Israel Tech. Inst. Report.

Ronen, D., Berkovitz, B. and J. van Rijn. 1994. *Conceptual Design of an Intermediate Scale Artificial Aquifer and Test Facilities.* Dept. of Env. Sci. and Energy Res. Scientific Report, Weizman Institute of Science.

Ronen, D., Magaritz, M. and I. Levy. 1986. A Multi-layer Sampler for the Study of Detailed Hydrochemical Profiles in Groundwater. *Water Res.,* 20:311–315.

Rosenberg, E., Legman, R., Kushmaro, A., Taube, R., Adler, E. and E.Z. Ron. 1992. Petroleum Bioremediation—A Multiphase Problem. *Biodegradation* 3:337–350.

Saltzman, S. and U. Mingelgrin. 1984. Nonbiological Degradation of Pesticides in the Unsaturated Zone. In Yaron, B., Dagan, G. and J. Goldshmid (eds.). *Pollutants in Porous Media.* Springer, Berlin, Heidelberg, New York, pp. 237–246.

Science (Editorial) 1992. Remediation of Hazardous Waste Sites, 255:901.

Sela, M., Garty, J. and E. Tel-Or. 1988. The Accumulation and the Effect of Heavy Metals on the Water Fern *Azolla filiculoides. New Phytol.,* 112:7–12.

Shevah, Y. and M. Waldman. 1989. Advances in Management of Fertilizer Application and the Resulting Effect on the Pollutant Load in Soil. In Bar-Yosef, B., Barrow, N.J. and J. Goldshmid (eds.). *Inorganic Contaminants in the Vadose Zone.* Springer-Verlag, Berlin, Heidelberg, pp. 179–189.

Shochat, E., Hermoni, I., Cohen, Z. and A. Abeliovich. 1993. Bromoalkane-Degrading *Pseudomonas* Strains. *Appl. Env. Microbiol.,* 59:1403–1409.

Skinner, J.H., Omdick, G.G. and T.L. Baugh. 1992. US EPA Bioremediation Research Program. In Hinchee, R.E. & R.E. Offenbuttel (eds.). *In-situ Bioreclamation,* Butterworth-Heineman, I: 1–15.

Soares, M.I.M., Belkin, S. and A. Abeliovich. 1989. Biological Groundwater Denitrification: Laboratory Studies. *Water Sci. Tech.,* 20 (3):185–195.

Soares, M.I.M., Breaster, C., Belkin, S. and A. Abeliovich. 1991. Denitrification in Laboratory Sand Columns: Carbon Regime, Gas Accumulation and Hydraulic Properties. *Water. Res.,* 25:325–332.

Steiber, M., Werner, P. and F.M. Frimmel. 1994. Investigations on the Microbial Degradation of Polynuclear Aromatic Hydrocarbons in Contaminated Soils. In Hinchee, R.H., Leeson, A., Semprin, C. and S.K. Ong (eds.). *Air Sparging for Site Remediation.* Lewis Publishers, CRC Press Inc., 164–171.

Swelt, G.M. 1992. Bioremediation: Myths vs. Realities. *Environmental Protection,* May, 23–26.

Tel-Or, E. 1995. Development of *Azolla filiculoides* Biofilter for Heavy Metal Removal; From the Lab to the Industry. The Otto Warburg Center and the Department of Agricultural Botany, The Hebrew University of Jerusalem, Rehovot, Israel.

USEPA. 1990. *Record of Decision. Remedial Alternative Selection,* USEPA, Region IV, Atlanta, Georgia.

Werner, P. 1985. A New Way for the Decontamination of Polluted Aquifers by Biodegradation. *Water Supply,* 3:41–43.

Wolman, S. and D. Ronen. 1992. *Nitrate Trends in the Coastal Plain Aquifer.* Israel Hydrological Service, Tech. Rep. No. 13.

Yaron, B. 1989. On the Behavior of Petroleum Hydrocarbons in the Unsaturated Zone: Abiotic Aspects. In Gerstl, Z., Chen, Y. and U. Mingelgrin (eds.). *Toxic Organic Chemicals in Porous Media,* Springer-Verlag, Berlin, Heidelberg, New York, pp. 211–230.

Zaidel, J. and D. Russo. 1993. Analytical Models of Vadose Zone Steady State Kinetically-Controlled Organic Transport. *Water Resources Res.,* 29 (10):3343–3356.

Zoller, U. 1992. Distribution and Survival of Nonionic Surfactants in the Surface, Sea and Groundwater. *J. Env. Sci. and Health,* 27 (6):521–1533.

Current Status of Soil Pollution and Bioremediation in Japan

YASUHIRO OKUBO
Nishihara Environmental Sanitation Research Corp., Ltd.
Shibaura 3-6-18 Minatoku
Tokyo 108, Japan

OSAMI YAGI
National Institute for Environmental Studies
Japan Environmental Agency
Onogawa, Tsukuba
Ibaraki 305, Japan

INTRODUCTION

Soil contamination is one of the oldest examples of environmental pollution in Japan. Although the first case of soil contamination was copper pollution on farmland in the Watarase River Basin early in the 1880s, it was finally recognized as a national problem in 1968 when agricultural land contaminated with cadmium was found to be the cause of Itai-Itai disease occurring in the Jinzu River Basin. As for urban land soil pollution, a former industrial site was found to be contaminated with hexavalent chromium in 1975 and became a social problem. However, soil contamination cases revealed at that time were limited to heavy metal pollution discovered in the course of pursuing the cause of damage that had come to the surface. More recently, results of investigations performed in the United States showed cases of groundwater polluted by organic chlorine compounds. Stirred by these reports, the Environmental Agency conducted an investigation in 1982 to determine the actual conditions of groundwater contamination in several areas for 18 items, focusing on organic chlorinated compounds. As a result, it was confirmed that organic chlorinated compound contamination was spread over unexpectedly wide areas.

To cope with these conditions, environmental quality standards (specifying desirable levels to be maintained in order to protect human health and preserve the living environment) for 10 soil quality items were established in August 1992. In February 1994, the original standards, which mainly specified heavy metal levels, were amended to add a further 15 items, including organic chlorine compounds as major targets. Thus, 25 compounds are currently subject to these standards.

Because of this amendment, the subjects to be investigated also increased to 25 items. This may result in an increased number of contaminated sites being revealed as further investigations are performed on soil and groundwater contamination.

These circumstances have compelled more and more investigators and companies to study toxic and harmful materials and have promoted research activities by newly established study groups.

So far, there is no instance in Japan of bioremediation being used to remove toxic and harmful materials. Bioremediation, however, has been increasingly studied because it is recognized as a promising technique for removing widespread, low concentrations of pollutants, soil contaminants, and other forms of pollution that are difficult to treat by conventional means.

The majority of bioremediation studies in Japan have been conducted to trace, isolate, and culture useful bacteria and to use such bacteria for treatment.

In this chapter, we describe (1) the current status of soil pollution in Japan, (2) the current activities of two study groups, and (3) a review of bioremediation studies in Japan.

CURRENT STATUS OF SOIL POLLUTION

Current Status of Soil Pollution and Control Measures

Soil contamination is one of the causes of soil deterioration and may be directly caused by the leakage of raw materials or permeated compounds from reclaimed wastes or secondarily by water and air pollution. Soil contamination is an accumulative type of pollution that exerts long-term effects on agricultural crops and groundwater.

The first reported farmland soil contamination was found in the basin of the Watarase River as early as 1877. After World War II, agricultural soil contaminated with cadmium was found in the basin of the Jinzu River. These are among the oldest cases of environmental pollution that caused serious health problems for residents. To prevent such pollution, various regulations, including the Water Pollution Prevention Law, were established, and measures to control pollution sources were implemented. In addition, standard levels were set for cadmium, copper, and arsenic in accordance with the Anti-Agricultural

Soil Pollution Law (Table 1). Farmland contaminated with the above elements at concentrations higher than the standard levels were made subject to control measures including soil dressing.

Contamination of nonfarm urban land has been increasingly revealed in all parts of the country. In particular, urban soil contamination has been frequently found in connection with the redevelopment of former factory sites. Investigations conducted in 1992 on the state of implementation of soil pollution control measures confirmed that a total of 177 cases have been identified since 1975. The number of cases has shown an increasing trend over the years (Table 2). In 75 of the 177 cases, soil contamination has affected the surrounding environment. In 63 of

Table 1. Specific harmful substances and requirements for designation as agricultural land soil pollution.

Specific Harmful Substance	Date Specified	Requirements for Designation
Cadmium and its compounds	24 June 1971	(1) Area where it is determined that the content of cadmium in rice is 1 ppm or more (2) Area adjacent to (1) where it is determined that the content of cadmium in rice is likely to be 1 ppm or more, since the quantity of cadmium in its soil is equal to or greater than (1) while the nature of its soil is roughly equal to (1)
Copper and its compounds	17 Oct. 1972	Area where it is determined that the content of copper in the soil is 125 ppm or more (0.1 NHCl extraction) (for paddy fields only)
Arsenic and its compounds	4 Apr. 1975	Area where it is determined that the content of arsenic in the soil is 15 ppm or more (1 NHCl extraction) (other values from 10–20 ppm to be specified by prefectural governor with approval of Director-General of Environment Agency in cases where this value is not feasible for special reasons peculiar to the area's natural condition) (for paddy fields only)

Table 2. The number of urban land soil contamination cases by causal industrial category and by contaminant.

Industrial Category	Number of Cases	%	Cad-mium	Cyano-gen	Lead	Hexa-valent Chromium	Arsenic	Mercury	PCBs	Tri-chloro-ethylene	Tetra-chloro-ethylene	Copper	Zinc
Textiles	1	0.6									1		
Timber/wood products	2	1.1				2	2					1	
Chemical	31	17.5	7	3	12	4	7	16	2	1	2	3	2
Plastic products	2	1.1	1			1							
Ceramics/earthenware/stoneware	5	2.8	2		2		1	1		1	1		
Iron and steel	2	1.1	1	1	2		1	1					1
Nonmetal products	7	4.0	3	1	4	1	3			1			2
Metal products	35	19.8	3	11	4	20	1			6	2		2
(electric plating)	(22)	(12.4)	(1)	(7)	(1)	(18)				(2)	(1)		(1)
General equipment	1	0.6											
Electric equipment	18	10.2	2	1	4	3		1	5	7	1		1
Transport equipment	6	3.4	1	1	2	3	2	1		1			1
Precision equipment	2	1.1								2	1		
Fuel gas	2	1.1		2									
Recycled resources wholesale	3	1.7			1				2				
Laundry	16	9.0								1	16		
Waste treatment	6	3.4	2		2	1			1	2	1		1
Natural science laboratory	7	4.0	3		5		2	7	1				3
Others	12	6.8	3				1		4		1	1	
Unknown	19	10.7	2		5	1	3	6	2	6	5		3
Total	177	100.0	30	20	43	36	23	33	17	28	31	5	16
		%	9.5	6.3	13.7	11.4	7.3	10.5	5.4	8.9	9.8	1.6	5.1
Results of investigation in 1987	58		13	8	14	19	10	16	2	1	1	4	7

Table 2. (continued).

Industrial Category	Oil	Fluorine	Nickel	1,1,1-Trichloroethylene	Phenol	Carbon Tetrachloride	*cis*-1,2-Dichloroethylene	1,2-Dichloroethane	Others	Total (Cumulative Number)
Textiles										1
Timber/wood products					1					6
Chemical	1	1	1	1	2				4	69
Plastic products										2
Ceramics/earthenware/stoneware										8
Iron and steel										7
Nonmetal products			1							16
Metal products				3						32
(electric plating)				(1)						(32)
General equipment				1						1
Electric equipment										25
Transport equipment										12
Precision equipment				1						4
Fuel gas										2
Recycled resources wholesale										3
Laundry										17
Waste treatment			1							11
Natural science laboratory										21
Others	3								1	14
Unknown	2			2	1	1	3	2		44
Total	6	1	3	8	4	1	3	2	5	315
	1.9	0.3	1.0	2.5	1.3	0.3	1.0	0.6	1.6	100.0
Results of investigation in 1987	1	—	2	1	2	—	—	—	—	101

these cases, soil contaminated with compounds such as trichloroethylene and tetrachloroethylene secondarily polluted the groundwater. Recently, trichloroethylene and tetrachloroethylene contaminants have shown a marked increase, in addition to heavy metals such as lead, hexavalent chromium, and mercury.

Soil pollution control measures currently in practice consist of the following.

ESTABLISHMENT OF ENVIRONMENTAL STANDARDS

The environmental standards for soil pollution, established in accordance with Section 16 of the Basic Law for Environmental Pollution Control, are the maximum permissible levels of contaminants allowed in any type of soil, including agricultural soil, to protect human health and preserve the living environment. It is intended that these standards will provide criteria for judging the presence of soil contamination and outline the goals to be accomplished by control measures against soil contamination.

In August 1991, soil contamination standard levels for 10 substances, including cadmium, were established. In February 1994, 15 items, consisting of 10 organic chlorine compounds including trichloroethylene and five agricultural chemicals such as Shimazine, were added in order to match the extended and enhanced environmental standards for water quality. At the same time, environmental restrictions on lead and arsenic were tightened (Table 3).

AGRICULTURAL SOIL POLLUTION CONTROL MEASURES

Among areas where specified compounds were detected at concentrations above the standard levels, 6230 ha (64 regions) have been designated as areas subject to control measures as of December 15, 1993. Control plans have been prepared for 6110 ha (63 of the 64 regions). For 4330 ha, control work has been completed under projects subsidized by the National Treasury, such as special land improvement projects for preventing environmental pollution. When adding another area of 390 ha, where control projects have been completed independently by local governments, control work has been completed for a total area of 4720 ha. This represents 66.1% of the total area where specified substances were detected at concentrations above standard levels.

To understand the general conditions of agricultural soil contamination by heavy metals, fixed-point basic investigations on the soil environment have been conducted. In addition, field tests have been carried out to evaluate the effects of improvement measures, including soil dressing and soil improvement in areas with polluted agricultural soil.

Table 3. Environmental quality standards for soil pollution (Prime Minister's office ordinance No. 46 of August 23, 1991; final amendment: No. 25 of February 21, 1994).

Substance	Target Levels of Soil Quality through Leaching Test and Content Test
Cadmium	0.01 mg/L or less in test liquid and less than 1 mg/kg in rice for agricultural land
Arsenic	0.01 mg/L or less in test liquid and less than 15 mg/kg in soil for agricultural land (paddy fields only)
Copper	Less than 125 mg/kg in soil for agricultural land (paddy fields only)
Lead	0.01 mg/L or less in test liquid
Hexavalent chromium	0.05 mg/L or less in test liquid
Total mercury	0.0005 mg/L or less in test liquid
Alkyl mercury	Not detectable in test liquid
PCBs	Not detectable in test liquid
Total cyanide	Not detectable in test liquid
Organic phosphorus	Not detectable in test liquid
Dichloromethane	0.02 mg/L or less in test liquid
Tetrachloro methane	0.002 mg/L or less in test liquid
1,2-Dichloro ethane	0.004 mg/L or less in test liquid
1,1-Dichloro ethylene	0.02 mg/L or less in test liquid
cis-1,2-Dichloro ethane	0.04 mg/L or less in test liquid
1,1,1-Trichloro ethane	1.0 mg/L or less in test liquid
1,1,2-Trichloro ethane	0.006 mg/L or less in test liquid
Trichloro ethylene	0.03 mg/L or less in test liquid
Tetrachloro ethylene	0.01 mg/L or less in test liquid
1,3-Dichloropropene	0.002 mg/L or less in test liquid
Thiuram	0.006 mg/L or less in test liquid
Simazine	0.003 mg/L or less in test liquid
Thiobencarb	0.02 mg/L or less in test liquid
Benzene	0.01 mg/L or less in test liquid
Selenium	0.01 mg/L or less in test liquid

Note: The above standards are not applicable to the following places: (1) places where natural toxic substances exist such as in the neighborhood of mineral veins and (2) places designated for the storage of toxic materials such as waste disposal sites.

URBAN SOIL POLLUTION CONTROL MEASURES

In order to maintain environmental standards for urban soil, investigations are to be conducted when soil contamination is found or suspected or at the time of changing land usage, using the standard levels as the basis. When soil contamination is confirmed by the investigation, the companies concerned are to be advised to voluntarily undertake the required corrective measures to ensure that environmental standards can be met as soon as possible. For the purpose of providing guidance and promoting the smooth implementation of control measures, the Environmental Pollution Control Service Corporation finances investigations by local public bodies on compliance with environmental standards and urban soil contamination control measures by private companies.

MINE POLLUTION CONTROL MEASURES

Metal and other types of mines are subject to mine pollution control measures in accordance with the Mine Safety Law. As metal mine facilities run the risk of causing mine pollution even after shutdown, efforts are being made to systematically implement mine pollution control projects in accordance with the Law for Special Measures to Control Metal and Other Mine Pollution and to establish an implementation system that enables reliable and permanent treatment of mine wastes. According to this law, in 1993, an amendment was made (effective from April 1, 1993) to the basic policy on mine pollution control projects for the coming 10 years to make the projects more systematic.

Current Status of Groundwater Pollution and Control Measures

About 30% of municipal water still comes from groundwater sources. Since the 1980s, however, groundwater pollution by substances such as trichloroethylene has been revealed.

Investigations on the actual groundwater pollution situation conducted by the Environment Agency in 1982 and 1983 revealed groundwater pollution by substances such as trichloroethylene in many parts of the country. Similar results were obtained from investigations on the general conditions of groundwater pollution performed by local public bodies during the 5-year period from 1984 to 1988. Trichloroethylene and tetrachloroethylene were detected in 2.7% and 4.1% of the approximately 27,000 wells investigated in various parts of the country, respectively, at concentration levels above those permitted by the water quality standards for drinking water.

Regular monitoring of groundwater pollution conditions started in 1989, in accordance with the Water Pollution Prevention Law. These groundwater quality measurement efforts were conducted by national and local public bodies according to the plans prepared by individual local governments.

In 1993, investigations on general groundwater pollution conditions were performed by 1506 municipal governments to determine groundwater quality in various areas. The percentage of wells where specific pollutants were detected at concentrations above the permissible levels were revealed by these investigations to be 0% (0/2625) for cadmium, 1.4% (37/2561) for arsenic, 0.1% (3/2626) for total mercury, 0.3% (15/4480) for trichloroethylene, 0.5% (24/4480) for tetrachloroethylene (Table 4). No wells were found to contain other specified pollutants at concentrations above permissible levels in any of the investigations.

Groundwater pollution by nitrates became an issue in the 1960s in Europe and the United States as large amounts of nitrogenous fertilizers were used for agricultural-chemical-intensive cultivation. However, because nitrites may cause cyanosis or methemoglobinemia, in Japan nitrogen from nitrites is regulated to ensure that it is less than the water quality standard levels for drinking water (10 mg/L) in accordance with the Drinking Water Law. According to investigations on the general condition of groundwater performed by 11 local governments in 1992, water from 5.4% of the wells contained nitrogen from nitrates at higher levels than the guideline level (10 mg/L), and this was specified as an item required to be monitored. Thus, continuous investigation is needed to keep track of the actual situation of groundwater pollution by nitrates.

Table 4. Results of the investigation on groundwater quality in 1993 (general conditions).

Toxic Pollutants	No. of Wells Investigated	No. of Wells Showing Excess Contaminants	Percentage of Wells Showing Excess Contaminants	Permissible Levels
Cadmium	2625	0	0	<0.01 mg/L
Arsenic	2561	37	1.4	<0.01 mg/L
Total mercury	2626	3	0.1	<0.0005 mg/L
Trichloroethylene	4480	15	0.3	<0.03 mg/L
Tetrachloroethylene	4480	24	0.5	<0.01 mg/L

Note: No wells were found to contain toxic chemicals or tentative control items other than ones shown above at concentrations higher than the permissible levels.

Once polluted, groundwater is difficult to clean. Thus, it is necessary to take preventive measures before the pollution becomes serious. From August 1984, the Environment Agency has given factories and operation sites tentative guidelines restricting the discharge of three compounds, including trichloroethylene. The Ministries of Health and Welfare and International Trade and Industry have also provided guidance to related industries, and the Ministry of Construction has given advice and conducted investigations on groundwater.

Subsequent investigations showed, however, that groundwater pollution by compounds such as trichloroethylene was still prevalent in many parts of the country. Thus, a partial amendment was made to the Water Pollution Prevention Law, effective October 1, 1989, with the aim of preventing further groundwater pollution by harmful substances.

According to the amendment, water containing one or more of 11 harmful substances, including trichloroethylene, was prohibited from permeating into the ground, and groundwater was to be regularly monitored for such pollution by the concerned prefectural governors (Figure 1).

In order to strengthen the regulations, 13 compounds, including 1,1,1-trichloroethane, were newly specified as harmful substances in December 1993.

APPROACHES OF STUDIES ON SOIL POLLUTION

Contaminated soil has usually been treated by soil replacement and soil dressing methods for agricultural land and by enclosing/isolating methods for urban land. It is clear, however, that these methods allow contaminated soil to remain unchanged semipermanently and possibly to cause secondary contamination in the future. Therefore, it is necessary to develop techniques in which the polluted soil itself can be decontaminated. Several study groups have been established for this purpose in Japan. An outline of the studies conducted by these groups is introduced in the following sections.

Studies by Japan Waste Research Foundation

Japan Waste Research Foundation was established in August 1989 as a center of research and development under the Ministry of Health and Welfare, with the aim of promoting research and development in waste management. The Ministry of Health and Welfare is a governmental agency responsible for the treatment and disposal of industrial

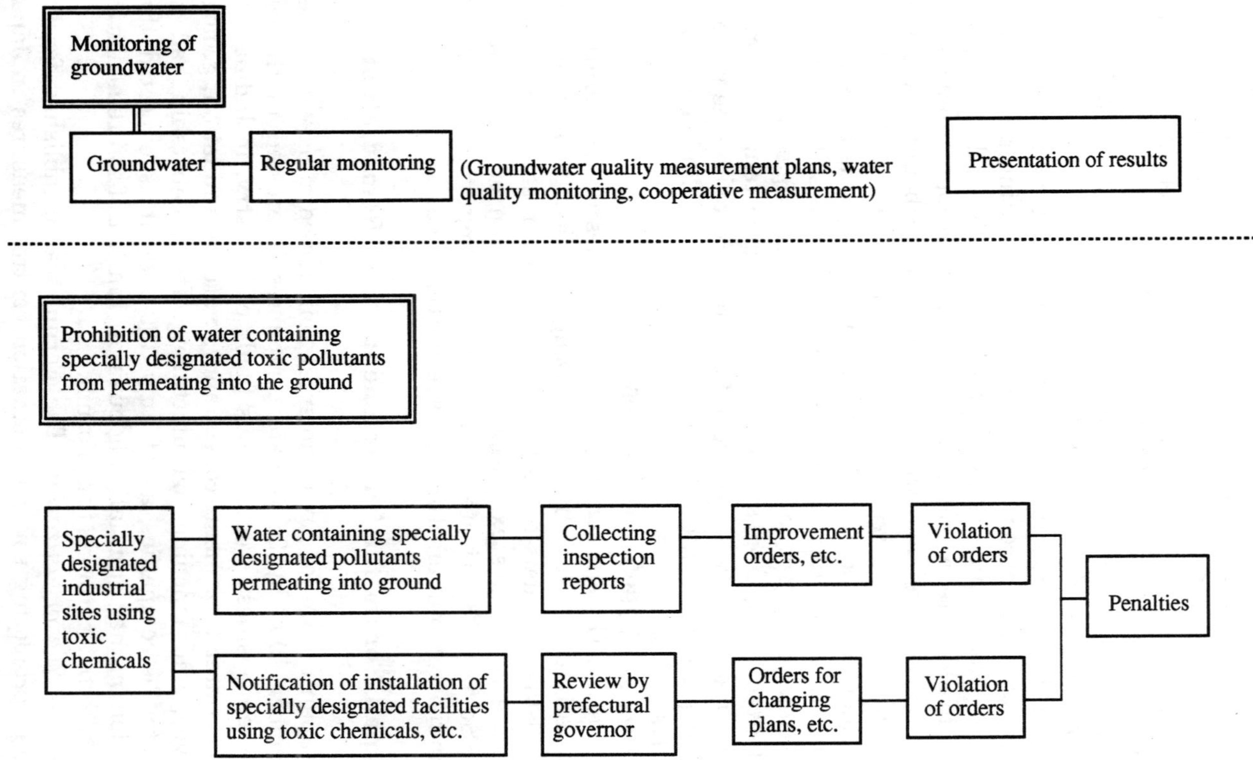

Figure 1. The system of regulations and other actions in accordance with the water pollution law (groundwater).

and domestic waste. Under the direction of the Ministry, which is also responsible for the treatment of leachate from waste reclamation sites, the Foundation started a research and development project targeting the removal of harmful materials expected to be present in such leachate. This project is outlined below.

Subject: Study of remediation techniques for soil and groundwater polluted by illegally disposed waste.

Objective: In addition to the fact that the capacity of final waste treatment plants is becoming inadequate for treating an increasing amount of waste, uncontrolled illegal waste disposal and serious cases of inadequate waste treatment are a serious problem that prevents safe and reliable waste treatment from being guaranteed.

If illegally disposed waste contains harmful materials such as organic chlorine compounds, which require a long time to decompose naturally, these harmful materials may affect the soil and subsoil. This is a major subject to be evaluated and investigated, and techniques for recovering from such damage are an important development objective. In this study, therefore, efforts are directed to the research and development of contamination survey methods and remediation techniques.

Duration: April 1993 to March 1997.

Project team: Governmental and private bodies are cooperating to conduct technical research by organizing a project team consisting of specialists in various fields from universities, research institutes, and private companies expressing their intention of joining in the project. So far, 18 private companies, including water treatment firms and plant manufacturers, have participated.

The Japan Forum for Environmental Soil Remediation

The Japan Forum for Environmental Soil Remediation was formed in December 1992 by 36 private companies interested in the soil remediation business from various fields of industry, including construction, water treatment, consultancy, water examination, geological survey, and machine manufacturing. Its membership had increased to 74 companies as of July 1994, as it is expected that information on the potentially big business market of soil remediation can be collected through this Forum.

Purpose of the organization: The Forum was established for the purpose of promoting soil contamination control measures in Japan by presenting its ideas to the government on the desirable measures to be taken. These ideas stem from cooperative research conducted between public and private bodies on the present state of pollution problems in Japan, including soil contamination, and on standards, systems, and techniques employed in foreign countries.

REVIEW OF STUDIES ON BIOREMEDIATION

There are two basic concepts in the treatment of toxic and harmful materials in soil by bioremediation:

1. Utilize bacteria naturally present in soil. In this method, nutrients and the required amount of oxygen are supplied to soil in order to improve growth conditions for useful bacteria so that they can grow and decompose toxic and harmful materials.
2. Trace and isolate bacteria are able to decompose particular toxic and harmful materials from soil and other substrates.

Since mainly the latter method is studied in Japan, studies with bacteria are reviewed here.

Study on Bioremediation with Strain M

Various bacteria are known to decompose trichloroethylene (TCE) under aerobic conditions. They include methane-utilizing bacteria, phenol-utilizing bacteria, nitrifying bacteria, and ethylene-utilizing bacteria. Among them, methane-utilizing bacteria and phenol-utilizing bacteria have been extensively investigated in various countries. All of these bacteria decompose TCE through so-called co-oxidation reactions in which enzymes decomposing carbon sources required for growth also catalyze oxidation of TCE. For example, methane oxygenase and toluene (phenol) oxygenase co-oxidize TCE in methane-utilizing bacteria and phenol-utilizing bacteria, respectively. These enzymes are produced only when corresponding compounds are present; methane and phenol must be added for methane-utilizing bacteria and phenol-utilizing bacteria, respectively, to evoke such co-oxidation. The following is a description of Strain M (*Mythylocystis* sp.), which has the highest TCE-decomposing activity among methane-utilizing bacteria isolated from soil by the National Institute for Environmental Studies.

DECOMPOSING ACTIVITY OF STRAIN

An attempt was made to isolate bacteria with superior ability to decompose TCE from soil and sludge in various parts of the country because such microorganisms are essential for the bioremediation of soil and groundwater contaminated with TCE. As a result, a mixed culture was found that can decompose TCE sufficiently in the presence of methane. This mixed culture decomposed 35 ppb of TCE almost completely in 3 days, simultaneously showing growth of bacteria and consumption of methane and oxygen. (This decomposition was observed only in the presence of methane under aerobic conditions, but not under anaerobic conditions.)

Soil bacterium "Strain M" with TCE-decomposing activity was successfully isolated as the result of efforts to isolate a methane-utilizing bacterium that can decompose trichloroethylene from the mixed culture by repeating pure culture using methane alone as the carbon source.

Figure 2 shows the disassociation curve at various concentrations of trichloroethylene. It indicates that 90% and 50% of trichloroethylene were decomposed in 4 and 12 days at 0.035 ppm and 10 ppm, respectively. Strain M showed sufficient growth even at a concentration of trichloroethylene as high as 100 ppm after adaptation culture containing increasing concentrations of trichloroethylene was added to the medium. This suggests that it is possible to increase the

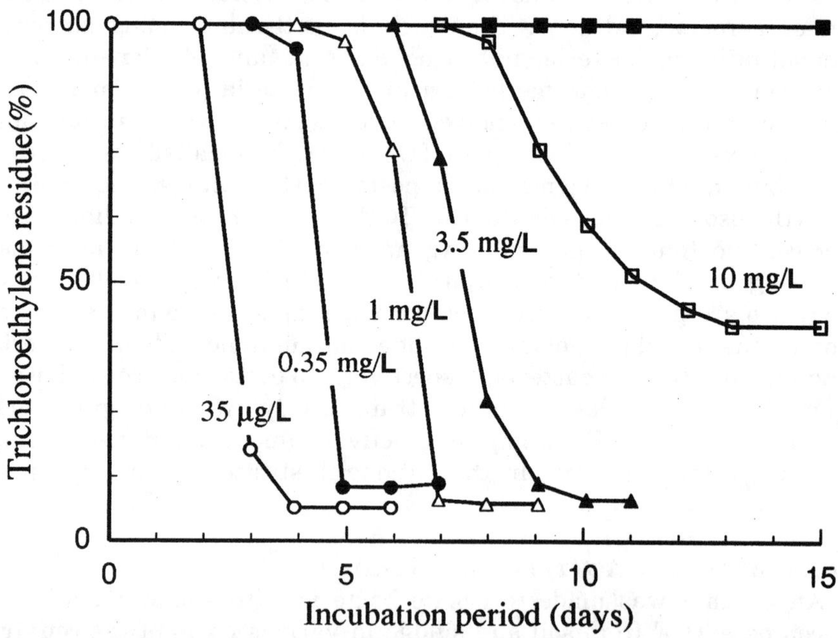

Figure 2. Decomposition of trichloroethylene at different concentrations.

decomposition rate of TCE considerably with Strain M after adaptation.

EXPERIMENTS ON THE TREATMENT OF TCE-CONTAMINATED WATER WITH IMMOBILIZED STRAIN M

The methods used to treat TCE-contaminated water are as follows:

1. Immobilization method—Strain M was added to 2% sodium alginate solution to prepare a cell suspension containing 900 mg/L of bacteria. This cell suspension was added dropwise to 0.1 mol/L $CaCl_2$ solution through a dropping nozzle. In the solution, the drops formed spherical gel particles of about 3 mm in diameter containing fixed bacteria.
2. Continuous treatment reactor—As shown in Figure 3, a glass column reactor with an effective capacity of 20 L was made and filled with about 2.2 L of fixed bacteria as described above, corresponding to 1.5 g/L average concentration of bacteria in the column. TCE added to tapwater was passed upwards through the column. At the same time, it was aerated with air/methane gas mixture. In order to determine TCE leakage from sources other than treated water and exhausted gas, a column filled with alginic acid gel without bacteria was operated in the same manner as for TCE treatment control.
3. Operation methods—As mentioned above, it is necessary to add methane for methane-utilizing bacteria so that methane oxygenase is induced and TCE-decomposing activity is maintained. When the added methane built up to excessively high concentrations, however, the problem of decreased efficiency arose because TCE decomposition was competitively inhibited. However, it was necessary to add methane even after the reaction was initiated because TCE-decomposing activity rapidly decreased when the addition of methane was interrupted. In the experiment, therefore, different conditions of methane concentration were established for maintaining TCE-decomposing activity and for bacteria-recovering activity. This was accomplished by adding a low concentration of methane during the TCE-decomposing operation and by adding a high concentration of methane along with the medium shown in Table 5 during activity-recovering operation. The latter operation was performed every other day during which the TCE supply was stopped. The operation conditions for the ordinary TCEdecomposing operation and activity recovering operation are shown in Table 6. In a separate experiment, changes in the efficiency of TCE decomposition and the stability of TCE-decomposing activ-

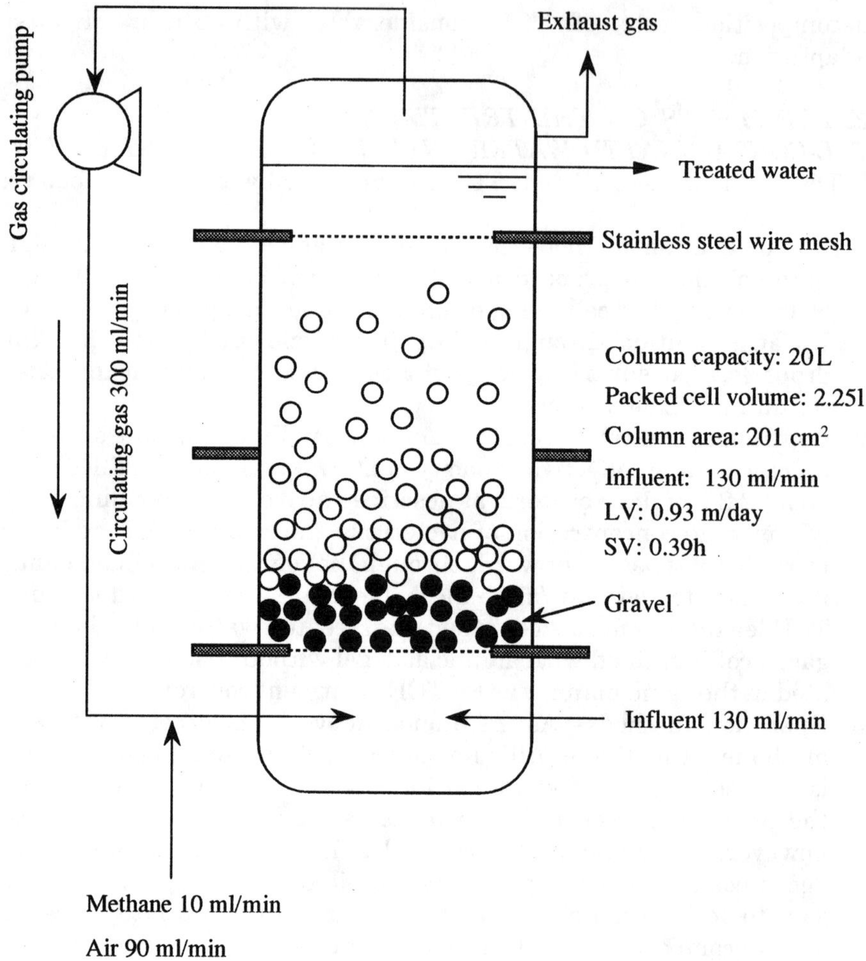

Figure 3. The 20-L bioreactor.

ity were evaluated with a reactor to which methane gas was added at different concentrations, ranging from 3% to 28%, under the ordinary operation conditions mentioned above.

Figure 4 shows the results of the experiment in which the TCE-decomposing operation and activity-recovering operation were performed alternately. During the period from Day 6 to Day 15, a stable quality of treated water with TCE concentrations of around 0.1–0.2 mg/L could be obtained by activity recovering every other day.

Table 5. Composition of media for activity recovery.

Component	Concentration (mmol/L)
$Ca(NO_3)_2$	1.2
KH_2PO_4	0.04
K_2HPO_4	0.08
$FeSO_4$ $4H_2O$	0.038

Study on Treatment of Hexavalent Chromium with Bacteria

One method for treating harmful heavy metals with bacteria is biological reduction of hexavalent chromium. Although its practical use for improving a natural environment polluted by hexavalent chromium has not yet been reported, we will report on this technique as one of the bioremediation methods that may be put to practical use in the future.

HEXAVALENT CHROMIUM-REDUCING BACTERIA

Since bacteria generally die out due to the toxicity of hexavalent chromium at higher concentrations, for the treatment of hexavalent chromium by biological reduction, it is necessary to find bacteria with both toxicity tolerance and high reducing activity. It is believed that the first reports on bacteria highly active in the reduction of hexavalent chromium were made by Russian microbiologists, Romanenko and Koren'kov (1977) and Lebedeva and Lyakova (1979). They found that hexavalent chromium was reduced to the trivalent form by a purely isolated

Table 6. Operating conditions of reactor.

	Flow Rate (ml/min)	TCE Concentration of Influent (mg/L)	Methane Flow Rate (ml/min)	Air Flow Rate (ml/min)	Gas Circulation Rate (ml/min)
Normal conditions	130	1.61–0.86	10	90	3000
Activity recovery conditions	260	0	100	400	4000

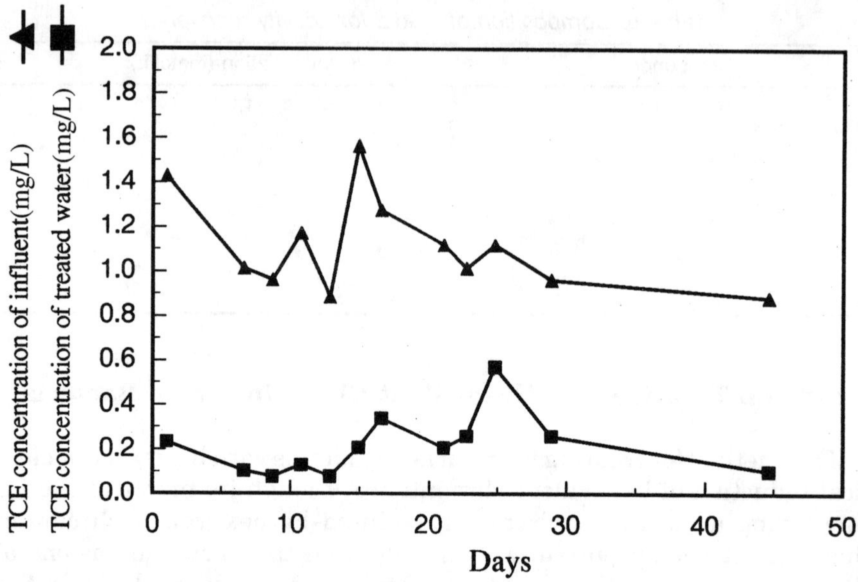

Figure 4. Results of tests of alternating TCE decomposition and activity recovery operations. Test days with points shown in the figure are those for normal operation; activity recovery operation was carried out on test days for which no points are shown.

reducing bacteria under anaerobic conditions, forming a precipitation of chromium hydroxide at a neutral pH. On the basis of these results, they assumed that the reduction of hexavalent chromium by the bacteria might be a sort of anaerobic respiration. However, there is no further research on the function of such bacteria.

More recently, in the 1980s, Shimada and Matsushima (1983) and Horitsu et al. (1983) in Japan and Bopp et al. (1983) in the United States have independently isolated bacteria resistant to hexavalent chromium with compound-reducing activity. In contrast with bacteria reported by the Russian investigators, these bacteria were reported to reduce hexavalent chromium under aerobic conditions and in the presence of intracellular redactase.

TREATMENT OF HEXAVALENT CHROMIUM USING STRAIN H01

Ohtake et al. (1989) tried to isolate a hexavalent chromium-reducing bacterium from various materials, including soil, in areas contaminated with hexavalent chromium. Although failing initially, they finally isolated a hexavalent chromium-reducing bacterium from activated sludge from a sewage treatment plant. This strain was identified as intestinal bacteria *Enterobacter cloacae* and designated Strain H01. Strain H01 is an ideal bacterium for treatment of hexavalent chromium because it has little intracellular accumulation of hexavalent chromium under both aerobic and anaerobic conditions. Strain H01 converts hexavalent chromium extracellularly into less toxic trivalent chromium to form an insoluble precipitation of chromium hydroxide, thus deactivating hexavalent chromium.

When determining the extent to which hexavalent chromium can be treated with Strain H01, both the range of hexavalent chromium concentrations that can be treated and the treatment rate must be considered. A test of treatment capacity with a synthetic medium showed that the rate of hexavalent chromium reduction by Strain H01 was proportional to cell density: 5 mM (260 ppm as Cr) and 10 mM (520 ppm) of hexavalent chromium could be treated completely at approximately 10^8 cells/ml in 1 day and at approximately 10^9 cells/ml in about 11 hours, respectively (Figure 5).

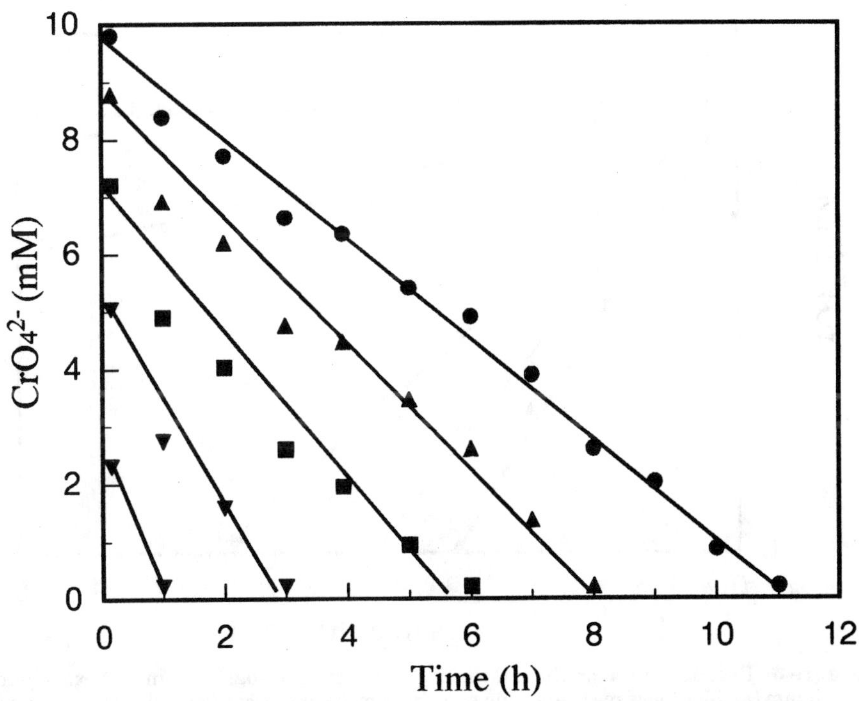

Figure 5. Reduction of hexavalent chromium in a synthetic medium.

A treatment test was then conducted with industrial wastewater containing hexavalent chromium. It was found that hexavalent chromium in industrial wastewater could also be reduced rapidly when adding an appropriate carbon source (Figure 6). Although insoluble chromium hydroxide was also formed in this case as a result of hexavalent chromium reduction, it was necessary to add a coagulant for rapid sedimentation. As hexavalent chromium (anion) is reduced to trivalent chromium (cation), it is possible to collect treated chromium almost completely by the dialysis culturing method with an ion exchange membrane. Moreover, a drop culturing method in which hexavalent chromium is added dropwise into a medium is useful for decreasing the toxic effects of hexavalent chromium in reduction treatment processes.

Removal of Phenol by Gene Recombination

It is possible to design a bacterial strain with enhanced ability and increased efficiency in phenol decomposition by means of gene recombination using genes from a purely isolated phenol-decomposing bacte-

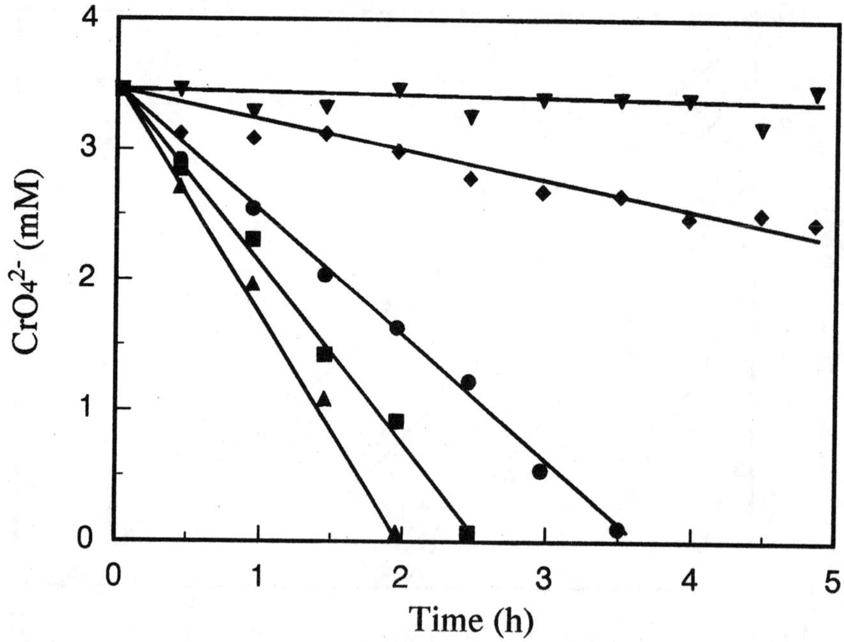

Figure 6. Results of tests on the treatment of industrial sewage containing hexavalent chromium (artificial sewage containing peptone compounds as carbon and energy sources was added to the industrial sewage).

rium. Phenol is converted to catechol via the path shown in Figure 7 and then to 2-hydroxymuconic semialdehyde (2-HMS) in the meta path and enters the TCA cycle via pyruvic acid. Accumulation of catechol has often been observed in phenol decomposition, suggesting that decomposition of catechol to 2-HMS is slower than that of phenol to catechol. Therefore, the entire decomposition rate of phenol to 2-HMS is expected to be improved by accelerating the reaction-decomposing catechol to 2-HMS. A recombinant strain was prepared successfully under the above design conception by isolating element genes of the phenol decomposition path shown in Figure 7 by shotgun cloning method using a cosmid vector, then taking the phe B gene from the isolated genes and inserting it into the parent strain. Figure 8 shows the phenol decomposition by the recombinant strain. This figure indicates that the recombinant strain is superior in decomposition and growth in the presence of 100 mg/L of phenol when compared with its parent strain. In addition, another recombinant strain was prepared by inserting the phe A gene into the parent strain in order to accelerate the first step of phenol decomposition, i.e., decomposition of phenol to catechol (data not shown). Although this recombinant was expected to be more effective in improving the efficiency of phenol decomposition, this strain showed considerable accumulation of catechol but no improvement in the phenol decomposition rate. This recombination, however, may be useful in preventing phenol from accumulating to a high concentration, which interferes with phenol decomposition in the parent strain. It is known that trichloroethylene (TCE) is decomposed by methane-utilizing bacteria, aromatic compounds decomposing bacteria, and ammonia-oxidizing bacteria. To evaluate TCE decomposition ability, a recombinant strain was prepared by inserting the phe A gene into *E. coli* along with a control gene. As shown in Figure 9, this recombinant strain rapidly decomposed TCE. In order to put this strain to practical use, it is necessary to eliminate the enzyme induction process by adding phenol as employed in this example. This may be accomplished by searching for less toxic, inexpensive phenol substitutes or by improving the recombinant strain to produce enzymes as its structural elements through gene engineering.

FUTURE DIRECTIONS

Environmental quality standards for soil pollution were set as the desirable levels for protection of human health and the living environment. This has impelled some local governments to plan or enforce means of control against soil contamination. Because of the lack of

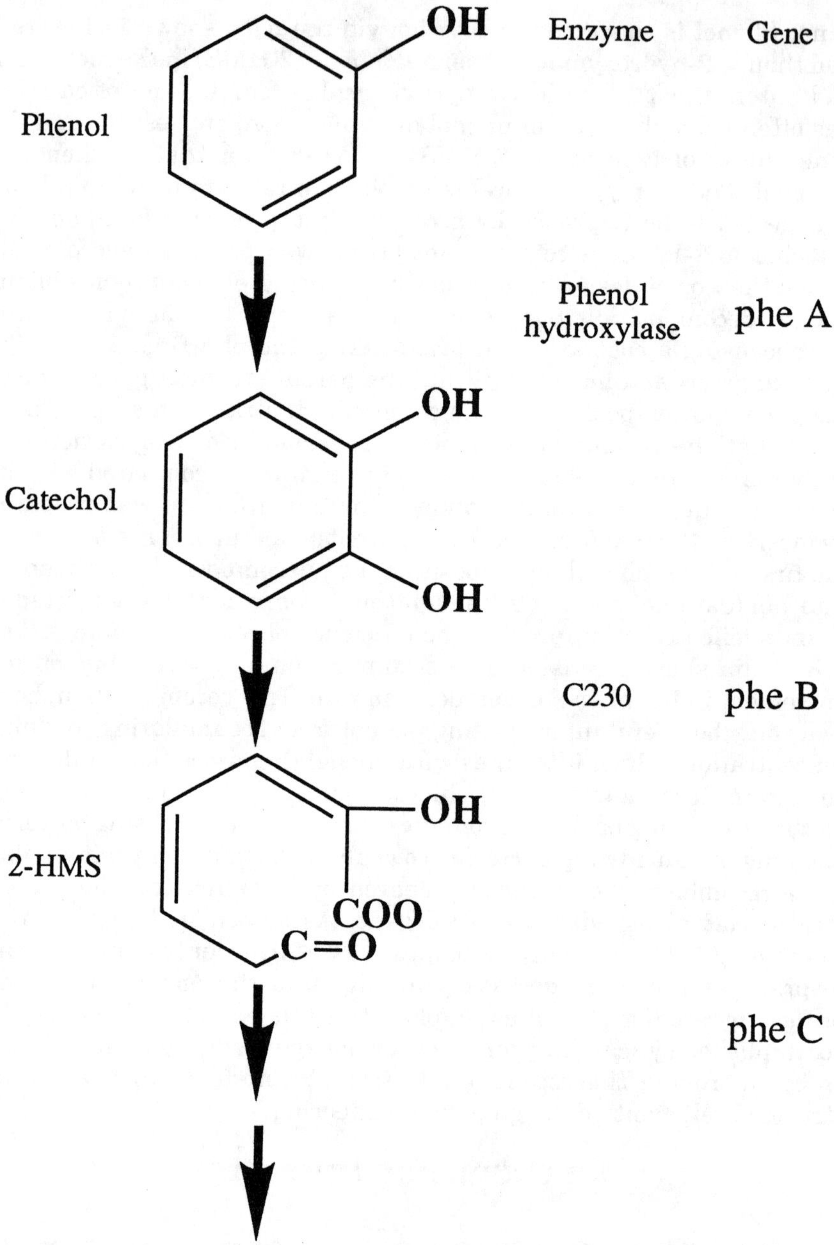

Figure 7. Mechanism of phenol decomposition by *P. putida* BH.

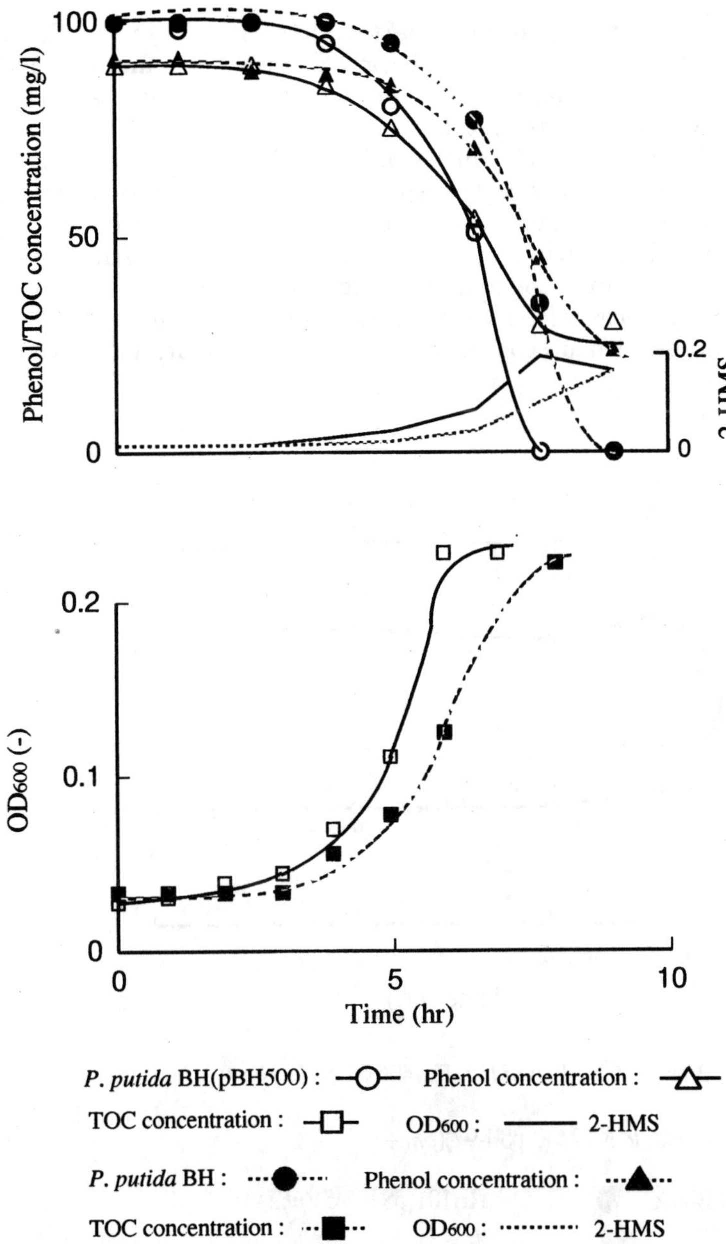

Figure 8. Results of tests on phenol decomposition with *P. putida* BH and *P. putida* BH (pBH500)(phenol concentration of the test medium = 100 mg/L).

laws and regulations, however, particularly regarding urban land soil contamination, actual means of control have been limited to those against soil contamination cases directly affecting residents. An urgent need today seems to be the establishment of a legal system for the prevention of soil contamination, which would rectify contaminated sites under such provisions that the party responsible for contamination would bear the expense of remediation. It is expected that an increasing number of contaminated sites will be revealed as further investigations are conducted to understand the actual conditions of soil contamination. Bioremediation is recognized as a promising technique to remedy soil widely contaminated with low concentrations of harmful materials. To use this

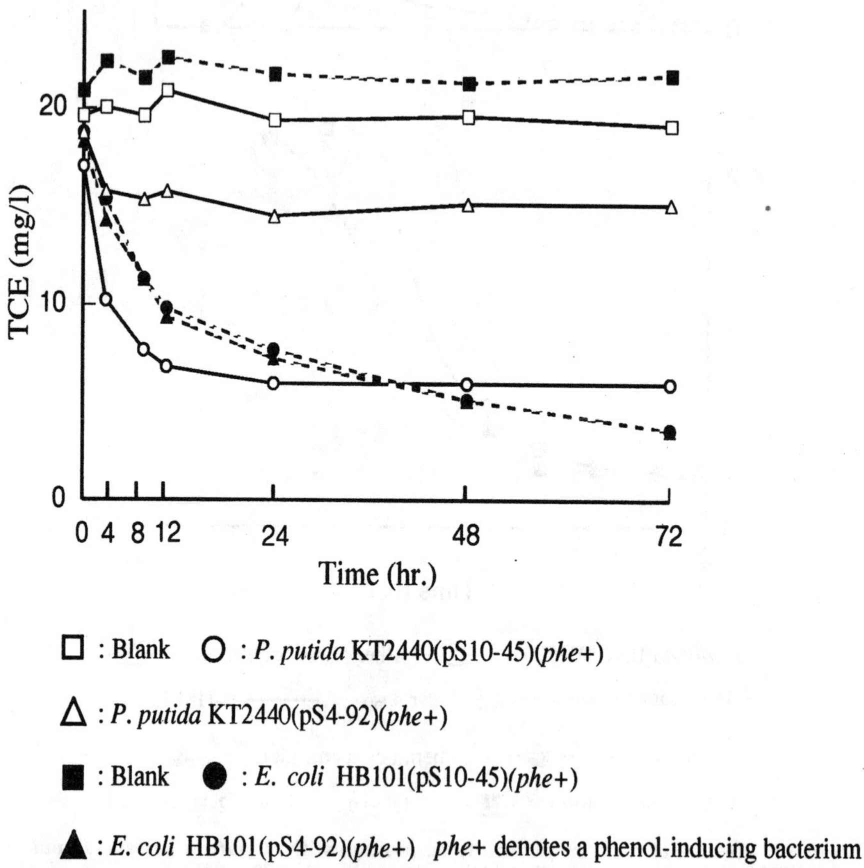

Figure 9. TCE decomposition with recombinant strains of *P. putida* KT2440 and *E. coli* HB101.

technique practically, early establishment of a method of evaluation is necessary to understand the effects of useful bacteria on the environment when they are used to remedy contaminated soil because this may provide safe application of useful bacteria, which is acceptable to residents.

REFERENCES

Bopp, L.H., Chakrabarty, A.M. and Ehrlich, H.L. 1983. "Chromate resistance plasmid in *Pseudomonas fluorescens," J. Bacteriol.,* 155:1105–1109.

Environment Agency. 1994. Quality of the Environment in Japan (written in Japanese. Japan: Environment Agency, Planning and Co-ordination Bureau.

Fujita, M., Ike, M. and Kamiya, T. 1993. "Accelerated phenol removal by amplifying the gene expression with a recombinant plasmid encoding catechol 2,3-oxygenase," *J. Wat. Res.* 27:9–13.

Fujita, M., Ike, M., Mori, K., Hioki, J. and Takeo, M. 1993. "Decomposition of trichloroethylene with a recombinant *E. coli* strain," Summary of Lectures of the 1993 Annual Meeting of The Research Society of Antibacterial and Antifungal Agents, Japan (written in Japanese).

Hinchee, R.E., ed. 1994. *Bioremediation of Chlorinated and Polycyclic Aromatic Hydrocarbon Compounds,* Chelsea, MI: Lewis Publishers, pp. 298–302.

Horitsu, H., Futo, S., Ozawa, K. and Kawai, K. 1983. "Comparison of characteristics of hexavalent chromium-tolerant bacterium *Pseudomonas ambigua* G-1, and its hexavalent chromium-sensitive mutant," *J. Agric. Biol. Chem.,* 47:2907–2908.

Komori, K., Rivas, A., Toda, K. and Ohtake, H. 1990. "Biological removal of toxic chromium using an *Enterobacter* cloacae strain that reduces chromate under unaerobic conditions," *J. Biotechnol. Bioeng.,* 35:951–954.

Lebedeva, E.V. and Lyakova, N.N. 1979. "Reduction of crocoite by *Pseudomonas chromatophilia* sp.," *Mikrobiologiya,* 48:517–522.

Ohtake, H., Cerbantes, C. and Silver, S. 1987. "Decreased chromate uptake in *Pseudomonas fluorescens* carrying a chromate resistance plasmid," *J. Bacteriol.,* 169:3853–3856.

Ohtake, H., Fujii, E. and Toda, K. 1989. "Reduction of toxic chromate in a chromate-reducing strain of *Enterobacter* cloacae," *J. Environ. Technol. Letters,* 11:663–668.

Romanenko, V.I., Koren'kov, V.N. 1977. "A pure culture of bacteria utilizing chromates and bichromates as hydrogen acceptors in growth under anaerobic conditions," *Mikrobiologiya,* 46:414–417.

Shimada, K., Matsushima, K. 1983. "Isolation of potassium chromate-resistant bacterium and reduction of hexavalent chromium by the bacterium," *Bull. Fac. Agric. Mie Univ.,* 67:101–106.

Shimomura, T., Okada, F., Uchiyama, H. and Yagi, O. 1994. "Experimental study on the continuous decomposition of sewage containing trichloroethylene with the M strain, a methane-utilizing bacterial strain," Summary of Lectures of the Third Study Meeting on Groundwater Contamination and Control Measures (written in Japanese).

Uchiyama, H., Nakajima, T., Yagi, O. and Tabuchi, T. 1989. "Aerobic degradation of trichloroethylene at high concentration by a methane utilizing mixed culture," *Agric. Bio. Chem.,* 53:1019–1024.

Wang, P., Mori, T., Komori, K., Sasatsu, M., Toda, K. and Ohtake, H. 1989. "Isolation and Characterization of an *Enterobactor* cloacae Strain That Reduces Hexavalent Chromium under Anaerobic Conditions," *J. Appl. Environ. Microbiol.,* 55:1665–1669.

Yamamoto, K., Kato, J., Yano, T. and Ohtake, H. 1992. "Kinetics and modeling of hexavalent chromium reduction in *Enterobacter* cloacae," *J. Biotechnol. Bioeng.,* 41:129–133.

Bioremediation of Soil and Groundwater Contaminated with Volatile Chlorinated Compounds by a Methane-Utilizing Bacterium

Osami Yagi, Hiroo Uchiyama and Kazuhiro Iwasaki
Water Quality Science Section
National Institute for Environmental Studies
Onogawa, Tsukuba
Ibaraki 305, Japan

INTRODUCTION

Volatile chlorinated aliphatic compounds, such as tetrachloroethylene, trichloroethylene, and 1,1,1-trichloroethane, have been detected in groundwater. More than one-third of the well water surveyed by the Japan Environment Agency was contaminated by these compounds (Kawasaki, 1985). This groundwater contamination is thought to be related to the widespread use of these compounds as cleaning solvents in dry-cleaning operations and semiconductor manufacturing plants. Leakage from hazardous waste sites also contributes to the groundwater contamination. Although PCE and TCE are resistant to biodegradation in the environment, they are considered to be carcinogens. As a result, the Japan Ministry of Health and Welfare established drinking water standards of 10, 30, and 300 μg/L for PCE, TCE, and TCA, respectively.

Various methods have been used to reduce groundwater pollution, such as air stripping and carbon adsorption, vacuum extraction, and excavation. Unfortunately, these processes do not degrade the contaminants; they only transfer the pollutant to other media, and many of the approaches can be quite costly. Bioremediation may be an effective way to clean soil and groundwater pollutants because of less expensive and in situ implementation technology. However, little information exists on in situ bioremediation technologies for TCE-contaminated soil and groundwater.

Recently, the microbial degradation of halogenated aliphatic compounds under anaerobic conditions has been studied to show that reductive dehalogenation of alkanes and alkenes are favored under methanogenic condition (Vogel and McCarty, 1985). Under aerobic conditions, chlorinated alkenes have been reported to resist in the past, but several recent studies have provided to the contrary. Little et al. (1988) reported aerobic TCE degradation. Nelson et al. (1990) reported that toluene dioxygenase was involved in TCE degradation by *Pseudomonas putida*.

Application of bioremediation technology to clean up groundwater contamination would be very useful because of the potential cost savings, destruction of the pollutant, and in the case of in situ, limited environmental disruption. Therefore, we isolated a TCE-degrading bacterium, Strain M, from soil and studied TCE degradation characteristics. Also, we determined the activity of its soluble methane monooxygenase(sMMO). Moreover, we modeled reactor-based and in situ approaches for TCE-contaminated soil and groundwater.

ISOLATION AND IDENTIFICATION OF TCE-DEGRADING BACTERIA

We collected many soil and activated sludge samples from various parts of Japan and screened them for TCE degradation activity. Soil or activated sludge (0.1 mg) was added to 155-ml serum bottles containing 30 ml of mineral salt medium. To this, 30 μg TCE and 20 ml methane were added as carbon sources. The culture was kept at 30°C with shaking for 7 days, Then, 0.5 ml of culture broth was transferred into a new medium and incubated for 7 days. After four repetitions of this procedure, the enrichment culture was spread onto mineral salt-agarose plates. The agarose plates were incubated in a desiccator under an atmosphere of methane:air (1:1, v/v) at 30°C. Isolated colonies were subcultured at least five times by passing onto new plates.

We isolated a TCE-degrading bacterium, Strain M, from one of the soil samples. This bacterium was rod-shaped, nonmotile, gram-negative, catalase(−), and oxidase(−), and it grew on methane and methanol. Strain M's predominant cellular fatty acid was C18:1. Its quinone type was Q-8, and 65.4% of its DNA bases were guanine and cytosine. Strain M was designated a new species that was identified as *Methylocystis* sp. M (Uchiyama et al., 1989).

CHLORINATED COMPOUND DEGRADATION CHARACTERISTICS OF STRAIN M

The time course of TCE degradation by Strain M was studied in serum bottles. More than 90% of initial TCE (35 ppb) was degraded in 3 days. TCE degradation was closely related to both cell growth and the consumption of methane and oxygen. Degradation at various TCE concentrations was investigated. Strain M degraded more than 90% of TCE at initial concentrations up to 3.5 ppm within 9 days. At 10 ppm, 60% of TCE was degraded within 13 days. Ten percent of TCE was also degraded at an initial concentration of 35 ppm. Controls that were run in parallel to these studies showed little loss of contaminant.

The activity of Strain M towards several structurally related compounds was examined (Table 1). TCE, 1,1-dichloroethylene, *cis*- and *trans*-1,2-dichloroethylene, 1,2-dibromoethylene,1,1,2-trichloroethane, 1,2-dichloroethane, and chloroform were all degraded significantly within 7 days, whereas PCE, TCE, carbon tetrachloride, and aromatic compounds were not degraded. The ability of strain M to degrade halogenated alkenes and alkanes was closely related to the chemical structure of the compounds; that is, incompletely halogenated compounds could be degraded, but completely halogenated could not.

AEROBIC TCE DEGRADATION PATHWAYS OF STRAIN M

There are a few reports describing the degradation of TCE by microorganisms. To characterize the metabolic transformations, the TCE

Table 1. Degradation of different compounds by strain M.

Compound	Initial Conc. (ppm)	Incubation Time (days)	Decrease[a] (%)
Tetrachloroethylene	0.15	7	2
TCE	0.36	7	96
1,1-Dichloroethylene	0.005	7	92
cis-1,2-Dichloroethylene	0.45	7	100
trans-1,2-Dichloroethylene	0.45	7	98
1,2-Dibromoethylene	0.5	7	93
1,1,1-Trichloroethane	0.55	7	6
1,1,2-Trichloroethane	0.85	7	82
1,2-Dichloroethane	0.85	7	100
Carbon tetrachloride	0.6	7	4
Chloroform	0.15	7	99
Benzene	10	10	6
Chlorobenzene	10	10	5
Phenol	10	10	9
Benzoic acid	10	10	10
p-Chlorobenzoic acid	10	10	4
p-Chlorobiphenyl	10	10	2

[a]Data represent average percent decreases compared with those in the cases of controls that were not inoculated with Strain M. Experiments were performed in duplicate.

breakdown products with ^{14}C-labeled TCE were determined. After 140-h incubations, the distributions of ^{14}C-labeled activity were analyzed. In Strain M, 32% of TCE was converted to CO and CO_2, 32% was associated in the water-soluble fraction, and 36% was retained by cell material. We also studied TCE degradation in a mixture of strains M and DA-4. Strain DA-4 was isolated and identified as *Xanthobacter autotrophicus*. In this case, 51% of the initial TCE was converted to CO and CO_2, 12% was in the water soluble fraction, and 37% was in the cells. The mixed culture of Strain M and DA-4 exhibited higher mineralization rate than that of Strain M (Uchiyama et al., 1992).

We used a high concentration of cells and 200 ppm of TCE to identify the TCE degradation products. Only low levels of the metabolites were detected. The water-soluble fraction was extracted with ether under acidic conditions at pH 2 and then methylated with diazomethane. Two distinct peaks (peaks 1 and 3) and a small peak (Peak 2) were detected by GC-MS analysis. Mass patterns of Peaks 1 and 3 corresponded to those of dichloroacetic acid and trichloroacetic acid, respectively. Peak 2 corresponded to 2,2,2-trichloroethanol. Dichloro- and trichloro-acetic acids and glyoxylic acid were determined by HPLC analysis of the water-soluble fraction. Moreover, the gaseous phase was verified to contain CO.

On the basis of these results, a hypothetical pathway of TCE degradation in strain M was developed (Figure 1). TCE is oxidized to TCE oxide by methane monooxygenase and TCE oxide spontaneously changes to carbon monoxide, formic acid, glyoxylic acid, dichloroacetic acid and CO_2. Trichloroacetic acid and 2,2,2-trichloroethanol, which are formed via chloral, are considered to be the substantial metabolites of TCE degradation (Nakajima et al. 1992a).

PURIFICATION AND PROPERTIES OF MMO FROM STRAIN M

To clarify the mechanisms controlling the TCE degradation, we isolated and purified the TCE-degrading enzyme. It is known that methanotrophs oxidize methane to methanol through the activity of MMO, which is a nonspecific enzyme capable of oxidizing a variety of aliphatic and aromatic compounds. MMO of *Methylococcus capsulatus* and *Methylosinus trichosporium* OB3b have already been purified. MMO is a multicomponent enzyme that consists of a hydroxylase, an NADH-dependent reductase and a regulatory protein. Two types of MMO are known. One is a soluble form (sMMO), and the other is a particulate membrane-bound form (pMMO). sMMO has high TCE degradation

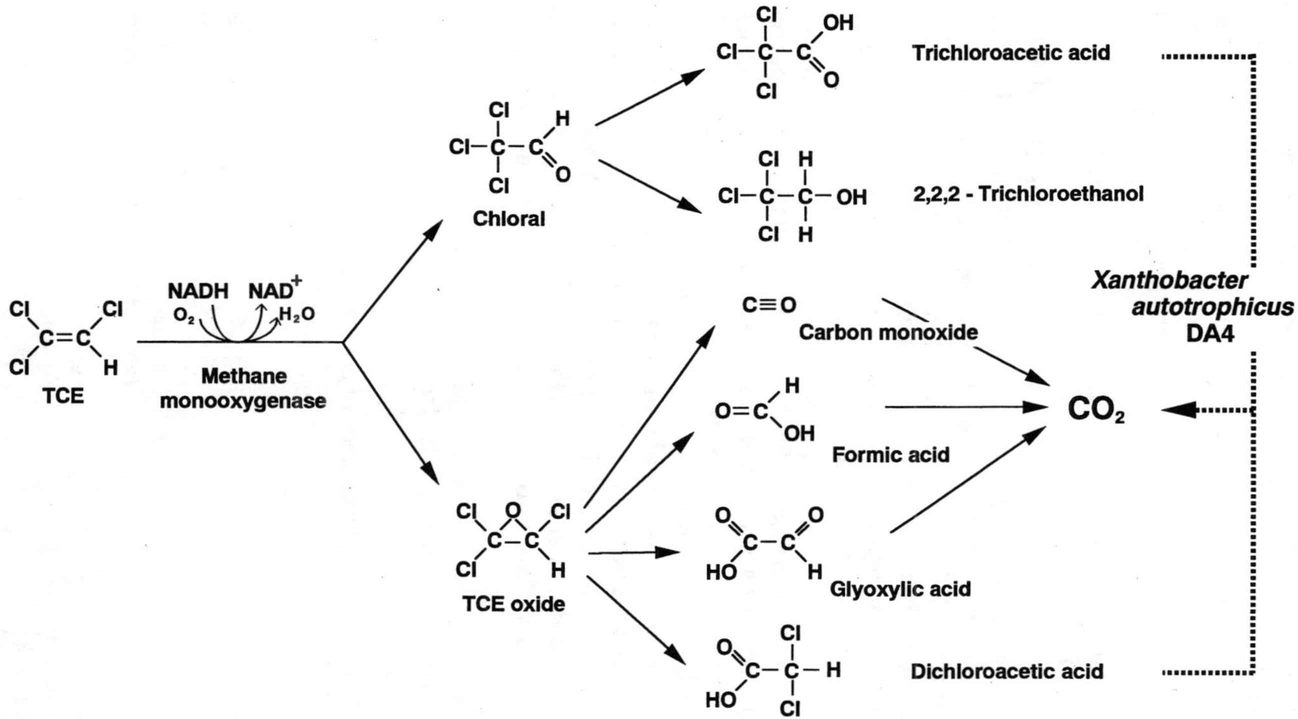

Figure 1. A hypothetical pathway of TCE degradation by strain M.

ability. Therefore, sMMO is important for the degradation of xenobiotic compounds (Koh, 1993).

We purified Strain M's MMO. Cell-free extract was prepared as follows. Frozen strain M's cells were broken by a French Pressure Cell at 30,000 psi, and the extract was centrifuged at 20,000g for 20 min. The supernatant was further centrifuged at 150,000g for 60 min. The cell-free extract supernatant was fractionated by HPLC Mono Q ion exchange column.

Fractions I, II, and III were separated. Fractions II and III were essential for MMO activity. Fraction I was not essential. Fraction II and III were further purified by gel chromatography and ion exchange chromatography (Table 2).

Component II from fraction II and component III from fraction III were purified to homogeneity and were essential for the sMMO activity. Components II and III had molecular masses of approximate 233,000 and 42,000, respectively. Component II consisted of three subunits, α, β, γ, with molecular masses of 55,000, 44,000, and 21,000, respectively. This suggested that structure of component II was $\alpha_2\beta_2\gamma_2$. Component II contained 4 mol of iron and was considered to be a hydroxylase. Component III was a flavoprotein, containing 1 mol of FAD, as well as 1–2 mol of iron, and was considered to be an NADH-dependent reductase. Molecular weights of the hydroxylase and the α, β, and γ subunits of reductase of Strain M were similar to those of other MMOs (Nakajima et al., 1992b; Nordlund et al., 1992) (Figure 2). Component I accelerated the MMO activity. Therefore, component I is considered as regulatory protein.

When Strain M was cultivated at high copper concentration with methane as a carbon source, TCE degradation activity decreased signif-

Table 2. Physical properties of purified sMMO components.

	Component II	Component III
Molecular mass (Da)		
Gel filtration	233,000	39,000
SDS/PAGE	α: 55,000 β: 44,000 γ: 21,000	42,000
Structure	$(\alpha\beta\gamma)_2$	Monomer
p*I*	5.4	5.5
Metal content (mol/mol protein)	Fe: 4.3 Zn: 0.2	Fe: 1.5
Prosthetic groups (mol/mol protein)	—	FAD: 0.82

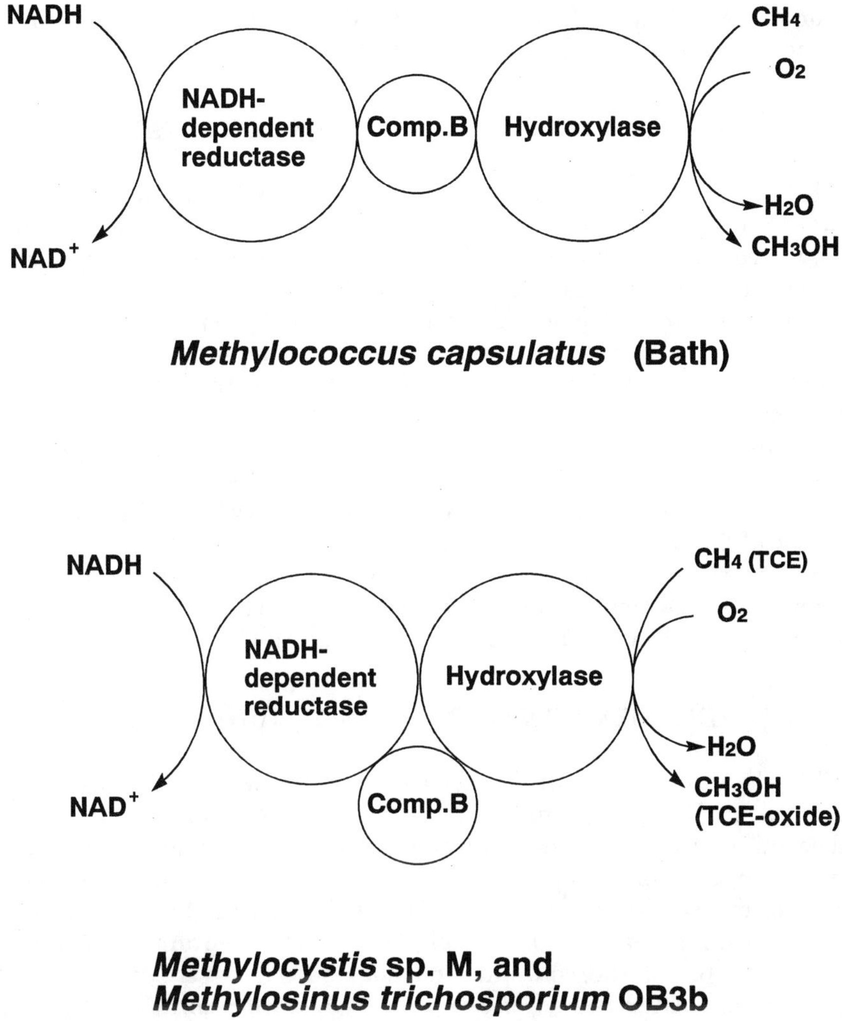

Figure 2. Structure of soluble methane monooxygenase (sMMO).

icantly. With methane and low copper, the amounts of α, β, and γ subunits were significantly high levels. In this case, TCE degradation activity was high, but bacterial growth was limited.

Molecular analysis of the MMO gene has begun. The restriction endonuclease map of chromosomal 12 Kb DNA of *Methylococcus capsulatus* encoding mmo X for the α subunit, mmo Y for the β subunit, mmo

B for protein B, mmo Z for the γ subunit, and mmo C for protein C (reductase) was determined (Cardy, 1991).

APPLICATION OF STRAIN M FOR TCE TREATMENT

Utilization of microbial TCE degradation activity can help to eliminate liquid and gaseous TCE released from wastewater and exhaust gas. We constructed bioreactors of immobilized cells to treat liquid and gaseous TCE. We studied the TCE degradation activity of immobilized Strain M in serum bottles. Cells immobilized by polyurethane, photocross-linkable resin, and a polyelectrolyte complex resulted in the loss of the TCE degradation activity. Alginate immobilized cells showed the highest degradation activity. More than 90% of 1 ppm liquid-phase TCE was degraded in 14 h, and more than 95% of 2 ppm gaseous PCE was degraded in 2 h in batch culture. We studied liquid TCE degradation in a continuous 5-L reactor using 230 g of the immobilized cells of alginate gel beads of 3 mm diameter (Figure 3). One ppm of TCE in water was continuously fed into the reactor. More than 70% of 1 ppm TCE was continuously degraded with a 12-h reactor retention time up to 100 h without carbon source (Figure 4).

BIOREMEDIATION OF TCE-CONTAMINATED SOIL

TCE degradation by Strain M was evaluated with two types of soil microcosms. In the first microcosm, 30 g of dry soil was put into a 70-ml serum bottle, and TCE-contaminated water was added to bring the final volume of water and soil mixture to 50 ml. In the second system, 560 g of dry soil were put into a 450-ml cylindrical glass column (40 mm inner diameter by 300 mm height), and TCE-contaminated water was added to fill the column. Andosol and sand were used as the soil samples. After the equilibrium of TCE adsorption to soil in the microcosm was established in 2 days, Strain M was added and turned up and down several times. The soil microcosms were incubated in the dark at 20°C. TCE concentration was periodically determined with a gas chromatography with ECD or FID detectors.

Strain M cell density had a marked effect on TCE degradation in the column microcosms with sand (Figure 5). We varied cell density between OD 0.005 and OD 0.1 at 660 nm, with OD 0.01 corresponding to 10^6 cells/ml. As the initial TCE concentration was 0.1 ppm, more than 50% and 95% of initial TCE were degraded in 1 day at OD 0.005 and 0.05, respectively. Only a small amount of TCE was decreased in

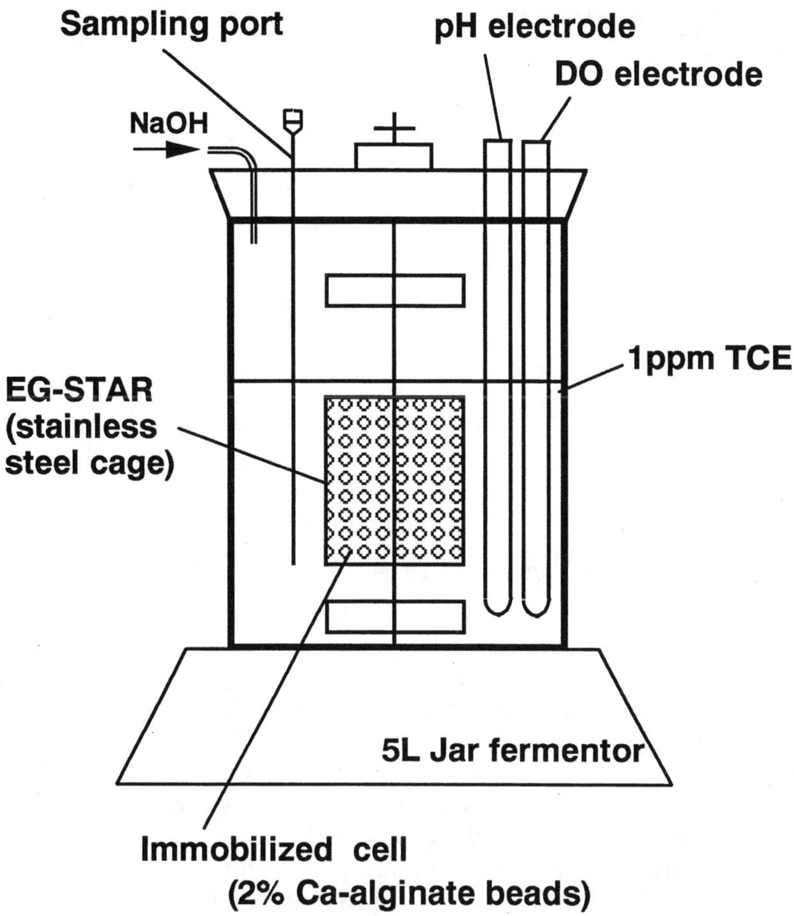

Figure 3. Bioreactor.

the strain M free control system. Clearly, strain M is able to degrade TCE in the soil environment at rates corresponding to cell density.

Degradation of 1,1-, *cis*-, and *trans*-DCE in serum bottle microcosms were all accelerated by the addition of Strain M (Figure 6). For *cis*- and *trans*-DCE, the degradation rate was highly dependent on cell density. Among DCEs, 1,1-DCE is the most toxic. For this reason 1,1-DCE is more difficult to degrade than *cis*- and *trans*-DCE.

TCE degradation rates in column microcosms filled with sand were affected by initial TCE concentrations (Figure 7). When Strain M concentrations corresponded to OD 0.1 in aquifer, about 20% of an initial

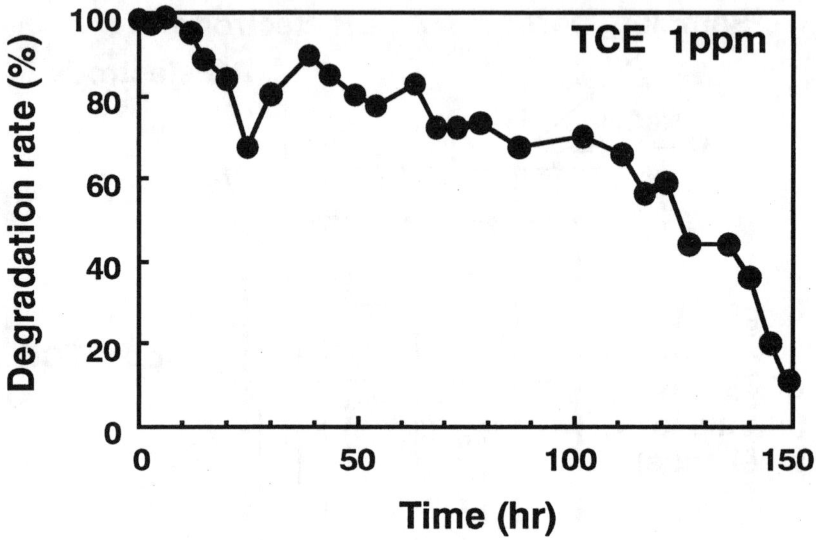

Figure 4. TCE degradation by immobilized Strain M.

1 ppm TCE concentration was degraded in 1 day without Strain M. However, more than 90% of initial TCE was degraded in 1 day with Strain M. About 30% of an initial 20 ppm TCE concentration was degraded in 1 day with Strain M; however, TCE degradation stopped after 1 or 2 days. This might be attributable to a lack of oxygen. Therefore, the effect of H_2O_2 on TCE degradation was studied in column microcosms with sand. The addition of H_2O_2 accelerated TCE degradation (Yagi et al., 1994). No degradation was observed when H_2O_2 alone was added. H_2O_2 seemed effective in increasing TCE degradation by Strain M.

CONCLUSION

A bioremediation technology has been developed to clean up TCE-contaminated soil and groundwater using a methane-utilizing bacterium, Strain M. It was demonstrated that Strain M was very effective in degrading TCE in contaminated soils and groundwater. Further research on topics such as the efficacy and persistence of Strain M in natural environments and the risk assessment of this bioremediation technology must be developed.

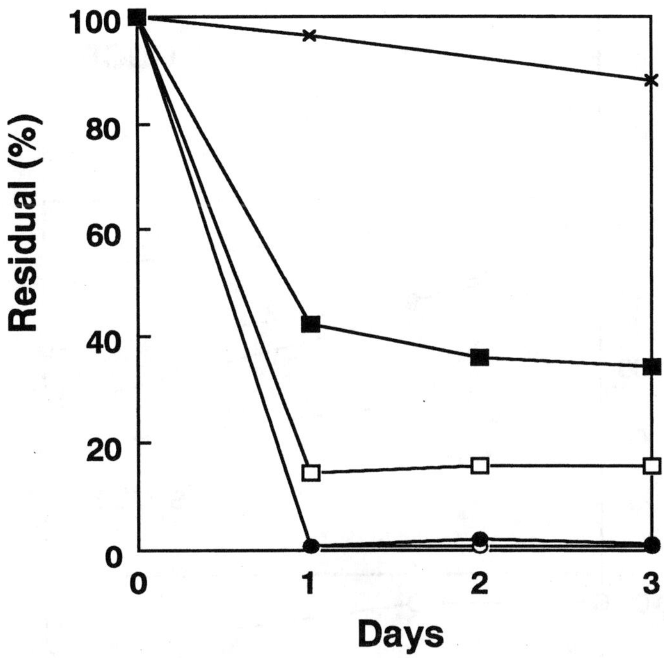

Figure 5. Effect of Strain M cell density on TCE degradation in column: X: without Strain M; ○: OD = 0.1; ●: OD = 0.05; □: OD = 0.01; ■: OD = 0.005.

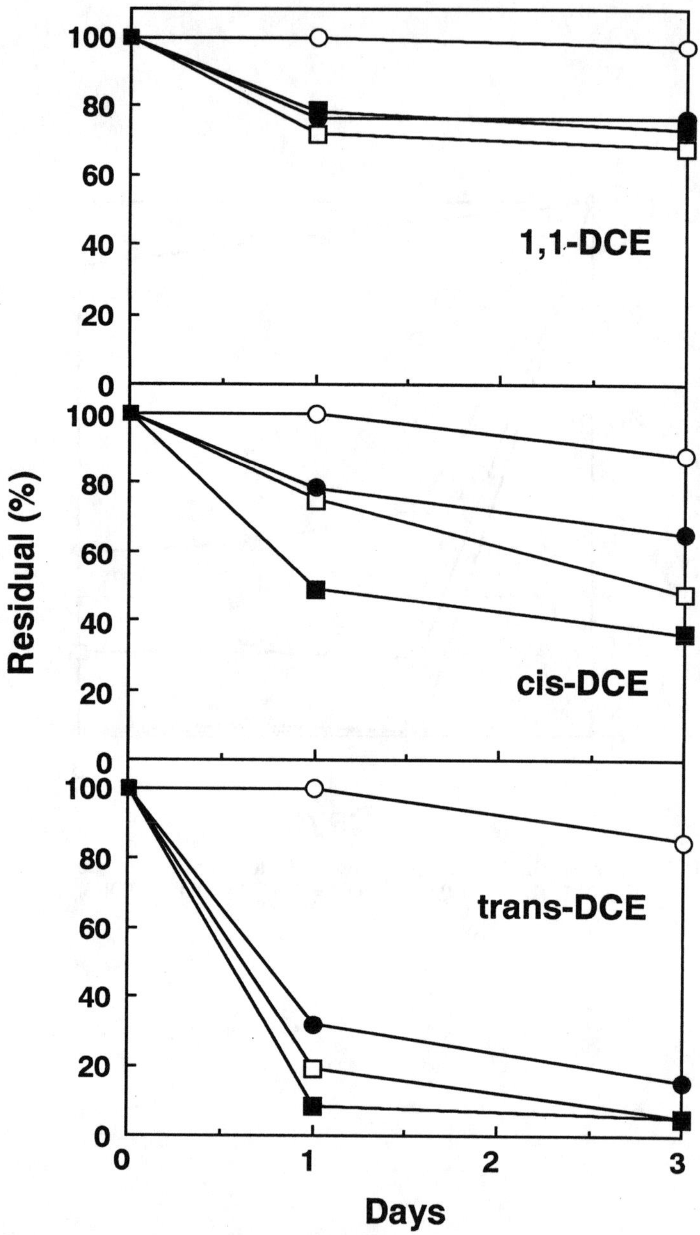

Figure 6. Effect of Strain M cell density on DCE degradation in bottle: O: without Strain M; ●: OD = 0.05; □: OD = 0.1; ■: OD = 0.2

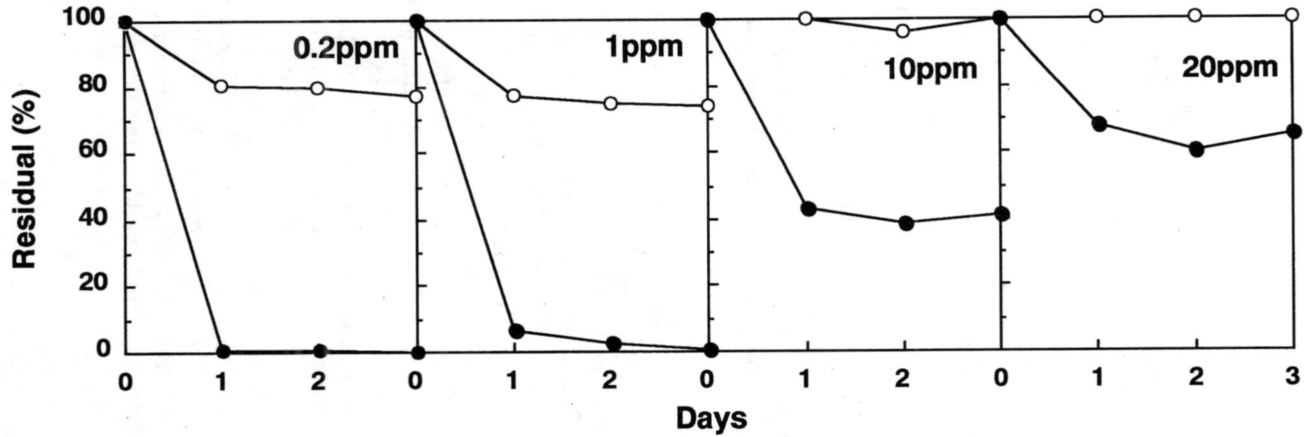

Figure 7. Effect of TCE concentration on TCE degradation by Strain M in column: ○: without Strain M; ●: with Strain M.

REFERENCES

Cardy, D. L. N., V. Laidler, P. C. Salmond and J. C. Murrell. (1991). Molecular Analysis of the Methane Monooxygenase (MMO) Gene Cluster of *Methylosinus trichosporium* OB3b, *Molecular Microbiology,* 5:335–342.

Kawasaki, T. (1985). Present Condition and Countermeasure with Groundwater Pollution, *Jpn. J. Water Pollut. Res.,* 8:264–268.

Koh, Sung-Cheol, J. P. Bowman and G. S. Sayler. (1993). Soluble Methane Monooxygenase Production and Trichloroethylene Degradation by a Type I Methanotroph, *Methylomonas methanica* 68-1, *Appl. Environ. Microbiol.,* 59:960–967.

Little, C. D., A. V. Palumbo, S. E. Herbes, M. E. Lidstrom, R. L. Tyndall and P. J. Gilmer. (1988). Trichloroethylene Biodegradation by a Methane-Oxidizing Bacterium, *Appl. Environ. Microbiol.,* 54:951–956.

Nakajima, T., H. Uchiyama, O. Yagi and T. Nakahara. (1992a) Novel Metabolite of Trichloroethylene in a Methanotrophic Bacterium, *Methylocystis* sp. M and Hypothetical Degradation Pathway, *Biosci. Biotech. Biochem.,* 56:486–489.

Nakajima, T., H. Uchiyama, O. Yagi and T. Nakahara. (1992b) Purification and Properties of a Soluble Methane Monooxygenase from *Methylocystis* sp. M., *Biosci. Biotech. Biochem.,* 56:736–740.

Nelson, M. J., J. V. Kinsella and T. Montoya. (1990) In situ Biodegradation of TCE contaminated Groundwater, *Environ. Prog.,* 9:190–196.

Nordlund, P., H. Dalton and H. Eklund. (1992) The Active Site Structure of Methane Monooxygenase Is Closely Related to the Binuclear Iron Center of Ribonucleotide Reductase, *FEBS Letters,* 307:257–262.

Uchiyama, H., T. Nakajima, O. Yagi and T. Tabuchi. (1989) Aerobic Degradation of Trichloroethylene by a New Methane-Utilizing Bacterium Strain M, Type II, *Agric. Biol. Chem.,* 53:2903–2907.

Uchiyama, H., T. Nakajima, O. Yagi and T. Nakahara. (1992) Role of Heterotrophic Bacteria in Complete Mineralization of Trichloroethylene by *Methylocystis* sp. M., *Appl. Environ. Microbiol.,* 58:3067–3071.

Vogel, T. M. and P. L. McCarty. (1985) Biotransformation of Tetrachloroethylene, Dichloroethylene, Vinyl Chloride and Carbon Dioxide under Methanogenic Conditions, *Appl. Environ. Microbiol.,* 49:1080–1083.

Yagi, O., H. Uchiyama and K. Iwasaki. (1994) Bioremediation of Trichloroethylene-Contaminated Soils by a Methane-Utilizing Bacterium *Methylocystis* sp. M., In: *Bioremediation of Chlorinated PHA Compounds,* Lewis Publishers, 28–36.

Status and Scope of Bioremediation in South Africa

KAREN PEARCE AND ROLF A. OELLERMANN
Division of Water Technology
CSIR
P.O. Box 395
Pretoria, 0001
South Africa

INTRODUCTION

Bioremediation is important universally as a means of reducing the negative impact of pollution by undesirable solids, liquids, and gases on the environment. Internationally, the technologies of bioremediation, viz., bioaugmentation and biostimulation, are proving to be economic methods for the effective treatment of effluents and reclamation of polluted sites. These technologies offer innovative methods to minimize health hazards and degrade pollutants to innocuous forms, resulting in a cleaner and healthier environment.

South Africa faces a number of challenges, some of which can be directly addressed through harnessing the process of bioremediation. A comparatively low, and highly variable, rainfall, averaging about 502 mm per annum, as compared to a world average of 802 mm per annum (Department of Water Affairs, 1986) makes South Africa a relatively arid country. It is thus not only important to develop water resources, but also crucial to protect the quality of water. Although use of groundwater is currently limited to 13% of total consumption, it is expected to increase in the future; hence, this valuable resource must be protected (Retief, 1994).

South Africa has a population of approximately 40 million. It experienced an annual population growth rate of 2.06% over the period 1985–1993 (Harber and Ludman, 1994). Population densities vary widely across the country. Low population densities over the majority of South Africa make it difficult to ensure an equal and adequate supply of safe drinking water and sanitation to all people. Thus, the level of existing services and infrastructure vary notably between regions within the country. Rural communities and unserved areas have to rely on small localized site-specific treatment plants for treatment of water to potable quality and for treatment or disposal of wastes. Here again, bioremediation has an important role to fulfill.

However, the differences exist not only between urban and rural communities. With a growing industrial sector and rapid urbanization, adequate housing with necessary services such as water supply, sanitation, and waste disposal are often lacking or unable to keep up with the demand. A large proportion, representing almost 66% of the people without adequate water, are living in informal settlements around the metropolitan areas (Water Research Commission, 1993a). Because of a high measure of unemployment, a low living standard is common. As can be expected, outbreaks of disease and other associated hazards prevail in such environments.

The Republic of South Africa has been through a period of change. Its geographical location and imposed sanctions in the past have contributed to its isolation from technological advances overseas. Despite the many associated disadvantages, South Africa has made considerable progress towards finding solutions, and in this regard, a relatively high standard of research was attained in South Africa. Recent changes have had far-reaching effects, many of which are as yet not well defined. With the restructuring of government, a change in policy, including that of the environment, is still in progress. It is uncertain to what extent South Africa will be able to carry a substantial research and development sector. Major research institutions have had to undergo a marked change, moving away from fundamental research to applied research and development. The emphasis is placed on the adaptation and optimization of available technologies to existing needs.

In order to contribute to and influence science and technology policy in South Africa, a Directorate for Science and Technology Policy was started. A survey conducted amongst business leaders and policymakers identified the following areas as the most critical for economic development: environmental technologies, networks and communications, software development, advanced materials and composites, and biotechnology (Foundation for Research and Development, 1994). Because bioremediation fits into both environmental technologies and biotechnology, it has a sound future.

Funding is appropriated by Parliament according to the Scientific Research Council Act of 1988 (*Government Gazette,* 22 April, 1988) and the Resource Development Act, No. 75 of 1990. Another major source of funding is through the Water Research Act, No. 34 of 1971, which provides for the promotion of research in connection with water affairs. The Water Research Commission was established to manage the fund whose resources were obtained from a levy charged on water supplied. However, due to the severe drought that necessitated water restrictions in a number of regions, less water was used, thus negatively impacting on this source of funding.

To meet these challenges, research is being aimed at two levels. Firstly, to eliminate pollution as far as possible, while on the other hand to determine which levels of pollution would be acceptable to consumers (Water Research Commission, 1993a).

The Department of Water Affairs and Forestry in South Africa aims to manage various factors determining water quality in an integrated fashion. They are seen as the custodians of water resources with the overall aim to maintain the fitness of use of the water resources on a sustained basis (Department of Water Affairs, 1993). With this approach, the relative importance of diffuse, as well as point source, pollution becomes apparent. These factors and other previously mentioned aspects are being reflected in changing research needs and in new research priorities.

Legislation on the use of water and discharge of wastewaters is contained in the Water Act (1956) Act 54 of 1956, as amended by the Water Amendment Act, Act 96 of 1984. The Act contains details on water usage and its disposal, and the discharge standards laid down are presented in Table 1.

In this chapter, a brief overview will be given on the current status of bioremediation in South Africa and how the technologies are applied. Needs and opportunities identified within South Africa with regards to bioremediation, as well as competing technologies, will be discussed. Recent and current research activities will be spotlighted and anticipated future developments indicated.

THE INDUSTRIAL AND DOMESTIC WASTEWATER SCENE

In most towns and cities of South Africa, industrial and domestic wastewaters are discharged and, in the process, become mixed before treatment by municipal water care works. This results in the dilution of industrial pollutants, which makes their breakdown less effective. On the other hand, the availability of a readily biodegradable carbon source in domestic sewage can be beneficial. In general, however, the diversity of industrial discharges and the possible toxicity of some makes control and efficient treatment of such mixed effluents difficult and results in relatively frequent operational disruptions.

Table 1. Standards for effluent discharge in terms of the South African Water Act (1956), Act 54 of 1956, as amended.

Parameter	General Standard Limits	Special Standard[a] Limits
Colour, odor, taste	NiL	Nil
pH	5.5–9.5	5.5–9.5
Dissolved oxygen (% saturation) (DO)	75	75
Temperature (°C)	35	25
Typical faecal coli (per 100 ml)	Nil	Nil
Chemical oxygen demand (COD)	75	30
Oxygen absorbed	10	5
Conductivity (mS/m)	250; 75 above intake	250; 15% above intake
Suspended solids	25	10
Sodium	90 above intake	90 above intake
Soap, oil and grease	2.5	Nil
Residual chloride (as Cl)	0.1	Nil
Free and saline ammonia (as N)	10	1
Nitrate (as N)	N.S.	1.5
Arsenic (as As)	0.5	0.1
Boron (as B)	1.0	0.5
Chromium (total, as Cr)	0.5	0.05
Copper (as Cu)	1.0	0.02
Phenolics (as phenol)	0.1	0.01
Lead (as Pb)	0.1	0.1
Sulphides (as S)	1.0	0.05
Fluoride (as F)	1.0	1.0
Zinc	5.0	0.3
Soluble orthophosphate[b]	N.S.[b]	1.0
Iron (as Fe)	N.S.	0.3
Manganese (as Mn)	0.4	0.1
Cyanides (as Cn)	0.5	0.5
Cadmium (as Cd)	0.05	0.05
Mercury (as Hg)	0.02	0.02
Selenium (as Se)	0.05	0.05
Hexavalent chromium (as Cr)	0.05	N.S.

[a]The Special Standard applies to certain selected rivers and catchments as specified in *Government Notice*. All units in mg/l unless specified otherwise; N.S. = Not specified.
[b]In terms of Government Notice No. 991 of 18 May 1984, effluents draining to certain sensitive areas must have soluble orthophosphate (as P) concentrations of <1.0 mg/l.

Insufficient on-site treatment of effluent is carried out by industry, and they tend to be complacent about environmental pollution. Non-compliance or discharge of effluent outside the legal limits may result in the imposition of excessive levies, and specific action may be taken by the Department of Water Affairs and Forestry to improve the quality of effluent discharged. This naturally does not apply to all companies, and a change towards proactive environmental protection is noticeable.

In biological treatment processes, the removal of contaminants is achieved by microbial biomass. Primarily, organic contaminants are removed, depending on their biodegradability. These processes, in the correct configuration, may also be used to remove nitrogen and phosphorus from wastewaters. The successful biodegradation of recalcitrant and xenobiotic compounds may require the supplementation by specially adapted microorganisms.

Biological treatment of industrial and domestic wastes by municipalities is well established in South Africa and ranges from full nutrient removal systems through activated sludge plants, to simpler, less operator-intensive systems such as constructed wetlands, ponds, and aerated lagoons. Some commonly used biological treatment processes are listed in Table 2.

Activated Sludge Treatment

In the 1920s, the first continuous-flow treatment plants were constructed using activated sludge to remove organic pollutants from wastewater. The term *activated sludge* originated from the fact that the biomass is efficiently aerated and highly active in removing soluble organics from liquid. A heterogeneous living microbial culture develops in the process and the microorganisms grow on the nutrients supplied in the influent. The composition and characteristics of the activated sludge are therefore determined, amongst others, by the composition of the influent, rate of aeration, and sludge age.

It has been widely recognized that South Africa is well established in the area of nutrient removal, and basic research in our laboratories resulted in the development of the Bardenpho system (Barnard, 1975). Since then, many activated sludge systems have been brought into operation using the Bardenpho, the Phoredox, or the UCT processes or modifications of these (Water Research Commission, 1984). Research is ongoing, and the systems are continually being improved and problem areas identified for further research. The Water Research Group at the University of Cape Town has undertaken extensive research in a number of areas, including the development of a hypothesis on the causes of bulking sludges at nutrient removal plants. The achievement of greater clarity on the causes of bulking should lead to practical strategies on the control of filamentous bulking with concomitant benefits in effluent quality, cost reduction, and ease of plant operation (Casey *et al.*, 1993; Water Research Commission, 1993a).

Table 2. Biological treatment processes for domestic and industrial wastewaters.

Type of System	Descriptive Name
Suspended growth systems	Aerobic systems Activated sludge Aerated lagoons Oxidation ditches
	Anaerobic digestion (conventional)
Immobilized biomass systems	Aerobic systems Biotowers (Trickling filters) Rotating biological contactors (RBCs)
	Anaerobic systems Upflow pellet reactors Fluidized-bed reactors Packed-bed reactors
Combined systems	Waste stabilization ponds (WSPs) Anaerobic WSP Facultative WSP Maturation WSP
	Wetlands

It is thought that the floc-forming bacteria denitrify NO_x to N, while the filamentous bacteria only denitrify NO_3 to NO_2. The less oxidized forms of NO_x, particularly NO, inhibit cell metabolism. The inhibitory effect of the accumulated intermediate denitrification compounds within the cells of the floc-formers may give the filamentous bacteria the growth advantage in nutrient removal activated sludge plants (Water Research Commission, 1993a).

Biological phosphate removal has an important role to play in limiting the amount of eutrophication occurring in natural water bodies. Research is ongoing to obtain a better understanding of the metabolic mechanisms responsible for the accumulation of excess phosphate by bacterial cells in the activated sludge process. Through this, the management of the phosphorus removal process can be refined. Research is being undertaken at the University of Pretoria to investigate possible connections between the presence of enzymes and the ability to remove phosphorus (Water Research Commission, 1993a—Report No. 314).

Treatment in Biotowers (Trickling Filters)

Biotowers are fixed film growth reactors with biomass attached to the surface of crushed rock or other suitable support medium. They represent an example of a closed heterogeneous continuous culture system in which the effluent permeates the biofilm on the support medium acting as a filter (biofilter). They have found application worldwide as a popular, rugged, and dependable means of wastewater purification. Recirculation of biotower effluent has become a standard operation and has been effectively implemented in many instances to remove toxic pollutants and improve tolerance to shock loads. Increasing the hydraulic loading reduces the chance of a dry surface during low-flow operations, thereby maintaining maximum treatment capability.

Based on the organic and hydraulic loading, the biotowers are classified into low-rate, intermediate-rate, high-rate, and roughing filters.

Biotowers at peak operational performance only infrequently result in chemical oxygen demand reductions of more than 80%. This has resulted in a reduced application also because of their configurational inflexibility and large volumes of support media required to achieve acceptable effluent quality. Research has therefore been done to develop design and operational criteria for two modifications of the conventional biofiltration process. This encompasses, firstly, increased recirculation directly around the biofilm and, secondly, to optimize the permeate-biofilm contact in the filtration process. Higher flow rates result in more efficient sloughing of excess biofilm, reducing the possibility of clogging of the system. Recirculation also ensures that organic matter, having permeated through on the first passage and missed exposure to the biofilm, will be degraded on the second. The enhancement of the permeate-biofilm contact may also improve the phosphorus removal efficiency (Water Research Commission, 1993a—Report No. 569).

Rotating Biological Contactors (RBCs)

RBCs consist of a series of closely spaced circular discs, usually of a suitable plastic material. A series of these mounted on a horizontal shaft are partially immersed in the effluent and rotate slowly. A film of biomass adheres to the rotating discs similar to that on the surface on the biotower support medium. Alternating exposure to air and wastewater results in mineralization of the pollutants. Excess biomass is sloughed from the discs and washed out in the process effluent and separated by settling.

Because of structural limitations and drive control, RBC units are most easily constructed as modular units. In most cases, two to six modules are placed in series, depending on the effluent strength and the degree of treatment required. Sloughing of biofilm is a continuous process, and secondary clarification is necessary. In some cases, intermediate clarifiers are placed between modules, and RBCs have found application in wastewater treatment for small communities in South Africa. As many as 500 installations have been installed during the past 20 years and are operational in the country.

Bioaugmentation

The use of specially selected and adapted microorganisms for the specific biodegradation of organic pollutants is an emerging technology (Meyer and Oellermann, 1993). Supplementing biological treatment systems with consortia of microorganisms is relatively new to South Africa, and its application is not on a sound footing. Research is being conducted to carefully evaluate the potential of the technology and establish the necessary basis for its implementation (Oellermann and Pearce, 1994).

Anaerobic Digestion

Anaerobic treatment of industrial effluents is not uncommon in South Africa. Suspended growth systems are most frequently employed. Full-scale applications include anaerobic digestion of brewery effluent where significant reductions in organic pollutants are demanded. Upflow anaerobic sludge blanket (UASB) digesters are used for high hydraulic loads. However, hydraulic or organic overload conditions or low influent alkalinity may result in process failure due to acidification as a result of a deviation from the fermentation balance (Perry, 1994). To obviate these problems, researchers have looked at the provision of a conditioning tank and the use of a liquid recycle (Dudley *et al.,* 1993). Generally, anaerobic reactor effluent requires further treatment to achieve the standards required for discharge.

Treatment by systems employing immobilized biomass are more tolerant of fluctuating load and toxic pollutants than the conventional digesters. Biofilm treatment technologies have resulted in more compact and high-rate reactors that generally proved expensive. Research has therefore been conducted to apply inexpensive support media. Such support media, although not unique to the world, include cinders, crushed brick, and clay pellets. Their practical application has been demonstrated by the treatment of abattoir effluent (Oellermann *et al.*, 1994a, 1994b).

To obtain a better understanding of the complexities of anaerobic digestion, Britz *et al.* (1994) have investigated the bacterial diversity required for the biodegradation of various substrates.

Treatment in Waste Stabilization Ponds (WSPs)

Waste stabilization ponds, lagoons, or oxidation ponds are shallow lakes in which wastewater is treated by natural processes based on the activities of both algae and bacteria. They are efficient in the destruction of pathogenic bacteria, and the systems are easy to operate and maintain.

In warmer climates, WSPs offer a suitable and economical treatment system for low- to medium-strength effluents. In South Africa, oxidation ponds have many advantages, not only in economic terms, but also in areas where only relatively unskilled labor is available. However, meeting the stringent legal discharge requirements has been a factor that has resulted in limited use of this technology. Research is currently being done on methods to upgrade pond effluent, through use of an integrated pond system developed in South Africa, known under the acronym PETRO, or Pond Enhanced Treatment and Operation (Meiring, 1993; Meiring and Oellermann, 1993). The system has the potential to remove algae and upgrade oxidation pond effluent satisfactorily. The PETRO system is adaptable to different configurations, which makes it suitable for a number of applications. Important advances have been made to improve the performance of an anaerobic pond as a pretreatment facility, both with regard to organic and ammonia removal, and reductions of up to 50% of the Kjeldahl nitrogen content of raw sewage have been achieved (Meiring and Hoffmann, 1994).

Treatment in Wetland Systems

In the South African context, constructed wetlands or reedbeds may offer an effective treatment option for a wide range of wastewaters (Wood, 1991). However, a lack of understanding of the mechanisms by which a reedbed functions may lead to an unacceptable performance. The wetlands can be designed to treat raw or secondary sewage, to upgrade septic tank and oxidation pond effluents, stormwater, and agricultural and aquacultural wastes, as well as a variety of industrial and mining wastewaters (Batchelor, 1994b; Wood, 1991). During 1991, approximately 50 full-scale systems were either at an operational, construction, or planning stage in the country.

Constructed wetlands are effective in the improvement of water quality, specifically in their ability to treat domestic wastewater. Their water purification functions are dependent upon the four principle components of wetlands—the vegetation, the water column, substrate, and their associated microbial populations. The principle function of vegetation in wetlands systems is to create additional environments for microbial populations. The stems and leaves in the water column obstruct flow and facilitate sedimentation and increase the surface area for attachment of microbes and constitute thin film reactive surfaces. This is in addition to the microbial environments associated with the water column found in stabilization and maturation ponds (Batchelor, 1994a).

Bacteria, fungi, algae, and protozoa degrade contaminant substances. In addition, many microbial groups are predatory and will forage on pathogenic organisms. The effectiveness of wetlands in water purification is dependent on developing and maintaining optimal environmental conditions for the desirable microbial populations. Most of the desirable microbes are ubiquitous and are likely to be found in most wastewaters with nutrients and energy; hence, inoculation of specific strains is generally not necessary (Batchelor, 1994a).

Combination Processes

Industrial wastewater treatment is very much dependent on the characteristics of the effluent. Optimum treatment may consist of a combined physical-chemical-biological approach. A physical or chemical pretreatment may render the water more suitable for a subsequent biological treatment. An example of this is the improved removal of phosphorus through the combination of a chemical process with that of a biological treatment.

A major problem is posed by phosphate pollution, which is an essential nutrient for microbes. Biological phosphate removal processes frequently cannot cope with the load, resulting in the discharge of treated wastewaters containing unacceptable concentrations of phosphorus. To alleviate the problem, research is being conducted on the integration of phosphate crystallization in a fluidized-bed reactor, in tandem, or in series with an activated sludge plant. Phosphate is removed as hydroxyapatite pellets by treating a sidestream of the effluent from an anaerobic digester containing high concentrations of phosphate released by the microorganisms. It represents a clean technology producing reusable calcium phosphate (Momberg and Oellermann, 1992; Saayman *et al.*, 1994).

Technological problems still exist in solids-liquid separation and biomass retention (Ross *et al.*, 1994). Here again, a combined treatment would enhance the efficiency of biological treatment. The South African-developed anaerobic digestion-ultrafiltration (ADUF) process is currently being optimized (Ross *et al.*, 1994) and is a system that combines biological treatment with a physical process. The use of a suitable filtration unit recycles biomass and effectively decouples the solids and liquid residence times within an anaerobic digester. Thus the volumetric throughput can be substantially increased without impairing biomass digestion. An advantage is that increases in flow can be accommodated, and the construction of additional costly digesters is delayed. A further benefit is that the downstream processes have a reduced volumetric loading on the sludge dewatering equipment. This process has been implemented at both pilot- and full-scale (Water Research Commission, 1993a—Project No. 560).

BIOREMEDIATION IN THE MINING SECTOR

Mining is an important industrial sector contributing a large percentage towards the GNP. However, mining activities lead to the production of acidic mine waters containing high concentrations of heavy metals and sulphates and can have pH values as low as 2.5. These acid mine waters are prohibited from being discharged to public streams and may negatively impact on groundwater. Apart from neutralization with lime, much research has also been done for the use of sulphate-reducing bacteria with notable success. A three-stage process (anaerobic–aerobic–anaerobic) employing upflow packed bed reactors (or sludge bed reactors) for anaerobic treatment, and an activated sludge system for aerobic treatment could be used for reducing sulphate concentrations from 2.5 g/l to 0.5 g/l with concomitant removal of H_2S, carbonates, complexed cyanides, phenols and heavy metals, using molasses as an energy source (Maree *et al.*, 1986). Maree (1989) has further been able to show that sulphuric acid-rich water can be treated biologically for sulphate removal without prior neutralization with an alkali such as lime, due to the alkalinity produced in the biological sulphate removal process. A major disadvantage when sulphate is removed biologically using an organic carbon source as substrate is the high residual organic carbon content. This can be overcome through the use of producer gas (a mixture of H_2, CO, and CO_2) as an alternative carbon source (du

Preez *et al.,* 1991). Further work by du Preez and Maree (1994) has demonstrated the technical feasibility of an integrated process, consisting of a number of stages: sulphate reduction, H_2S stripping, H_2S absorption in an Fe(III) solution for sulphur production, and biological oxidation of Fe(II).

The potential of constructed wetlands to ameliorate acid mine drainage is currently being investigated at a number of sites within southern Africa (Wood, 1991). The combination of pond systems with wetlands is perceived to provide additional benefits. Water loss is enhanced through evaporation in the ponds, as well as nutrient uptake through algae. In addition, pH and alkalinity are increased before the effluent is discharged to a reedbed. Expectations are that, in the long term, the degradation of plant biomass will provide the organic carbon content needed for sulphate reduction. Further applications being investigated within the mining sector are the use of reedbeds for the treatment of [slimes dams decant water] (Wood, 1991).

RESOURCE RECOVERY

The University of Pretoria has been engaged for a number of years in purposefully cultivating filamentous microorganisms to purify effluents, while simultaneously producing single-cell protein. It has been shown that each effluent has its own specific microorganism yielding best results, due to the specific chemical composition of the effluent. In this regard, the fungus *Geotrichum* was best suited for petroleum effluent, while *Aspergillus* shows promise for the treatment of paper effluent (Water Research Commission, 1993a—Report No. 535). In some cases, bioremediation is used as a treatment of the effluent and for valuable resource recovery. Research in production of single-cell protein as a possible food source has shown that it could be an economically viable proposition (Pretorius, 1993). Many different potential sources of raw materials are available for microbial biomass protein, including petrochemical, pulp mill (Pretorius, 1993), and abattoir effluents (Anon., 1994a). Evaluation of the protein recovery from abattoir effluent was done at three levels, namely, system stabilization and operation, sludge characteristics, and treatment effectiveness. As a batch process is used, a biological steady state is not attainable; however, stabilization of the system is achievable. The protein content of the product is 40%; the sludge has a sludge volume index of 80 ml/g, which settles well. Through use of a belt press, a solids content of 31% is achievable. A 99% reduction of suspended solids was obtained, with a final effluent quality of 45 mg/l suspended solids (Anon., 1994a).

APPROPRIATE TECHNOLOGIES FOR DEVELOPING COMMUNITIES

Background

Inadequate access and supply of potable water and proper sanitation to a large proportion of our urban and rural communities (Water Research Commission, 1993b) is due to the lack of reliable surface and groundwater sources, long distances between dwellings and communal distribution points, and the cost of water purchased from water vendors. These communities use much less than the 50 L/capita/day prescribed by the World Health Organization as the minimum health standard.

The supply of water to the rural areas in South Africa is managed through a diversity of arrangements, and a strategic plan to provide the necessary infrastructure is vital (Anon., 1994b). Appropriate technologies to provide water and sanitation to these communities have to solve the immediate water supply and sanitation needs, be sustainable in the long term, and hence should also aim to reduce pollution of soil and water. This is required in view of the changing priorities in South Africa, although they may appear insignificant compared to the large scale of operations in the South African mining and manufacturing industries.

The construction and maintenance of these systems should aim to create employment and, subsequently, channel funds back into the communities. Low capital cost and maintenance technologies will therefore be the most appropriate. The approach of self-help is the hallmark of success, and projects should go beyond the simple provision of water. Improving the understanding of water and sanitation will raise the quality of life in the local population with respect to hygiene, sanitation, and water storage.

A brief description of the current status is given, together with technologies available to address these issues.

Existing Systems Used for Production of Potable Water in Developing Communities

Aquifers underlying rural and urban informal and semi-formal settlements often serve as both a source of cheap, readily available drinking water and as a repository for waste emanating from the sanitation systems and waste disposal activities. Systems need to be researched and designed to adequately treat groundwater to potable quality. A problem associated with groundwater is high salinity, a contributing factor being the nitrates in the water, which may cause adverse effects in humans and animals. Biological denitrification has been shown to be a less expensive and simpler method than nitrate removal by ion exchange and thus better suited towards rural areas with a lack of skilled operators. Research using a packedbed reactor indicated that a mixed culture of activated sludge performed poorly when compared to a pure culture of denitrifiers, namely the *Alcaligenes* species. It was shown that an initial concentration of 60 mg NO_3-N/l could be treated at a design capacity of $20m^3$/d at R1.11/m^3 and R0.95m^3 for ion exchange and biological denitrification respectively (Letimela, 1993).

Relatively little research is being conducted on groundwater, and South Africa generally takes leads from technological research elsewhere. The major users of groundwater are the mining and agricultural sectors for process or irrigation purposes.

Groundwater is not often used as a source for drinking water, except in some rural areas. Perhaps of greater concern in areas of informal settlements is groundwater contamination, not with chemical pollutants, but with fecal coli, making it a hazardous source of drinking water without proper treatment (Fey, 1995).

Existing Systems Used for Sanitation in Developing Communities

The primary objectives of sanitation are to reduce the occurrence of wasteborne diseases and to minimize the impact of sewage on the natural environment, in particular nutrient loading of ground- and surface waters. Sanitation systems commonly used in South Africa include

- full waterborne sanitation systems
- septic tank systems in association with small-volume flush toilets and small bore sewer systems for applications in sanitation
- low-volume flush, with on-site anaerobic digester linked to an on-site soakaway or "solids-free" reticulation system (LOFLOS)
- ventilated improved pit latrines (VIPs)
- bucket collection systems
- constructed wetlands and ponds for pollutant removal
- biodigesters to recover energy in the form of methane gas for domestic use
- package plants for small communities

The percentage of the different systems in operation is shown in Table 3.

Frequently, the bucket collection systems for disposal of sewage were initially installed as an interim measure until funds were secured for a full waterborne sanitation system but have now become an almost permanent feature of the urban landscape in many areas. The night soil is typically disposed of in lagoons, which, by necessity, have to be situated some distance from the community. Malodors, fly breeding, and disease-causing pathogens are a direct result of these lagoons, which have an adverse impact on the population and the environment. The current situation clearly indicates the need for research in this regard. However, a lack of funds poses difficulties in implementing research findings.

Systems that appear to have potential in the treatment of domestic waste, where no waterborne systems exist, are wetlands and the composting of sewage sludge. Research has shown that forced aeration co-composting of night soil with unpulverized refuse as a bulking and filtering agent is a cost-effective process for integrated stabilization, disinfection, and resource recovery (Water Research Commission, 1993a—Project No. 341).

The treatment options for wastewater from low-volume flush systems have been discussed elsewhere (Oellermann and Batchelor, 1993). Constructed wetlands for use in small and medium communities have a particular advantage in that they can be used where flows are irregular and where mechanical systems are perceived to be both expensive and unreliable (Wood, 1991). Not only can they be constructed for the treatment of raw and secondary sewage effluents, but they can also protect nearby rivers from untreated stormwater runoff. However, most rural communities do not have a water supply on tap and wastewater treatment is currently therefore of secondary importance. Appropriate

Table 3. Sanitation systems commonly in use in South Africa (Water Research Commission, 1993b).

System	Percentage in Operation
Waterborne	64
Ventilated improved pit latrines (VIP)	1.4
Septic tanks	2
Pit latrines	21
Buckets	8
Other	1.6
None at all	2

technologies are available and timely planning for their implementation is important to preempt unnecessary pollution.

BIOGAS RECOVERY

Implementation of relatively low-cost biogas technology for utilization by the lower-income groups in the rural areas has been investigated by installing four plants in rural areas. The operational conditions required were established. The estimated costs of biogas in rural areas appear to compare favorably with the costs of paraffin and liquid petroleum gas, especially if a fixed-dome plant is used. Important aspects to consider include its viability in terms of numbers of cattle needed to sustain the biogas plant with manure, water, the skills needed for the maintenance of the plant, and the acceptance of the system on a sociocultural level (Thom, 1994).

In a further study, the potential quantity and quality of biogas from 85 anaerobic household digesters is being investigated by the municipality of Grahamstown. This technology, if successful, could be of great benefit to rural communities throughout South Africa. Although much research has been documented on the production and capture of methane gas from large anaerobic reactors and sewage works, information is lacking on the collection and utilization of biogas from digesters serving individual facilities (Water Research Commission, 1993a—Project No. 551).

BIOREMEDIATION OF SOILS AND SOLID WASTES

Bioremediation of Soils

Bioremediation of contaminated soil and groundwater is less well established. Although not foreign to South Africa, it has had limited application at full scale. This section highlights the current status of bioremediation of contaminated soils within South Africa, and an indication is given to what extent it has been used as a cleanup technology.

Due to the large petrochemical industrial sector in South Africa, one of the main contaminants of soils are the petrochemical products. Such contamination is often economically treatable through bioremediation.

The South African oil industry agreed (Camp, 1994a) that

- "No liquid hydrocarbons or soil polluted with liquid hydrocarbons shall be disposed of to a landfill site, waste dump or class 1 or A site."

- "Liquid hydrocarbons shall be recovered and be reused whether before or after reprocessing as conditions require."
- "Polluted soil shall be bioremediated, whether in situ or at an approved location, or treated in another acceptable manner so as to render it acceptable to the environment."
- "Hydrocarbon polluted waters shall have all free hydrocarbons removed and the polluted water treated to remove or render harmless the dissolved petroleum hydrocarbon fractions."

Guidelines are available for the determination of the extent of the problem, soil sampling, analyses required, and acceptable goals to which to bioremediate (Camp, 1994b). There is no national legislation with respect to target levels of total petroleum hydrocarbons in South Africa at present. Each site is assessed individually, and targets are set in conjunction with the Department of Water Affairs and Forestry. Where there is a danger of contamination of water, be it surface or groundwater, these levels are understandably lower than where this is not a factor. The South African Soil and Water Pollution Manual to the South African oil industry (Camp, 1994b) states that South Africa will strive to work towards a level equal to or less than 1000 ppm total petroleum hydrocarbons (TPH) above that of levels occurring in surrounding soils, although a level of 5000 ppm on a dry mass:mass TPH basis will support plant growth. Further work needs to be done to provide legislative guidelines on target limits. It is recognized that, although a site-specific approach is needed, the risks, both to the environment and to man, should be taken into account.

A recent survey (van Heerden, 1994) has shown that there is a growing awareness of bioremediation as a cleanup technology for contaminated soils. Full-scale application is increasing; however, it is not utilized to its full potential. The technology is seen generally as being cost-effective in the treatment of petroleum products, where it has most widely been used. It has potential for growth and is seen to be effective to reach cleanup target goals.

Various techniques for soil bioremediation are being investigated. In the majority of applications in South Africa, landfarming, being the most economical, has been used. Both *in situ* and off-site treatment at full scale have found application. *In situ* bioremediation has obvious economic advantages; however, this may not always be possible to apply (Ginster, 1994). Landfarming of petrochemical products is usually done in the open, because legislation requirements for the treatment of less hazardous volatile fractions are not as stringent as in the developed countries.

The first documented use of bioremediation in South Africa was approximately in 1980, when a refinery established an area to treat their oily wastes through "landfarming." Since that time, various refineries have followed suit, as did the storage and handling sections of the oil industry. However, the majority of contaminated soils are still excavated and landfilled, usually in first-class landfill sites, of which there are only six in the country. This can be ascribed in most cases to ignorance of alternatives and the relatively low charges of landfilling. The shortsightedness of these actions is exacerbated by an "out-of-sight, out-of-mind" philosophy.

To date, contaminated soil from approximately 30 service stations, 40 vehicle accident spills, and 11 bulk storage facilities have been treated in South Africa using bioremediation. The largest environmental cleanup operation undertaken by the South African oil industry recently came to a conclusion, with the flooding of the former oil tank farm in the Cape Town harbor to create a yachting marina and land for waterside housing. The tank farm had been established in 1914, when there was little concern about environmental pollution. The tanks stored a variety of chemicals, from oils and fuels to toluene and xylene. Cleanup actions started in 1987 and involved the excavation of old tanks and pipes and the remediation of the contaminated soil. The soil was remediated to a level of 1000 ppm or less (Camp, 1994a).

A further risk of oil contamination occurs on the beaches since South Africa lies on one of the world's major shipping routes. South Africa has a coastline of about 3000 km, which exacerbates the risk. It was estimated in 1993 that approximately 120 million tons of Middle East exports passed the Cape of Good Hope. Although this is a considerable decrease compared to the 635 million tons that were shipped in 1977 [due mainly to alternative shipping routes opening up (Moldan, 1994)], it still represents a major risk to ecologically sensitive coastal areas. This was highlighted earlier in June 1994, when the Apollo Sea lost 2500 tons of heavy fuel oil, just off the Western Cape coastline. Apart from the many scenic beaches that were contaminated with oil, the marine life and penguins were also adversely affected.

In this respect, the Department of Sea Fisheries has an oil spill response strategy for the whole coastline, where the coast has been divided into 23 coastal zones (Moldan, 1994). These give details, amongst others, as to how to react to a spill, what resources are available in the area, and the cleanup procedures, as well as the legal aspects of who has jurisdiction over the area.

A case study was done on oil-contaminated beach sand following the Petingo oil spill on our northeastern seaboard in 1990, when approximately 1650 tonnes of heavy fuel oil contaminated 5500 m^3 of sand. Landfarming was used; however, rates of degradation were lower than expected. Problems hindering bioremediation that were identified included the remoteness of the spill, lack of nutrients in the sand, and the heterogenous nature of the contaminated sand, which made it difficult to quantitatively monitor the degradation of the oil (Dehrmann, 1991).

Research currently being conducted on contaminated soils focuses mainly on petroleum products, such as oil-contaminated soil (Pearce *et al.,* 1994; Lees and Senior, 1992). Rapid breakdown of river sediment contaminated with heavy oils to the order of 94% was shown over a period of 10 weeks (Pearce *et al.,* 1994). Creosote has also been investigated (Bichara, 1991, cited by Lees and Senior, 1992), as have other polyaromatic hydrocarbons such as PCBs. Preliminary research has also been undertaken on a laboratory scale for treating phenol contaminated soils using slurry reactors and soil columns.

Research is also being done using thermal bioventing (Snyman, 1994), a process that uses similar principles to composting. Heat is provided through the microbial activity, while low levels of aeration increase the oxygen available to the microorganisms, keeping volatilization down to a minimum. Application for this technique is anticipated for soils with a high clay content, soils contaminated with very high concentrations of dangerous volatiles that need to be treated.

The University of Natal has also been engaged in various research projects involving contaminated soil. The studies focus on the rate of biodegradation of petroleum hydrocarbons and metals. It was found that a C:N ratio of 5:1 caused microbial inhibition. Adjustment of the soil pH (4.6–8.2) had no beneficial effect. Addition of a coarse ash from an industrial process as bulking agent was effective in improving hydrocarbon degradation. An additional benefit was a reduction in heavy metal concentration in the leachate resulting from the increased soil pH. A biodegradation index was developed to assess the relative concentration of C (<20) and C (>20) hydrocarbons, since not all hydrocarbons pose an environmental risk. This index is believed to more accurately assess biodegradation and environmental risk since the heavier hydrocarbon fractions are less prone to degradation and leaching. Research is currently ongoing to find common soil characteristics that affect pollutant migration and microbial degradation, so that accurate predictions can be made about the fate of organic soil pollutants. Soil surfaces were found to significantly affect contaminant adsorption and surface attachment of microorganisms. Porosity of the soils has also been found to exert an influence on pollutant attenuation and on microbial migration. It is believed that the ratio of micro- to macropores in the soils is important because the growth rate of organisms occupying the pores is often different. Migration of pollutants into the pores must also play a role (du Plessis, 1994).

Evaluation of the potential application of various species of fungi for degradation of petroleum hydrocarbon contaminated soils is being investigated. This research has involved an extensive isolation program from contaminated soil, as well as characterization of fungal isolates in terms of their hydrocarbon degradation capabilities (du Plessis, 1994).

Another technology seen as an option for decontamination of sites is phytoremediation, a process that can be defined as the use of plants to remove, contain, or render harmless an environmental contaminant (Ginster, 1994). Despite the historical use of this technology for treatment of wastewaters in wetlands, it may also find an application in the treatment of contaminated soils.

Further research is needed to establish the level of nutrients required in the soil and how best to achieve this. Currently, mono ammonium phosphate is commonly used, but this is applied mainly to improve the soil for agricultural purposes and is not necessarily optimal for encouraging bacterial growth. It may be rewarding to investigate the use of slow nutrient release pellets versus liquid application.

There is a need to improve the accuracy and sensitivity of analyses to be able to monitor pollutant volatilization and breakdown.

These goals can probably best be achieved by careful coordination of local research and networking worldwide. It is essential to conduct bioremediation on a more scientific basis, rather than by rules of thumb.

Landfill Sites

Novella *et al.* (1994) demonstrated that co-disposal of wastewater sludge with domestic refuse in a sanitary landfill could be a viable option, provided proper control over the operation is exercised. Alternatively, wastewater sludge in South Africa, with its dry climate, is frequently disposed of in drying beds. Furthermore, in the Western Cape, wastewater sludge from virtually any source is used as a soil conditioner by farmers (Novella and Fawcett, 1992, cited in Novella *et al.,* 1994). However, its future use will be prescribed by stricter legislation.

Biogas utilization from landfills has been demonstrated in South Africa. Methane is also used as a feedstock by a chemical manufacturing industry for the production of cyanide in the vicinity of the Robinson Deep landfill site in Johannesburg. However, the cost of electricity is still considered to be one of the lowest in the world (Letcher *et al.,* 1994).

BIOREMEDIATION OF OFF-GASES AND ODORS

Types of Atmospheric Discharge

Atmospheric discharges are generally classified as

- specific gaseous chemicals that are potentially toxic or environmentally damaging (atmospheric pollution)
- complex mixtures of indeterminate chemical constitution of minimal risk in terms of direct health or environmental damage (odor nuisance)

A significant difference exists between the two categories in South Africa in terms of legislation, monitoring, and enforcement. Pollution limits can easily be defined in terms of specific substances and their maximum concentrations. Odor nuisances, on the other hand, can only provide objective data by a complex and expensive means of what has been defined as computer-controlled dynamic dilution olfactometry. It is not in use in South Africa. As a consequence, it is difficult to control the nuisance of odors through local legislation. Fortunately, many responsible organizations voluntarily control malodor emissions.

Biological Methods of Odor Control

At the very low concentrations encountered, it is difficult to use physico-chemical systems in a cost-effective manner to address odor control. However, biological treatment of odor emissions are technically simpler, require a minimum of consumable input materials, do not generate problem waste products, and can operate with very little or no supervision and at acceptable cost.

Since the properties of aerobic catabolism are employed, these systems all require the presence of air (oxygen) for operation. The microbes utilize the problem pollutant as a substrate for their continued survival. Provided there is a suitable selection of microbial species at the start, the system will adapt to that pollutant composition and concentration presented to it. Solid-phase materials, such as dusts or smokes, can create problems with accumulation and blockage, so they should be removed by physical methods, such as prefiltration, wet scrubbing, or electrostatic precipitation, before putting the odor-laden airstream through a device such as a biofilter. Recalcitrant materials or substances at or above toxic concentrations will kill or inhibit the microbial activity, as will extremes of heat or cold, leading to functional failure of the system.

Although proven technology is often available overseas, it was more cost-effective to develop technologies for odor control in South Africa. Consequently, a local program of development and implementation of systems that are suited to the types of odor emissions and climatic conditions that prevail in our country has been embarked upon (Clark, 1994).

Types of Technology in Biological Odor Control

Biological methods of odor control fall roughly into three technological categories, as has been depicted in Table 4. Odor biofilters contain an air-pervious matrix of controlled moisture content, in which the microbial consortium is entrained. They handle very low pollutant concentrations with high efficiency but have the disadvantage of a relatively large size of the device in relation to its load capacity. A biofilter can run for as much as 5 years without any supervision or maintenance.

Odor biotricklers resemble their namesakes in the wastewater treatment field, except that they are designed to accept an airstream as the influent pollutant carrier. Unlike biofilters, there is a free flow of water over the biofilm, which is attached as a fixed growth on a suitable support matrix. Biotricklers cannot achieve the degree of "polish" to the exit airstream that can be achieved with biofilters, but biotricklers can accept significantly larger concentrations of influent material and are thus more compact devices in relation to biofilters of equivalent capacity. Having moving parts such as recirculation pumps and microbial nutrient dosing equipment, they require some degree of skilled supervision.

Odor bioscrubbers resemble conventional wet scrubbers, except that the water phase contains a suspended biomass that circulates with the water. The scrubber tower may or may not contain a matrix to aid mass transfer from gas to aqueous phase, but no biofilm is allowed to attach to such packing because it would impair air flow and reduce operating efficiency. The bioscrubber has a wet sump, in which the bulk of the suspended biomass is retained. As in an activated sludge

Table 4. Main types of technologies in odor control.

Phase	Biofilters	Biotricklers	Bioscrubbers
Aqueous phase	Fixed	Flowing	Flowing
Microbial phase	Fixed	Fixed	Flowing

basin, the pollutants scrubbed from the air in the scrubber tower are metabolized in the aqueous suspended biomass to innocuous breakdown products. Bioscrubbers represent the most technically complex form of system of this series and can deal with the largest concentrations of pollutant, as well as large air flows, in relation to the size of this equipment. "Polish" of emergent air from bioscrubbers is not as high as that from biotricklers. Being a complex plant, they require skilled supervision for reliable service.

These various forms of biological odor control can be used in tandem combinations, if necessary, where both high capacity and high final "polish" is sought.

Development and Implementation of Biological Odor Control Technology in South Africa

Nuisance odor emissions from sewer pump stations have been successfully controlled by means of biofilters that embody design details that have been filed in patent applications on behalf of the CSIR (Clark, 1991, 1993a). These units range from a small device (50 m^3/h) designed to drop into the aperture of a standard 500 mm diameter manhole, to a twin 6-m transport container system for the treatment of 3500 m^3/h, emanating from a large sewer pump station.

The CSIR in its Division of Water Technology has also developed bioscrubber technology (Clark, 1993b) that has been purpose-designed for the control of odor emissions from sewer inlet works, screen houses, and other points of odor emission at wastewater treatment plants.

RESEARCH AND DEVELOPMENT: TOWARDS THE FUTURE

No formal research and development plan exists as such for bioremediation, although a master plan has been drawn up for sanitation (Mitchell and Offringa, 1991). The master plan identified the main thrust areas for future research on sewage effluent treatment and sludge management. The main research aims were to improve the utilization of the country's scarce funding, material and skilled manpower resources to manage sanitation related effluents, wastes and runoffs in the most efficient and cost-effective way, and for the best interest of the Republic of South Africa and all its inhabitants.

It is generally thought that the impetus of research will have to be maintained by increasing support from industry. However, current industry involvement is low but could change by the introduction of incentive legislation. It is difficult to obtain support for the implementation of research outputs, and industry could play a more direct role in these efforts. Industry should be encouraged to fund research in fields beyond their immediate line of business to ensure continuing development in the country.

Insufficient planning results in problem solving being pursued on a reactive, rather than proactive, basis.

Research teams at universities and other organizations are often small. This, together with the lack of monetary support, points towards the need that research in the interim be slanted towards application, rather than fundamental research. A positive move is the trend to prevent duplication and to apply and develop technologies towards specific needs in South Africa.

Upliftment and the improvement of the quality of life, particularly in the rural areas and informal settlements, by providing affordable potable water and appropriate sanitation, will remain a focus activity in the years to come. However, a balance with industrial development and associated research in environmental protection, resource development, and pollution control will be vital for the continued development and sustainable growth of the subcontinent.

ACKNOWLEDGEMENTS

We sincerely thank Mr. C. R. Camp of the Oil Industry Environment Committee, Cape Town, for the valuable information provided and his time for numerous meetings; Dr. S. A. Mitchell of the Water Research Commission, Pretoria, for his important input in the compilation of this manuscript; Mr. A. Gerber, Programme Manager of the Water, Waste and Effluent Treatment Programme of the Division of Water Technology, CSIR, Pretoria, for his continued interest and personal support of our work; all industries, organizations, companies, and individuals for their participation in the surveys carried out and sharing of information in the compilation of this manuscript; and our colleagues in the Division of Water Technology, CSIR, Pretoria, for their assistance in research and the preparation of this manuscript.

REFERENCES

Anon. 1994a. Eerste resultate van proteinherwinning uit abattoiruitvloeisel lyk positief. *SA Waterbulletin,* 20:4, 11.

Anon. 1994b. Institutional options for water supply and sanitation. Executive summary for discussion document. *Water Sewage and Effluent,* 14:28–30.

Barnard, J.L. 1975. Biological nutrient removal without the addition of chemicals. *Water Research,* 9:485–490.

Batchelor, A. 1994a. Personal communication. Division of Water Technology, CSIR, P.O. Box 395, Pretoria, 0001, South Africa.

Batchelor, A. 1994b. Recent advances in constructed wetland technology. *Water Sewage and Effluent,* 14:9–13, 15–16.

Britz, T.J., Spangenberg, G. and Venter, C.A. 1994. Acidogenic microbial species diversity in anaerobic digesters treating different substrates. *IAWQ Seventh International Symposium on Anaerobic Digestion,* Cape Town, South Africa, 23–27 January 1994. Oral Paper Preprints, pp. 74–79.

Camp, C.R. 1994a. Personal communication. Oil Industry Environment Committee, P.O. Box 6094, Roggebaai, Cape Town, 8012, South Africa.

Camp, C.R. 1994b. *The South African Soil and Water Pollution Manual.* June 1994, In preparation for the Oil Industry of South Africa.

Casey, T.G., Ekama, G.A., Wentzel, M.C. and Marais, G v R. 1993. Causes and control of low F/M filamentous bulking in nutrient removal activated sludge systems. *Water Sewage and Effluent,* 13:10–26.

Clark, R.C. 1991. "Biofilter". South African Patent No 91/8139. Patentee, CSIR.

Clark, R.C. 1993a. "Filter Matrix". South African Final Patent Application No 93/8720. Patentee, CSIR.

Clark, R.C. 1993b. "Odour Treatment". South African Final Patent Application No 93/3181. Patentee, CSIR.

Clark, R.C. 1994. Personal communication. Division of Water Technology, CSIR, P.O. Box 395, Pretoria, 0001, South Africa.

Dehrmann, A.A. 1991. Landfarming: An option for disposal of oily waste following a marine oil spill off the South African coast. *Biennial Ground Water Convention,* Midrand, South Africa, 21–23 August 1991. Paper No 25.

Department of Water Affairs. 1986. *Management of the water resources of the Republic of South Africa.* Department of Water Affairs and Forestry, Private Bag X313, Pretoria.

Department of Water Affairs. 1993. *South African Water Quality Guidelines.* Department of Water Affairs and Forestry, Private Bag X313, Pretoria.

Dudley, B.T., Taylor, M.A. and Howgrave-Graham, A.R. 1993. Process Performance investigations on two laboratory-scale and one full-scale UASB digesters treating brewery waste water. *Proceedings of the Third Biennial WISA Conference,* Durban, South Africa, 24–26 May, 1993, Abridged Papers, pp. 160–161.

du Plessis, C.A. 1994. Personal communication. Department of Microbiology and Plant Pathology, University of Natal, Private Bag X10, Scottsville, Pietermaritzburg, 3209, South Africa.

du Preez, L.A. and Maree, J.P. 1994. Pilot-scale biological sulphate and nitrate removal utilizing producer gas as an energy source. *IAWQ Seventh International Symposium on Anaerobic Digestion,* Cape Town, South Africa, 23–27 January 1994. Oral Paper Preprints, pp. 190–204.

du Preez, L.A., Maree, J.A. and Jackson-Moss, C.A. 1991. Biological sulphate removal form mining effluents by utilising producer gas as an energy source, *Proceedings of the Fourth International Mine Water Congress,* Ljubljana, Yugoslavia, 25–30 September, 1991, 2:255–264.

Foundation for Research Development. 1994. *Annual Report 1993 / 1994.* (Joubert, M. Ed), Words'Worth Publishing Consultants, p. 35.

Fey, M. 1995. Personal communication. Department of Geological Sciences, University of Cape Town, Rondebosch, 7700, South Africa.

Ginster, M. 1994. In-situ remediation options for petroleum hydrocarbon contaminated soils. *Proceedings of the Wastecon'94 Conference,* Somerset West, South Africa, 27–29 September 1994. Oral Papers, pp. 254–261.

Harber, A. and Ludman, B. (ed). 1994. *A–Z of South African Politics: The essential handbook.* The Weekly Mail and Guardian. Penguin Books.

Lees, Z.M. and Senior, E. 1992. Bioremediation: A novel technology for South Africa. *Proceedings Third Southern Africa Anaerobic Digestion Symposium,* Pietermaritzburg, South Africa, 13–16 July, 1992, pp. 269–275.

Letcher, T.M., Jarmain, D. and Daneel, R. 1994. Exploitation of methane gas from landfill in South Africa. *IAWQ Seventh International Symposium on Anaerobic Digestion,* Cape Town, South Africa, 23–27 January 1994. Oral Paper Preprints, pp. 448–459.

Letimela, O.N. 1993. *Denitrification of groundwater for potable purposes.* Water Research Commission, Pretoria, South Africa. Report No. 403/1/93, pp. i–ii.

Maree, J.P. 1989. Biological treatment of water. United States Patent No. 4,839,052, 13 June. (Filed: 23 June 1988).

Maree, J.P., Gerber, A. and Strydom, W.F. 1986. A biological process for sulphate removal from industrial effluent. *Water SA,* 12:139–144.

Meiring, P.G.J., 1993. Integrating oxidation ponds and biological trickling filters. *Proceedings of the Third Biennial WISA Conference,* Durban, South Africa, 24–26 May, 1993, 2:182–193.

Meiring, P.G.J. and Hoffmann, J.R. 1994. Anaerobic pond reactor in-line with biological removal of algae. *IAWQ Seventh International Symposium on Anaerobic Digestion,* Cape Town, South Africa, 23–27 January 1994. Oral Paper Preprints, pp. 385–395.

Meiring, P.G.J and Oellermann, R.A. 1993. Biological removal of algae in an integrated pond system. *2nd IAWQ International Specialist Conference on "Waste Stabilization Ponds and Reuse of Pond Effluents,"* Berkeley, California, 30 November–3 December 1993.

Meyer, V. and Oellermann, R.A. 1993. Biosupplementation: An emerging technology for the degradation of selected pollutants in wastewaters. *Proceedings of the Second Biennial WISA Conference,* Durban, South Africa, 24–26 May 1993, 2:194–201.

Mitchell, S.A. and Offringa, G. 1991. *Master plan for future research on sanitation in South Africa.* Water Research Commission, Pretoria, South Africa. September 1991, AV 2/91.

Moldan, A. 1994. Personal communication. Sea Fisheries Research Institute, Private Bag X2, Roggebaai, 8012, South Africa.

Momberg, G.A. and Oellermann, R.A. 1992. The removal of phosphate by hyroxyapatite and struvite crystallisation in South Africa. *Water Sci Tech,* 26:987–996.

Novella, P.H., Ross, W.H., Lord, G.E., Stow, J.G., Fawcett, K.S. and Greenhalgh, M.A. 1994. The simultaneous co-disposal of waste-water sludge liquor in a sanitary landfill bioreactor. *Proceedings of the Wastecon'94 Conference,* Somerset West, South Africa, 27–29 September 1994. Poster Papers, pp. 93–101.

Oellermann, R.A. and Batchelor, A. 1993. Waste water treatment options for small bore sewer systems. Paper presented at the *Seminar on design guidelines for small bore sewer systems under Southern African Conditions,* Cape Provincial Administration, Port Elizabeth, South Africa, 30 September 1993.

Oellermann, R.A. and Pearce, K. 1994. Bioaugmentation technology for wastewater treatment in South Africa. *Water Research Commission Project K5 / 429,* 1993–1994.

Oellermann, R.A., Meyer, V. and Carlsson, F.H.H. 1994a. A laboratory comparison of support media for biomass immobilization in anaerobic wastewater treatment. *IAWQ Seventh International Symposium on Anaerobic Digestion,* Cape Town, South Africa, 23–27 January 1994. Poster Paper Preprints, pp. 287–290.

Oellermann, R.A., Meyer, V. and Carlsson, F.H.H. 1994b. Pilot-scale evaluation of anaerobic treatment of abattoir effluent using a fixed film reactor. *IAWQ Seventh International Symposium on Anaerobic Digestion,* Cape Town, South Africa, 23–27 January 1994. Poster Paper Preprints, pp. 353–355.

Pearce, K., Snyman, H. and Oellermann, R.A. 1994. Bioremediation of petroleum contaminated soil using landfarming. *Proceedings of the Wastecon'94 Conference,* Somerset West, South Africa, 27–29 September 1994. Poster Papers, pp. 102–111.

Perry, M. 1994. Personal communication. South African Breweries, P.O. Box 782178, Sandton, 2146, South Africa.

Pretorius, W.A. 1993. Unconventional protein production on industrial effluents: Present situation in South Africa. *Proceedings of the Third Biennial WISA Conference,* Durban, South Africa, 24–26 May, 1993, 2:308–314.

Retief, E.A. 1994. Petroleum pollution and the environment: a review of the current status of a serious potential threat to groundwater resources in SA. *Municipal Engineer,* September 1994, pp. 45, 47–48.

Ross, W.R., Strohwald, N.K.H., Grobler, C.J. and Sanetra, J. 1994. Membrane Assisted anaerobic treatment for industrial effluents: The South African ADUF Process. *IAWQ Seventh International Symposium on Anaerobic Digestion,* Cape Town, South Africa, 23–27 January 1994. Oral Paper Preprints, pp. 550–559.

Saayman, G.B., Loots, P.A., Oellermann, R.A. and Pearce, K. 1994. *Pilot studies on phosphate crystallization in biological wastewater treatment systems.* Final report to the Water Research Commission on Project K5/366, Pretoria, South Africa.

Snyman, H. 1994. Unpublished data. Division of Water Technology, CSIR, P.O. Box 395, Pretoria, 0001, South Africa.

Thom, C. 1994. *A socio-technical feasibility study to develop biogas as an energy source for rural South Africa.* Final report to the Department of Mineral and Energy Affairs by the Division of Water Technology, CSIR, Pretoria, South Africa.

van Heerden, H. 1994. Unpublished data. Division of Water Technology, CSIR, P.O. Box 395, Pretoria, 0001, South Africa.

Water Research Commission. 1984. *Theory, design and operation of nutrient removal activated sludge processes.* (Contributors: G.A. Ekama, G. v R Marais, I.P. Siebritz, A.R. Pitman, G.F.P. Keay, L. Buchan, A. Gerber and M. Smollen). Water Research Commission, Pretoria, South Africa.

Water Research Commission, 1993a. *Annual Report.* Water Research Commission, Pretoria, South Africa.

Water Research Commission, 1993b. *Technical, socio-economic and environmental evaluation of sanitation systems for developing urban areas in South Africa.* Report number 385/1/93, Pretoria, South Africa.

Wood, A., 1991. Application of constructed reedbeds for wastewater treatment in South Africa. *Proceedings of the Second WISA Biennial Conference,* Kempton Park, South Africa, 13–15 May, 1991, 2:702–712.

Bioremediation R&D Planning in the United Kingdom

PHILIP CAUNT AND MICHAEL L. HOLROYD
Biotal Ltd.
5 Chiltern Close
Cardiff Industrial Park
Cardiff CF4 5DL Wales, UK

INTRODUCTION

The History of Bioremediation in the United Kingdom

The remediation of contaminated land is an activity still very much in its infancy in the United Kingdom. The myriad problems left by Britain's history as an industrialized nation were only realized in the 1970s and 1980s, and development and implementation of remedial activities only really began in the 1980s.

In the early days, bioremediation research was mostly carried out by a small number of specialist companies. Only very few groups were able to assemble the combination of skills required to develop bioremediation, namely a thorough understanding of the processes of biological degradation and their translation to on-site conditions; how the behavior of the contamination is affected by the soil type, geology, and hydrogeology of the site; and the application skills, more usually associated with the civil and process engineering industry.

However, the increase in requirement for remediation of contaminated land and the commercial opportunities offered had a noticeable effect. Bioremediation was accepted as an option for the treatment of contaminated sites and much of the early mystique associated with the process was removed, resulting in an increase in the number of organizations offering remedial services, including bioremediation. Many of these new organizations were either specialist environmental consultancies or were associated with the large civil engineering contracting companies. Unfortunately, other less reputable players entered the market, offering bioremediation as a panacea for the treatment of a wide range of contaminated sites. Many offered services not backed by all the necessary expertise, including the sale of microbial preparations that were claimed to clean contaminated land and were often used without the necessary associated site work to optimize the biodegradation process. As a result of the general mood of optimism at that time, the capabilities of bioremediation were tested to, and eventually beyond, the limits, in many cases without sufficient research and development or understanding of the difficulties before going to the full scale. As a result, bioremediation gained the reputation of being unreliable, slow, and relatively expensive when compared to the other available technologies.

This change in attitude towards bioremediation also coincided with a period of economic change in the United Kingdom. The prosperous years of the early 1980s were followed by a very deep economic recession in the later years of the decade and into the 1990s. One of the industries worst hit by the recession was property development and construction, and as a result, the demand for remediation of contaminated land fell sharply. What remedial activity remained and the approach taken during those recession years were more often decided solely by cost. Remedial technologies became a routine part of the civil engineering industry, with projects put out to tender in the normal way. U.K. legislation permits the disposal of contaminated material to a licensed landfill, and off-site disposal or on-site encapsulation became very popular, rapid, and cost-effective choices for site cleanup. This requirement to compete with other technologies resulted in a considerable sharpening of attitudes towards alternative technologies, including bioremediation.

More recently, changing attitudes to the environment have included the realization that the use of techniques that do not solve the problem in the long term are not necessarily acceptable. There is also increasing pressure on landfill space and a greater awareness of future liabilities. Bioremediation research is having something of a renaissance, with a significant number of groups involved, especially a larger number of universities. However, the growth in activity in bioremediation is largely research driven and is not necessarily shared by the industry that actually carries out the cleanup projects. The contracting industry in the United Kingdom is conservative in nature and slow to adopt new technologies, and it is an extremely competitive market. It is likely that the cheapest option will remain favored, and until the legislative position changes, this remains either off-site disposal or encapsulation.

However, this new activity in bioremediation research is likely to result in greater understanding of the fundamentals and limitations of the process, which in turn should result in less applications of the technologies in unrealistic or unsuitable projects. In addition, new technologies are emerging that overcome some of the perceived shortcomings of bioremediation. For example, groups are working on some of the problems previously not treatable by biological methods, such as metal contamination. In addition, some of the newer technologies offer promise as being more rapid than the original methods and able to reach much lower targets than before.

Now, more than ever, there is a crucial role for the U.K. government in terms of research and development planning in land reclamation generally and in bioremediation in particular. Some of that role is in the funding of appropriate projects, while in other cases, it is by introducing lines of communication to set up project groups with common interest. Overall, there is a third role, which is perhaps the most difficult, that of steering and coordination of the U.K. bioremediation effort.

U.K. GOVERNMENT ATTITUDES TO CONTAMINATED LAND

As a general principle, the U.K. government has confirmed that the concept of multifunctionality; i.e., the cleanup of contaminated land to a state suitable for every kind of use should be rejected as the costs are likely to far outweigh the benefits. As an alternative, the government has proposed a "suitable for use" approach. Under this methodology, the remedial operations should deal with any unacceptable risks to health or the environment, taking into account the costs and the intended use of the site.

Historically, the U.K. land remediation industry is less developed than that in the United States, both in terms of the number of companies involved and the number of sites successfully treated. On a practical level, results from American research and field work have given U.K. groups confidence to attempt remediations based on the techniques and experience from the U.S. projects. Perhaps more importantly, the U.S. experience has helped in the formation of U.K. government policy, in that an outsider's view of the American legislative position regarding contaminated land has helped to steer the United Kingdom towards the fit for purpose concept. The overall aim is to encourage cleanup by avoiding burdening business and industry with excessive costs, and so reducing the pressure to develop the United Kingdom's precious green field areas. Thus, R&D work on remedial technologies for contaminated land in the United Kingdom are very closely related to this policy.

The Department of the Environment has a division with specific responsibility for contaminated land and liabilities (CLL). Within this section is the contaminated land branch (CLL1), which provides policy and technical support. Among the responsibilities is the supplementary credits approval program, which allows local authorities to borrow money to investigate and carry out remedial work on closed landfill sites and contaminated land that they are responsible for. The Department of the Environment sponsors research in order to assist in ministerial decision making and policy and to promote innovation and the spread of good practice and technology. The contaminated land branch administers a contaminated land research program with an annual budget of approximately £1.5 million. This finite resource has to cover a very wide range of requirements, and priority has been given to projects in the following areas:

1. The development of standards for dealing with contaminated land
2. A better understanding of contaminant behavior where this is relevant to setting standards and determining cost-effective solutions
3. Information requirements: the identification of contaminated sites related to past industrial usage. "Blight" of contaminated land and sale and transfer issues
4. Risk assessment: frameworks for the assessment of risk, site inspection and sampling methodologies, analytical methods and impact of contaminated land on water quality, effects of landfill gas and contamination on development
5. Remedial methods and monitoring: review of available technologies and selection strategies, review of research, soil separation techniques, use of microbiological tests to assess toxicity of substances released during biological treatment

The use of this finite resource may be taken as a fair indication of government attitude towards bioremediation and other technologies for the treatment of contaminated land. Most of the R&D programs funded so far have been directed toward two areas. The first is the accurate and consistent measurement of the real risks associated with the many types of land contamination. The second is the review of remedial technologies that have already been developed worldwide and particularly in the United States. The actual development and support of bioremediation and other remedial technologies has been seen as a lower priority than these fundamental issues.

At the present time, there is a general change towards funding of technology developments. The government intends to support research into a range of environmental technologies, including the remediation of contaminated land under its general research program. The various pieces of work are normally brought together in a number of schemes.

At the time of writing, several schemes of this type were in operation, and three past and three current schemes are described more fully below.

PUBLIC FUNDING SCHEMES FOR ENVIRONMENTAL RESEARCH

Environmental Technology Innovation Scheme (ETIS)

This scheme was run jointly by the Department of the Environment (DoE) and the Department of Trade and Industry (DTI) and promoted the development of technologies through to the proof of concept stage. The scheme was designed to cover research leading to new techniques, processes, materials, or equipment in four broad categories:

- cleaner technologies
- recycling
- waste or effluent treatment and disposal
- environmental monitoring

Within this framework, one of the priority areas identified was the treatment of organic contamination in soils. Launched in precursor form in 1988 and in full form in 1990, with a projected time span to 1993, the scheme received over 500 applications, and 35 projects were supported, with an input of £6.6 million. Biologically based projects included the use of anaerobic digestion for the treatment of wool scouring effluent, fungal biosorption of heavy metals from effluent, and the bioleaching of iron-stained silica sands for glass production.

Euroenviron

Euroenviron is the Department of Trade and Industry's environmental network within the EUREKA initiative, which is funded by the European Community. In broad terms, this scheme supported a wide range of environmental research, including European collaborations between 1990 and 1992. The projects involved the U.K. organization collaborating with at least one other European partner in the Euroenviron scheme. Thirteen full-scale projects were approved, along with six smaller feasibility studies. In total, £4.6 million of support were given. Although no bioremediation projects are included, one of the projects involved the treatment of contaminated soils in a slurry phase, and another was on wood preservation technologies using biological agents.

DEMOS

DEMOS is an acronym for the Department of Trade and Industry's Environmental Management Options Scheme, which again is closed to further applications. This scheme promoted the demonstration of environmental best practice techniques and technologies. The 18 supported projects included the anaerobic digestion of municipal solid waste fractions, co-composting of sewage sludge and municipal waste, and the establishment of an effluent processing club. Although bioremediation projects are noticeable in their absence from DEMOS, the various activities on waste minimization and recycling will have a marked effect on the future applications of bioremediation technologies.

The LINK Schemes

At the time of writing, 39 different LINK programs have been set up by the Department of Trade and Industry and administered by the Office of Science and Technology. LINK resources are concentrated in a series of programs, each addressing a particular area of science and technology and aiming to build up a portfolio of collaborative projects. The LINK programs have the aim of accelerating the commercial exploitation of government-funded research and focus on advances in science and engineering with particular commercial promise. The initiative encourages collaboration between partners in industry and the research communities (i.e., universities) and provides up to 50% government funding for the eligible costs of the project.

Nine of the LINK programs fall into the area of biological sciences, and most recently, a program has been set up to promote research on biological treatment of soil and water (known as BSTW). The scheme is jointly sponsored by the Department of Trade and Industry; The Ministry of Agriculture, Fisheries, and Food; and the Office of Science and Technology via the Research Councils. The scheme is scheduled to last 5 years with total funds of £4.6 million. In the first year of the program (from April 1993), four projects were funded through this route. The LINK program is also important in the organization of conferences on bioremediation and communication and flow of ideas that such events encourage.

The areas that have been identified for high priority in the LINK scheme are these:

1. Reclamation of polluted land, both long-term procedures and rapid measures that can restrict major harmful environmental effects

2. Monitoring and environmental assessment, including the development of analytical techniques and means of tracking pollutants and degradative organisms in the field
3. Waste treatment, the improved digestion and detoxification of domestic industrial and agricultural liquid and solid wastes
4. Agriproduction systems and the reduction of groundwater and river pollution by nutrients and pesticides, and treatment of animal slurries

LINK projects approved so far as part of the biological treatment of soil and water program include the development of a combined technology for the removal of heavy metals from soil and water, microbial treatment of spent oxides at gas works sites, and the use of biologically active carbon filters for groundwater treatment.

Fourth Framework Program

This is a European Community program of research and development that offers a source of funding for U.K. organizations in environmental technologies. Participants must work with at least one partner from another European Economic Area country, and the work must be precompetitive. The EC will contribute up to 50% of the research costs, and the participation of small- and medium-sized companies is actively encouraged. U.K. applications are to be coordinated by the DTI, and one of the areas identified as specifically of interest in the environment and climate part of the program is the in situ treatment of polluted soils.

Environmental Technology Best Practice Program

This is a very new scheme, once again run jointly by the DoE and DTI. It is more involved in technology transfer to promote better environmental performance and increase the competitiveness of U.K. industry and commerce, through the routes of waste minimization and cost-effective cleaner technologies. Special attention is to be focused on particular areas, which have initially been identified as the problems of volatile organic compounds, foundries, and the metals finishing industry. Although not directly relevant to bioremediation research, other areas will be added during the life of the program. Under the future practice part, the program is also intended to support research and development for techniques which will become future opportunities for improved environmental performance.

BIOREMEDIATION RESEARCH AT U.K. UNIVERSITIES

Research in U.K. universities is predominantly funded through the Government Office of Science and Technology via six research councils. Applications for grants are made by the various university departments to the research councils who then assess the merits of the projects together before making the funding decisions. These grants then fund a number of research assistants or postdoctoral scientists, as well as the associated capital and consumable costs of the research. Research studentships are awarded in a separate competitive exercise.

Three research councils are currently funding bioremediation research at British universities. The most active of these is the Biotechnology and Biological Science Research Council (BBSRC), although projects are also funded by the Engineering and Physical Sciences Research Council (EPSRC), and the Natural Environment Research Council (NERC). The projects range from basic, long-term R&D into the fundamentals and mechanisms of the bioremediation process, to the specific and applied short-term work on a particular problem. This is often, but not always, as a result of varying levels of input from an industrial partner.

A second important source of funding for university research is via private industry, and although the funding of biological science research is very well established in areas such as pharmaceutical sciences, this route is not as well established in the field of bioremediation. There are two schemes worthy of note that combine funding from research councils and industry. The first is industrial sponsorship of research students through special cooperative awards (known as CASE studentships) where the industrial partner supplies some financial assistance, as well as facilities and equipment for the student.

A second program is the teaching company scheme, funded jointly by DTI and the research councils, among others, where a student carries out Ph.D. research at the industrial partner, thus transferring knowledge and developing people from the academic environment with industrially relevant skills. Of the 500 teaching company schemes presently in the United Kingdom, some are in biological sciences and some in the area of the environment, but as yet, no bioremediation research projects have been carried out in this way.

THE NATO/CCMS PILOT STUDY

The North Atlantic Treaty Organization alliance civil structure includes a Committee for Challenges to Modern Society, which has among its terms of reference the improvement of the exchange of views and experience among allied countries to create a better environment for their societies. The CCMS operates through pilot studies proposed by member states, which last up to 5 years and then report their conclusions. A relevant pilot study has been established for the development and evaluation of technologies for the treatment of contaminated soil and groundwater. Although there is no assistance with project costs from NATO funds, the alliance will fund Fellowships to enable participation in international meetings of the study, in addition to study visit funds and cooperative research grants to encourage international collaboration on research projects. The United Kingdom has submitted a number of projects as part of the study, and five have been accepted, all of which were using chemical, physical, or integrated technologies. Biological projects from other countries form a major part of the program, and participation in the scheme is an important way of ensuring the international relevance of U.K. land remediation research. It is also of considerable value in international commercialization of the technologies once developed.

CONCLUSIONS

Despite the slightly checkered history of bioremediation in the United Kingdom, with many things promised and not always delivered by the technology, the situation at the present time is very positive. The full extent of the problems of contaminated land are only recently being realized, and under the steerage of the various U.K. government departments, a logical program of research is being pursued. The first stages were to identify the extent of the problem and to ensure a reasonable degree of consistency of approach to various sites. Next, the technologies developed elsewhere, in other countries and, to some extent, other industries, were reviewed. Now we have reached the stage where public funding is available for bioremediation projects, and activity in the universities is higher than ever before.

In the United Kingdom, any innovative technology must compete with the existing technologies for site remediation. There is no attitude where innovative technologies are favored, as appears to exist in other European countries, and the legislative position over landfilling of contaminated soil also works against the more permanent, but potentially more expensive, technologies. There is obviously hope within groups developing these innovative technologies that this will change. However, the timescale for these changes is not known. Thus, there are two challenges for bioremediation in the future. The first is to show that the bioremediation is in fact reliable, can be rapid, and is competitive with other technologies. The second, in the longer term, is to have all the fundamental work carried out and all the variables understood, so that when the legislative position changes, bioremediation is ready to challenge competitively with other innovative technologies that have been developed in the same period.

SECTION II

Pilot, Full-Scale, and Commercial Demonstration

Regulatory Issues Applying to Bioremediation as a Risk Reduction Technology

KATE DEVINE
DEVO Enterprises, Inc.
1003 K Street, NW, Ste. 501
Washington, D.C. 2001

PETER LAGOY
LaGoy Risk Analysis, Inc.
11 Grove Street
Hopkinton, MA 01748

INTRODUCTION

Laws and regulations drive the supply of and the demand for many commodities, including cleanup technologies. Like any cleanup technology, bioremediation is employed to reduce the risk from contaminant exposure at a site. Several environmental statutes, as well as the regulations and policies developed under these laws, have affected the use of bioremediation. This is due, in part, to the fact that the cleanup technology potentially applicable to the site-specific risk has not been a key parameter in the development of many of these regulations. However, there is a growing national financial crisis in which there are too many sites still to be cleaned up with inadequate financial reserves to continue with the current cleanup standard regime. Cost-effective technologies, including bioremediation, are being considered more often as the concept of risk-based cleanup standards established on a site-specific basis gains some prominence.

The first part of this chapter gives a historical overview of aspects of federal regulations of noted importance to those in the cleanup community, the role of risk-related factors in the creation of these regulations, and the effect of these regulations on the use of bioremediation as a risk reduction technology. This overview is followed by a section on the potential economic value of cleanup standard development that uses a risk-based approach considering site-specific contaminant availability. Lastly, a more in-depth discussion of risk assessment in remediation is given and exemplifies, through a case study, how risk factors can play a major role in the process by which a technology is chosen for cleanup. The case, concerning a Mississippi wood-preserving site originally slated for cleanup by incineration, illustrates how a risk-based approach aided in the determination that using bioremediation, instead of the conventional technology of incineration, was an appropriate action at this site.

SELECT FEDERAL REGULATIONS AND IMPLEMENTATION OF BIOREMEDIATION AS A RISK REDUCTION TECHNOLOGY

The Resource Conservation and Recovery Act and the Land Disposal Restrictions

The Resource Conservation and Recovery Act (RCRA) of 1976 provides a basic framework for regulating waste generators, waste transporters, and waste management facilities under Subtitle C (USEPA, 1990). 1984 amendments to RCRA, termed the Hazardous and Solid Waste Amendments (HSWA), then restricted the land disposal of hazardous wastes unless these wastes had been treated according to treatment standards. HSWA directed the U.S. Environmental Protection Agency (EPA) to develop these treatment standards for all hazardous wastes. Wastes meeting these standards, promulgated through a series of rules known as the land disposal restrictions (LDRs), would not be subject to this restriction of land disposal of wastes (USEPA, 1990c).

EPA promulgated regulations specifying either levels or methods of treatment that substantially reduced the toxicity or mobility of RCRA hazardous wastes and their hazardous constituents (USEPA, 1990c). According to the EPA, a recurring debate during the Agency's development of the land disposal restrictions was whether treatment standards should be technology-based or risk-based (*Biotreatment News,* 1993). EPA had begun developing risk-based treatment standards before HSWA's passage and continued to promote a risk-based approach after HSWA. EPA states that, through hearings and written comments on the proposed LDRs, Congress made clear its desire for technology-based treatment standards. Therefore, EPA reversed its initial standard development decision and issued technology-based treatment standards instead (USEPA, 1990a).

These standards, which required a method of treatment or a specific concentration level for the target contaminant in the treated waste, were based on the concept of "best, demonstrated technology," or BDAT, which is defined as the following:

- A treatment technology is DEMONSTRATED for a given waste if it is currently used to treat the waste; bench- and pilot-scale technologies are not demonstrated.
- A technology is AVAILABLE if it can purchased or leased. In addition, the technology must substantially diminish the toxicity of the waste or substantially reduce the likelihood of migration of hazardous constituents from the waste.
- After determining the technologies that are demonstrated and available, a statistical analysis is used that incorporates a variability factor to account for normal process variations in well-designed and well-operated treatment systems to determine the BEST technology (Jonesi, 1989).

At the time of the LDR promulgation (1986–1990), some technologies, including bioremediation, had not been commercially implemented at full scale for a significant period of time. The EPA felt that insufficient data were available to consider using these types of technologies as a basis for treatment standards. The BDAT levels established were usually based on data from treatment of pure industrial process wastes, which are typically less difficult to treat than those that are highly variable (USEPA, 1990b). Therefore, while bioremediation offered significant risk reduction potential for many hazardous wastes, it could not be utilized as a treatment, either in instances where other technology had been explicitly dictated or in instances where the standard was a concentration level based on that attainable by incineration. Additionally, because states can choose to implement standards more stringent than those mandated at the federal level, concentrations as restrictive as nondetectable (ND) or level of detection (LOD) can be required for cleanup under the RCRA program. Such non-risk-based standards also could preclude use of bioremediation at a given site.

RCRA'S HAZARDOUS WASTE IDENTIFICATION RULE

The EPA has indicated that its ultimate policy preference is to establish risk-based levels that represent minimal threat levels and, in doing so, cap the extent of hazardous waste treatment. EPA claims that, due to the generally low concentrations of hazardous constituents in soil, incineration is not uniformly appropriate for hazardous soil (USEPA, 1993a). The Agency is working on the Hazardous Waste Identification Rule (HWIR), which will define hazardous constituent concentration levels below which a waste is no longer considered hazardous. A "bright line" will be used to distinguish highly contaminated media (i.e., hot spots) from other less contaminated media. Media above the bright line would be subject to Subtitle C requirements. Media below the bright line would be subject to more flexible site-specific management requirements, as specified by approved state or federal cleanup programs (*Biotreatment News,* 1994a). Such regulatory action could positively affect commercial implementation of bioremediation at an RCRA site. This proposed rule was scheduled for release for public comment before the end of 1995.

RCRA UNDERGROUND STORAGE TANK PROGRAM

In 1984, Congress enacted Subtitle I of RCRA as a means of controlling and preventing leaks from underground storage tanks (USTs) (USEPA, 1990c). EPA and states under cooperative agreements with EPA were given the authority to clean up releases from UST systems or require their owners and operators to do so. Later, as part of the amendments to the Comprehensive Environmental Response, Compensation, and Liability Act (CERCLA) in 1986, Congress added a trust fund to Subtitle I to aid in cleanup of select UST sites. The Leaking Underground Storage Tank Fund is financed through a tax on gasoline, diesel, and aviation fuels and is used under certain conditions, such as when cleanup costs exceed coverage requirements of the financially responsible party, when the owner or operator refuses to comply with a corrective action order, when a solvent owner or operator cannot be found, or when an emergency exists (USEPA, 1990c).

Although bioremediation is a proven, full-scale, cost-effective technology, particularly for petroleum-related contaminants, the most recent EPA UST data collected (1990 and 1991) indicated bioremediation was being used at less than 10% of the UST sites in the United States (USEPA, 1992a). Regulatory factors have contributed to this less than maximum implementation of the technology in the UST program. For example, a few years ago, many states based their cleanup goals specifically for benzene, toluene, ethylbenzene, and xylene (BTEX) constituents and total petroleum hydrocarbon (TPH) compounds on the Leaching Potential Analysis Method, which led to typical parameters of 1 ppm BTEX or 100 ppm TPH for soils at petroleum-contaminated sites. Because few scientific aspects of subsurface contamination are known with absolute certainty, agencies may rely on policy to support decisions about soil and groundwater cleanup levels (Walsh, 1990). In the early days of the UST program, states set these UST site standards without a scientific approach or guidelines, and these numbers were adopted across the country in an effort to quickly establish cleanup goals for each state (Liptak and Lombardo, 1994). Although from a risk reduction perspective, bioremediation may be an acceptable means of cleanup at a given UST site, it could be precluded from use because bioremediation cannot universally achieve these levels of 1 ppm BTEX or 100 ppm TPH.

Comprehensive Environmental Response, Compensation, and Liability Act (Superfund)

In 1980, Congress passed the Comprehensive Environmental Response, Compensation, and Liability Act (CERCLA) in response to public concern over the large number of abandoned, leaking, hazardous waste sites threatening human health and the environmental (USEPA, 1992b). A major difference in CERCLA and RCRA is that CERCLA was designed to remedy mistakes made in hazardous waste management in the past, while RCRA is focused on avoiding such mistakes through proper management in the present and future (USEPA, 1990c).

CERCLA created a trust fund to pay for cleanup and imposed cleanup liability on those responsible. This "superfund" was comprised primarily of tax assessment on oil and designated chemicals (USEPA, 1992b). CERCLA response is triggered by a release or substantial threat of release of a dangerous substance. CERCLA authorizes removal action and remedial actions. Removals are short-term cleanup actions conducted in response to an emergency situation, such as cleaning up a spill. Remedial actions are intended to provide permanent solutions to hazardous substance threats. EPA takes remedial action at sites on the National Priorities List (NPL), which were placed on this list after being evaluated through the hazard ranking system (HRS), a model that determines the relative risks to public health and the environment posed by hazardous substances (USEPA, 1990c). See Section 5 for a case study involving a Superfund remedial action and the use of risk assessment in remedy selection.

The Superfund process consists of several steps briefly outlined as follows. After site discovery, a site assessment, beginning with a preliminary assessment (PA) is conducted to determine if the site poses a potential hazard. If the PA shows a contamination problem, EPA will do a site inspection and use the HRS to evaluate the relative risks to human health and the environment. For remedial responses, that is, longer term actions to eliminate or substantially reduce threat from contamination, there are two main phases: remedial investigation and feasibility study (RI/FS) and remedial design and remedial action (RD/RA). During the RI, EPA, the state, or the potentially responsible party (PRP) collects and analyzes information to determine the nature and extent of the contamination. This information is used in a site-specific risk assessment, prepared following procedures outlined in EPA's manual entitled *Risk Assessment Guidance for Superfund.* The *Human Health Evaluation Manual* provides guidance for developing health risk and information at Superfund sites while the *Environmental Evaluation Manual* provides guidance for environmental assessment. These manuals were developed to be used in the RI/FS process at Superfund sites, although the analytical framework and specific methods described in the manuals may also be applicable to other assessments of hazardous materials and wastes. The human health evaluation manual is addressed primarily to the individuals actually conducting health risk assessments for sites, who frequently are contractors to EPA, other federal agencies, and potentially responsible parties (PRPs). The *Environmental Evaluation Manual* is addressed primarily to remedial project managers and on-scene coordinators who are responsible for ensuring a thorough evaluation of potential environmental effects at sites (USEPA, 1989b).

If a site poses a risk, remediation will probably be needed and an FS is prepared. During the FS, specific alternative remedies are evaluated by EPA. A preferred remedy is identified from this list of remedies in what is termed a record of decision (ROD). In the remedial design and action stage, the recommended cleanup is undertaken (USEPA, 1992b).

CERCLA APPLICABLE OR RELEVANT AND APPROPRIATE REQUIREMENTS

Under the Superfund Amendments and Reauthorization Act (SARA) of 1986, Congress stipulated that response actions conducted under CERCLA must at least attain all "applicable or relevant and appropriate requirements" (ARARs) of other federal environmental laws and more stringent state environmental laws. ARARs are identified on a site-by-site basis for all on-site response actions where CERCLA authority is the basis for cleanup (USEPA, 1992). However, the use of ARARs as the dominant cleanup standard, particularly the reliance on drinking water standards, makes it difficult to implement a technology such as bioremediation. ARARs can raise the threshold of acceptance so high that bioremediation is excluded in favor of more familiar technologies (Brown, 1995).

CERCLA REMEDY SELECTION

Proponents of bioremediation have often criticized the lack of flexibility in the Superfund remedy selection process. As in the RCRA program, there is a predisposition to familiar technologies, with pump and treat and incineration often the remedies selected. Often, when a ROD is finalized, it specifies performance criteria, such as attaining acceptable levels for drinking water standards. If these standards are not met, the ROD frequently specifies that the site would have to be addressed with another technology, such as excavation or incineration. Additionally, the length of time that elapses between the selection and implementation of the remedy is often long enough that significant innovation in a technology can occur, yet the current system does not allow for a systematic reopening of the ROD so that new or improved technologies can be evaluated and implemented (Brown, 1995).

Additionally, the applicable regulatory agency focus often seems to place as much emphasis on the volume of the contaminant as on its mobility. One of the benefits of bioremediation is that it removes the more mobile constituents of a waste, first reducing the mobility of and, consequently, the potential for exposure to the contaminant. Therefore, as a risk reduction technology, bioremediation would be an appropriate means of cleanup to employ at many Superfund sites.

TSCA

INTERGENERIC MICROORGANISMS

Section 5 of the Toxic Substances Control Act (TSCA) of 1976 authorizes EPA to review new chemicals not under the scope of other legislative jurisdiction (e.g., the Food, Drug, and Cosmetics Act) for potential human health and environmental effects prior to manufacture or import. It further dictates to limit any unreasonable risk that the new chemical may present by taking action (USEPA, 1987b). In the summer of 1994, the EPA published a proposed rule, "Microbial Products of Biotechnology; Proposed Regulations under the TSCA." This rule proposes, among other things, that persons intending to manufacture, import, or process intergeneric microorganisms for commercial or field experimentation purposes in the United States must notify the EPA, providing certain information for review, before such activity (*Biotreatment News,* 1994b). In doing so, the EPA has made the determination of intergeneric microorganisms as new chemical substances. Thus, any genetically engineered microorganism used for bioremedial field research or commercial purposes will be subject to reporting requirements not imposed on those utilizing naturally occurring microorganisms.

PCBs

Under TSCA Section 6, Congress singled out PCBs for immediate regulation and phased withdrawal from the market (USEPA, 1987a). TSCA mandated that approved disposal methods be used and that Agency approval be sought by obtaining a permit. In 1986, the Agency issued guidelines for persons applying to the EPA for approval of PCB disposal by methods alternative to incineration because under the PCB regulations, incineration is the standard for PCB destruction. EPA approval using technologies other than incineration or alternative to incineration can be sought, but equivalency to incineration must be demonstrated (Blake, 1994). Therefore, as with RCRA and CERCLA programs, a familiar, traditional method of treatment is given preference in remedy selection although from a cost-effective, risk-based perspective, other technologies could be prime candidates for use.

In order to issue a permit for bioremediation as an alternative disposal method, the EPA must make the finding that the process not only destroys PCBs, but produces no toxic by-products or toxic emissions, and that any microorganisms used as inoculum pose no unreasonable risk to human health or the environment. At this time, incineration is the only PCB disposal process approved to remove PCBs from soils and sediments (Blake, 1994).

FINANCIAL CRISIS RESOLUTION: SITE-SPECIFIC RISK-BASED APPROACH

Financial Crisis

The United States is spending billions of dollars annually on environmental cleanup of hazardous and industrial wastes. Estimates of the cost for cleanup now run as high as being in excess of $1 trillion (Zaborsky, 1995). In the UST programs, many states have been faced with a growing financial crisis as the cost of cleanup increases faster than the growth in petroleum restoration funds (see Section 2.1 for more on Leaking Underground Storage Tank Trust Fund). The EPA passed regulations in 1988 that required petroleum UST owners, predominantly commercial gas stations and operators of car and truck fleets, to upgrade USTs by December 1998. In doing so, an estimated 460,000 leaking tanks will be uncovered and may require remediation. With an estimated 75% of the sites contaminated, many states are recognizing the need to move beyond traditional approaches (Environmental Information Ltd., 1994).

Environmentally Acceptable Endpoints

One concept that would save billions in cleanup costs for private industry and federal and state governments, if utilized, is that of environmentally acceptable endpoints. An "environmentally acceptable endpoint" (EAE) is defined as the concentration of chemical substances in soil or other medium that does not adversely affect human health or the environment (Zaborsky, 1995). One of the most important site-specific factors for risk evaluation of contaminated soil is the availability of the chemicals held within the soil. It has been reported that substantial reduction in the toxicity of soil contaminants has been achieved with varying degrees of chemical removal. These results support the concept that the risk associated with a contaminated soil is independent of the total concentration of specific chemicals and depends on the fraction of the total concentration that is available to either human or ecological receptors (Gas Research Institute, 1995). Therefore, a scientifically based approach to determine "How clean is clean?" must not rely solely on the analytical detection limit of each contaminant; rather, it should consist of a combination of toxicological and biological tests that can directly assess the concentration of chemicals in soil or water that will not adversely affect human health or the environment (Gas Research Institute, 1995).

Over the past decade, numerous laboratory and field studies and actual cleanup operations have been conducted on hydrocarbons in soil. The results of these efforts indicate that (1) hydrocarbons are biodegraded by indigenous soil microorganisms to a plateau concentration, i.e., a concentration that no longer decreases or decreases very slowly with continued treatment; (2) reductions below the plateau concentration are limited by the availability of the hydrocarbons to the microorganisms; (3) residual hydrocarbons that remain after biological treatment, regardless of the extent of treatment that has occurred exhibit significantly reduced leachability to the aqueous phase and significantly reduce toxicity; and (4) residual hydrocarbons that are present in an aged and/or weathered soil, depending upon the environmental condition to which they were exposed, also exhibit a reduction in leachability and toxicity compared to virgin hydrocarbons (Gas Research Institute, 1995).

It is believed that the hydrocarbons that remain in the treated soil are no longer available to the environment or ecological and human receptors and represent an EAE. Therefore, bioremediation that can reduce hydrocarbon concentrations to a level where they no longer pose an unacceptable risk to the environment, i.e., they no longer leach to the groundwater or harm ecological receptors or human health, could be employed as a risk reduction technology in many instances (Gas Research Institute, 1995).

Site-Specific Risk-Based Approach

As mentioned previously, many states adopted the UST site standard of 100 ppm for TPH and 1 ppm for BTEX in previous years. However, states are beginning to realize that the concept of one size fits all (i.e., one standard for all sites) is not a cost-effective means of site cleanup. The growing environmental financial crisis has led states to begin to consider alternative measures for cleanup of sites within their respective state boundaries. In fact, earlier this year, the EPA issued a directive, "Use of Risk-Based Decision-Making in UST Corrective Action Programs" (USEPA, 1995). In a 1995 survey of state UST programs, many states indicated that a risk-based approach was now being considered, and some specifically mentioned the possibility of following the state of Texas's lead in its risk-based approach that was adopted into regulation in late 1993 (Devine and Graham, 1995). Additionally, according to a recent survey of state cleanup standards, 33 states may use a risk/health parameter and/or a site-specific approach in determining cleanup standards for soil and/or groundwater for certain contaminants in certain instances. Moreover, 31 states have requested training on the American Society for Testing and Materials (ASTM) *Emergency Standard Guide for Risk-Based Corrective Action Applied at Petroleum Release Sites* (ASTM ES 38–94), more commonly referred to as RBCA (pronounced "rebecca") (Devine and Graham, 1995). This guidance, compiled by an ASTM subcommittee, standardizes existing Superfund risk assessment methods and promotes the concentration of resources for sites that pose the greatest threat to human health and the environment. The RBCA approach recognizes the diversity of parameters affecting risk, such as contaminant complexity and physical and chemical characteristics, and uses a tiered approach that tailors assessment and remediation activities to site-specific conditions and risks (ASTM, 1994). (See Section 4 for further details on the RBCA strategy.)

Based on a survey of states' UST programs, only the state of Wisconsin has actually specified bioremediation in a regulation that mandates that intrinsic bioremediation be considered as a cleanup alternative at UST sites (Devine and Graham, 1995). Other states, such as California, have exhibited a significant interest in RBCA, having studied the effect of RBCA on its UST program (McNeely, 1995). Also in California, a "plume-a-thon" is being conducted that began in the fall of 1994 and is expected to continue until the summer of 1995. Data is being collected on 1500 UST sites to ascertain the extent of contamination and migration. The state also has drafted a policy on the concept of nonattainment zones whereby a site with groundwater that cannot be feasibly treated is monitored or closed. This policy, which considers both organoleptic criteria and likely future use and potential exposure to the water, would raise the state cleanup standards. For example, a current benzene standard of "nondetect," or 1 ppb, would increase 10 to 100 times, to 10 to 100 ppb (Dooher, 1995). Such policy would allow for use of certain bioremedial containment techniques, such as a "microbial fence" (Swett, 1995).

A few other states are considering or implementing programs that are favorable for utilization of innovative technologies, including bioremediation. Based on a 1995 UST program survey, the states viewed as the most conducive for full-scale bioremediation, particularly on petroleum hydrocarbon-contaminated soil sites regulated under a state UST program, include California, New York, New Jersey, Michigan, New Hampshire, Nevada, Texas, Minnesota, and Wisconsin (Devine and Graham, 1995). On a U.S. Environmental Protection Agency regional basis, therefore, Regions 2, 5, and 9 were considered as having the most favorable regulatory climate for innovative technologies, including bioremediation (see Figure 1).

RISK ASSESSMENT IN REMEDIATION

The ultimate purpose of hazardous waste site remediation is to render a site safe for human or ecological populations. However, historically, cleanup goals often have not explicitly addressed if the remedial action results in a safe site. Previous values have been based on limits of detection (LODs), natural background chemical levels, or the technical ability

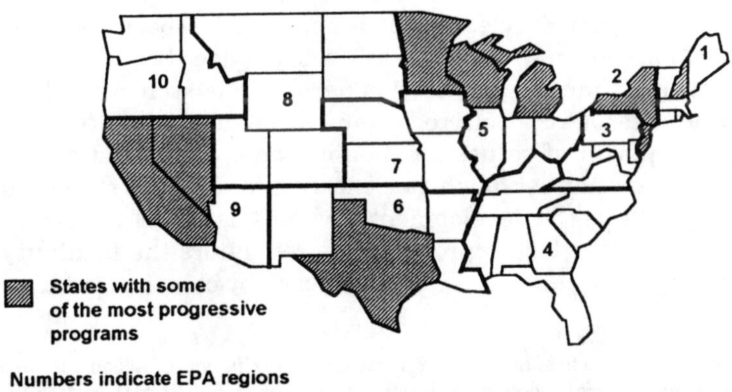

Figure 1. States (within EPA regions) with some of the most progressive cleanup programs (Devine and Graham, 1995).

to destroy the waste (best demonstrated available technology, or BDAT; see Section 2.1). The use of background as a criterion is appropriate in that it is unreasonable to expect remediation for levels below naturally occurring levels, even if these levels may pose a health risk.[1] However, selecting the appropriate value to represent background concentrations can be problematic in that there is a substantial amount of variability in the natural chemical composition of soils, and many chemicals have both natural and anthropogenic sources (LaGoy and Schulz, 1993). The use of BDAT or LOD as criteria, on the other hand, is inappropriate if the goal of remediation is to protect human health or the environment. In either case, cleanup goals may be either overprotective, leading to unnecessary remediation and costs, or underprotective, allowing health or environmental risks to exist. And, as illustrated in the example at the end of this chapter, BDAT-based criteria can be both under- and overprotective, preventing the use of a more cost-effective remedial technology (bioremediation) while, in some cases, focusing on a subset of chemicals of limited toxicity, making it impossible to determine if remedial action removed the health threat posed by other chemicals in the mixture.

Risk assessment is a tool that can be used to measure the risks to public health or the environment posed by a site and, therefore, can be used to determine whether exposure to chemicals at a site is "safe." Risk assessment-based remediation goals can be a useful tool in encouraging and justifying the use of innovative technologies such as bioremediation, since risk-based goals tend to be less stringent than non-risk-based remediation goals. The use of risk-based goals and the risk assessment process in the remediation of hazardous waste site work is gaining acceptance on the federal level and has been adopted by several states. In general, these states have developed similar approaches that incorporate a three-tiered strategy for site cleanup. This approach has been adopted recently by ASTM and is known as "RBCA" (rebecca) (see Section 3.3). The first two tiers are similar, with tier 1 generally consisting of fairly stringent numeric criteria for a selected group of chemicals, and tier 2 involving the development of values for chemicals not on the tier one list using the formulas that establish tier one values. In most states, the tier 3 generally allows the development of values based on a site-specific risk assessment. This last approach gives regulators the flexibility to consider factors that decrease the risks posed by chemicals at a particular

[1] In some cases, background levels of naturally occurring materials do pose health risks. For example, naturally occurring radon gas has been linked with excess lung cancer risks, and high natural selenium levels have been shown to cause adverse effects on livestock.

site, such as known future site use, chemical behavior at the site, and soil cover.

Although, in theory, bioremediation can totally destroy all measurable levels of many different chemicals at a site, in reality, degradation at most sites is limited to some extent by factors that prevent contact between microorganisms and chemicals. For such non-totally destructive technologies, risk-based cleanup levels are critical to justify leaving residual levels of chemicals on-site. Either of two types of risk-based values can be used, depending on needs for a particular site and regulatory requirements. First, generic (non-site-specific; e.g., the tier 1 and 2 values discussed above), risk-based values may be useful in situations where only low levels of chemicals remain. These values are fairly stringent, but for many biodegradable compounds, these values are still higher (less stringent) than non-risk-based cleanup levels. Second, site-specific, risk-based values (e.g., tier 3 values) that incorporate a consideration of site use may show that even higher chemical levels can be left in place without posing an unacceptable health risk. In some cases, deed restrictions may be needed to clearly limit future land use.

Risk assessment is an inherently uncertain science due to the nature of the subjects being evaluated. Humans and ecosystems exhibit a wide range of behavior patterns and, therefore, different individuals will have exposures and consequent risks that cover a large interval. The same chemical level at a site may pose no risk to one person yet pose a serious risk to another. Regulators, who are required to make decisions on appropriate cleanup levels, were initially understandably reluctant to embrace risk assessment in the face of this uncertainty. However, in the past 10 years, substantial advances have been made in the ability to characterize exposure and risk, and approaches have been developed to start to address some of the inherent uncertainty in the risk assessment process. Based on this fact and on an increasing awareness of the costs associated with non-risk-based values that may be over- or underprotective, regulators have begun to consider and use risk-based remediation goals to establish cleanup levels at hazardous waste sites.

"Generic" remediation goals, that is, cleanup levels that can apply to any site, are designed to be conservative (health protective) in order to ensure that public health and the environment are protected. These generic values are developed using reasonable maximum exposure (RME) estimates that are unlikely to underestimate true exposure levels for almost all individuals. The basic function of generic remediation goals is to provide a bounding estimate of risk (i.e., risks are unlikely to be higher than estimated but could be considerably lower). In general, these generic cleanup levels are developed based on the assumption that a hazardous waste site could be used for the construction of residential homes in the future. Another future use scenario that is also sometimes considered is continued use of an area for industrial purposes. In either case, cleanup levels are derived using an approach that is highly and possibly unrealistically conservative but that almost[2] ensures that any use of a site is safe.

Generic risk-based cleanup goals are an advance over non-risk-based remediation goals but are still not always appropriate for a particular site. Site-specific factors can greatly influence the potential for exposure, and the use of generic values means that these factors are not considered. Consequently, the site may be cleaned up to meet standards or guidance that are inappropriate for the particular site. The most familiar example is the use of criteria based on residential use of a property for a site that cannot be used for this purpose. Another factor that is more commonly overlooked is that chemicals in the environment generally exist in both bound and unbound forms, with the amount that is unbound (and available) dependent on environmental and chemical characteristics. Many non-totally destructive technologies remove all of the mobile fraction of a chemical and, therefore, significantly decrease the environmental mobility of target chemicals. In addition, natural attenuation, including intrinsic biodegradation, natural chemical transformation processes, and adsorption, may significantly influence the potential for a chemical to pose risks at a particular site.

More accurate and less conservative, but still health-protective, risk-based cleanup goals can be developed to allow consideration of site-specific factors such as future site use, chemical availability, and natural attenuation. Increased use of risk assessment and true, site-specific, risk-based cleanup levels should lead to more appropriate and efficacious remediation. And for non-fully destructive technologies, such as bioremediation, risk-based criteria may be required to allow the use of the technology.

CASE STUDY

An example of the use of risk assessment to assist in the use of bioremediation as a remedial option is provided by a project completed

[2] There is always a possibility that a greater exposure could occur under rare circumstances. For example, EPA assumes that people ingest 100 mg/day of soil, but women in some cultures consume soils (primarily clay) on a regular basis at much higher levels. Standard risk estimates would not be adequate for this subgroup.

by OHM Remediation Services Corp. (OHM) at the Southeastern Wood Preserving Site in Canton, Mississippi. This site initially was the subject of an emergency removal action by the USEPA (Region VI), with soils stockpiled on site for treatment. OHM was hired to treat the soil using bioremediation but had difficulty achieving the incinerator-based remedial target levels. Risk assessment was used as one factor in establishing achievable target goals for the site.

Introduction

The Southeastern Wood Preserving Company located in Canton, Mississippi, operated as a wood treatment facility from 1928 to 1979 (Jerger et al., 1994). The site is approximately 10 acres in size and is located in a mixed industrial and residential area. Creosote was the primary preservative used in the wood-treating operations, with small amounts of pentachlorophenol also used.

EPA performed an emergency response action at the site in 1986 to stabilize three on-site unlined surface impoundments and prevent release of creosote-containing material into Bachelor Creek, which runs along the northern side of the site. As part of this action, over 10,500 cubic yards of contaminated material were excavated. EPA was also concerned about the potential for the polycyclic aromatic hydrocarbons (PAHs) in the creosote to leach to shallow groundwater in the vicinity of the site. Wastes with PAH levels exceeding 1000 mg/kg were to be treated on-site using the alternative treatment of bioremediation. Soils containing between 100 and 1000 mg/kg of PAHs were to be buried on-site.

Cleanup goals for the creosote wastes at the site were based on the RCRA LDR standard for K001[3] waste specified in sections 268.41 and 268.43 of 40 CFR *(Code of Federal Regulations).* The cleanup goals for the K001 wastes (bottom sediment sludges from the treatment of wastewaters from wood-preserving processes that use creosote and/or pentachlorophenol) are 1.5 mg/kg for naphthalene, phenanthrene, and pyrene; 7.4 mg/kg for pentachlorophenol; 28 mg/kg for toluene; and 33 mg/kg for xylene. These cleanup goals were developed based on the ability of the best demonstrated available technology (BDAT), which is incineration to destroy these compounds (see Section 2.1 for more on BDAT). Under the EPA's LDR guidance, if cleanup goals can be achieved by an alternative treatment technology, that technology can be used for remediation. EPA's Gulf Breeze Research Laboratory (Florida)

[3] Under RCRA, certain hazardous wastes are classified by number and letter categories.

was conducting a field treatability study to confirm that bioremediation was an appropriate option for creosote that had been stabilized with kiln dust, but based on the data available at that time, EPA indicated that LDR standards could be achieved with bioremediation. Consequently, bioremediation was selected as the treatment technology for the site.

Now, more recent information from both the OHM lab and from other researchers suggested that meeting the RCRA LDR requirements for K001 wastes using bioremediation was, if not impossible, at least highly impractical for soils containing high concentrations of PAHs. Based on this information, OHM determined that it was necessary either to evaluate alternative treatment goals or to consider other treatment technologies, namely incineration. Because the state of Mississippi indicated that they were opposed to incineration, OHM focused on developing alternative cleanup goals.

Alternative Cleanup Goals

As mentioned previously, RCRA LDRs are established based on BDAT and do not consider risk issues. As an example, the K001 standards do not address (i.e., require analysis for) the component of the waste, namely the carcinogenic PAHs, that is likely to pose the greatest health risk to individuals exposed to the wastes. In this case, a cleanup action that achieved LDRs might not meet the CERCLA criteria of ensuring that remediation protects human health and the environment. Several approaches are available for more accurately ensuring that remediation at a Superfund site meets the EPA's overall goal of protecting public health, welfare, and the environment. These approaches and allowable cleanup goals are presented in this section.

SUPERFUND LDR GUIDE #6A

For Superfund sites contaminated with RCRA wastes covered by LDRs, EPA has developed guidelines for obtaining what is termed a "treatability variance."[4] These guidelines are presented in *Superfund LDR Guide #6A, Obtaining a Soil and Debris Treatability Variance for Remedial Actions.* Under these guidelines, RCRA wastes containing polynuclear aromatics (PAHs) must be treated to within a range of 0.5–20 mg/kg if initial concentrations are less than 400 mg/kg, or concentrations must be reduced by 95–99.9% if initial concentrations are

[4] A treatability variance is available if a waste cannot be treated in compliance with a concentration level or by using a specific treatment method.

above 400 mg/kg. At the Southeastern Wood Preserving Site, initial PAH concentrations were well above 400 mg/kg, and consequently, the percent reduction guidelines were applicable. The guidelines specify biological treatment, low-temperature stripping, soil washing, and thermal destruction as treatment options but note that other technologies may be used if they can achieve the necessary percent reduction range. The basis for the guidance appears to be BDAT, but the #6A guidance has an advantage over the LDR regulations for K001 in that Guide #6A considers all the PAHs and not simply a subset of three chemicals (naphthalene, phenanthrene, and pyrene).

As noted previously, initial information from EPA's laboratories where PAHs in sandy soils were being studied suggested that 95% reduction was achievable using bioremediation. However, based on the results of laboratory, pilot, and field studies, achieving between 95% and 99.9% reduction of PAHs at the Southeastern Wood Preserving Site by biodegradation was difficult or impossible. High levels of fine particulate matter and wood chips were present in the mixture and appeared to decrease the ability of the soil microorganisms to contact and degrade the PAHs. Consequently, although biodegradation is capable, under certain situations, of achieving PAH degradation in the desired range, it was difficult to achieve 95% reduction for all materials at the Southeastern Wood Preserving Site because of the very high initial concentrations and problems of matrix interference.

OHM established that, if Guide #6A was to be used as a criterion for the site, cleanup levels on the order of 250–500 mg/kg total PAHs would be required based on the PAH concentrations in the initial material ranging from approximately 5000–10,000 mg/kg. One problem with these levels is that they do not specify the PAHs to be remediated. Consequently, it is not possible to ensure that meeting the #6A guidance is protective of human health and the environment.

RISK-BASED CLEANUP GOALS (PRG APPROACH)

The 1991 RAGS' Volume 1, *Human Health Evaluation Manual Part B, Development of Risk-Based Preliminary Remediation Goals,* has established procedures for determining preliminary remediation goals (PRGs) for chemicals at hazardous waste sites. These PRGs are not site-specific and use maximum exposure scenarios and conservative assumptions so that it is unlikely that contact with the chemical-affected media of concern at the PRG levels would pose a significant health risk under almost any condition. The PRGs are designed primarily to be used as a screening tool; if all levels of a chemical are below the PRGs, no further work is necessary. However, if the PRGs are exceeded, further analysis of the cleanup criteria, including evaluation of site-specific factors, is warranted and remediation *may* be needed.

This EPA *Human Health Evaluation Manual Part B* presents both detailed and summary (called reduced) equations for calculating PRGs for chemicals. These equations are presented for both residential use scenarios (which assume a site will be used for residential purposes) and nonresidential use scenarios (which generally assume that a site will continue to be used for industrial purposes). For nonresidential (industrial/commercial) land use, the reduced equation for nonvolatile compounds like the carcinogenic PAHs is

$$\mathrm{PRG} = 5.8/\mathrm{SF_o}$$

where SF_o is the oral cancer slope factor (also called the cancer potency factor), a value that expresses the relationship between the probability of developing cancer and the dose.

Based on this equation and the cancer slope factor for benzo[*a*]pyrene (B[a]P) of 7.3 (mg/kg/day)$^{-1}$, a PRG for B[a]P, or for PAHs that are equivalent in their toxic potency to B[a]P, in soils in an industrial area is 1 mg/kg. This value equates to a one in one million (10^{-6}) cancer risk, assuming long-term exposure. It should be noted that EPA regulations generally dictate a risk range of 10^{-4}–10^{-6}.

For noncarcinogens, the PRG equation for commercial/industrial exposure is

$$\mathrm{PRG} = 2 \times 10^{6}\ (\mathrm{RfD_o})$$

where RfD_o is the oral reference dose (formerly referred to as the allowable daily intake, or ADI, a value at which exposure is considered unlikely to pose a health concern).

Using this equation, PRGs can be determined for two of the three PAHs considered in the K001 standards. No toxicity values (RfDs) are available for phenanthrene, so no PRG can be established. Based on the PRG approach, concentrations of naphthalene below 2000 mg/kg and pyrene below 10,000 mg/kg are considered unlikely to pose health risks by direct contact exposure.

As noted in the PRG guidance, "The recommended approach for developing remediation goals is to identify PRGs at scoping [project inception], modify them as needed at the end of the remedial investigation [RI] or during the feasibility study [FS] based on site-specific information from the baseline risk assessment, and ultimately select remediation levels in the Record of Decision (ROD)." At the Southeastern Wood Preserving Site, a consideration of several factors suggested that final remediation goals that were less stringent would still suitably protect human health. Key factors that would influence risk at the site include

- Biodegradation and weathering will remove the more mobile PAHs and other chemicals that could act as carriers for the less mobile and more toxic PAHs. These more toxic PAHs, with high octanol water partition coefficients, indicate a high binding potential with organic matter in soils and, therefore, are quite immobile in the environment. Direct contact with contaminated material is the only route of exposure that is likely to pose a health risk at the site. OHM did some limited work looking at leachable compounds before and after treatment that supported this assumption.
- As a practical matter, a combination of deed restrictions and cover soils will limit potential exposure to levels well below levels assumed in calculating the PRG.

RISK-BASED CLEANUP GOALS (CDC/ATSDR APPROACH)

At a Superfund site in Louisiana, the Bayou Bonfouca site in Slidell, the Agency for Toxic Substances and Disease Registry (ATSDR) evaluated the potential effects of PAH contamination on local residents. Based on a comparison with published cleanup criteria for a more toxic compound (2,3,7,8-tetrachlorodibenzo-*p*-dioxin, or TCDD), ATSDR (1986) estimated that levels less than 100 mg/kg of B(a)P would not pose a health hazard.

The approach used by ATSDR was to base the PAH cleanup criteria on the relative difference in carcinogenic potency between B[a]P and TCDD and on allowable levels of TCDD that had been established and used for Superfund site soils across the country. The relationship, which is reciprocal because higher toxic potency requires lower target levels, can be expressed as

$$\frac{SF_o \text{ for TCDD}}{SF_o \text{ for B[a]P}} = \frac{\text{Criteria for B[a]P}}{\text{Criteria for TCDD}}$$

A TCDD cleanup criteria of 1 μg/kg in soils was developed by the Center for Disease Control (CDC) during work at Times Beach, Missouri, and has been used widely as a cleanup goal for residential soils. Multiplying the ratio of the cancer slope factor for TCDD [1.5×10^5 $(mg/kg/day)^{-1}$] to the former cancer slope factor for B[a]P of 11.5 $(mg/kg/day)^{-1}$ by the allowable TCDD soil criteria of 1 μg/kg gives an allowable B[a]P concentration of approximately 10 mg/kg. (ATSDR arrived at a value of 100 mg/kg due to slight differences in approach and rounding.) However, as noted above, the B[a]P cancer slope factor has been lowered to 7.3 $(mg/kg/day)^{-1}$, or approximately in half. In addition, EPA commonly regulates in a risk range of 10^{-4}–10^{-6} with 10^{-6} used as the target, but approximately 10-fold higher rates often are allowed for industrial exposure then are allowed for residential exposure. Consequently, a higher cleanup criteria, in the range of 150–200 mg/kg for B[a]P or PAHs of equivalent potency would be consistent with the ATSDR cleanup criteria.

CRITERIA FOR PAHS AND OTHER CHEMICALS AT OTHER SUPERFUND SITES

A review of EPA's record of decisions (RODs) for Superfund sites contaminated with creosote indicates large differences in the cleanup goals (Table 1), with no clear rationale for the differences. Cleanup goals have been established for total polycyclic aromatic hydrocarbons (PAHs), carcinogenic PAHs, or individual PAH compounds, with allowable levels ranging from <1 to 100 mg/kg of the carcinogenic PAHs and from 100–2450 mg/kg total PAHs. The variability in these cleanup criteria stems in part from the complexity of the PAHs as a class and from several unresolved issues regarding the risks posed by these compounds (LaGoy and Quirk, 1994; Table 2).

At the Southeastern Wood Preserving Site, cleanup was completed to a level of 1000 mg/kg total PAHs. Based on our evaluation of PAH data from the site, the material that remains in place (<1000 mg/kg PAHs) probably contains slightly over 100 mg/kg of carcinogenic PAHs. These levels do not necessarily represent allowable PAH levels in soils, but they should be considered along with other site-specific background samples in establishing cleanup goals for the removed material.

EPA has developed soil remediation levels at other sites based on a consideration of cover material. For example, at the Whitewood Creek Superfund Site in South Dakota, EPA used a soil level of 100 mg/kg of arsenic as a soil remediation objective, with soil exceeding 100 mg/kg of arsenic in residential areas either being removed or covered with at least 12 inches of clean fill. Based on their environmental behavior, the potential for exposure to arsenic and carcinogenic PAHs generally would be similar and the cancer slope factors for the materials are similar. Considering that at the Southeastern Wood Preserving Site, soils will be treated to below 200 mg/kg B(a)P equivalent PAH concentrations, covered with 2 feet of clean material, and that the site is not in a residential area, it seems likely that the potential for health risks after completion of remediation is probably lower for the Southeastern Wood Preserving Site than at the Whitewood Creek Site.

Table 1. Soil cleanup criteria for selected EPA Superfund sites.

Site	Location	Cleanup Criteria	Comments
Bayou Bonfouca	Slidell, LA	100 mg/kg carcinogenic PAHs	Level based on an ATSDR memo comparing carc. PAHs with dioxins and the dioxin cleanup criteria of 1 ppb in soils
Burlington and Northern	Somers, MT	3.6 mg/kg carc. PAH 1875 mg/kg total PAH	Based on potential residential use and a 10^{-5} risk level
North and Cavalcade	Houston, TX	1 mg/kg carc. PAH	Based on potential residential use and a 10^{-5} risk level
South Cavalcade	Houston, TX	700 mg/kg carc. PAH; no leaching potential in a TCLP[a] analysis	Based on commercial use and a 10^{-5} risk level; TCLP designed to protect groundwater and may be more stringent
Texarkana Wood Preserving	Texarkana, TX	3 mg/kg B(a)P equiv. 2450 mg/kg total PAH	Based on industrial exposure scenario and a 10^{-5} risk level

[a]Toxicity Characteristic Leaching Procedure; 40 CFR 261.24.
Source: EPA Records of Decisions (RODs) for the sites; USEPA, 1987c; USEPA, 1988a; USEPA, 1988b; USEPA, 1990; and USEPA, 1989a.

Table 2. Key factors affecting uncertainty in risk assessments for the polycyclic aromatic hydrocarbons.

Toxicity Factors

- The lack of a dose-response estimate for site-of-contact tumors caused by dermal exposure to the individual PAHs
- Questions regarding the accuracy of the available cancer slope factors for oral exposures
- The lack of an adequate approach for addressing the potency of mixtures of PAHs

Exposure Factors

- Questions regarding the effect of the environmental matrix on the availability of the chemicals to a biological receptor
- The lack of information on levels of those PAHs that are not detected using standard analytical procedures
- The lack of a comprehensive understanding of naturally occurring and anthropogenic background PAH levels

Discussion

Biodegradation was determined to be unlikely to be able to achieve 95% or greater reduction in PAH levels at the site, and the more toxic PAHs (the high molecular weight compounds) would be degraded to a lesser degree than other PAHs. However, biodegradation will remove the more mobile of the PAHs and other associated hydrocarbons in the creosote, and consequently, residual material is likely to be quite immobile in soils. In addition, the matrix effects that prevent reduction of soil PAH concentrations by 95% are also likely to limit mobility and possible bioavailability of the PAHs.

Several site-specific factors were considered likely to further limit the potential for biotreated PAH-containing material to pose health risks at the site. As noted above, the residual PAHs present at the Southeastern Wood Preserving Site are quite tightly bound by the soil and wood chip matrix. Since these materials are not readily available for degradation by soil bacteria, it also seems likely that they are relatively unavailable to pose risks to other biological organisms. Consequently, direct contact with the residual biotreated PAH-containing material is less likely to pose risks than equal concentrations of carcinogenic PAHs in untreated material.

The residual treated material will be covered with 2 feet of clean soil, graded, and seeded. In addition, the site is in a mostly industrial area. Consequently, the potential for contact with PAH-containing residual materials is limited.

PAH LEVELS AT THE SITE

Although achieving 95–99.9% degradation seemed unlikely, PAH levels are substantially reduced by biodegradation in an OHM bioslurry reactor. Table 3 summarizes information on the starting and ending PAH levels. As can be seen from this table, PAH levels were degraded substantially by the slurry bioreactors, but the carcinogenic PAHs were less affected than the noncarcinogenic compounds. The noncarcinogenic PAHs tend to be lower molecular weight compounds, while the carcinogenic PAHs tend to be higher molecular weight chemicals. In addition to the PAHs, other compounds were also reduced, as shown by the substantial reduction in limits of detection. The biodegradation is likely to have a substantial influence on actual site risks:

- The more soluble compounds, such as the lower molecular weight aliphatic and aromatic petroleum hydrocarbons, can act as carriers for the carcinogenic PAHs, which have very low water solubility. Consequently, destruction of these soluble compounds should greatly reduce the mobility of the carcinogenic PAHs.
- Low molecular weight PAHs that could act as carriers for dermal or oral absorption will be reduced, possibly decreasing the potential bioavailability of the carcinogenic PAHs.

Table 3. Summary of PAH concentrations in a full-scale, slurry-phase bioreactor at Southeastern Wood Preserving Site, Canton, Mississippi (all values in mg/kg).

	Initial Conc.	Final Conc.	Percent Reduction
3-Ring PAHs[a]	4650	290	(94%)
4-Ring PAHs[a]	2420	110	(95%)
5- and 6-Ring PAHs[a]	730	350	(52%)
Total PAH Detected	7800	750	(90%)
CPAHs[b] (EPA carcinogenic)	1200	400	(65%)
B[a]P-equivalent	290	140	(50%)

[a]3-ring PAHs include acenaphthene, acenaphthylene, anthracene, fluorene, and phenanthrene; 4-ring compounds include benzo[*a*]anthracene, chrysene, fluoranthene, and pyrene; and 5- and 6-ring PAHs include benzo[*b*]fluoranthene, benzo[*k*]fluoranthene, benzo[*ghi*]perylene, benzo[*a*]pyrene, dibenzo[*ah*]anthracene, and indeno[2,3-cd]pyrene.

[b]The value for CPAHs is the sum of all carcinogenic PAHs, which include benzo[*a*]anthracene, chrysene, benzo[*b*]fluoranthene, benzo[*k*]fluoranthene, benzo[*a*]pyrene, dibenzo[*ah*]anthracene, and indeno[2,3-cd]pyrene. The value for B[*a*]P-equivalent adjusts for differences in potency between different compounds (Nisbet and LaGoy, 1992; USEPA, 1993b).

- Some reduction in carcinogenic PAH levels is being achieved; the residual material is apparently not readily available for biological action and may be less bioavailable as well.

CONCLUSIONS

A risk-based decision framework for site prioritization and remedy selection, which incorporates the concept of environmentally acceptable endpoints (EAE), should bode well for risk reduction technologies, such as bioremediation. It is anticipated that, as risk-based site management becomes accepted, more sites will employ bioremediation, leading to a more maximally utilized risk reduction technology.

On a case study basis, the final action levels for treatment of creosote wastes from the Southeastern Wood Preserving Site were established based on a consideration of both technical capability and risk. Target remediation goals included values for both total PAHs (950 mg/kg was the target level) and B[a]P-equivalents (B[a]P) or any combination of PAHs of equivalent toxic potency (180 mg/kg of B[a]P-equivalent PAHs was the target level). In general, final levels were well below these values. Unlike either the RCRA LDR values or the LDR Guide #6A values discussed previously, the target levels used for the Southeastern Wood Preserving Site allow a consideration of the effect of treatment on site risks. B[a]P-equivalent levels of PAHs at the site were reduced by bioremediation to slightly below 100 mg/kg. This level is in the upper end of the range allowed by EPA at various Superfund sites and for similar situations. Using standard EPA approaches to calculating allowable preliminary remediation goals, or PRGs, the level of 100 mg/kg is above the EPA target risk level of a 10^{-6} excess cancer risk but is at the upper end of the range of risks allowed by EPA in regulatory decisions (10^{-4}–10^{-6}). In addition, site-specific factors suggest that actual risks would be lower than predicted using the PRG approach, particularly if direct contact exposure were to be controlled.

There are currently several unresolved issues regarding the risks posed by the PAHs (LaGoy and Quirk, 1994; Table 2). These unresolved issues substantially increase the uncertainty in a risk assessment involving PAHs and do not allow the direction of the uncertainty (i.e., whether risks are being under- or overestimated) to be known. The lack of a cancer slope factor for dermal exposure, questions about the oral cancer slope factor, and testing for only 17 of the numerous PAHs are all likely to lead to underestimating risks. On the other hand, not accounting for differences in the relative potency of individual PAHs and ignoring matrix effects may lead to overestimating risks. Despite these uncertainties, the approach outlined here of using risk assessment to aid in developing appropriate target levels gives more favorable target levels (from the viewpoint of encouraging biologic treatment) and provides more information to a regulator tasked with making a decision on a particular site.

Considerable work remains to be done to decrease the uncertainties in risk assessment, and some uncertainties are inherent to the process and will persist. However, risk assessment is a valuable tool that can be used to encourage the use of innovative, but often non-totally destructive remediation technologies, by providing regulators with a method to judge the adequacy of a cleanup action. Totally destructive actions such as excavation and off-site incineration are certainly effective treatment technologies, but their costs are substantial. Alternative treatment methods, such as bioremediation, can offer a cost-effective (and politically advantageous) alternative to these more destructive technologies.

REFERENCES

ASTM. 1994. *Emergency Standard Guide for Risk-Based Corrective Action Applied at Petroleum Release Sites, ES 38-94.* American Society for Testing and Materials. Philadelphia, PA.

ATSDR. 1986. Memorandum from J.A. Lybarger, Acting Director, Office of Health Assessment of C.R. Hickem, Public Health Advisor, EPA Region VI Regarding Bayou Bonfouca, SI-86-035B. Agency for Toxic Substances Disease Registry. Dated Sept. 11, 1986.

Bakst, J. and K. Devine. 1994. "Bioremediation: Environmental Regulations and Resulting Market Opportunities." In P.E. Flathman, D.E. Jerger and J.H. Exner (Eds.), *Bioremediation Field Experience.* pp. 11–48. Lewis Publishers, Boca Raton, FL.

Biotreatment News. 1993. "EPA Proposes Long-Awaited Contaminated Soils Rule." *3*(11):7, 8 and 10.

Biotreatment News. 1994a. "EPA Rolls Contaminated Soil Treatment Standards into HWIR." *4*(7):4 and 19.

Biotreatment News. 1994b. Second Phase of LDRs Issued, Superfund and CWA DIe, TSCA Rule Proposed. *4*(12):10–12.

Biotreatment News. 1995. "Second Phase of LDRs Issued, Superfund and CWA Die, TSCA Rule Proposed. *4*(12):10, 11, 12.

Blake, J. 1993. "Approach to the Regulation of Bioremediation of Polychlorinated Biphenyls." In R.E. Hinchee, A. Leeson, L. Semprini and S.K. Ong. (eds.), *Bioremediation of Chlorinated and Polycyclic Aromatic Hydrocarbon Compounds.* pp. 432–435. Lewis Publishers, Boca Raton, FL.

Brown, R.A. 1995. "Regarding the Advantages of Using Bioremediation to Clean Up Superfund Sites." Testimony of Dr. Richard A. Brown, Vice President, Remediation Technology, Groundwater Technology, Inc. Testifying on Behalf of the Biotechnology Industry Organization, Inc. before the Senate Environment and Public Works Committee, Subcommittee on Superfund, Waste Control and Risk Assessment. Washington, D.C.

Devine, K. and L.L. Graham. 1995. "States' Attitudes on the Use of Bioremediation." To be published in conference proceedings of the *Third in situ and On-Site Bioreclamation International Symposium*. Battelle Press, Columbus, OH.

Devine, K. 1992. *Bioremediation Case Studies: An Analysis of Vendor Supplied Data.* EPA/600/R-92/043. U.S. Environmental Protection Agency, Office of Research and Development. Washington, D.C.

Dooher, B. 1995. Personal communication with K. Devine. University of California at Los Angeles, School of Engineering. Los Angeles, CA.

Environmental Information Ltd. 1994. "Petroleum UST Cleanup Initiatives Show Strengths and Weaknesses of State-Based Programs." Press release. Minneapolis, MN.

Gas Research Institute. 1995. *Environmentally Acceptable Endpoints in Soil. Risk-Based Approach to Contaminated Site Management Based on Availability of Chemicals in Soil.* Draft report. Chicago, IL.

Jerger, D.E., Erickson, S.A. and Rigger, R.D. 1994. *Full Scale Slurry-Phase Biological Treatment of Wood Preserving Wastes at a Superfund Site.* OHM Remediation Services Corporation, Findlay, Ohio.

Jonesi, G. 1989. "Impact of RCRA Land Disposal Restrictions on Bioremediation." USEPA Office of Enforcement and Compliance Monitoring. Unpublished paper.

LaGoy, P.K. and Quirk, T.C. 1994. "Establishing Generic Remediation Goals for the Polycyclic Aromatic Hydrocarbons: Critical Issues." *Environ. Health Perspective,* 102(4):348–352.

LaGoy, P.K. and Schulz, C.O. 1993. "Background Sampling: An Example of the Need for Reasonableness in Risk Assessment (Letter to the Editor)." *Risk Analysis.* 13:483–484.

Liptak, J.F. and Lombardo, G. 1994. *The Development of Chemical-Specific Risk-Based Soil Cleanup Guidelines Results in Timely and Cost Effective Remediation.* New Hampshire Department of Environmental Services, Groundwater Protection Bureau. Concord, NH.

McNeely, S. 1995. U.S. Environmental Protection Agency, Office of Solid Waste and Emergency Response, Office of Underground Storage Tanks. Personal communication with K. Devine. Washington, D.C.

Nisbet, I.C.T. and LaGoy, P.K. n.d. Toxic Equivalency Factors (TEFs) for Polycyclic Aromatic Hydrocarbons (PAHs). In preparation.

Swett, G. 1995. Personal communications with K. Devine. RETEC, Tucson, AZ.

USEPA. 1987a. *The Layman's Guide to the Toxic Substances Control Act.* U.S. Environmental Protection Agency, Office of Pesticides and Toxic Substances.

USEPA. 1987b. *The PMN Process: A Discussion and Analysis of the New Chemicals Program in the Office of Toxic Substances.* EPA 560/1-87-011. U.S. Environmental Protection Agency, Office of Toxic Substances. Washington, D.C.

USEPA. 1987c. *Record of Decision (ROD)—Remedial Alternative Selection—Bayou Bonfouca Site, Slidell, Louisiana.* EPA Region VI, Dallas, Texas. March.

USEPA. 1988a. *Record of Decision (ROD)—Summary of Remedial Alternative Selection—North Cavalcade Street Site, Houston, Texas.* EPA Region VI, Dallas, Texas. June.

USEPA. 1988b. *Record of Decision (ROD)—Summary of Remedial Alternative Selection—South Cavalcade Street Site, Houston, Texas.* EPA Region VI, Dallas, Texas. September.

USEPA. 1989a. *Record of Decision (ROD)—Burlington Northern (Somers Plant) Superfund Site, Flathead County, Montana.* EPA Region VIII, Montana Operations Office. September.

USEPA. 1989b. *Risk Assessment Guidance for Superfund. Volume I. Human Health Evaluation Manual (Part A).* Interim final. EPA/540/1-89/002. December. Office of Emergency and Remedial Response.

USEPA. 1990a. *The Nation's Hazardous Waste Management Program at a Crossroads. The RCRA Implementation Study.* EPA/530-SW-90-069. U.S. Environmental Protection Agency, Office of Solid Waste. Washington, D.C.

USEPA. 1990b. Quality Assurance Project Plan for Characterization Sampling and Treatment Tests Conducted for the Contaminated Soil and Debris (CS&D) Program. Office of Solid Waste.

USEPA. 1990c. *RCRA Orientation Manual.* EPA/530-SW-90-036. U.S. Environmental Protection Agency, Office of Solid Waste. Washington, D.C.

USEPA. 1990d. *Record of Decision (ROD)—Texarkana Wood Preserving Company Superfund Site, Texarkana, Texas.* EPA Region VI, Dallas, Texas. September.

USEPA. 1991. *Integrated Risk Information System (IRIS).* Chemical-specific reference doses and cancer potency factors and EPA Toxicology Background Documents. Office of Health and Environmental Assessment, Environmental Criteria and Assessment Office. Cincinnati, Ohio.

USEPA. 1992a. *CERCLA / Superfund Orientation Manual.* EPA/542/R-92/005. U.S. Environmental Protection Agency, Office of Solid Waste and Emergency Response. Washington, D.C.

USEPA. 1992b. *Technologies and Options for UST Corrective Actions: Overview of Current Practice.* EPA/542/R-92/010. U.S. Environmental Protection

Agency, Office of Solid Waste and Emergency Response Technology Innovation Office and Office of Underground Storage Tanks. Washington, D.C.

USEPA. 1993a. 58 *Fed. Reg.* 48092-48204. Sept. 14, 1993. Office of Solid Waste.

USEPA. 1993b. *Provisional Guidance for Quantitative Risk Assessment of Polycyclic Aromatic Hydrocarbons.* Office of Research and Development, Washington, D.C. EPA 600/R-93/089.

USEPA. 1995. "OSWER Directive 9610.17: Use of Risk-Based Decision-Making in UST Corrective Action Programs." Memorandum from U.S. Environmental Protection Agency, Office of Solid Waste and Emergency Response Assistant Administrator to Regional Administrators. Washington, D.C.

Walsh, W.J. (1990). "Making Science, Policy, and Public Perception Compatible: A Legal/Policy Summary." *In Groundwater and Soil Contamination Remediation: Toward Compatible Science, Policy, and Public Perception,* pp. 206–249. National Academy Press, Washington, D.C.

Zaborsky, O.R. (1995). "Environmentally Acceptable Endpoints: The Scientific Approach to Cleanup Levels." *Biotreatment News.* 5(6): 17, 19, 20, 24.

Landfarming: A Practical Guide

ROBERT L. KELLEY, BILL LIU AND VIPUL SRIVASTAVA
Institute of Gas Technology
Environmental Processes
1700 S. Mt. Prospect Road
Des Plaines, IL 60018

INTRODUCTION

Landfarming or land treatment involves the use of natural processes, either biological, chemical, and/or physical, to transform organic contaminants in soil. Solid-phase bioremediation denotes a landfarming operation that complies with revised environmental regulations. These operations can be "engineered" with various control and collection systems. Landfarming operations began more than 20 years ago, primarily for the treatment of petroleum hydrocarbon wastes. By 1984 there were more than 150 landfarming operations, ranging in size up to 1500 acres. In recent years, there has been an enormous growth of interest in bioremediation in general, specifically landfarming techniques. Since numerous books [1,8,21,26,32], reports [27,31], and review papers [16,33] have been written about landfarming, the purpose of this chapter is to briefly summarize the technology, describe the important factors, and give a case study that describes the field testing and the resulting conceptual design for the remediation of former manufactured gas plant (MGP) wastes. This chapter will emphasize both scientific and engineering aspects of this technology.

The advantages of landfarming are (1) less time and money than other remedial operations, (2) complete destruction of waste material, (3) limited disruption of site activities, (4) capable of handling large volumes, and (5) general acceptance by the public. Bioremediation technologies will reduce health and environmental risks of hazardous waste sites because these systems will result in environmentally benign products (CO_2 and H_2O) from both halogenated and nonhalogenated organics. Near-term risk reduction includes the reduction or elimination of the problem of hazardous volatile organic compound (VOC) emissions seen in some current waste elimination technologies.

The microorganisms used in the biological treatment are normally obtained from nature and have not been genetically modified. Landfarming processes use microorganisms that occur naturally at the site or microorganisms derived from similar contaminated wastes. Public and scientific concerns regarding the use of recombinant strains of microorganisms are obviated. Acclimation of these cultures is the only step that may be necessary and is best determined on a site-specific basis. For example, some sites may contain organisms that are already carrying out or that are capable of carrying out waste degradation steps. These organisms that have arisen by natural selection processes are potential candidates for use in the biological process.

Parameters can be readily controlled during biological treatment. This is a primary advantage over other processes where engineering and process control are expensive and/or difficult. Because process parameters can be manipulated as appropriate in the chemical treatment component of a land treatment process, optimal oxidation of the wastes can be readily achieved. Similarly, incubation temperature, oxygen, nutrient, and substrate concentrations can be controlled to attain extensive biodegradation in a timely and economic manner. This flexibility is beneficial because it provides treatment options if site-specific problems arise.

Commercially available waste elimination technologies, such as incineration, and other new technologies currently under investigation have high capital and operating costs. Significant cost savings are possible using a landfarming process because it combines technologies that are much less mechanically complex and have been shown to require low capital investment requirements and to have low operating costs. Costs are projected to be lower for the proposed system because only simple and well-established equipment and simple process controls are required, and no additional pollution control systems are needed. In a case history of a field test conducted by the Institute of Gas Technology (IGT), comparing an integrated chemical/biological treatment process and a conventional bioremediation treatment in land treatment mode, the total treatment costs were $60 and $150 per ton, respectively [17,19]. A landfarming treatment system is especially cost-effective on large-volume, low-concentration wastes but can also be used to efficiently treat high-concentration wastes. The process is also relatively simple and safe to operate when compared to high-pressure, high-temperature processes.

There are disadvantages and limitations to landfarming. Not all contaminants are degraded quickly and completely. Contaminants may have inhibitory or toxic effects on the biological process. Also, heterogenously dispersed wastes may cause concentration effects in which there is either too much or too little of the waste material. Because landfarming relies on biological processes, operational parameters, such as pH and temperature, may be somewhat more narrow than in typical industrial processes. Once degradation has begun, daughter products may be produced that may be accumulated and/or may be more toxic. The best example of this phenomenon is the anaerobic degradation of trichloroethylene (TCE), which results in the accumulation of vinyl chloride. In fact, regulators of landfarming activities have recently become very interested in the production of intermediates during a bioremedial process. If treating organic wastes mixed with heavy metals or radioactive materials, landfarming may effectively deal with the organic portion and have no effect on the other hazardous materials.

From an engineering point of view, landfarming poses some challenges and must be properly designed and operated to avoid adverse environmental impact. Landfarming or solid-phase bioremediation can involve an extensive amount of land and require intensive management. The design of the treatment facility should include proper control and collection systems to avoid air emissions, including odors, leaching, and lateral runoff.

BASIC PRINCIPLES OF "ENGINEERED" LANDFARMING

As mentioned previously, landfarming relies on natural processes that can be biological, chemical, or physical. Physical processes include sorption to waste matrices and bacterial cells, as well as volatilization. Chemical processes include UV degradation, as well as chemical reactive waste material causing oxidation, reduction, hydrolysis, precipitation, and polymerization. Soil chemistry can contribute to processes such as humification.

Dragun [8] has documented the biodegradation rates for almost 300 chemicals. In landfarming, degradation is probably not the result of a single bacterial strain but is caused by a consortia of bacteria, fungi, and other single-celled organisms such as protozoa. Usually, the microorganisms derive carbon and/or energy from the chemicals that they degrade. This degradative process could be straightforward, with the contaminant being transported into the bacteria, which degrades it to CO_2 and H_2O, or it could incorporate the intermediates into cellular components. Degradation of the contaminant may also be done by complementary or sequential reactions by a variety of organisms. Finally, a contaminant can be degraded by a gratuitous reaction in which the microorganism gains nothing. This type of metabolism is referred to as co-metabolism. Table 1 outlines critical factors for microbial activity.

Landfarming techniques were originally developed for treating petroleum, municipal sludge, and agricultural wastes (Table 2). The advent of "engineered" landfarming came with the desire to treat more hazardous wastes, such as wood-preserving wastes, chlorinated solvents, manufactured gas plant (MGP) wastes, explosives, and propellants (Table 3). The major use of landfarming today, however, continues to be petroleum-derived wastes, such as gasoline, diesel fuel, and jet fuel. It has also been widely used to reduce the volume of exploration and production wastes associated with the petroleum and natural gas industries. In fact, the American Petroleum Institute has developed a manual of practice for landfarming petroleum wastes [2]. Chemical components such as C_4–C_{12} hydrocarbons, paraffins, olefins, naphthenes, and polyaromatic hydrocarbons are biodegradable. Soils containing up to 25% of petroleum wastes can be treated to 100–200 ppm, with typical soils contaminated with 5–8% wastes.

Wood-preserving wastes containing pentachlorophenol (PCP), creosote, and polynuclear aromatic hydrocarbons (PAHs) have been effectively treated by landfarming techniques. Solvents such as acetone, butanol, ethylene glycol, methylene chloride, and TCE are biodegradable. Finally, in a composting operation, 99% of a 76,000 ppm mixture of explosives, such as trinitrotoluene (TNT), RDX (Royal Dutch explosive),

Table 1. Critical environmental factors for microbial activity

Available soil water	25%–85% of water holding capacity; 0.01 MPa
Oxygen	Aerobic metabolism: Greater than 0.2 mg/L dissolved oxygen, minimum air-filled pore space of 10%; Anaerobic metabolism: O_2 concentrations <1%
Redox potential	Aerobes and facultative anaerobes: greater than 50 millivolts; Anaerobes: less than 50 millivolts
pH	5.5–8.5
Nutrients	Sufficient nitrogen, phosphorus, and other nutrients so not limiting to microbial growth
Temperature	15/C-45/C (Mesophiles)

Suggested C:N:P ratio of 100:10:1.

Table 2. Land treatment usage by major industry group.

Description	Number of Units
Petroleum refining and related industries	105
Chemicals and allied products	30
Electric, gas, and sanitary services	16
Fabricated metal products, except machinery and transportation equipment	12
National security and international affairs	9
Lumber and wood products, except furniture	97
Electrical and electronic machinery, equipment, and supplies	5
Food and kindred products	4
Textile mill products	4
Miscellaneous manufacturing industries	3
Machinery, except electrical	3
Paper and allied products	3
Oil and gas extraction	2
Water transportation	2
Miscellaneous repair services	2
Agricultural production—livestock	1
Rubber and miscellaneous plastics products	1
Primary metal industries	1
Transportation equipment	1
Wholesale trade—nondurable goods	1
Educational services	1

Source: K.W. Brown and Associates, Inc. (1981).

and HMX (Her Majesty's explosive), and propellants, such as dinitrotoluene (DNT), have been treated in 22 weeks [38].

Landfarming hazardous contaminants do require more control collection systems to comply with current environmental regulations. To minimize the potential adverse environmental impacts of volatilization, leaching, and runoff, the following are necessary components:

- impermeable clay or chemically resistant liner
- leachate control system
- surface runoff collection system
- vapor control system

Table 3. Land treatability of the six main groups of hazardous materials migrating from disposal sites.[a]

Hazardous Material Group	Land Treatability
(1) Solvents and related organics such as trichloroethylene, chloroform, and toluene	High
(2) PCBs and PBBs	Limited
(3) Pesticides High	4
(4) Inorganic chemicals such as ammonia, cyanide, acids, and bases	Limited
(5) Heavy metals	Limited
(6) Waste oils and greases	High

[a]High land treatability does not infer immunity from environmental damage. Only through proper design and management of land treatment unit can the desired level of treatment be obtained and the migration of hazardous materials be prevented (from Reference [6]).
Source: EPA (1980).

These land treatment or solid-phase bioremediation units are usually called prepared bed units. They require physical treatment of the soil in which the soil is crushed and sieved to an appropriate size. Soil characteristics are very important. For example, a clayey soil or soil with poor tilth properties will limit O_2 transfer.

CASE STUDY: FIELD TEST TO CONCEPTUAL DESIGN

Manufactured gas plants have produced large quantities of tar and related by-products. Although there are some similarities among these sites, the sites often vary significantly in the specific types and quantities of wastes present, depending upon the types of processes used (coal, water, or oil gas) and the era in which the plant operated. Polynuclear aromatic hydrocarbons are components of all kinds of tars, and tars are a by-product of all of these plants. The physical and chemical characteristics of these tars vary, however, according to the process employed [13,18,19]. Oxide wastes from purifier boxes may contain ferrocyanide and varying amounts of arsenic, chromium, copper, lead, nickel, and zinc (all of which appear on the Environmental Protection Agency priority pollutants list). Lamp black, clinker, cinders, and ash may also be present [4,5,18]. The MGP wastes can be categorized into five major types:

- free tars, oils, and lamp black
- organic waste or tar-contaminated soils
- organic waste or tar/oil-contaminated waters
- purifier box (or spent oxide) wastes
- mixed wastes and fills

The organic-contaminated, or PAH-containing, soils represent the largest waste type at most sites. Some of the PAHs are suspected or potential carcinogens and, therefore, must be removed or treated to reduce their concentrations so that they do not pose a risk to humans or animals. These risks may be due to the direct contact with soils (ingestion, dermal contact, inhalation, etc.) as well as direct and/or indirect effects associated with the groundwater and surface water.

The Institute of Gas Technology (IGT) has been developing processes to treat soils and water at former MGP sites that are contaminated with wastes such as PAHs, cyanides, and metals [6–10]. Remediation technologies are also being developed for other gas industry wastes such as halogenated hydrocarbons, including polychlorinated biphenyls (PCBs) and perchloroethylene (PCE). The ultimate goal of this technology development program is to provide a cost-effective waste treatment technology alternative to conventional options that include containment, land filling, chemical fixation, soil washing, thermal treatment, conventional bioremediation, etc. The limitations of the conventional options include either limited/incomplete/insufficient waste degradation, considerable expense, or both. IGT's approach has been to identify the treatment-limiting steps and then to develop approaches to overcome those limiting steps. As a result of extensive bench-scale studies conducted since 1987, IGT has developed and demonstrated processes for PAH-contaminated soils that are a combination of biological treatment and chemical treatment, the integrated Chemical/Biological Treatment (CBT) or MGP-REM.

The MGP-REM Process combines two complementary remedial techniques: (1) chemical oxidation as the pretreatment for hard-to-degrade contaminants and (2) biological treatment using an aerobic biosystem. The CBT Process uses mild chemical treatment with Fenton's reagent (H_2O_2 + Fe^{++}), which produces hydroxyl radicals that start the chain reaction with the organic contaminants, resulting in modification and degradation of organics to biodegradable and environmentally benign products. These products and other organics are later degraded in the biological step. Results from bench-scale studies conducted with approximately 25 MGP soil samples show that the CBT Process is capable of enhancing the rate, as well as the extent of PAH degradation.

The results indicate that, when a soil is dominated by 4- to 6-ring PAHs, chemical treatment is best performed as a pretreatment step. When initial soil screening indicates that a soil contains 2- and 3-ring PAHs in significant amounts, biological treatment is recommended as the initial step to remove as many of these compounds as possible. Chemical treatment using Fenton's reaction is then used to reduce the remaining persistent PAHs.

Bench-Scale Soil Treatability

A treatability protocol developed for the Gas Research Institute (GRI) and IGT's Sustaining Membership Program (SMP) by IGT scientists and other GRI contractors has been successfully evaluated using coal tar-laden soils. Some of the soil was used as a source of inoculum in a basal mineral salts nutrient medium and was incubated aerobically to determine if indigenous microorganisms exist that can grow at the expense of organic contaminants. Following this soil characterization phase, more of the same soil was evaluated for contaminant mass transfer using a method called "solids desorption," which was performed to estimate to what extent, if any, adsorbed wastes would passively desorb from the soil.

The next step of the treatability study evaluated biodegradation of the organic wastes. The rates and extents of biodegradation were determined in liquid-culture experiments and soil slurry reactor studies, which were conducted for 30–60 days. Microbiol cultures obtained from site enrichment or cultures maintained in IGT's stock culture collection were used as inoculum if needed. The results of these studies provided an indication of how effective a bioremediation process was likely to be for a given soil sample.

Simulated land treatment options were evaluated using soil-pan microcosms. Typically, 1 to 2 kg of contaminated soil was placed into Pyrex glass pans and tested under a variety of treatment conditions. A typical aerobic soil microcosm study takes a minimum of 4 to 6 months to perform even for aerobic conditions. Soil microcosm treatability studies are usually incubated under optimal conditions of pH, moisture, tilling, and temperature. Usually the most degradable pollutants are removed within a period of one growing season, the length of which depends on geographical location and climate conditions, normally 80 to 120 days.

The effectiveness of the MGP-REM Process was demonstrated using the highly contaminated soil. In an example of this effectiveness, an MGP soil was used in experiments involving Fenton's reaction applied after an initial period of aerobic biological treatment. Figure 1 shows that the

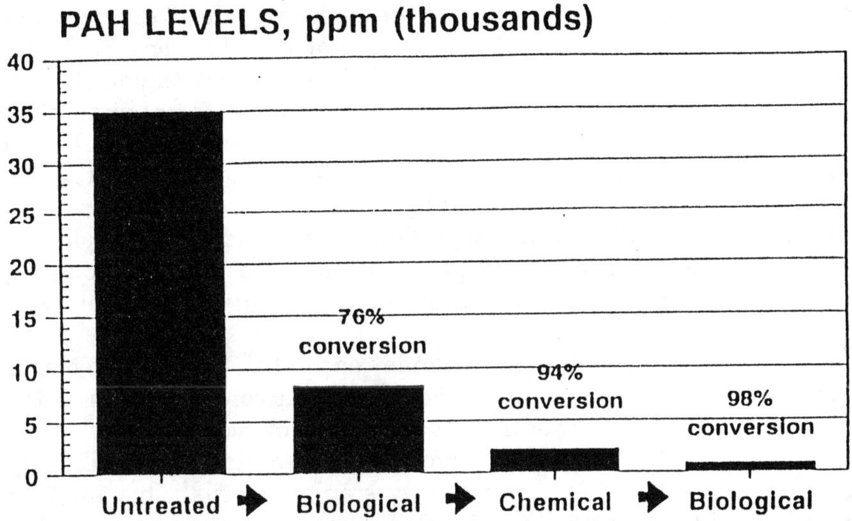

Figure 1. Total PAHs in TGS-7 soil after sequential biological-chemical-biological treatment.

initial period of aerobic biological treatment reduced the total PAHs by 76%, from 35,000 to about 8400 ppm. When Fenton's reaction was used as a posttreatment, PAH levels dropped by an additional 18% to 2200 ppm, and when coupled with a second round of biological treatment, the total PAH reduction approached 98%. Added cycles of combined chemical and biological treatments could be implemented as deemed necessary to reduce the end point further. The aerobic biological treatment alone could reduce PAHs to only just under 80% after a long treatment duration.

Chemical Treatment Protocol

Typically, chemical treatment consists of 1% Fe^{++} (16.87 g of ferrous sulfate/kilogram of soil) and 1% hydrogen peroxide (31.4 ml of 30% technical grade H_2O_2/kilogram of soil). Due to the exothermic nature of the chemical reactions, the temperature is maintained by slow addition of the hydrogen peroxide to the soil or soil slurry containing the Fe^{++}.

Field Design and Operating Protocol

Based upon the treatability results on the soil from the site in Dubuque, Iowa, it was decided to demonstrate this technology in the field. Several experimental designs were evaluated. One treatment plot design included four plots with dimensions of 4′ by 12′ each. Each plot was to employ a different land treatment technology with one of them acting as a control. Another design included eight treatment plots of the same dimension. The purpose of this design was to provide repetition in treatment and analysis, and thus reduce random error. The third design included a total of sixteen treatment plots of the same dimension. Four plots were to be dedicated to one of four land treatment technologies. The sixteen plots, however, were to be loaded in a random sequence, thus minimizing errors associated with loading, treatment, and analysis.

A liner system was established under each of the treatment blocks to facilitate water management and to prevent contamination of the underlying soil. Above this liner was a layer of rock or gravel, then clean sand, and finally 15 to 20 cm of the contaminated soil. Each of the treatment plots were bounded by similar boards. This made it possible to till the various plots using a garden tractor fitted with a rototiller on a weekly or biweekly basis. Decontamination was performed between treatment blocks. A water management system was established to hold water derived from each of the treatment blocks and the decontamination zone in an on-site impoundment. Collected water was used for irrigation during the experiment. Meteorological data were collected for ambient temperature, wind speed and direction, and precipitation at the site. Soil samples were collected at the beginning of the experiment and after 1 day and 1, 2, 4, 6, 8, and 10 weeks of treatment. Parameters measured included PAHs, nutrient levels, microbial population levels, pH, and moisture.

In the first field study, the main conditions evaluated were Treatment 1—unamended control; Treatment 2—amendment with nutrients, tilling, and irrigation; Treatment 3—amendment with nutrients, tilling, moisture, and Fenton's reagents at the beginning of the study and on a periodic basis by adding hydrogen peroxide directly to the soil during irrigation. During field studies, analytical measurements monitored include PAH concentration, soil pH, and total heterotrophic bacterial counts. Other data collected, including air and soil temperature, rainfall, and relative humidity, are shown in Table 4.

Random samples were collected from each of the test plots over the duration of the experiment. Composite samples were collected for each sampling event using a soil corer or auger. Eight to ten cores were taken from each plot and mixed together to form a composite sample of about 500 g. On occasions where samples were sent to laboratories for validation, 1 kg was collected as a composite. This 1-kg composite

Table 4. Summary of field weather data.

Parameter	1991 Field Test 8/14–9/18
Air temperature, °F	
Average	73.6 + 12.2
Median	72.6
Soil temperature, °F	
Average	81.9 + 10.3
Median	82.5
Relative humidity, %	
Average	74.9 + 19.2
Median	79.6
Total rainfall, inch	3.11

was divided, with 500 g being sent to the laboratory for analysis. Because test methods for Soxhlet extraction require only 10 g of soil, this provided a sufficient sample for dry soil, PAH analysis, and other tests.

PAHs were determined in the composite sample from the site using a modified EPA SW-846 Method 8100 analysis of Soxhlet extracts [1,2,20,21]. Soxhlet extraction was performed by EPA SW-846 Method 3540 using a 1 : 1 mixture of acetone and hexane as the extraction solvent. A suite of PAHs was analyzed in this treatability study and is summarized in Table 5 as IGT total PAHs. These 23 compounds are subdivided into two groups: IGT 2- and 3-ring PAHs and IGT 4- to 6-ring PAHs, based on the number of rings represented by the various

Table 5. Total PAH reduction after 42 days of treatment.

Treatment	Explanation	PAH Reduction, %	
		Total PAHs	Carcinogenic PAHs
1	Control—not managed	41 ± 7	28 ± 5
2	Conventional bioremediation Addition of N + P at Day 21	49 ± 6	29 ± 3
3	Integrated chemical—biological treatment Chemical addition as cotreatment (Day 1 and Day 29)	72 ± 6	56 ± 4

constituent PAHs. IGT's previous experience has revealed that 2- and 3-ring PAHs are most readily biodegradable, whereas the 4- to 6-ring PAHs are more persistent.

Field Evaluation Data

The merits of the MGP-REM (or CBT) Process were verified in a field experiment conducted with the soils from an MGP site that is also a Superfund site in the state of Iowa [10–17]. The primary objective of this field experiment was to compare PAH degradation as a direct consequence of microorganisms or from chemical treatment using Fenton's reaction as a pretreatment or as a co-treatment in conjunction with biological processes. Figure 2 presents the results of the field test conducted in the landfarming mode of the CBT Process. In this figure, the residual PAH concentrations in the soil are compared with control plots that did not receive any nutrients, conventional bioremediation plots that received nutrients, and CBT treatment plots. Because this particular soil naturally contained nutrients, the control plots also exhibited some degradation of PAHs. Therefore, this figure is intended to show the PAH reductions over and above those observed in control plots. Table 5 shows percent

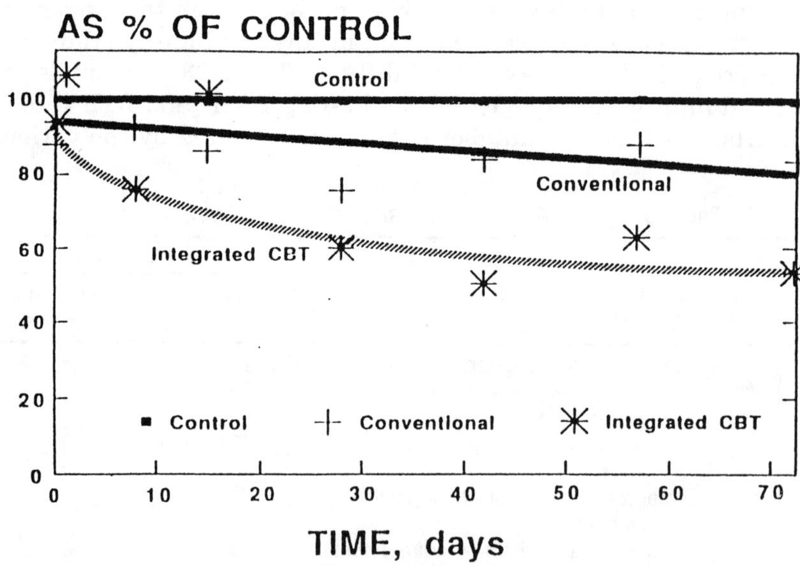

Figure 2. PAH concentrations in midwest gas soil during field test in 1991.

degradation of PAHs in each treatment plot group as total EPA PAHs and carcinogenic EPA PAHs (as defined by USEPA and presented in Table 6) after 42 days. Initial land treatment of 1% chemical treatment on Day 0 and additional 1% chemical treatment on Day 29 appears to give the best results with higher PAH degradation and at a faster rate of degradation. The integrated treatment reduced PAHs at a higher rate and to a greater extent than the conventional bioremediation. The treatment goals for this soil were met within the first 28 days when using the CBT Process.

Meteorological data were collected so that ambient temperature (both air and soil), wind speed and direction, precipitation, and relative humidity were known at the site. Nighttime temperatures below 32°F were first seen after 31 days of operation, and soil temperatures were below 45°F after 40 days of operation (Figure 3). Table 4 summarizes the field weather data.

Bacterial-soil titer for all the treatment remained relatively constant at 10^8 cells/gram of soil (Figure 4). An apparent loss in bacterial titer, of approximately one log coincides with the chemical treatments; however, the significance of this decline appears to be inconsequential.

Table 7 summarizes the preliminary economic evaluations for this site [11]. Results show that this site can be cleaned at $60/cu yd or less when using the landfarming mode of soil treatment using the CBT Process. Process optimization results, as well as the initial PAH concentration and final treatment goals, would affect the soil remediation costs for other sites.

In summary, the manufactured gas plant sites are nonhomogeneous in soil and waste types and concentrations of contaminants [4]. PAH-contaminated soils represent the largest fraction of the contamination at the MGP sites [4]. Studies conducted by IGT show that

- Bioremediation is effective in removing PAHs from MGP soils.
- Integrated chemical/biological treatment improves the rate, as well as the extent, of PAH removal.
- MGP soils can be effectively cleaned in the landfarming mode of the CBT Process.
- Soils with high sand content are easier to clean. Also, soil-slurry reactors are an effective and efficient system for contaminated soil bioremediation.
- Integrated chemical/biological treatment is also effective for soils with high silts and clay contents.

Finally, IGT also continues to develop, optimize, and demonstrate innovative and accelerated contaminated soil treatment technologies. A

Table 6. Polynuclear aromatic hydrocarbons assayed by IGT chromatography and EPA SW-846 method.

PAH Analyte	Ring Number	IGT Total PAHs	IGT2- and 3-Ring PAHs	IGT 4- to 6-Ring PAHs
Naphthalene	2	Naphthalene	Naphthalene	
1-Methylnaphthalene	2	1-Methylnaphthalene	1-Methylnaphthalene	
2-Methylnaphthalene[a]	2			
Biphenyl	2	Biphenyl	Biphenyl	
Acenaphthylene	3	Acenaphthylene	Acenaphthylene	
Acenaphthene	3	Acenaphthene	Acenaphthene	
Dibenzofuran[a]	3	Dibenzofuran	Dibenzofuran	
Fluorene	3	Fluorene	Fluorene	
Phenanthrene	3	Phenanthrene	Phenanthrene	
Anthracene	3	Anthracene	Anthracene	
1-Phenylnaphthalene	3	1-Phenylnaphthalene	1-Phenylnaphthalene	
2-Methylphenanthrene	3	2-Methylphenanthrene	2-Methylphenanthrene	
1-Methylphenanthrene	3	1-Methylphenanthrene	1-Methylphenanthrene	
Fluoranthene	4	Fluoranthene		Fluoranthene
Pyrene	4	Pyrene		Pyrene
Benz(*a*)anthracene	4	Benz(*a*)anthracene		Benz(*a*)anthracene
Chrysene[b]	4	Chrysene[b]		Chrysene[b]
Triphenylene[b]	4	Triphenylene[b]		Triphenylene[b]
Benzo(*b*)fluoranthene[c]	5	Benzo(*b*)fluoranthene[c]		Benzo(*b*)fluoranthene[c]
Benzo(*k*)fluoranthene[c]	5	Benzo(*k*)fluoranthene[c]		Benzo(*k*)fluoranthene[c]
Benzo(*a*)pyrene	5	Benzo(*a*)pyrene		Benzo(*a*)pyrene
Indeno(1,2,3-cd)pyrene	5	Indeno(1,2,3-cd)pyrene		Indeno(1,2,3-cd)pyrene
Dibenzo(a,h)anthracene	5	Dibenzo(a,h)anthracene		Dibenzo(a,h)anthracene
Benzo(g,h,i)perylene	6	Benzo(g,h,i)perylene		Benzo(g,h,i)perylene

Table 6. (continued).

PAH Analyte	EPA Total PAHs	EPA Noncarcinogenic PAHs	EPA Carcinogenic PAHs
Naphthalene	Naphthalene	Naphthalene	
1-Methylnaphthalene			
2-Methylnaphthalene[a]	2-Methylnaphthalene[a]	2-Methylnaphthalene[a]	
Biphenyl			
Acenaphthylene	Acenaphthylene	Acenaphthylene	
Acenaphthene	Acenaphthene	Acenaphthene	
Dibenzofuran[a]	Dibenzofuran[a]	Dibenzofuran[a]	
Fluorene	Fluorene	Fluorene	
Phenanthrene	Phenanthrene	Phenanthrene	
Anthracene	Anthracene	Anthracene	
1-Phenylnaphthalene			
2-Methylphenanthrene			
1-Methylphenanthrene			
Fluoranthene	Fluoranthene	Fluoranthene	
Pyrene	Pyrene	Pyrene	
Benz(*a*)anthracene	Benz(*a*)anthracene		Benz(*a*)anthracene
Chrysene[b]	Chrysene[b]		Chrysene[b]
Triphenylene[b]			
Benzo(*b*)fluoranthene[c]	Benzo(*b*)fluoranthene[c]		Benzo(*b*)fluoranthene[c]
Benzo(*k*)fluoranthene[c]	Benzo(*k*)fluoranthene[c]		Benzo(*k*)fluoranthene[c]
Benzo(*a*)pyrene	Benzo(*a*)pyrene		Benzo(*a*)pyrene
Indeno(1,2,3-cd)pyrene	Indeno(1,2,3-cd)pyrene		Indeno(1,2,3-cd)pyrene
Dibenzo(a,h)anthracene	Dibenzo(a,h)anthracene		Dibenzo(a,h)anthracene
Benzo(g,h,i)perylene	Benzo(g,h,i)perylene	Benzo(g,h,i)perylene	

[a]2-Methylnaphthalene and dibenzofuran are PAHs listed as EPA PAHs for consistency with previous testing performed by Midwest Gas. These PAHs are not normally analyzed by EPA SW-846 Method 8100 but had been selected from the suite of compounds listed in EPA SW-846 Method 8270. Also 2-methylnaphthalene is not an analyte included in the IGT chromatography protocol. All other PAHs listed as EPA PAHs were analyzed for this treatability study.
[b]Chrysene and triphenylene coelute during chromatography.
[c]Benzo(*b*)fluoranthene and benzo(*k*)fluoranthene coelute during chromatography.

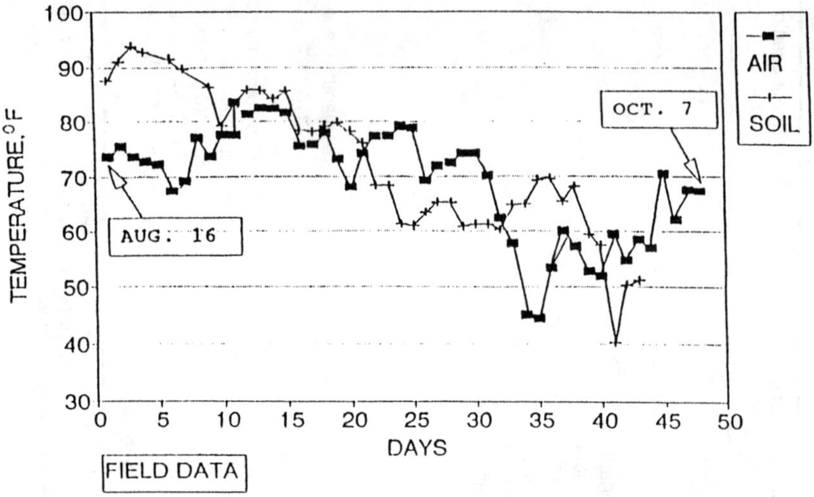

Figure 3. Temperature profiles for soil and air during the field demonstration period.

field evaluation of the integrated treatment in the soil-slurry treatment system has been completed.

Conceptual Design of a Land Treatment Unit

Based on the results of the treatability study and field test, a conceptual design can then be developed. This section will explain the basics of designing a solid-phase bioremediation system. The site used for the conceptual design is an ex-MGP site, contaminated with PAHs and requiring remediation. The sample site contains approximately 8000 cu yd of contaminated soil, of which 6000 cu yd is highly contaminated and is not suitable for bioremediation. Since there was no immediate

Table 7. Preliminary economic analysis: landfarming at Iowa Site 1991 [11].[a]

Chemical Concentration (wt %)	Treatment Cost ($ per cu yd)
0.5	48
1.0	60
1.5	72
2.0	85
3.0	110

[a]A 1% chemical addition was the minimum dosage that resulted in the successful attainment of the treatment goals. The 2% and 3% additions also reached the goals.

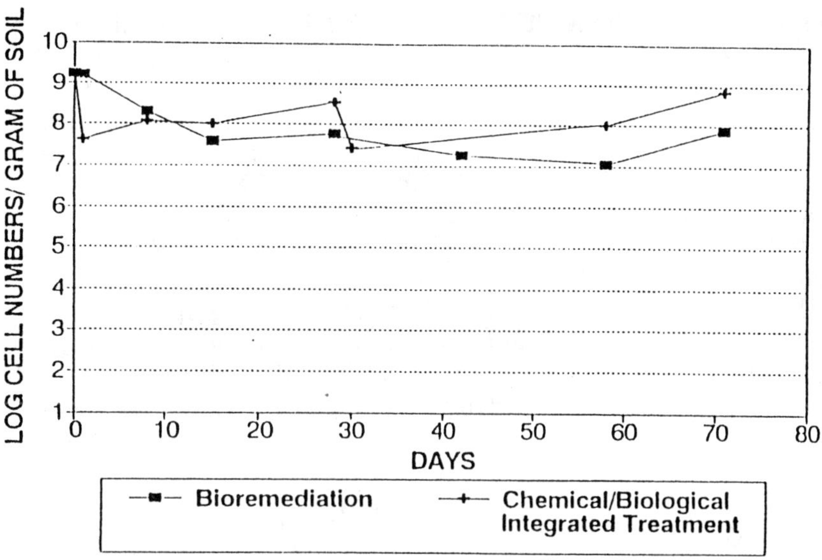

Figure 4. Bacterial soil titer for each treatment during the field demonstration.

plan for reuse, it was determined that an integrated chemical/biological treatment of landfarming would be most suitable because of its lower remediation cost. The background information of the site is summarized in Table 8.

The conceptual design starts with a modular land treatment unit (MLTU), which is the building block of the entire land treatment sys-

Table 8. Background information for the Iowa MGP Site.

Parameter	Value	Unit
Area available for landfarming	100 × 100	ft^2
Estimated total contaminated soil	8000	yd^3
Estimated highly contaminated soil	6000	yd^3
Water table depth	15–17	ft
Average t-PAHs concentration	1400	mg/kg
Average c-PAHs concentration	300	mg/kg
Cleanup standards—t-PAHs	500	mg/kg
Cleanup standards—c-PAHs	100	mg/kg
First-order kinetics for t-PAHs degradation	0.02	day^{-1}
First-order kinetics for c-PAHs degradation	0.0086	day^{-1}
Average soil TOC	16,700	mg/kg
Soil density	1.5	g/cm^3

tem. If the lift of the MLTU is 1 ft, six MLTU will be required for the remediation.

The second step of the design is to determine the amount of time required to remediate the site to desired treatment standards, i.e., 300 mg/kg of t-PAHs and 100 mg/kg of c-PAHs. Since the literature has demonstrated that PAH degradation follows a first-order kinetics, the treatment times needed to reach the treatment end point are 128 days for t-PAHs and 52 days for c-PAHs. Consequently, the desired treatment time for this MLTU will be 128 days.

The other parameters significant to landfarming operation are the chemicals for nutrient requirement of nitrogen and phosphorus. The total carbon source in soil can be represented by the amount of TOC and the ideal carbon to nitrogen and phosphorus ratio should be 100:5:1.

CONCLUSIONS

Landfarming technologies have been shown to be effective in reducing the environmental and hazardous risks associated with various waste materials and matrices. Technical and political impediments exist. There is a lack of knowledge of biological methods among field engineers and other decision makers. For a full-scale landfarming operation to work, it requires detailed data from lab-scale treatability and field pilots to be generated and validated. Once operational, landfarming operations lack reliable, real-time, on-site monitoring methods for biodegradation or biological activity. Finally, local, state, and federal laws and rules have to incorporate realistic treatment goals that are not so low as to eliminate every other technology but incineration.

ACKNOWLEDGEMENT

The research program mentioned in this Chapter is jointly funded by the Gas Research Institute, IGT's Sustaining Membership Program, USEPA, and several gas companies.

The author acknowledges the significant technical contribution of Dr. T. D. Hayes and Mr. D. G. Linz of GRI; Ms. Norma Lewis, Ms. Annette Gatchett, and Ms. Naomi Barkley of the USEPA; Drs. R. L. Kelley, J. J. Kilbane, J. R. Paterek, and W. K. Gauger (now with Radian Corp.) of IGT; Mr. Sam Nelson of Midwest Gas Company; and Dr. Johanshir Golchin of the state of Iowa.

REFERENCES

1. Alexander, M. (1994). *Biodegradation and Bioremediation.* Academic Press, New York.
2. American Petroleum Institute. (1984). *The Land Treatability of Appendix VIII Constituents Present in Petroleum Industry Wastes.*
3. Anonymous, *Status Report: Bioremediation of Organic-Contaminated Soils / Sludges from Manufactured Gas Plant Sites,* Gas Research Institute, GRI-90/0117, 1990.
4. Anonymous, *Test Methods for Evaluating Solid Waste, Physical / Chemical Methods,* SW-836, 1986.
5. Anonymous, *Standard Test Methods for Carbon and Hydrogen in the Analysis of Coal and Coke.* Am. Soc. Test. Mat., ASTM D-3178-84, 1985.
6. Brown K.W., Evans, Jr., Frentrup B.D. (1983). *Hazardous Waste Land Treatment.* Butterworth Publishers, Boston, MA.
7. Pope, D.F., Matthews, J.E. (1995). *Bioremediation Using the Land Treatment Concept.* EPA/600/15:1–28.
8. Dragun J. (1988). *The Soil Chemistry of Hazardous Materials.* Silver Spring, MD, Hazardous Materials Research Institute.
9. Eng, R. *Survey of Town Gas and By-Product Production and Locations in the U.S. (880–1950).* Report No. EPA/600/7-85/004, 468, Office of Research and Development, EPA, Research Triangle Park, North Carolina, 1985.
10. Environmental Research and Technology and Kippers Co., Inc., *Handbook on Manufactured Gas Plant Sites.* Utility Solid Waste Activities Group, Superfund Committee, Washington, D.C., 1984.
11. Gauger, W.K. and Srivastava, V.J. "Bioremediation of Gas Industry Wastes: Current Status and New Directions." Paper presented at the *Symposium, Hazardous Waste and Environmental Management in the Gas Industry,* Institute of Gas Technology, Chicago, June 1990.
12. Gauger, W.K., Srivastava, V.J., Hayes, T.D. and Linz, D.G. "Enhanced Biodegradation of Polyaromatic Hydrocarbons in Manufactured Gas Plant Wastes." Presented at *IGT's Third International Symposium on Gas, Oil, Coal, and Environmental Biotechnology,* New Orleans, LA, December 3–5, 1990.
13. Gauger, W.K., Kilbane, J.J., Kelley, R.L. and Srivastava, V.J. "Enhancement of Microbial Degradation of Hydrocarbons in Soil and Water." Paper presented at *IGT's Second International Symposium on Gas, Oil, Coal, and Environmental Biotechnology,* New Orleans, Louisiana, December 1989.
14. Golueke, C.G., Diaz, L.F. (1989). Biological Treatment for Hazardous Wastes. *Biocycle,* 58–63.
15. Hatzinger, P.B., Alexander, M. (1995). Effect of aging chemicals in soil on their biodegradability and extractability. *Environ Sci & Technol,* 29:537–545.

16. Heitzer, A, Sayler, G.S. (1993). Monitoring the efficacy of bioremediation. *TIBTECH,* 11:334–343.
17. Kelley, R.L. and Srivastava, V.J. "Field-Scale Evaluation of an Integrated Treatment for Remediation of PAHs in Manufactured Gas Plant Soils." Paper presented at *1992 Spring National A.I.Ch.E. Meeting,* New Orleans, Louisiana, March 1992.
18. Kelley, R.L., Gauger, W.K. and Srivastava, V.J. "Application of Fenton's Reagent as a Pretreatment Step in Biological Degradation of Aromatic Compounds." Paper presented at *IGT's Third International Symposium on Gas, Oil, Coal, and Environmental Biotechnology,* New Orleans, Louisiana, December 3–5, 1990.
19. Kelley, R.L., Moeleer, M.W., Lenschner, A.D. and Srivastava, V.J. *Land Treatment of MGP Soils Field Scale Demonstration: Engineering Assessment and Economic Evaluation,* Topical Report for Gas Research Institute, In press.
20. Kilroy, A., Gray, N.F. (1995). Treatability, toxicity and biodegradability test methods. *Biol Rev Cambridge Phil Soc,* 70:243–275.
21. King, R.B. Long, G.M., Sheldon, J.K. (1992). *Practical Environmental Bioremediation.* Lewis Publishers, Boca Raton, LA.
22. Klute, A. *Methods of Soil Analysis Part 1. Physical and Mineralogical Methods.* Madison, Wisconsin: Am. Agron, Soc., Soil Sci. Sci. Am., 1986.
23. Liebs, L.H. "Town Gas—An Overview," in *American Gas Association Operational Section Proceedings,* 369–79. Arlington, Virginia American Gas Association, 1985.
24. Mueller, J.G., Lin, J.E., Lantz, S.E., Pritchard, P.H. (1993). *Recent Developments in Cleanup Technologies,* EPA/600/J-93/390:369–380.
25. Page, A.L. *Methods of Soil Analysis: Part 2. Chemical and Microbiological Properties.* Madison, Wisconsin: Am. Agron. Soc., Soil Sci. Sci. Am., 1986.
26. Parr, J.F., Marsh, P.B., Kla, J.M. (1983). *Land Treatment of Hazardous Wastes.* Noyes Data Corp. Park Ridge, N.J.
27. Peyton, T.O. (1984). Biological disposal of hazardous waste. *Enzyme Microb Technol,* 6:146–154.
28. Ramaswami, A., Groshal, S., Luthy, R.G. (1994). Mass transfer and biodegradation of PAH compounds from coal tar. Water Sci Technol., 30:61–70.
29. Remediation Technologies, Inc., *Remediation Alternatives and Costs for the Restoration of MGP Sites.* Final Report to Gas Research Institute, Chicago, March 1990.
30. Rhodes, E.O. (1974). "Water-Gas Tars and Oil-Gas Tars," in Hoiberg, A. J., Ed., *Bituminous Materials: Asphalts, Tars, and Pitches. Volume III: Coal Tars and Pitches,* 1–31. Huntington, New York: Robert E. Krioeger Publishing Co.
31. Sims, J.L., Sims, R.C., Matthews J.E. (1989). *Bioremediation of Contaminated Surface Soils,* EPA-600/9-89/73:1–23.

32. Skladany, G.J. Metting, F.B. (1992). Bioremediation of Contaminated Soil. In: Metting F.B., Jr. (ed) *Soil Microbial Ecology: Application in Agricultural and Environmental Management.* Marcel Dekker, Inc. New York, USA, pp. 483–513.
33. Srivastava, V.J. (1993). "Manufactured Gas Plant Sites Characterization of Wastes and IGT's Innovative Redmediation Alternatives," paper presented at the *Symposium for Hazardous and Environmentally Sensitive Waste Management in the Gas Industry,* Albuquerque, New Mexico.
34. Thompson, S.N., Burgess, A.S. and Ada, D. (1983). "Coal Tar Containment and Cleanup: Plattsburgh, New York," in *Proceedings of the National Conference on Management of Uncontrolled Hazardous Waste Sites,* 331–37. Washington, D.C.
35. U.S. Environmental Protection Agency, *Test Methods for Evaluating Solid Waste, Vol. 1,* Sections A, B, and C, 3rd ed., SW-846, November 1986.
36. U.S. Environmental Protection Agency, *Superfund Public Health Evaluation Manual,* Government Printing Office, Washington, D.C., EPA/540/1-86/060, October 1986.
37. Villaume, J.F. (1984). "Coal, Tar Wastes: Their Environmental Fate and Effects," in Majumdar, S. K. and Miller, E.W., Eds., *Hazardous and Toxic Wastes: Technology Management and Health Effects.* Easton, Pennsylvania: The Pennsylvania Academy of Sciences.
38. Williams, A.J., Ziegenfuss, P.S., Mohrman, G.B., Sisk, W.E. (1989). Composting of explosives and propellant contaminated sediments. *Hazardous and Industrial Wastes: Proceedings of the 21st Mid-Atlantic Industrial Waste Conference.* Technomic Publishing Co., Inc., Lancaster, PA, 269–281.

Field-Scale Use of White Rot Fungi for Soil Remediation in Finland

MICHAEL L. HOLROYD AND PHILIP CAUNT
Biotal Ltd.
5 Chiltern Close
Cardiff Industrial Park
Cardiff CF4 5DL
Wales, UK

INTRODUCTION

The problem of chemically polluted soil is becoming a major environmental issue in industrialized countries. Biological processes are potentially attractive as cost-effective, permanent, and environmentally acceptable solutions to the risk associated with contaminated soil. Microorganisms are known to have the ability to detoxify an increasingly wide variety of environmental pollutants and, by doing so, can provide a permanent and destructive solution in contrast to some other treatment methods, for example, encapsulation.

Traditionally, bioremediation projects have been conducted in a number of ways:

- The activity of degradative microorganisms already present in the environment has been enhanced by the addition of suitable nutrients and by improving other physico-chemical conditions.
- Suitable microorganisms or their enzymes have been introduced to the contaminated material.
- Reactors have been used, whereby the microbial activity takes place in a specially designed vessel under ideal conditions.

However, bacterial bioremediation has been found to have a number of potential drawbacks. The ability to degrade the compounds of interest in the laboratory does not ensure success on the actual site. Lack of knowledge on ecological competitiveness is one of the main obstacles in assessing the likely success of a bioremediation project. Acea and Alexander (1988) reported populations of several genera of bacteria declining by one or two orders of magnitude per week in natural loamy soil. The main reason for this decline was thought to be predation by protozoa although other factors related to physical conditions also play a role. The type of soil, temperature, water content, oxygen, and pH levels will all have an influence on the microbial activity, and their effects are not fully understood.

The specificity of most bacteria towards a particular chemical or closely related chemical groups generally limits their effectiveness against the varied profile of contaminants found in polluted soils. In these situations, different types of chemical contaminants may be targeted sequentially by a range of bacterial species. The contaminated soil may contain sources of carbon that are utilized in preference to the targeted compound. On the other hand, a chemical may only be degraded co-metabolically, that is, in the presence of a suitable alternative carbon source. This is often the case with xenobiotic pollutants, towards which potential degrading microorganisms have had relatively little time to adapt. Finally, the effects of the target compounds present in the polluted soil on microbial activity need to be determined at concentrations likely to be encountered on the actual site in question. Many wastes are toxic to metabolic activity, and it may only be possible to bioremediate these pollutants when present at concentrations below a certain threshold limit.

In addition, factors relating to the pollutants themselves are very important, not only in terms of biodegradability, but also accessibility to the microorganisms, which is essential before degradation can occur. Many common environmental pollutants have been found to be resistant to biodegradation, for example, high molecular weight polyaromatic hydrocarbons (PAHs) and some halogenated aromatics. Some of these compounds, especially PAHs with more than three fused aromatic rings, tend to be very stable and may only be degraded in the presence of another source of carbon and energy (Cerniglia and Heitkamp, 1989). Additionally, higher molecular weight PAHs are much less water soluble than the lower molecular weight species, tending to adsorb tightly to soil humic materials, thereby substantially reducing their bioavailability.

Advantages of Fungal Bioremediation

The use of fungal bioremediation technology has been developed to overcome some of the problems associated with bacterial treatments. The species *Phanerochaete chrysosporium* is a wood-degrading Basidiomycete that has been shown to degrade a wide range of hazardous chemicals under laboratory conditions (Glaser, 1990). The normal habitat of *P. chrysosporium* is decaying wood and, due to its ease of cultivation and rapid life cycles, has been used as a model system to understand the process of lignin biodegradation. Lignin is one of the most recalcitrant natural substrates, a nonrepeating aromatic heteropolymer that is resistant to attack by most microorganisms. Microorganisms capable of metabolizing lignin include some species of fungi and a relatively small number of species of bacteria (Kirk, 1984).

P. chrysosporium has been shown to generate a carbon-centered free radical enzyme system that allows it to catalyze numerous nonspecific cleavage reactions, resulting in breakdown of the lignin macrostructure into smaller fragments capable of being metabolized by more conventional enzymes. The ability to produce these potent oxidative enzymes explains its ability to degrade aromatic environmental pollutants. In fact, the enzymes have been demonstrated to exhibit very broad substrate specificity and act on a wide range of aromatics including PAHs, PCBs, and dioxins (Bumpus *et al.,* 1985).

Although the ability of *P. chrysosporium* to degrade a wide range of chemicals of environmental concern is now well established, attempts to remediate contaminated soil by direct inoculation with these organisms have produced little success mainly because soil is not their normal habitat (Glaser, 1990). Competition from other soil microorganisms severely limits growth; however, both growth and survival are poor even in nonsterile soils (Loske *et al.,* 1990).

The research presented here demonstrates the degradation of chlorinated phenols in nonsterile soil by *P. chrysosporium* LA1, a naturally isolated lignin degrader. The process has been scaled up to produce a pilot scheme for the remediation of 6400 m^3 of soil at a former timber treatment facility in Finland. The site was contaminated down to a depth of 8 m due to the use of KY-5, a crude chlorophenol preparation used to prevent fungal attack on timber products, consisting of 81% 2,3,4,6-tetrachlorophenol (TECP), 11% 2,4,6-trichlorophenol (TCP), 7% pentachlorophenol (PCP), and 1% other chlorophenols. The treatment process has a number of interesting features that distinguish it from more conventional (bacterial) bioremediation schemes:

- The microbial inoculum is pregrown on a suitable lignin-based substrate before it is combined with the contaminated soil. The established mycelium is then able to withstand competition from the native microflora and penetrate the surrounding soil.

- Degradation of the chemical contaminants is not dependent on their concentration since the process is co-metabolic; that is, degradation occurs as a consequence of growth on the lignocellulosic primary substrate.
- The enzyme system is nonspecific, and a range of aromatic compounds is degraded. This is especially important in many contaminated soils that contain more than one toxic compound of interest.
- Concentrations of pollutants normally thought of as being too high for biological processing can be attempted due to the major growth being on a support material offering some protection against toxic effects of the chemical contaminants.
- The support material is ultimately degraded, and the introduced microorganisms gradually decline in numbers due to competition from the natural population.

The microorganisms used are naturally occurring isolates that have not been manipulated in any way to produce unusual enzymes.

Degradation of Chlorophenol by *P. chrysosporium*

Although chlorinated phenols, like many chlorinated compounds, are somewhat resistant to biodegradation, pure strains of bacteria have been isolated capable of mineralizing even the most chlorinated of these compounds, PCP (Middeldorp *et al.,* 1990). However, attempts to use these strains to bioremediate soils have generally resulted in treatment times of several years to reach the desired target concentration. For example, Valo and Hakulinen (1991) reported treatment times of over 3 years, using windrows (total volume 50 m^3) inoculated with degrading bacteria, to reduce chlorophenols from 215 mg kg^{-1} to less than 10 mg kg^{-1}. At larger scale (200 m^3), these workers report 4 years to reach 10 mg kg^{-1} from a starting level of 450 mg kg^{-1}. Due to government guidelines, target concentrations are now lower than they have been in the past, and it has become necessary to develop new biological treatment methods to meet these new targets within a reasonable time frame.

Laboratory studies by several research groups have suggested that *P. chrysosporium* is capable of metabolizing a range of chlorinated phenols within accelerated time spans. Valli and Gold (1991), working on mechanisms of chlorophenol degradation, found that the degradation pathway for 2,4-dichlorophenol (2,4-DCP) by *P. chrysosporium* involved removal of both chlorine atoms from the aromatic ring of

2,4-DCP before ring cleavage and mineralization. The degradation of PCP by *P. chrysosporium* has also been reported to occur by this mechanism (Mileski *et al.*, 1988; Hammel and Tardone, 1988; Lin *et al.*, 1990)

In the kinetic model by Lin *et al.* (1990), PCP is mineralized to carbon dioxide by two possible mechanisms:

- PCP is degraded by extracellular enzymes to some intermediate, with the intermediate being subsequently mineralized to carbon dioxide by cell mass or cell bound enzymes.
- Cell mass (or cell bound enzymes) can directly mineralize PCP to carbon dioxide without requiring the initial reaction mediated by extracellular enzymes.

Mileski *et al.* (1988) state that the requirements for lignin degradation are that the initial step must be extracellular and nonspecific. The same requirements are also true for degradation of toxic chemicals with high molecular weight and complex structure. However, for compounds such as PCP, with low molecular weight, entry of these toxic compounds into the cell also makes them accessible to cell bound or intracellular degradative enzymes. The mechanism of degradation of these compounds may therefore be multiple, rather than only being degraded by a sequence of reactions initiated by extracellular enzymes.

LABORATORY PROGRAM

Before the start of any bioremediation scheme, it is desirable to undertake careful laboratory tests on soil samples from the site itself. From these tests, it should be possible to determine the rate-limiting factors affecting biodegradation of contaminants within a particular soil. The major factors affecting biological treatment of soils include temperature, pH, moisture content, oxygen availability, particle size distribution, soil type, nutrient concentrations, the presence of inhibitory compounds, and the availability of the contaminants to microorganisms. Some of these factors, which are often interrelated, can be altered to improve microbial activity. The effects of these adjustments can be judged by performing small-scale simulation experiments in the laboratory under carefully controlled conditions.

These experiments, known as microcosm tests, were performed using nonsterile, chlorophenol-contaminated soil from a former timber treatment site in Finland. A suitable carrier substrate consisting of straw, pine woodchips, and sawdust was inoculated with *P. chrysosporium* LA1, incubated for 3 weeks at 30°C, and then mixed with the screened (2 mm mesh) contaminated, soil. The soil was characterized as a silty sand, pH 7.5, with a total organic matter content of 5.1% w/w. At various time intervals, five microcosm jars were removed from the incubator and the contents solvent extracted by the method of Veikko (1990). Analysis was then carried out using a gas chromatograph fitted with an electron capture detector. The results (Figure 1) show an 80% reduction in total chlorophenol after 15 weeks treatment. The main contaminants, tetrachlorophenol (TECP) and PCP were reduced by 83% and 73%, respectively. Degradation was extremely rapid in the first few weeks, followed by a second phase during which very little degradation occurred. Further experiments were required to understand this phenomenon, and to control it in order to reach the treatment

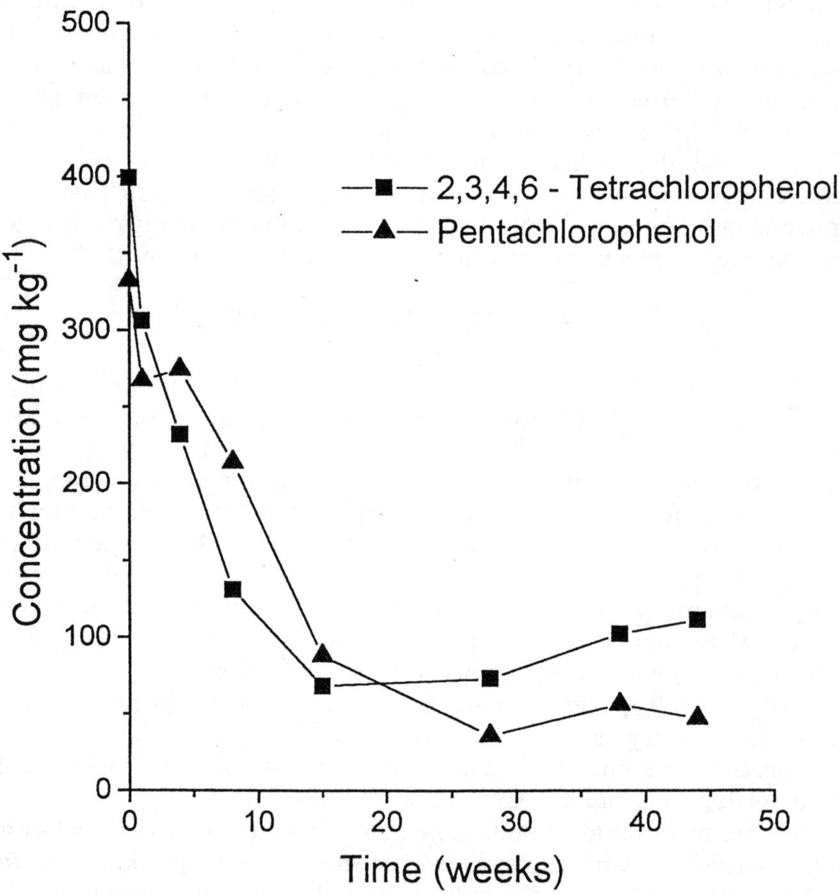

Figure 1. Degradation of chlorinated phenols in laboratory microcosms.

targets at the full scale. Although the starting concentration was about 750 mg kg^{-1} total chlorophenol, visual inspection indicated no apparent inhibition of growth.

FIELD-SCALE OPERATIONS IN FINLAND

The laboratory simulation had shown that treatment of chlorophenol-contaminated soil was feasible, but the more difficult challenge was to transfer the technology from the laboratory to the field application. In order to ensure success in the field, process engineering must be carefully designed and planned to meet the parameters determined during laboratory tests.

A site investigation was carried out which gave information on the levels of chlorophenols present and the approximate volume of contaminated soil. The top half meter of soil was found to contain dioxins and furans at levels unacceptably high (to local environmental authorities) to be incorporated into treatment beds. This material was therefore removed for disposal off-site. Analysis results revealed that metals in the soil were not likely to present a toxic hazard to a biological treatment method.

Before excavation of the soil could begin, it was necessary to pregrow the fungal inoculum on a mixture of lignocellulosic materials, including straw, woodchips, sawdust and pine bark. The mixture was moistened with clean water and then inoculated with *P. chrysosporium* LA1. Inoculation was made using both solid-phase and liquid-grown fungal material which was prepared at the laboratory in Cardiff, South Wales before shipping to the site. The solid-phase inoculum was grown in 200-L barrels at 30°C, containing a mixture of straw and sugar beet pulp. The liquid phase inoculum was grown in bulk in sporulation broth at 30°C before harvesting, concentrating, and formulating into a stable paste. This was diluted to 16 m^3 before applying to the woody materials through a spray nozzle. There were problems with mixing the solid-phase inoculum into the prepared material, and it was concluded that distribution of the diluted liquid inoculum was more even at the large scale. After the material had been inoculated, it was formed into a suitable heap and covered with plastic sheeting. Careful checks were subsequently made on moisture content and more water was added when necessary.

After several weeks, when the *P. chrysosporium* was seen to be well established, this primary inoculum pile was used to inoculate an even larger volume of woody matter. Further liquid inoculum was added, and the material formed into a heap, which was again covered. Several weeks later, when the fungus was clearly seen to be well established, the material was ready to be incorporated into the contaminated soil.

The contaminated soil was excavated immediately prior to forming the treatment beds. Screening of the soil and mixing with the inoculated organic material was carried out in a one-step process using a rotating drum screen. Processed soil was then transported to the treatment area and laid out over a gravel base enclosed by an HDPE liner. This allowed for control of any leachate that could be removed to a collection tank. Due to the depth of excavation (up to 8 m), some contaminated run-off water was encountered. Peat was used to soak up the contaminated water, and this was also added to the treatment beds.

Incorporated into the beds was a network of perforated pipes, which, when connected to a pump, allow forced aeration. This was necessary due to the depth of the beds (2–2.5 m) and to avoid turning the bed material during the 2-year treatment period. A single blower unit was used, switched between the four beds for equal times using a solenoid valve. Heavy machinery was not allowed to travel onto the beds to avoid unnecessary compaction, which would restrict airflow. Each treatment bed (approximate size 50 m × 30 m × 2 m) was finally covered with HDPE sheeting. Due to some delays as the processes were being developed and used for the first time, the first two treatment beds were completed in September 1992, while the final two were constructed and finished in the following spring of 1993.

MONITORING AND RESULTS

Initial Results for the Measured Parameters

Table 1 shows results obtained for the British Standard Sedimentation Test on two soil samples taken from the contaminated area

Table 1. Results for the British Standard Sedimentation Test on soil from the contaminated area.[a]

		Silt			Sand		
Sample	Clay	Fine	Medium	Course	Fine	Medium	Course
a	15	15	45	18	4	3	0
b	17	13	44	15	3	6	2

[a]Shows the percentage distribution of soil particle size for two samples from the contaminated area prior to excavation and treatment. Sample a was taken from a depth of 1.5 m and Sample b was taken from 3.0 m. Particle size definitions are as follows: clay, <0.002 mm; silt, 0.002–0.05 mm; sand, 0.05–2.0 mm.

Table 2. Analytical results for various parameters in treatment beds A to D.[a]

	Sampling Date				
	23.09.92	05.04.93	04.08.93	26.10.93	26.04.94
Bed A					
pH	n/a	7.5	7.2	7.3	7.2
TOL_x (mg kg^{-1})	2727	2191	1521	1614	n/a
TOM (%)	n/a	5.1	4.8	5.6	n/a
Bed B					
pH	n/a	7.7	7.2	7.3	7.2
TOL_x (mg kg^{-1})	1323	1194	813	750	n/a
TOM (%)	n/a	3.3	2.7	2.9	n/a
Bed C					
pH			7.6	7.7	7.3
TOL_x (mg kg^{-1})			816	734	n/a
TOM (%)			2.1	2.7	n/a
Bed D					
pH			7.1	6.8	7.2
TOL_x (mg kg^{-1})			839	465	n/a
TOM (%)			3.6	4.1	n/a

[a]Shows analytical results, where available, at different time points during treatment. Abbreviations are as follows: TOL_x, toluene extractable matter; TOM, total organic matter; n/a, results not available. Each value is average analysis result of 10 samples. Each sample analyzed was a composite of five subsamples.

prior to excavation and treatment. Samples A and B were taken from depths of 1.5 and 3 m, the results illustrating the silty nature of the soil at depth. The top 1 m of the area was mainly course sand, the top 0.5 m containing high levels of dioxins and furans. These compounds were present in the original wood preservative, as by-products in the manufacture of the chlorinated phenols. Since these compounds are very hydrophobic, binding tightly to soil, only the top 0.5 m of the contaminated area had to be removed for alternative treatment.

Once the soil had been screened, mixed with the inoculated lignocellulosic materials, and formed into treatment beds, samples were taken for chemical analysis, for a number of parameters, as shown in Table 2. Toluene extractable material (TOL_x), giving a measure of the hydrocarbon content including extractable aromatic compounds from the added lignocellulosic materials, had initial values ranging from 2727 mg kg^{-1} for Bed A to 816 mg kg^{-1} for Bed C. Total organic matter (TOM) initially varied from greater than 5% for Bed A to 2.1% for Bed

C. The starting pH levels were all slightly higher than neutral, varying between 7.1 for Bed D and 7.7 for Bed B.

Ongoing Monitoring of Process Parameters

The treatment beds were regularly monitored throughout the treatment period. Bed temperatures were measured at 1.5 m depth using a portable probe. Gas monitoring points were installed in the treatment beds for on-site measurement of oxygen and carbon dioxide using a portable infrared gas monitoring system.

Gas measurements showed the importance of efficient aeration to maintain high oxygen levels in deep treatment beds, which is necessary for the highly oxidative fungal process to work optimally. However, the number of hours of aeration per day had to be carefully adjusted, depending on the outside air temperature. This was to avoid unnecessary cooling, or even freezing, of the beds during the colder months. Additionally, since there was no irrigation system, aeration was reduced to the lowest possible level during the summer months to minimize loss of water vapor. Previous mass-balance studies by Lamar and Dietrich (1990) have shown that volatilization is not a significant mechanism for the loss of PCP from soil. Since the levels of aeration in the present project were minimal, air stripping was unlikely to play a noticeable part in the removal of chlorinated phenols.

Bed temperatures were generally higher than the typical midday air temperature by around 5°C in the winter and up to 15°C in the summer. This elevated temperature was probably due to nonspecific biological activity in the bed, yet assisted in the year-round activity of the fungus, thus overcoming the seasonal nature of some bacterial bioremediation systems in cold climates.

TOL_x values fell substantially during treatment, indicating a large reduction in the amount of toluene extractable organic material present in the beds. The results in Table 2 show a 41% reduction in Bed A and a 43% reduction in Bed B during the first year. Substantial reductions were also evident in Beds C and D. Results for TOM were variable, and no real conclusions could be drawn regarding trends during treatment, while pH values also fell slightly and stabilized at about 7.2.

Results for Chlorinated Phenols and Related Compounds

Each treatment bed was sampled using a hand auger. The sampling regime generated 10 composite samples from each bed, five from 0.5

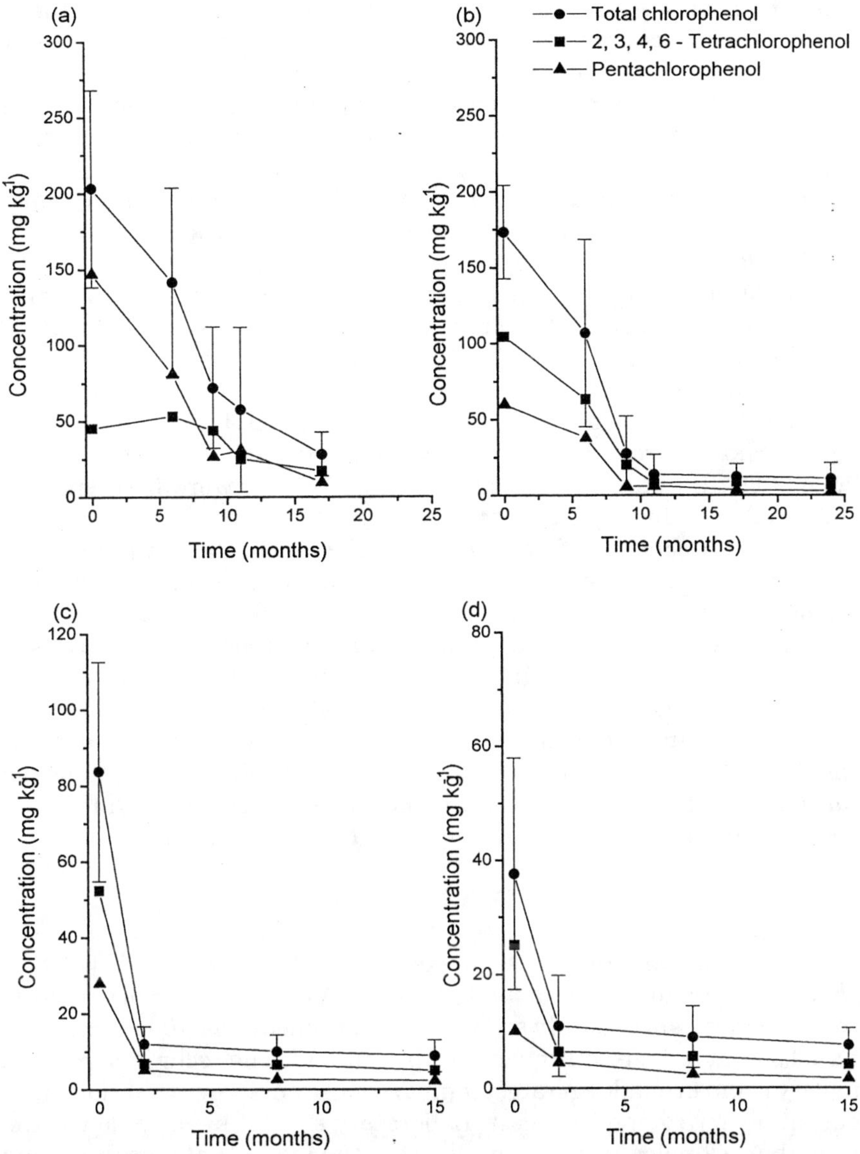

Figure 2. The degradation of chlorinated phenols in Beds A–D. For clarity, error bars were only included for the total chlorophenol data points: (a) Bed A; (b) Bed B; (c) Bed C; (d) Bed D.

m and five from 1.5 m depth. This was carried out as soon as possible after the beds were set up and at regular intervals during treatment.

The starting concentrations for total chlorinated phenol varied between 203.1 mg kg^{-1} for Bed A and 37.6 mg kg^{-1} for Bed D. However, Beds C and D were not sampled until approximately 2 months after construction. It is therefore likely that the starting levels were significantly higher than those reported. Figure 2 shows reduction in total chlorophenol in Beds A to D. There was a considerable drop in all four treatment beds, the reduction continuing through the winter, despite long periods of subzero temperatures on the site. This was due to the elevated temperatures in the beds, which were inherent to their design. Beds C and D reached the target concentration of 10 mg kg^{-1} after 9 months of treatment, while in October 1994, the mean concentration in Bed B was 11.0 mg kg^{-1}. At the time of writing, the October results for Bed A were not available; however, in April 1994, the level was 28.0 mg kg^{-1}. From Figure 2, it can be seen that the levels of both 2,3,4,6-TECP and PCP, the main contaminants, were both substantially reduced during the treatment.

There were concerns, from various parties, about the possibility of the formation of toxic intermediates during the breakdown of chlorinated phenols, especially anisoles and veratroles, which have a high mammalian toxicity. For this reason, samples were subjected to analysis by GC/MS, specifically targeting these compounds. However, in contrast to results obtained by Lamar and Dietrich (1990), there was no indication of the formation of these compounds, the levels remaining below the detection limit of 0.02 mg kg^{-1} throughout. Similarly, dioxins and furans were also monitored in the beds during treatment. GC/MS analysis showed no evidence for an increase in these compounds.

CONCLUSIONS

Small-scale laboratory tests demonstrating the ability of *P. chrysosporium* to degrade a mixture of chlorinated phenols in nonsterile soil have been successfully scaled up to bioremediate 6400 m^3 contaminated soil at a disused timber yard in Finland. The technique resulted in very rapid breakdown rates, compared to other bioremediation methods, at a very competitive cost. Laboratory studies have demonstrated that the technology can be applied to other pollutants, such as high molecular weight polyaromatic hydrocarbons. Work ongoing at the moment, funded by the U.K. Department of Trade and Industry, is aimed at evaluating the possibility of using this technique for the remediation of PCB- and dioxin-contaminated soils.

REFERENCES

Acea, M.J. and Alexander, M. (1988). Growth and survival of bacteria introduced into carbon-amended soil. *Soil Biol Biochem,* 20 (4):703–709.

Bumpus, J.A., Tien, M., Wright, D. and Aust, S.D. (1985). Oxidation of persistent environmental pollutants by a white rot fungus. *Science,* 228.

Cerniglia, C.E. and Heitkamp, M.A. (1989). In *Metabolism of Polycyclic Aromatic Hydrocarbons in the Aqueous Environment.* Ed. U. Varanasa. CRC Press.

Glaser, J.A. (1990). Hazardous waste degradation by wood degrading fungi. In *Biotechnology and Biodegradation (Advances in Applied Biotechnology Series, Vol 4).* Eds. D. Kamely, A. Chakrabarty and G.S. Ommen. Gulf Publishing Co.

Hammel, K.E. and Tardone P.J. (1988). The oxidative 4-dechorination of polychlorinated phenols is catalysed by extracellular fungal peroxidases. *Biochemistry,* 27:6563–6568.

Kirk, T.K. (1984). In *Microbial Degradation of Organic Compounds.* Ed. Gibson, D.T. Dekker, New York.

Lamar, R.T. and Dietrich, D.M. (1990). In situ depletion of pentachlorophenol from contaminated soil by *Phanerochaete* spp. *Appl. Environ. Microbiol.,* 56:3039–3100.

Lin, J., Wang, H.Y. and Hickey, R.F. (1990). Degradation kinetics of pentachlorophenol by *Phanerochaete chrysosporium. Biotechnology and Bioengineering,* 35:1125–1134.

Loske, D., Hutterman, A. Majcherczyk, A., Zadrazil, F., Lorsen, H. and Waldinger, P. (1990). In: *Advances in Biological Treatment of Lignocellulosic Materials.* Ed. Coughlan, M.P. Elsevier Science.

Middeldorp, P.J.M., Briglia, M. and Salkinoja-Salonen, M.S. (1990). Biodegradation of pentachlorophenol in natural soil by inoculated *Rhodococcus phenolicus. Microb. Ecol.,* 20:123–139.

Mileski, G.J., Bumpus, J.A., Jurek, M.A. and Aust, M.D. (1988). Biodegradation of pentachlorophenol by the white rot fungus *Phanerochate chrysosporium. Appl. Environ. Microbiol.,* 54:2885–2889.

Valli, K. and Gold, M.H. Degradation of 2,4-dichlorophenol by The lignin-degrading fungus *Phanerochaete chrysosporium. J. Bacteriol.,* 173:345–352.

Valo, R. and Hakulinen, R. (1991). *Proc. Environmental Conference* (Book 1). Hilton Palacio/San Antonio Convention Center, San Antonio, TX. April 7–10, 1991. Tappi Press, Atlanta.

Veikko, K. (1990). The use and formation of CPs, PCPPs and PCDDs/PCDFs in mechanical and chemical wood processing industries. Doctoral thesis, University of Helsinki.

REFERENCES

[illegible]

Oil Spill Bioremediation on Coastal Shorelines: A Critique

Albert D. Venosa
U.S. Environmental Protection Agency
Risk Reduction Engineering Laboratory
26 W. Martin Luther King Drive
Cincinnati, OH 45268

BACKGROUND

The purpose of this chapter is not to provide an extensive review of the literature on oil spill bioremediation. For that, the reader is referred to Swannell *et al.* (1996), who have conducted the most exhaustive review I have yet to come across. Other reviews are also available (Prince, 1993; Leahy and Colwell, 1990; Owen, 1991; Lee and Levy, 1989a). Rather, it is my intent to critique how investigators have been misled by the improper use of inferential statistics and experimental design to support their contention that bioremediation enhances the removal of crude oil contaminating an impacted shoreline and to offer an approach that improves upon one's ability to convince the scientific community of the veracity and authenticity of field data when recommending response action. I am using as the basis of my entire argument the classic treatise on pseudoreplication published over 10 years ago by Hurlbert (1984). In that article, Hurlbert argues forcefully for the use of true replication, randomization, and treatment interspersion when conducting any kind of field investigation in which the researcher wishes to test the effectiveness of various treatments through the use of manipulative experiments. He pointed out quite effectively how often ecologists misuse replicate sampling of unreplicated plots as a substitute for replicated and interspersed treatment layouts. There is nothing inherently wrong with using a pseudoreplicate design when gross effects of a treatment are anticipated (such as, for example, treatment of an oil slick by *in situ* burning). It *is* inherently wrong, however, to conclude that a manipulated treatment effect is significant from a pseudoreplicated experiment when precise differences need to be determined or when the cost-effectiveness of a technology under study needs to be established. The problem arises when the investigator attempts to use inferential statistics to back up his interpretations, which is scientifically indefensible and mathematically invalid. Usually, field conditions preclude knowledge of all important variables, which is why true replicates are needed. This chapter builds on Hurlbert's assertions and extends the ideas to field experiments involving oil spill bioremediation on shorelines and beaches. For that matter, the statistical concepts discussed here apply to any field experiment regardless of the research objectives.

In the following sections, I will attempt to show how replete the oil spill bioremediation literature is with respect to pseudoreplicated field studies and why, because of this, skepticism reigns among those charged with the responsibility to deploy appropriate countermeasures (among which bioremediation may be a reasonable choice) to clean up the impacted environment. Because of the extremely high cost of oil spill response, we can no longer afford to conduct field studies lacking a statistically sound and fundamentally credible basis supporting recommendations for implementation of bioremediation technology. The intent of this chapter is to provide intellectual guidance that will aid responders, on-scene coordinators, regional response teams, and the general engineering and scientific communities in focusing on the proper means of evaluating and assessing bioremediation field trials.

Definitions of Terms

To clarify the meaning of terms that might not be familiar to the reader, the following definitions are appropriate. For a much more detailed discussion of this terminology, the reader is referred to Hurlbert (1984).

1. *Pseudoreplication* refers to the use of inferential statistics to test for treatment effects (e.g., does bioremediation through nutrient enrichment enhance the rate of loss of crude oil from a beach compared to the inherent loss?) with data from experiments where either treatments are not replicated (though samples may be) or replicates are not statistically independent.
2. *Controls* are treatments that are used for comparison against manipulated treatments. They are used to control temporal changes (inherent biodegradative and abiotic loss of oil components) and procedural effects of the experimenter (addition of fertilizer or microbial inocula for enhanced remediation).

3. *Randomization* refers to the spatial assignment of treatments (experimental units) on the experimental plane in such a way that mitigates or eliminates systematic natural bias or experimenter bias (e.g., effects of longshore currents, prevailing winds, underground water flow, light gradients, etc.). Its purpose is to eliminate biased placement of plots that would impose one or more systematic effects on some plots but not others.
4. *Replication* refers to the repeated use of treatments and controls (two or more) to obtain an estimate of variability, either experimenter-induced variability (random error) or initial or inherent variability among experimental units [e.g., hydrocarbon concentration differences or geomorphological differences (grain size, hydraulic conductivity, permeability, soil structure) throughout the study area].
5. *Interspersion* refers to the act of mitigating or eliminating the regular spatial variation in properties of experimental units through statistical randomization procedures (unrestricted or restricted) and segregation of treatments. Placement of four control plots in a row followed by four treated plots in a row fails the criterion of proper interspersion of treatments.
6. *Type I error* is the conclusion that a treatment effect is significant when indeed it is not (e.g., bioremediation enhanced the disappearance of oil from a beach by several-fold when indeed it did not). The probability of making a Type I error is usually denoted as α, and we try to set this value as low as *reasonably* possible (typically about 0.05).
7. *Type II error* is the conclusion that a treatment effect is not significant when indeed it is. The probability of making a Type II error is denoted as β. If we subtract β from 1, we obtain the probability of calling a difference significant when indeed it is. This is called the *power of the test,* and we set this probability to be moderately high (usually about 0.80).

PAST FIELD STUDIES

Although other examples exist in the literature, I have limited discussion to recent field trials dating back into the 1980s (i.e., those conducted in Norway, Canada, Israel, and Alaska). It should be pointed out, however, that most of the studies reported in the literature prior to 1980 suffered from the same statistical design problems that are discussed in the following section. Two examples of studies that indeed did make proper use of replication and randomization to demonstrate significant treatment effects were Westlake *et al.* (1978) and Jobson *et al.* (1974).

Spitzbergen, Norway

Several reports were published in the 1980s describing field studies carried out in the cold climate of Norway (Sendstad, 1980; Halmo, 1985; Sveum and Ladousse, 1989; Sveum, 1987). The first (Sendstad, 1980) involved experimentally releasing 10 L/m^2 Forcados unweathered crude oil on each of two 10 m^2 plots on a seashore at Spitzbergen. One plot served as a "control" to which no amendments were added. The other was treated with a commercial fertilizer at a concentration of 0.1 kg/m^2, although it was not clear whether the fertilizer was applied once or several times over the course of the study. A third plot served as an unoiled control against which to compare soil heterotrophic activity. The investigators demonstrated that the "oil alone" site showed a small, but significant ($p < 0.1$), decrease in respiration compared to the unoiled control plot but that the oiled/fertilized plot elicited a marked increase over that of both plots. They attributed this to the effect of the fertilizer. Further, the cumulative carbon release from the fertilized plot was substantially greater than the unfertilized plot. With respect to total oil disappearance, only 430 g oil/m^2 was remaining in the unfertilized plot after 4 years versus 320 g/m^2 in the fertilized plot, having started with 10 L/m^2. Sendstad attributed the difference to biodegradation, and from these data he computed an oil decomposition rate of 55–96 g/m^2/yr. Although it was not clear if inferential statistics were used to support the conclusions, two flaws were apparent in this design: (1) an inadequate control and (2) use of unreplicated plots. A proper control, in addition to controlling temporal changes, must account for procedural effects of the experimenter. Hurlbert draws an excellent analogy from the pharmaceutical industry: to control for mice injected with a drug dissolved in saline, control mice are injected with saline alone. In the Spitzbergen case, nothing was applied to the control plot. Thus, the control plot was not "treated" in exactly the same way as the fertilized plot (i.e., seawater with no nutrients should have been applied to the untreated plot). The results from the field study are equivocal because (1) no estimation of experimental error was possible due to the lack of adequate replicate plots, and (2) rigorous care was not taken to treat both plots in precisely the same way. Is it possible, however remotely, that the method of applying fertilizer to the treated plot or that unobserved differences in the experimental matrix might have caused the differences noted?

Halmo (1985) conducted another shoreline bioremediation experiment in Spitzbergen in 1983. In this experiment, three plots were set up in the supralittoral zone (above the normal high tide line), which contained developed vegetation and was moistened occasionally by wave action and spring tides. A 20 L/m^2 water-in-oil emulsion was applied to the three plots: One plot was treated with a single application of an oleophilic fertilizer (1:10 ratio of fertilizer to oil) consisting of urea in a microemulsion of oleic acid; one was treated with a water-soluble commercial grade fertilizer formulation (same carbon-to-nitrogen ratio as the oleophilic fertilizer) consisting of salts of NH_4^+ and NO_3^- (also in a single application); and one was not treated with anything. Each plot was divided into 200 subplots that were sampled at random over a 1-year period. Oil degrader populations were found to increase in all plots and remained high in number for the duration of the experiment. A statistically significant difference in total number of bacteria was recorded between the oil-soluble and the water-soluble fertilizer, the oil-soluble giving the higher number. The low to medium weight *n*-alkanes were observed to decrease in the period June–September 1983 by about 45–85% in the fertilizer treated plots compared to a 10–25% decline in the unfertilized plot. In June 1984, the paraffins had completely degraded in both fertilized plots, while 20–30% of the paraffins were still present in the unfertilized site. The n-C_{17}/pristane and n-C_{18}/phytane ratios were also shown to be lower in the fertilized plots than in the untreated control plot, but no statistical significance tests were reported. The authors did conclude that no statistically significant difference in oil biodegradation occurred between the two fertilizer types. Again, this experiment suffered from the same flaws as the previous Spitzbergen studies. Since two different types of fertilizers were used in the Halmo study, two controls should have been used: one receiving the oleophilic material without the urea and phosphate, the other receiving seawater without water-soluble nutrients. With respect to the replicate sampling layout, it did not matter how many subplots were established for sampling. Since they were all within the same plot, they were not independent of each other, so they could not be used to estimate experimental error. "Replicate" samples in the same plot are merely used to determine within-plot variability, which measures precision but does not have anything whatever to do with evaluation of treatment effects, the main hypothesis under consideration. The authors used statistical analysis to support their conclusions only on selected portions of the data. Because of the inconsistency in the reporting of the experimental results, the lack of adequate information on the details of the treatment methodology used in the field study and the flawed use of a pseudoreplicate design for demonstrating effectiveness, the report is unconvincing as to the utility of bioremediation as an oil spill response countermeasure.

Sveum and Ladousse (1989) conducted similar studies with the oleophilic fertilizer Inipol EAP 22 (Elf Aquitaine) in Kings Bay, Spitzbergen, on different types of beach material. One experiment used sandy sediments as the contaminated matrix; the other involved fine-grained lagoon sediments. The sandy soil experiment was conducted in two plastic containers, one of which was treated with Inipol while the other was untreated. An identical, but separate, experiment at a different time was conducted with the lagoon sediment. In each case, the containers were located in the intertidal zone and experimentally contaminated with diesel oil at a concentration of 6 L/m^2. Only one application of the oleophilic fertilizer was used in each experiment. Results showed that, after 120 days, the Inipol-treated container on the sandy sediment gave reductions in the n-C_{17}/pristane and n-C_{18}/phytane ratios (about 30%) compared to the untreated plot, whereas no differences were noted in the fine-grained experiment. Examination of the figures displaying these data reveals that the differences were only apparent on the last day of sampling. They attributed this difference in biodegradation behavior in the two soil types to the fact that the Inipol might have partitioned better to the coarse-grained material, thus having less tendency to wash away with incoming and outgoing tides. However, no attempt was made to estimate error by any means (no standard deviation error bars were shown in the figures). Consequently, it is not possible for the reader to dispute the interpretations of the authors. In this instance, the experimental conclusions were hampered not only by the lack of replicate treatments and adequate controls, but also by the fact that two separate experiments, each using a different sediment type, were carried out at different times. If one wishes to study the difference between fertilizer and no fertilizer *and* the difference between two sediment types and if the experimenter wants to use inferential statistics to report significant differences among these various types of treatments, the experiment must incorporate all treatment types at the same time. This is the only way one can make certain that comparisons were made under identical conditions. This, coupled with the fact that the differences noted between fertilized and untreated plots were not that impressive to begin with, points out the necessity to conduct substantially more careful investigations with proper controls and extensive descriptions of materials, methods, and assumptions. Again, I would argue that results reported in this study were equivocal and unconvincing for the reasons cited.

Sveum (1987) studied bioremediation of an accidental spill of 88,000 L of diesel fuel into Kings Bay, Spitzbergen, in November 1985. The oil weathered for 2 months on the coarse sediment prior to conducting the experiment. Four unreplicated plots were set up, three treated with Inipol, the other untreated (reference plot 1). One of the fertilized plots was treated with Inipol 2 months after application of the diesel fuel (experimental plot 1). A second was treated with Inipol 6 months later (experimental plot 2), and a third 1 month after that (experimental plot 3). In addition, two more unreplicated plots were set up during the summer months, one serving as an unfertilized control (reference plot 2) and the other fertilized with Inipol (experimental plot 4). Biodegradation was again followed by observing changes in the n-C_{17}/pristane and n-C_{18}/phytane ratios over time. From the two unreplicated plots in the winter months, the authors observed a 60% decrease in the normal/branched alkane ratios in the Inipol-treated plot compared to the untreated plot. Biostimulation was not as pronounced from the summer data (i.e., about a 25% comparative decrease). Statistical analyses were not reported in the paper. Again, because of lack of plot replication during both the winter and the summer months, the experimental design failed to account for procedure effects, experimenter bias, experimenter-generated variability, and inherent variability. It is extremely important to replicate all treatments because failure to do so precludes estimation of overall experimental error. Without the ability to estimate error, one cannot conclude that a treatment effect is significant with any degree of confidence. Of course, the evidence was suggestive that a stimulatory effect might have occurred, but without authoritative statistical support to back up a claim, it would be very difficult to convince an on-scene coordinator (OSC) to adopt bioremediation as a countermeasure for an oil spill cleanup.

Nova Scotia, Canada

The Lee group in Canada's Department of Fisheries and Oceans presented and published a series of papers in the *International Oil Spill Conference Proceedings* every biennium since 1987 (Lee and Levy, 1987, 1989b, 1991; Lee *et al.*, 1993). The experimental designs reported in each paper were all very similar. They consisted of establishing replicate oiled and unoiled "plots" on a low energy, sandy beach in Nova Scotia [or salt marsh in one instance (Lee *et al.*, 1993)], applying a certain type of fertilizer or microbial culture to some, and no amendment to others, depending on the experimental objective desired. They achieved isolation of plots by burying sand contained either in stainless steel perforated trays or fine mesh Nitex bags (both holding 7 L of sand) below the surface of the sandy experimental plane. Independent grab samples from each experimental unit were collected at different times and analyzed for microbial growth, biodegradative activity (^{3}H-thymidine uptake or $^{14}CO_2$ evolution from radiolabeled hexadecane or glutamate), most probable number (MPN) estimation of oleoclastic bacteria, and decline in *n*-heptadecane/pristane ratio or individual oil components. Unfortunately, none of the papers presented enough experimental design information to enable another group to reproduce or confirm the findings. The papers did not discuss how many replicates of each treatment were set up, how the treatments (trays or bags) were interspersed on the beach (random or systematic?), whether composite sampling was conducted, or whether the data were even subjected to objective inferential statistical analysis (although they did state in one paper that "statistically significant differences between the oiled control and the experimental treatments were not obtained during the experiment"). [Lee later informed me via personal communication that two replicates of each treatment were used in these studies, that usually 10 treatments were set out (although not all results were reported), that the treatments were randomly situated on the beach in randomized block form, but that no analysis of variance (ANOVA) was ever performed on the data.] The conclusions drawn by the investigators were based on their best scientific judgment and common sense and were not supported by statistical analysis. Although this is perfectly acceptable, it nonetheless represents an instance where researchers who *apparently* might have used proper experimental design procedures could have elevated their conclusions to a higher level of rigor and achieved heightened credibility through the use of appropriate statistical analysis techniques. Unfortunately, they did not do this, and the resultant effect is the furtherance of the controversy with respect to effectiveness of bioremediation.

At this point, it is important to paraphrase Hurlbert's plea for a detailed and complete description of the experimental approach used in any field study (Hurlbert, 1984). Full descriptions should be provided for each of the following properties of the experiment:

- objectives
- the nature of the experimental units
- the number and kinds of treatments and controls and their measurement properties
- the manner in which treatments are assigned to experimental units

- the physical arrangement of the plots
- the temporal sequence in which treatments are applied to and measurements made on the different experimental units
- a thorough description of exactly what statistical techniques were used to analyze the data
- a complete description of the models used to test the hypothesis and their underlying assumptions

The above admonishments apply to all the studies reviewed in this chapter. A common mistake made by numerous field researchers is to assume that the reader can infer the properties listed above from cursory descriptions without realizing they have left out critical information. I have singled out the Lee group to cite as examples of such oversights only because their experimental approach was, in almost every instance, proper (albeit incomplete) from a statistical standpoint. The lack of detail describing fully the experimental approach is especially unfortunate because the Canadian group has come closer than any other in designing definitive field experiments involving replicate treatments.

Lee and Levy's work with Scotian Shelf Condensate (SSC) biodegradation in the intertidal zone of sandy beaches starts off with the hypothesis that nutrient enrichment and microbial seeding enhance the biodegradation of SSC (Lee and Levy, 1987). In this case, the nutrient used was the oleophilic fertilizer Inipol. They conducted both a laboratory study in beakers and a field study with submerged trays. Results from the laboratory study suggested that Inipol stimulated the biodegradation of the SSC, as measured by the decline in the n-C_{17}/pristane ratio. When they went to the field, they made a one-time application of Inipol to the plots equivalent to the level used in the lab (10% v/v) and collected samples over time. The main hypothesis being tested was that a single application of the oleophilic fertilizer, which was purported to maintain a long-lasting residual of nutrients at the oil-water interface, would enhance hydrocarbon biodegradation over long periods without the need to reapply. However, since they found that the level of nitrogen fell to background within 2 days following Inipol application, it is not surprising that no enhancement was observed. When they added a commercial inoculum of bacteria to similar plots, they observed a slight decline in the n-C_{17}/pristane ratio. However, the authors did not account for variability through analysis of variance or other statistical tests of significance, so the results presented were not a *definitive* test of the hypothesis. Apparently, inadequate replication was provided as suggested by the authors' statement that "owing to the higher inter-

and intra-tray variability resulting from the tidal-driven migration of unoiled sand into, and of contaminated sand out of, the trays, no significant difference in the rate of condensate loss was observed among the oiled trays." Another plausible explanation for the lack of observed treatment effects was the failure of the trays to contain the oiled sand for a sufficiently long time period to permit a treatment effect to be elicited. In any event, a more complete description of the experimental approach might have shed some further light on the conclusions reached. How did the investigators physically arrange the experimental units on the beach? (The random order should be reported.) How many replicates were established? (Lee informed me he used duplicates, but he did not state that in the paper.) Did they subject their data to statistical tests of significance? Answers to these questions could have suggested a better subsequent approach.

A major recommendation emanating from the latter study was that proper enhancement of biodegradation of oil-contaminated beaches could be achieved by periodic application of fertilizer rather than a one-time application. The same authors adopted this concept in the next study reported (Lee and Levy, 1989a). One improvement over the last experiment was the use of fine mesh Nitex bags as enclosures for the experimental units. This prevented the migration of oiled sand into and out of the enclosures. Another improvement was the repeated nutrient application, which provided the degrading populations with renewed levels needed for continued breakdown of the hydrocarbons. Two types of oil (SSC and Hibernia crude) and two types of fertilizer (Inipol and water-soluble mineral nutrients) were used. No details were given as to the random arrangement (interspersion) of bags on the beach, number of replicates used, and how the data were analyzed. Fertilizer was added after each sampling event, the frequency of which was approximately monthly from viewing the figures. The authors found that periodic addition of Inipol might have actually suppressed the biodegradation of SSC and Hibernia crude, while corresponding applications of inorganic agricultural fertilizer accelerated it. Gas chromatographic analysis of the aliphatic fraction of SSC showed that components as light as n-C_{13} persisted after 165 days in untreated sediments and sediments treated with Inipol, whereas components with carbon number up to n-C_{24} were absent after 178 days in the sediments amended with agricultural fertilizer. The same trend was observed with the Hibernia crude oil experiment: components as light as n-C_{11} persisted in the untreated and Inipol-treated bags after 353 days, whereas components below n-C_{21} were absent in the enclosure treated with inorganic fertilizer. Unfortunately, the only time series data presented in the paper were the ratios of n-C_{17}/pristane and n-C_{18}/phytane. It would have been useful to have seen component analysis results at different times during the course of the experiments. The conclusions reached by Lee and Levy were acceptable because the authors did not attempt to use inferential statistics to support their interpretations. Without statistically sound evidence backing up the data, however, conclusions can only be qualitative. The experiment could have been greatly improved had the investigators been able to study both SSC and Hibernia crude oil at the same time using a randomized block design with several replicates of every treatment. Using triplicate treatments, 18 bags would have been required [(1 control and 2 fertilizers) × 2 oil types × 3 replicates]. With such a design coupled with a higher frequency of sampling and fertilizer application, the data would likely have had far greater impact.

The next study by the same authors involved periodic addition of water-soluble agricultural fertilizer granules (ammonium nitrate and triple super phosphate) in an attempt to enhance biodegradation of waxy crude oil in a low-energy, sandy beach and in a salt marsh (Lee and Levy, 1991). Two concentrations of the NH_4NO_3 were tested (0.34 and 1.36 g/L sediment). The oil used was Terra Nova crude at two different levels (0.3 and 3.0%), and isolation of the plots was again achieved by use of Nitex containment bags secured below the surface of the environmental plane, each bag holding 7 L oiled sand. Results from the sandy beach showed that at the lower level of oil contamination, no enhancement by fertilizer was evident (i.e., declines in the n-C_{17}/pristane ratio were equivalent for untreated and treated bags). However, at the higher oil contamination level, substantial declines in the n-C_{17}/pristane ratio occurred in the fertilized enclosures compared to the unfertilized, and a fourfold increase in nitrogen concentration did not further enhance the biodegradation. Results in the salt marsh were the exact opposite. Enhancement by fertilizer was significant at the 0.3% contamination level, but no enhancement occurred at the 3% oil contamination, where better penetration into the anaerobic zone occurred.

Although the above results were moderately convincing, the same lack of detail that was characteristic of the last two studies was also evident in this one. No measurements of interstitial nitrogen and phosphorus residuals were reported, so it is difficult to calculate nutrient needs to achieve optimum biodegradative loss of hydrocarbons for future studies or for an actual cleanup. Very little detailed description of the experimental layout (number of replicates, method of plot allocation, etc.) was provided to allow one to make judgments on the significance of the findings. Inferential statistics were not used to support the conclusions, and no error bars were shown on the oil depletion figures. Thus, quantitative estimates of significant differences were precluded, based on the findings from the paper.

The most complex experiment reported by the Lee group (Lee *et al.*, 1993) concerned the comparison of water-soluble fertilizer (NH_4NO_3 and triple super phosphate) with a slow-release granular fertilizer (sulfur-coated urea) on the same sandy beach as in previous studies (Lee and Levy, 1987, 1989b, 1991). No less than 14 treatments were set up involving various combinations and concentrations of quick-release and slow-release fertilizers. Nitex bags again were used to achieve plot isolation. Treatments were not replicated in this study, which is a major flaw in the experimental design. This is the first study in which the authors reported changes in the concentrations of both the aliphatic *and* the aromatic components of the oil. Only data involving the lowest concentrations of nutrients were reported in the paper, because results at higher concentrations (two- to fourfold) were "not significantly different" in terms of enhanced oil biodegradation. Significant reductions in the normal alkanes were achieved in the nutrient-treated plots compared to the untreated controls. Further, increased reductions in both the n-C_{17}/pristane ratio and the relative concentrations of normal alkanes were observed in the plots treated with slow-release nutrients, suggesting an operational advantage in the use of sulfur-coated urea granules. With respect to the aromatic fraction of the crude oil, which is the environmentally significant fraction because of its toxic properties, the investigators found no statistically significant differences between the oiled control and experimental treatments. However, they concluded on subjective grounds that the most consistent reductions occurred in the enclosures treated with the slow-release fertilizer. Since no replicates were used in the experimental design, no statistical comparisons were possible. Thus, it is difficult to accept statements of statistical significance as accurate reflections of the findings. The curves in Figure 5 of the paper show similar results among the various treatments, suggesting that fertilizer addition, regardless of the type, did not enhance removal of the aromatic components of the oil. The authors stated that "statistically significant differences between the oiled control and experimental treatments were not obtained during the experiment." How could the latter statement be made if the experiment did not include replicate treatments, which proper statistical analysis requires? In their conclusions, Lee *et al.* (1993) remark that bioremediation remains controversial, but they further point out that application of fertilizers has been shown from their field microcosm results to enhance the rate of removal of crude oil spilled in the marine environment. I would contend that conclusive statistical evidence on rate *and* extent of oil biodegradation in the field is still sorely needed. We need to end the controversy with carefully designed experiments that yield quantifiable results from statistically objective comparisons.

Israel

Rosenburg *et al.* (1992) described a recent field experiment involving bioaugmentation, biostimulation, and tilling all within the same treatment. The proprietary fertilizer used, which was developed by the authors, consisted of a polymerized ureaformaldehyde called F-1 that had oleophilic properties. By enriching certain oleoclastic bacteria on F-1 in the laboratory, they were able to develop cultures that could depolymerize the F-1 for use in degrading hydrocarbons at the oil/water interface. Thus, these bacteria had a putative selective advantage over other indigenous bacteria unable to utilize F-1 as a nitrogen source. The approach used to test the effectiveness of their F-1/bacterial consortium amendment was to establish two identical 50 m^2 plots on a sandy beach in Israel that had recently been accidentally contaminated by a 100-ton spill of heavy crude oil. One plot was treated with the Rosenburg amendment (38 kg of F-1 and 20 L of a mixed culture of *Gluconobacter* sp. and two pseudomonads), tilled manually once every day with a rake to a depth of 5 cm, and watered with 1.5 m^3 seawater. The control plot was left untreated. Oil removal rates were measured by following the disappearance of TPH over time. Results indicated that 84.5% of the oil was removed from the treated plot within 25 days, compared to only 15.6% from the control plot. Surprisingly, the intermediate weight alkanes (n-C_{20} to n-C_{32}) were more extensively degraded than the lower molecular weight fraction (n-C_{14} to n-C_{18}).

Clearly, several flaws are evident in the experimental design described above. First, the plots were not replicated, so the only truly scientific conclusion that can be reached is that significant differences existed between the two plots after 25 days. The difference cannot be ascribed to the bioremediation treatment for reasons that should be patently obvious by now (see early discussion on pseudoreplication). A second flaw is the different ways the two plots were treated by the experimenters. The bioremediated plot was watered daily and tilled; the control plot was not. To be a true control, the untreated plot(s) must receive the exact same treatment (aside from the bioremediation amendments) as the treated plot(s); i.e., it (they) must be watered with the same volume of seawater and tilled in exactly the same way and at exactly the same frequency. According to Hurlbert's standards, this was an uncontrolled experiment, not only because its lack of replication did not control for the stochastic factor, but also because the "control" did not control for procedural effects (tilling and watering). In their review of this article, Swannell *et al.* (1996) recognized the second flaw, yet they still concluded that the authors successfully demonstrated bioremediation in a warm climate. This type of biased interpretation only exacerbates the skepticism pervading the bioremediation field today.

Prince William Sound, Alaska

Another example where an experimental design involving pseudoreplication prevented an unequivocal interpretation of important bioremediation information is the Alaska oil spill project (Pritchard and Costa, 1991; Pritchard *et al.,* 1991; Pritchard *et al.,* 1992). In 1989, following the grounding of the supertanker *T/V Exxon Valdez* on Bligh Reef in Prince William Sound, EPA, in conjunction with the Exxon Corporation and the state of Alaska, embarked on the largest oil spill bioremediation project ever attempted. Two sites on Knight Island were selected for conducting the experiments: Snug Harbor, located on the southeast corner, and Passage Cove, located on the northwest edge of the island.

SNUG HARBOR

At the Snug Harbor site, six plots were laid out. Three were positioned on a cobble surface and three on a mixed sand and gravel surface. The plots varied in linear dimensions, ranging from 21 to 35 m wide and 8 to 12 m in length (high tide to low tide). Three treatments were compared: oleophilic fertilizer addition (Inipol EAP 22) (two applications, the second 9 days after the first); slow-release solid fertilizer briquettes (Woodace™ comprised of isobutylidine diurea) contained in herring seine bags tethered to the beach surface at the intersection of the midpoint of each horizontal section with each of the seven lines demarcating the longitudinal sections (repositioning took place 10 days after startup because the upper bags were not being submerged by incoming tides); and a no amendment control. The three treatments were repeated on each of the two terrains. Buffer zones of at least 5 m separated the treated plots. To minimize cross-contamination, larger buffer zones (~20 m) were established between the control and treated plots. Each plot was divided into 21 subplots (7 horizontal × 3 longitudinal array). Each subplot was further subdivided into 1 m × 1 m grid cells for sampling purposes. At each sampling event, an individual sample was collected from one grid cell within each subplot according to a predetermined random number sequence. All 21 subplots were sampled at the same random location. Thus, a total of 126 samples were collected (21 × 6 plots) at each sampling event and analyzed individually for oil residue weight (gravimetric), normal and branched alkanes, nutrients, and oleoclastic bacteria by most probable number (MPN).

The concentration of Inipol-N added to the Snug Harbor plots was 10.9 g/m^2 (first application) and 11.5 g/m^2 (second application). The total concentration of Woodace™ nitrogen applied was 141 g/m^2 after repositioning.

Although the Snug Harbor experiment did contain two replicates of the three treatments, the data were analyzed by linear regression analysis separately for each terrain type (after log transforming the gravimetric residue weight data). With respect to the cobble terrain, semi-logarithmic plots of the residue weights as a function of time were highly scattered, the data varying by as much as two orders of magnitude around the median. The authors concluded that test plots behaved differently during the first 30 days of the experiment. Use of the nonparametric Mann-Whitney test to compare the Inipol plot with the control revealed significance after the third sampling event, suggesting more oil was lost from the treated plot, which also agreed with visual observations. The opposite was true for the briquette-treated plot. Regressions were performed over the first 30 days of the experiment. Decay rates were fastest for the Inipol plot, slowest for the briquette plot, and intermediate for the control plot. They reported that "only the rate on the Inipol-treated beach . . . was significantly different from a slope of zero, suggesting the Inipol fertilizer increased oil removal rates by at least three-fold." Rates after the 30-day initial period were all significantly different from zero but not different from each other. This was interpreted to mean that Inipol application initiated a faster rate of removal that was maintained over the entire test period, but the effect of the Inipol was short-lived. One important observation made by the investigators was the fact that many samples from the untreated control plot had undetectable concentrations of oil, "indicating that the oil on this plot was much more heterogeneously distributed than on the other plots."

With respect to the sand and gravel part of the Snug Harbor beach, regression analysis comparing the briquette-treated plot with the control plot showed that only the slope of the control plot was significantly different from zero, suggesting inhibition by the briquette fertilizer. Although the Inipol-treated plot had substantially less oil at T_0 than the other two plots, the slope of the regression curve was significantly different from zero. Thus, Inipol caused a slightly greater removal rate of oil compared to the control plot.

Several points need to be highlighted when considering the overall experimental design of the Snug Harbor experiment and the subsequent data analysis and interpretation. First, with respect to the experimental design, two replicates of the three treatments were set up on the beach, one on a cobble surface, the second on a mixed sand and gravel surface. This is perfectly acceptable, because a randomized complete block design does not require that all blocks must share exactly the same environmental conditions to enable the treatments within them to be true replicates. Effects due to differences in terrain will be accounted for by the ANOVA in the block effect. The design does require, however, that the treatments contained within a block share as close to identical geomorphology as possible, which in this case they did. Second, the treatments were not randomized as to arrangement on the experimental plane: the oleophilic plot was always in the middle between the control and the water-soluble plots, and both water-soluble plots were adjacent to a running stream. This could have caused systematic errors that might have skewed the results in one direction or another. Third, even though a complete block design was used (albeit nonrandomized), the results were not analyzed by ANOVA. Rather, the investigators decided to treat each section of beach individually, using the 21 nonindependent data points from each plot at each sampling event as pseudoreplicates in support of their regression analyses. When analyzing the results of an experiment, the data from *all* plots must be analyzed simultaneously to estimate error variability leading to proper conclusions on treatment effects. It does not matter how many data points were collected from within each plot (21, 121, or 1021); the ANOVA only considers each plot datum when computing significant effects. It would have been more correct and acceptable if plot means had been calculated from the 21 measurements from each of the six plots and the means used in an ANOVA to infer treatment differences. Granted, the findings would likely have been the same (i.e., no significant differences), but the seemingly endless search for significant differences through the overuse and even misuse of regression analysis and nonparametric tests such as the Mann-Whitney test would have been averted. Finally, the plots were not identical as to size. Some plots were 21 m long, others 28 m, still others 35 m. The concentration of applied nitrogen was not the same for the Inipol and the Woodace plots.

All plots should be as nearly identical as possible in area and linear dimensions so that systematic fluctuations in environmental or climatic effects do not affect certain plots differently. Furthermore, if a valid comparison of nitrogen source is desired, all treated plots should receive the same dose of nitrogen so that the only difference is the composition of the chemical being applied.

The extreme variability encountered in oil chemistry results strongly suggests, even demands, the use of more independent replicates when conducting these types of field studies. Pritchard *et al.* (1991, p. 48) stated in their report: "Measured effects were attributable to both nutrient treatment effects and beach effects. It could not be determined statistically whether an increased bioremediation rate at a site was due to either the treatment or to a fortuitously good location, since the two variables were confounded." The only way to reduce such confounding is to increase the number of independent replicate plots, randomly intersperse them on the experimental plane, and analyze the resultant data using appropriate statistical testing procedures. Since the linear stretch of beach available at Snug Harbor approximated 200 m, the increased number of replicate plots could have been achieved by designing smaller plots (e.g., 5 m by 10 m) with buffer zones of 10 m between them. This arrangement would have easily allowed four replicates of the three treatments and enabled better control of the manipulated variables.

PASSAGE COVE

At the Passage Cove site, four plots of similar dimensions were laid out. One served as a control, two were treated on a one-time basis with a combination of Inipol and a slow-release granular fertilizer called Customblen™, and the fourth received daily applications of water-soluble fertilizer via a sprinkler system. Samples (21 from each plot) were collected approximately every 2 weeks for 8 weeks (i.e., four sampling events including T_0). The concentration of nitrogen added to the sprinkler plot (NH_4NO_3) was 3.7 g N/m^2 per day. The concentrations of nitrogen added to the two plots receiving Inipol and Customblen™ were 21.5 g/m^2 and 247 g/m^2, respectively. The nutrients were added to the latter plots once at the beginning; the sprinkler plot received daily applications, except on occasion when weather hampered the operation. The control plot received no treatment. Data were analyzed in exactly the same way as the Snug Harbor experiment, the primary dependent variable being residue weight.

Regression analysis of log-transformed data indicated that the plot receiving daily application of water-soluble nutrients via the sprinkler system gave the best removal of TPH. Comparisons were made individually between a treated plot and the control plot, rather than conducting an overall ANOVA. After concluding that the sprinkler plot was the only plot giving statistically significant loss rates compared to the control, the investigators then reexamined the nontransformed data by regression analysis and calculated a zero-order residue weight decay rate for each of the three treatments. In their analysis, they decided to exclude some of the data that did not conform to their prescribed linear decay pattern. Only the first three sampling days were included in the regression. From this, they concluded that application of fertilizer solution promoted the loss of oil residue four to five times faster than the intrinsic rate.

The Passage Cove experiment is another classic example of pseudoreplication and the misuse of statistical analysis to add a "veneer of rigor" (Hurlbert, 1984) to a nonrigorous experimental design. Not only were the control and sprinkler plots not replicated, but the control plot was not treated the same way as the treated plot. It should have been sprayed with seawater alone to control for the procedural effects of the experimenters. Is it possible that some of the enhanced oil removal *could* have been promoted by the action of the water being sprayed onto the plots? Although, admittedly, this possibility is somewhat remote, nonetheless, it certainly weakens the argument that spray irrigation provided "definitive proof that biodegradation (and not chemical washing) was responsible for the oil removal" (Pritchard *et al.,* 1991).

When conducting any kind of quantitative analysis of experimental data, the analyst must include all the data and not assume any kind of reaction order *a priori.* Once the analysis is complete, the analyst or statistician can then examine the data set as a whole and perform tests (e.g., residuals analysis) to determine if outlyers exist or expose other trends that may negate the reaction order presumption. By excluding data that do not properly conform to an assumed order, the analyst departs from the realm of scientific objectivity and promotes a lack of confidence on the part of the scientific community in the veracity of all the data.

Several conclusions can be addressed concerning the Alaska Oil Spill Bioremediation Project. In my opinion, the only treatment that appears to have been successful in stimulating indigenous populations to degrade the spilled oil was the sprinkler plot on Passage Cove. Because the plots were not replicated, however, only qualitative conclusions are possible. Recommendations can be made to on-scene coordinators that bioremediation with inorganic mineral nutrients applied by sprinkler can be pursued. The extent to which biological removal of oil contaminating cobble or mixed sand and gravel surfaces and the rates at which such removals take place cannot be ascertained from the data presented. All the statistical analyses presented in the report, with the hundreds of degrees of freedom shown in the data tables, were conducted on nonindependent data and thus cannot be relied upon in support of any quantitative estimates of hydrocarbon removal.

DISK ISLAND

In the summer of 1990, an experiment was conducted on Disk Island in Prince William Sound to test the effectiveness of mineral nutrients and two commercial microbial products in enhancing bioremediation of the Prudhoe Bay crude oil that had been spilled by the *Exxon Valdez* tanker more than 1 year before (Venosa *et al.,* 1992). A randomized block design was set up in which four blocks of four treatments were laid out in the intertidal zone of the beach. The four treatments consisted of a no-nutrient control, nutrients alone, and two commercial products made by Sybron and ERI Waste Microbes. Each plot measured 2 m × 5 m, and a 3-m buffer zone separated the plots within the blocks. The plots were randomly placed in line within the blocks. Each plot was subdivided horizontally into three equal sectors. Each of the sectors contained four small fiberglass screen bags buried 5 cm below the surface containing approximately 1 kg of previously sieved, oiled sand from the high intertidal zone of the beach. These bags served as samples that were collected on predetermined sampling days (0, 9, 18, and 27 days). On every sampling day, three bags were removed from each of the 16 plots, one from each sector of each plot. Thus, 48 samples were collected on four different days for capillary gas chromatography analysis of oil constituents. Nutrients (NH_4NO_3 and a commercial phosphate fertilizer called Upstart™ with an N-P-K content of 3-10-3) were added at time 0 and every 4 days for the duration of the experiment, which was 27 days. The nutrients were dissolved in seawater and immediately applied with a sprinkling can to the appropriate plots. The microbial products were applied in the same way. Results, which were analyzed by analysis of variance, indicated no significant differences in resolved alkane and residue weight disappearance among the four treatments after 27 days of exposure. Variability was extremely high, and this, coupled with the short duration of the experiment and the fact that the oil had been severely weathered prior to startup (the n-C_{17}/pristane and n-C_{18}/phytane ratios were 0.18 and 0.27 at day 0), was attributed to the lack of effectiveness encountered. A statistical power test was performed to determine if a Type II error had been made. The probability of rejecting the null hypothesis was found to be only 41%, meaning that, if the true means had differed by the same amount as the sample means, there would have been only a 41% chance of detecting such a difference. Thus, the variability was too high to enable a difference to be detected.

One error in the above experiment, as rigorous as it was from a randomization and replication standpoint, was the fact that the control plots were not treated exactly the same way as the treated plots (i.e., no seawater was applied to the control plots on nutrient/product application days). Since no differences were observed, however, the error had no effect on the results. It points out the great care that must be exercised in designing field studies properly. It is unfortunate that the above study had not been conducted a year earlier on a more representative Prince William Sound beach; results might have been more definitive.

The Exxon Study

Exxon researchers implemented a field monitoring program in 1990 to address the problem of sample heterogeneity and to provide more convincing evidence in support of bioremediation (Bragg *et al.,* 1994). They correctly reasoned that measuring changes in oil composition relative to the biomarker hopane [a nonbiodegradable pentacyclic triterpane (Peters and Moldowan, 1993)] would eliminate much of the extreme heterogeneity that confounded previous statistical analyses. Tests were carried out on three separate beaches on Knight Island (KN-135, a low-energy beach; KN-211, a high-energy beach; and KN-132, another low-energy beach). Each site was subdivided, half receiving fertilizer (a combination of Inipol and Customblen), the other receiving no treatment. Samples were collected periodically over a period of 109 days (the control for only 72 days). Three composite samples in the general vicinity of three water wells from each half of each site were collected and analyzed separately by gas chromatography and gas chromatography/mass spectrometry. Thus, each sampling event resulted in three control values and three fertilizer values from each site. Measured changes in oil chemistry were analyzed by regression analysis, and relationships were modeled by a multiparameter mathematical model. The investigators concluded that changes in hydrocarbon/hopane ratios were most influenced by the oil load, the concentration of the compounds comprising the polar fraction, and cumulative nitrogen measured in the interstitial water, the latter being the most important variable. With respect to KN-135, the model predicted that the fertilized plot gave a rate enhancement 8.6 times greater than the unfertilized control over the 109-day period.

Although at first glance one might be impressed with the enhanced biodegradation predicted by the model and actually measured, the final outcome was derived from a classical pseudoreplicate design. Whenever a research team reports that a field plot was divided in half, one side treated and the other not, and makes no mention that randomization was attempted (even though only two treatments were investigated), it is usually a good sign that a pseudoreplicate design was used to test the underlying hypothesis. In the Exxon case, three treated and three control samples were analyzed at each sampling event; however, since those samples were collected from the same plot, they were not independent. If an inference on treatment effects is desired, the triplicates do not improve the power and should not be used to artificially increase the degrees of freedom of the regression. A regression analysis makes the same assumptions as an ANOVA: all samples are independent and normally distributed. It would have been just as powerful to have tested the enhancement effect after pooling the three nonindependent samples.

A second point is the fact that the nitrogen data used in the multicomponent model were cumulative and that the disappearance of nitrogen was assumed to be linear between measurements (i.e., data points were connected by a straight line via extrapolation even though many days separated the actual measurements). If a spike of nitrogen is added at T_0 and an incoming tide totally washes it out, do the degrading populations still need nitrogen to continue the biodegradation process? The answer is yes. If a second addition is made 40 days later and again the nitrogen is diluted away by the first or second tidal wash, will the biodegradation rate continue unimpeded? The answer is no, although some nitrogen recycling might support a slightly enhanced rate over background. Therefore, what is the relevance of cumulative nitrogen in the overall scheme of bioremediation and how is the information used to design a bioremediation strategy?

The Exxon researchers came closer than the Norwegian and Israeli scientists in setting up a true replicate design, because they conducted their testing on three separate, independent sites. However, they failed the criterion that all treatments be treated alike (fertilizer application frequencies were substantially different among the three sites, and the control was fertilized after 72 days, thereby negating any possible further statistical treatment comparison beyond that time). Although they regressed the data from all sites, not all the data were independent for reasons mentioned before regarding the sampling strategy used.

Figure 3 of their paper is puzzling. It shows plots of observed versus predicted data with a 45° line drawn through the data and 95% confidence bands surrounding the line. This is a nonconventional way of presenting regression data and is very confusing. Usually, 95% confidence bands are shown on *x-y* plots, not observed versus predicted plots. The only way Figure 3 could have been generated is to have first fitted a model to the measured data (i.e., using a standard *x-y* plot) and then used the predicted data as *x*-values for regressing against the observed data in a second regression. Naturally, this technique will yield high r^2 values and narrow confidence bands. The best way for the authors to have presented the data from the multiple regression analysis was to have shown in a table the parameter estimates for the untreated and the treated plots along with the standard errors and r^2 values side by side.

Finally, examining the observed versus predicted plots in the same Figure 3, the vast majority of the data fell outside the 95% confidence limits shown on the figure. No explanation was given for this. Thus, we are still left with skepticism regarding statistically credible support for the effectiveness of bioremediation on oiled shorelines.

Future oil spill bioremediation studies must involve replicate plots and proper interspersion of treatments. Plots do not need to be large; they need only be representative of the terrain in which they preside and provide a reasonable representation of the conditions imposing on the test site (tidal inundation, longshore currents, daylight exposure, stream flow gradients, length of intertidal zone affected, etc.). If the intertidal zone is small, 3 m × 3 m or 4 m × 4 m plots may be all that are needed. If the intertidal zone is larger, and one wishes to determine the effect of tidal wash on performance within the plots, perhaps a 4 m × 10 m plot (the longer dimension being in the seaward to landward direction) might be designed. In either case, several repeat plots must be set out in random configuration, and control plots should be treated in exactly the same way as the treated plots. The rest of this chapter describes an experimental design that was used recently on the shoreline of Delaware Bay that takes into consideration all the points raised above.

THE DELAWARE PROJECT

In 1991, the Congressional Office of Technology Assessment (OTA) published a background paper on the technology of bioremediation applied to marine oil spills (Office of Technology Assessment, 1991). OTA discussed two approaches to bioremediation: biostimulation (nutrient enrichment) and bioaugmentation (addition of oil-degrading microorganisms). Biostimulation is viewed by the scientific community as more promising than bioaugmentation. Among its conclusions was the statement that the "usefulness of bioremediation for marine oil spills is still being evaluated, and its ultimate importance relative to other oil spill response technologies remains uncertain." The OTA report further pointed out that little detailed analysis or performance data are yet available using bioremediation on noncobble beaches, for different types of oil, or involving different climatic and marine conditions.

On June 1 and 2, 1992, the EPA Science Advisory Board convened in Crystal City, Virginia, to review the findings of the bioremediation efforts in Prince William Sound during 1989 and 1990. Their findings were published several months later (Science Advisory Board, 1992). They concurred with some of the conclusions of the report that biodegradation was enhanced by fertilizer application. They noted, however, that biodegradation was not enhanced in every case and pointed out some deficiencies in the research approach that call into question the generalized conclusions reached in the report. For example, none of the plots on any of the beaches was replicated, and thus statistical proof that bioremediation-enhanced oil removal was insufficient. Use of nonparametric statistical analysis of highly variable and nonreplicated data was questioned, and the attempt to conduct too many different types of experiments on such a limited number of sites was too confounding to the proper statistical treatment of experimental data. Future experiments should be kept simple, the experimental design should be statistically based *a priori,* and data should be treated by parametric statistical analysis that provides confidence intervals on rates and endpoints of bioremediation, if at all possible. The SAB correctly pointed out that the physical heterogeneity of the sites in Prince William Sound is likely to be encountered often in the future. Thus, alternative statistical designs should be considered, ranging from stratified analysis of variance with suitable replications to multiple regression, which maximizes the range of environmental characteristics.

The experimental oil spill project on Delaware Bay was first conceived during the winter of 1993, partially as a result of the needs cited by the OTA and SAB reports. Discussions began in March between officials of the Delaware Department of Natural Resources and Environmental Control (DNREC) and EPA's Risk Reduction Engineering Laboratory (RREL), Cincinnati. The DNREC was especially interested in studying bioremediation for several reasons:

1. The ports of the Delaware River receive the greatest quantities of crude oil, petroleum products, and break-bulk cargo on the eastern seaboard. Shipments in 1990 amounted to 1.25 billion barrels. A supertanker spill would be catastrophic to the rich shoreline of the Bay, and alternative remedial measures are conceived by DNREC to be greatly needed for a proper response action.
2. Every spring, millions of horseshoe crabs (*Limulus polyphemus*) come ashore on the Delaware and New Jersey coastlines to mate and deposit their eggs in the sand. These shorelines are a major feeding and congregation area for over 20 species of migratory shorebirds during the height of the migratory season. The main attraction for the more than 1 million spring migrant shorebirds is the horseshoe crab eggs that are deposited in the sands near the high tide line during the spring tides of that month. Because of the large proportion of populations involved, a failed migration and breeding season due to a catastrophic oil spill might require years for recovery.

The goals of the project were to obtain sufficient statistical and scientifically credible evidence to determine if bioremediation with inorganic mineral nutrients and/or microbial inoculation enhances the removal of crude oil contaminating mixed sand and gravel beaches, to compute the rate at which such enhancement takes place, and to establish engineering guidelines on how to bioremediate an oil-contaminated sandy shoreline. Other goals were to quantify biological effects of the oil and the comparative recovery in the various treated plots. Sediment assays included solid-phase Microtox, amphipod mortality, and Ames mutagenicity assays of oil extracts. Pore water assays included Microtox, grass shrimp mortality (Fisher and Foss, 1993), and echinoderm fertilization bioassays.

Approach

The plan was to maximize the effectiveness of bioremediation by maintaining a certain level of nitrogen (and phosphorus) in contact with the degrading populations so that they would be able to grow at their maximal rate at all times. A previous study (Venosa *et al.*, 1994a) had shown that the minimum nitrate-N concentration needed by oil degraders to grow on hydrocarbons at an accelerated rate under continuous flow conditions was 1.5 mg/L. It was also known that, if the incoming tide completely submerged the plot, the levels of nitrate in the interstitial pore water diluted to undetectable limits (Venosa *et al.,* 1994b). Thus, to maximize bioremediation during spring tides, we reasoned that nutrients would have to be applied every day. To achieve the target 1.5 mg/L interstitial pore water concentrations, we assumed a 100-fold safety factor to account for dilution. The amount of sodium nitrate needed under these circumstances was thus calculated to be 2 kg/plot (330 g nitrogen/plot).

The approach used to assess treatment effects in the field study involved a randomized complete block (RCB) design with repeated measures. This statistically based design uses randomization and replication to obtain unbiased estimates of treatment effects on oil biodegradation. Five areas of beach were selected based on the homogeneity of geomorphology within each area. Each area ("block") was large enough to accommodate four experimental units or test plots. The blocks were situated in a row on the beach parallel to the shoreline. Three treatments were tested on oiled plots: a no-nutrient control, water-soluble nutrients, and water-soluble nutrients supplemented with a natural microbial inoculum from the site. A fourth treatment, an unoiled and untreated plot, served as a background control for the biological repopulation studies. The four treatments were randomized in each of the five blocks.

BLOCK LAYOUT

A schematic diagram of the block layout is summarized in Figure 1. The five blocks of four treatments were located on one long beach as shown. Each treatment plot measured 36 m^2 in area (4 m × 9 m). The minimum distance between plots was 10 m. The distance between blocks was also 10 m, although 104 m separated blocks I and II and 154 m separated blocks IV and V. The plots were laid out using standard surveying equipment. Each plot was positioned at precisely the same elevation relative to known benchmarks (fence posts) placed strategically on the high dune area so that all plots would experience the same levels of submersion and exposure. Three steel fenceposts of known elevation were driven in the middle of each plot on a line bisecting the longitudinal axis of the plot. The purpose was to measure the change in beach topography with time as tide and wave action cause erosion and accretion of the sand within the plots. Such topographical alterations could artificially affect oil concentrations measured. The beach that had been selected (Fowler Beach) consisted of small gravel (about 25 mm or less in diameter) laid atop a mixed coarse sand and gravel matrix in a low to medium wave energy environment.

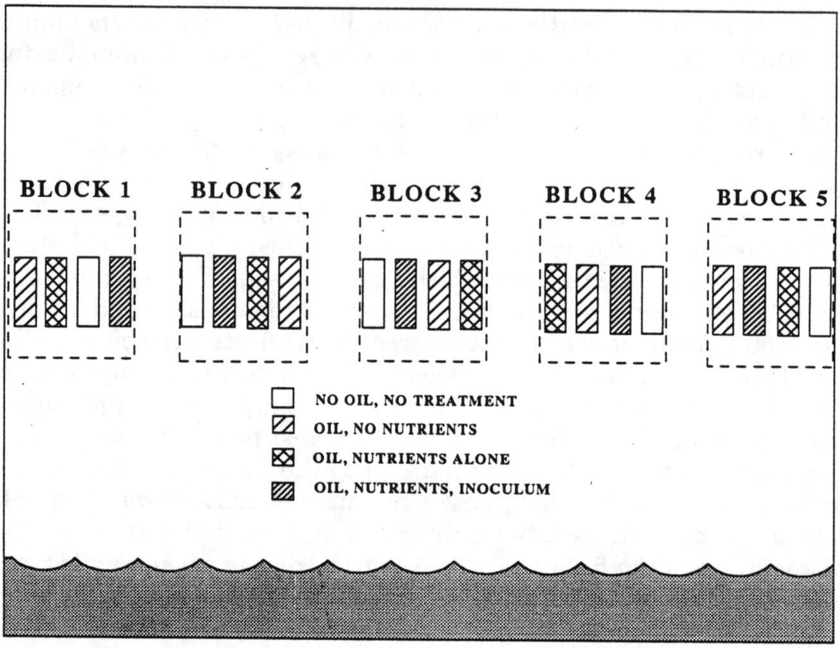

Figure 1. Randomized complete block design showing actual layout of treatment and control plots in five replicate blocks in the high-intertidal zone of Fowler Beach, DE (not to scale).

DETAILED DESCRIPTION OF BLOCKS

Each block had its own gasoline-powered generator to supply power to four electric pumps, one for each of the four plots. The electric pumps each had a pumping capacity of 4200 L/h and supplied seawater with or without nutrients and/or microorganisms to their respective plots. Adjacent to each pump was a Nalgene reservoir, capacity 850 L, holding the 800 L of water containing the appropriate amendments for application to the plots. Once every day, the field crew pumped the contents of the four reservoirs to the respective plots until emptied.

Each plot had six sprinkler heads, each rated at 400 L/h, connected by PVC piping. The contents of the reservoirs were pumped through the sprinkler heads for approximately 30 minutes (or until all the water had been applied). Water was applied to all four plots, regardless of whether or not a plot received nutrients and/or inoculum. For the nutrient plot, the fertilizer was previously dissolved in the reservoir. For the inoculum plot, the reservoir contained both predissolved nutrients and a suspended mixed population of microorganisms previously prepared (see below). For the no-nutrient control plot, only seawater was applied through the sprinkler system. Thus, each plot was treated in an identical fashion, the only differences being the contents of the water being applied.

CONTAINMENT BOOMS

The Marine Spill Response Corporation (MSRC) provided Slickbar booms for each plot to contain the oil and free-moving oiled sand. These booms were 98 ft in length and contained 2-ft long plastic floats molded into a plastic skirted sheet extending about 24 in. below the water surface. The booms were held in place between pairs of black iron pipe driven several feet below the surface of the sand and spaced every 5–6 ft around the perimeter of each plot. The booms floated on the water during high tide to contain floating oil and served as dams containing oiled sand when the tide went out. During the first several days after applying the oil, a second containment boom made of sorbent material was placed around the Slickbar boom as an added safety precaution in the event of an accidental release beyond the primary containment.

APPLICATION OF OIL TO PLOTS

Location of the oiled plots was crucial to the success of the project, because edge effects associated with small plots could seriously jeopardize the integrity of the plots. In other words, tidal and wave action can easily upwash and redistribute the oil from the plots to other parts of the beach, thereby precluding the study of bioremediation effectiveness. Indeed, this occurred during the Baffin Island Oil Spill (BIOS) project on two separate beaches, Crude Oil Point and Bay 106 (Owens and Robson, 1987), where it was observed that, 8 days after application of crude oil to 40 m^2 plots, the role of sediment transport and dispersion away from the plots by waves became a significant factor. On the Bay 106 intertidal plots, the mean oil-in-sediment concentration decreased from 9400 mg/kg immediately after application of the oil to 2220 mg/kg after 24 h, with the oil redistributing above the plots by the tides.

Oil was applied to those plots at approximately midtide on a flooding tide. Ambient temperature was approximately 7°C (Baffin Island is located above the Arctic Circle in Canada's Northwest Territories). Investigators wanted to simulate real-world Arctic spill conditions, where oil becomes stranded from the adjacent contaminated water surface, and to study fate and weathering of the stranded oil over time. However, in another part of the Baffin Island study, Owens *et al.* (1987) observed that, when oil is spilled on water and then allowed to contaminate the shoreline, most of the stranding takes place in the upper third of the intertidal zone, near the high water line. The researchers observed that, within 3 days after oiling had taken place, oil that had come into direct contact with sediment material was much more resistant to refloating. They had found this consistent with observations made in the *Amoco Cadiz* incident (Hayes *et al.,* 1989). Indeed, this was even consistent to some extent with observations from the *Exxon Valdez* spill in Prince William Sound.

From the foregoing, it was clear that to minimize edge effects in a small plot study such as this, oil should be applied near the spring high tide line, which it was. The crude oil (provided by Sun Oil Company) used was Bonny Light (Escravos), which is imported from Nigeria. It has a specific gravity of 0.848, API Gravity of 35.3, Reed Vapor Pressure of 6.9, viscosity of 5.6 CST at 70°F, a pour point of 27°F, and a sulfur content of 0.11% by weight. It was weathered prior to application by placing about 800 gal in a 12-ft diameter "fast tank," connecting a pump and hose, and continuously spraying and recirculating the oil within the tank for 2 days to evaporate the light fraction. Following this, the weathered oil was placed into 55-gal drums (36 gal per drum) to await application. Oil was applied on July 1, 1994. It was applied manually by two four-membered teams. Two persons held onto each end of a 13-ft pipe holding a hose connected to a pump that led into the 55-gal drum containing the oil. The oil was pumped through the hose out of four atomizer nozzles attached to the pipe manifold. The persons applying the oil walked slowly up and down the plot while a third person lifted the trailing hose over the black iron pipe holding the containment booms in place. The fourth person held a stopwatch and timed the walking pace of the appliers. The oil was applied uniformly in 14 sweeps of the plot over a period of about 7 min.

INOCULUM PREPARATION

The inoculum was prepared by growing the culture in two stainless steel 55-gal drums, each containing 45 gal of fresh seawater from Delaware Bay, the weathered Bonny Light crude oil (600 ml), and the same nutrients as used on the beach. Inoculation times of the two drums were staggered by 1 week to allow weekly inoculation of the appropriate plots. The original culture consisted of a mixed consortium isolated from the same beach several months prior to the experiment and growing it in the laboratory on Bonny Light crude oil. On a given inoculation day (every Wednesday), the contents of the stainless steel drum were transferred into a plastic drum and transported to the beach via ATV. The five Nalgene reservoirs designated to receive the inoculum were each inoculated with 8 gal of the culture along with the separate nutrient mix.

NUTRIENT MIX

The nutrients added to the 10 designated reservoirs consisted of technical grade sodium nitrate and sodium tripolyphosphate. The latter has been found in our laboratory to be more soluble in seawater than orthophosphate. Approximately 2 kg $NaNO_3$ and 128 g $Na_5P_3O_{10}$ were placed into plastic Ziploc® bags and screw cap plastic bottles, respectively, and transported to the beach where they were poured into the appropriate reservoirs just prior to sprinkling. Each reservoir was mixed by recirculating the seawater in the reservoir with a garden hose for 20 min, using the designated electric pump.

QUANTITY OF OIL RELEASED

The amount of crude oil released was computed as follows. Each plot measured 36 m^2 in area. Assuming that penetration of the spilled oil occurred to a depth of approximately 30 cm, a penetration depth commonly observed during the summer of 1989 in Prince William Sound, the volume of beach material to be contaminated was 10.8 m^3. Using a specific gravity of sand to be in the approximate range of 2650 kg/m^3 and assuming a void volume of 32%, the mass of beach material receiving oil treatment was 19,462 kg. The typical concentration of oil receiving bioremediation treatment in Prince William Sound was approximately 5 g/kg. Assuming an initial weight loss of 20% due to evaporation and sand redistribution, the starting concentration of oil would be 6.25 g/kg. Multiplying the desired initial oil concentration (6.25 g/kg) by the mass of beach material to be oiled (19,462 kg) and assuming a specific gravity of the oil to be about 0.9 g/cm^3 resulted in an oil volume requirement of 135 L/plot (approximately 36 gal per plot). Thus, the total volume of oil to be released in the study computed to 2025 L or 540 gal (36 gal/plot × 15 oiled plots).

STATISTICAL BASIS FOR NUMBER OF REPLICATES

To calculate the number of replicate treatments needed to detect treatment differences, if they were present, with a high probability (power), the following information was needed:

- concentration of target hydrocarbons in the crude oil at time 0
- expected average decrease in target hydrocarbons in 120 days (the minimum duration of the field study) in both treatments and controls

- expected difference in target hydrocarbons between treatments and controls after 120 days
- expected variability and the risks of an incorrect decision

Concentration of Target Hydrocarbons in Light Crude Oil

According to data accumulated at the EPA-Cincinnati laboratory, the average concentrations of total target alkanes (n-C_{10} to n-C_{35} plus pristane and phytane) and total aromatics (two-, three-, four-, and five-ring PAHs and heterocyclics) in Arabian Light crude oil (Bonny Light was presumed to be similar) as determined by GC/MS analysis of hexane extracts, were 127.5 mg/g oil and 10.7 mg/g oil, respectively. Multiplying these figures by the target concentration of oil after evaporative loss (5 g/kg) resulted in initial concentrations of 637.5 mg alkanes/kg beach material and 53.5 mg aromatics/kg. (The actual concentrations measured at T_0 were quite close: 488.0 mg alkanes/kg sand and 54.3 mg aromatics/kg sand.)

Expected Average Decrease in 120 Days

According to Pritchard and Costa (1991), the disappearance of oil residue weight on the control plot at Passage Cove (Passage Cove was the beach that was treated with water-soluble fertilizer via a sprinkler system in the summer of 1989) was 7.2 mg/kg/day. The disappearance of oil residue weight from the nutrient-treated plot on Passage Cove was 21.3 mg/kg/day. Assuming the same rates of disappearance as the Passage Cove observations, the total amount of residue weight loss expected after 120 days was 0.9 g residue weight/kg beach material in the control plots and 2.6 g/kg in the treated plots. This represented a 17% loss in the control plots and 51% loss in the nutrient-amended plots. Thus, the expected average concentration of alkanes and aromatics remaining on the control plots after 120 days, assuming conservatively that degradation of the target hydrocarbons proceeds at the same rate as the total residue weight, was 529.1 mg/kg (637.5 × 0.83) and 44.4 mg/kg (53.5 × 0.83), respectively. Expected average alkanes and aromatics in the treated plots were 312.4 mg/kg (637.5 × 0.49) and 26.2 mg/kg (53.5 × 0.49), respectively.

Expected Difference between Treatments and Controls after 120 days

Using the data above, the expected difference in alkanes and aromatics between the control plots and the treatment plots after 120 days was 216.7 mg/kg (529.1 – 312.4) for the alkanes and 18.2 mg/kg (44.4 – 26.2) for the aromatics.

Variance Estimates

This was the most difficult variable to estimate because it presumes knowledge of the variance *a priori*. We expected to find less variability than in a spill of opportunity because of the precisely measured amount of crude oil applied to the plots at time 0 and its uniformity of application. Also, measurement and reporting of oil and analyte concentrations normalized against the internal marker hopane should further reduce the expected variance. Nevertheless, to be conservative, we assumed an expected range of values of $\pm 50\%$ of the expected means in the control plots after 120 days. For alkanes, the range of values was 529.1 $\pm$ 265 mg/kg, and for aromatics, the expected range of values was 44.4 $\pm$ 22.2 mg/kg.

The estimated standard deviation of the data was computed by dividing the range by 6, which corresponded to ± 3 standard deviation units about the mean. This gives a 99.9% accountability of the data, assuming they are normally distributed. Thus, for alkanes, using a range of 530 (2×265), the estimated standard deviation was approximately 88.3 mg/kg ($530 \div 6$). For aromatics, the estimated standard deviation was 7.4 mg/kg ($22.2 \times 2 \div 6$).

The number of replicate treatments was then calculated (Montgomery, 1991). Using an estimated standard deviation of 88.3 for alkanes and 7.4 for aromatics and an expected average difference between treatments and controls of 216.7 mg alkanes/kg and 18.2 mg aromatics/kg after 120 days, the number of field replicates needed to detect the difference, if it was in fact present, 80% of the time ($\beta = 0.20$) at the 5% significance level ($\alpha = 0.05$) was five. Thus, five replicates of each treatment were used in the study.

SAMPLE SIZE, FREQUENCY, AND HANDLING

Samples of sediment were collected every 14 days from all of the oiled plots (15 in all). Each plot was divided into four equal sectors (labeled α, β, γ, and δ, landward to seaward, respectively). Subsamples were collected from each sector to determine if differences existed in oil biodegradation according to the length of time under water. Two subsamples from each sector were collected according to a predetermined random number sequence that disallowed subsequent resampling from the same hole, composited into one sample, frozen, and shipped to Cincinnati for extraction and analysis by gas chromatography/mass spectrometry (GC/MS). Thus, a total of 60 samples, stratified within plots according to location within the intertidal zone and spread among 15 independent plots, were collected at each sampling event.

The subsamples consisted of the following: two cores 7 cm in diameter × 15 cm deep, one on top of the other, from one hole, and two identical cores from another hole of the same sector, composited and mixed by a trowel in a galvanized bucket, and placed in a prelabeled 1-gal paint can. After the composited samples were thoroughly mixed, subsamples were taken from the mixture for other types of analyses (microbiological and toxicological). A portion of each composited sample was frozen and archived in the event that a reanalysis was needed at a later time.

MICROBIOLOGICAL ANALYSIS

Sediment subsamples from each sector of each plot, including the five unoiled plots (80 altogether), were placed in Whirlpak bags, brought back to the mobile laboratory trailer under ice, and immediately processed for MPN analysis. Approximately 10 g wet weight (exact weight was recorded after weighing on a top-loading balance) was placed in a dilution bottle containing 90 ml of sterile detachment solution (1 g/L disodium pyrophosphate and 20 g/L NaCl) and shaken for 1 h at 300 rpm. The samples were then placed onto a Beckman Biomek 1000 Laboratory Workstation for automated serial 10-fold dilution in 96-well microtiter MPN plates. The growth medium was Bushnell-Haas (Bushnell and Haas, 1941), supplemented with 2% sodium chloride to maintain the same salinity as the Delaware Bay water. The substrates used to grow the oleoclastic bacteria were hexadecane for alkane degraders and a combination of phenanthrene, anthracene, fluorene, and dibenzothiophene for the aromatic degraders. Plates were incubated for 2 weeks (alkane degraders) to 3 weeks (aromatic degraders) at 20°C prior to analysis. Alkane degraders were enumerated by adding iodonitrotetrazolium violet (INT) to the wells and scoring positive wells by observing the red color that had developed. For aromatic degraders, no dye was needed. Positive wells were scored by observing the bright yellow color developed from the cleavage product of fluorene and dibenzothiophene biodegradation.

OIL CHEMISTRY

After receiving the frozen sand samples from the field, lab personnel in Cincinnati processed the samples by extracting 100 g or 500 g of sand with dichloromethane (DCM), drying on $NaSO_4$, determining gravimetric TPH in an aliquot of the extract by evaporating off the DCM and weighing the residue, and injecting another portion of the extract into Hewlett-Packard 5890 gas chromatograph equipped with an HP 5971 mass selective detector for quantifying the oil components (both aliphatic and aromatic). All analyte data were normalized to the conservative, nonbiodegradable biomarker hopane [C_{30} 17α(H),21β(H)-hopane] (Peters and Moldowan, 1993).

NUTRIENT ANALYSIS

Concentrations of nitrate-N in the interstitial pore water of oiled plots were measured each day by collecting sand cores from each of the four sectors of a randomly selected test plot and the middle of a nonnutrient control plot, extracting the contents with 1.0 L of deionized water on a shaker, and quantifying the concentration on an autoanalyzer. The method used was based on cadmium reduction of nitrate to nitrite (Technicon Industrial Systems, 1972). Results were reported as milligrams of nitrate-N/L interstitial water after determining the water content of the sample gravimetrically.

STATISTICAL ANALYSIS

At the time of this writing, GC/MS data from only the first eight sampling events had been analyzed statistically. Thus, only preliminary results can be presented in this chapter. Statistical procedures used to analyze the results are described below.

Since the treatment plots were remeasured at prespecified sampling times during the course of the investigation, repeated measures analysis of variance (RMANOVA) (Freund *et al.,* 1986) were used to analyze the response variables (MPN and oil analytes). When the repeated measures ANOVA indicated significant differences, univariate ANOVAs were run on data at each time point. Where significant differences were indicated at a specific time point, protected least significant difference (LSD) mean separations were used to assess treatment differences. Nonlinear regression analysis will be used to estimate the rate of oil degradation for each of the three treatments. The rates (slopes) and intercepts will be compared statistically. Statistically valid variability estimates and power curves will be obtained that will aid the design of other oil bioremediation studies.

The above statistical treatment of the data assumes that all samples are independent and have the same variability properties. However, it is known that samples closer to each other in space are more autocorrelated (have less variability) than those separated by wider distances. If such spatial autocorrelation were confirmed, the nonindependence assumption would be invalid; thus, the ANOVA results would be questionable. Recently, powerful statistical techniques have been developed to address the effect of spatial variability (Cressie, 1991). This spatial analysis will be used, if sufficient data exist, to model variability relative to distance between plots, and, if necessary, new ANOVAs will be conducted on the data corrected for this spatial autocorrelation.

Preliminary Results

The microbiology data are preliminary because they are reported in terms of $\log_{10}$ cells/gram wet weight rather than dry weight. Dry weight data were still incomplete at the time of this writing. GC/MS data from the first 12 sampling events are shown in this chapter, but ANOVAs had only been completed for the first eight sampling weeks.

NUTRIENT ANALYSIS

Table 1 summarizes the daily nitrate data from the oiled plots by block. The plots receiving only seawater with no nutrients had measurable concentrations of nitrate (mean of 0.82 mg/L), while the concentrations in the nutrient- and inoculum-treated plots were substantially higher. Note the high variability of the data, especially on a block basis. This was due to variability in submersion caused by the occurrence of spring and neap tides.

The concentrations of nitrate-N measured in the control plots were close to the approximate 1.5 mg/L target level desired for maximum biodegradation. The Fowler Beach area of Delaware Bay was close to farmland, where runoff could possibly account for the high natural levels found. Also, the relatively high organic content of both the Delaware Bay seawater and the underlying geology of the site and the presence of a saltwater marsh several hundred meters landward from the beach could account for the high nitrogen levels encountered. It is not unusual for measurable levels of nitrogen to be found in

Table 1. Summary of nitrate concentrations by block (mg nitrate-N/L interstitial water).

	No Nutrients		Nutrients		Inoculum	
Block	Mean	St. Dev.	Mean	St. Dev.	Mean	St. Dev.
I	0.7	0.7	3.9	1.4	2.6	2.3
II	0.8	0.9	9.5	3.3	3.9	3.2
III	1.3	1.1	6.2	3.8	1.1	0.5
IV	0.6	0.5	3.6	2.1	5.5	5.5
V	0.8	1.3	8.6	7.9	4.2	2.4
Mean	0.8		6.3		3.5	
St. Dev.	0.3		2.7		1.7	

the vicinity of a saltwater marsh because of the significant vegetative decay that occurs there. It cannot be ruled out that some carryover of nutrients might have occurred from the sprinkler systems of adjacent plots, especially on windy days. However, samples along the shoreline of Delaware Bay were collected late in the project at considerable distances south and north of our experimental location and even in the Chesapeake Bay area, and relatively high levels of nitrate-N were measured in the interstitial waters. This adds strength to the assertion that agricultural runoff or the high indigenous organic matter characteristic of the Delaware and Chesapeake Bay areas might have been the major contributors to the high natural levels of nitrate on Fowler Beach.

MICROBIOLOGICAL ANALYSIS

Figure 2 summarizes the results of the aliphatic and aromatic degrader population data for all sampling events. ANOVAs were conducted on the data at each sampling event, and no statistically significant differences were evident among the three oiled treatments at any sampling event ($p > 0.05$) (although the populations were always significantly higher than the unoiled controls). Several observations are noteworthy from the graphs:

1. At time 0, corresponding to the first day nutrient application began, which was 4 days after oil was applied, the levels of alkane and aromatic degraders in the oiled treatments were two orders of magnitude higher than those in the unoiled controls. This suggests that considerable and rapid increases in degrader populations occurred within 4 days due solely to exposure to crude oil.
2. Aliphatic degraders in both unoiled and oiled plots slowly decreased over time, while aromatic degraders increased almost three orders of magnitude in the first 2 weeks, leveled off, and then decreased. The aromatic degraders even increased in the unoiled plots, suggesting that enough oil may have escaped from the oiled plots onto the unoiled plots to cause a stimulation in growth. Several plausible explanations could account for no change being observed in the alkane degrader populations:

- They were already at their maximum field capacity at T_0.
- Continual tidal inundations could have resulted in an artifactual constancy of population numbers due to semidiurnal growth and washout.

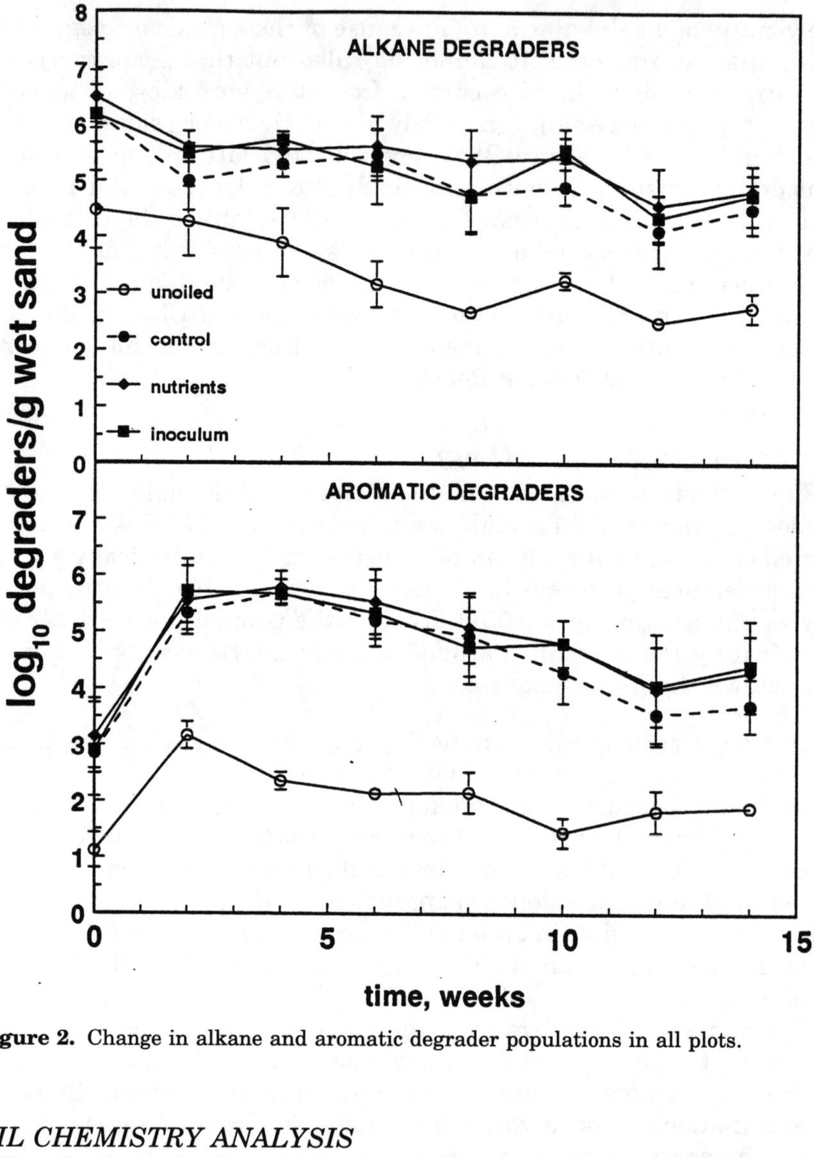

Figure 2. Change in alkane and aromatic degrader populations in all plots.

OIL CHEMISTRY ANALYSIS

Results from the first 12 weeks of the study period are summarized in Figure 3(a) and 3(b). The figures show the first-order hopane-normalized declines in alkanes [Figure 3(a)] and aromatics [Figure 3(b)] in the nutrient-treated, inoculum-treated, and control plots as a function of time. Each data point is the mean of five independent replicates, and the error bars represent one standard deviation unit above and

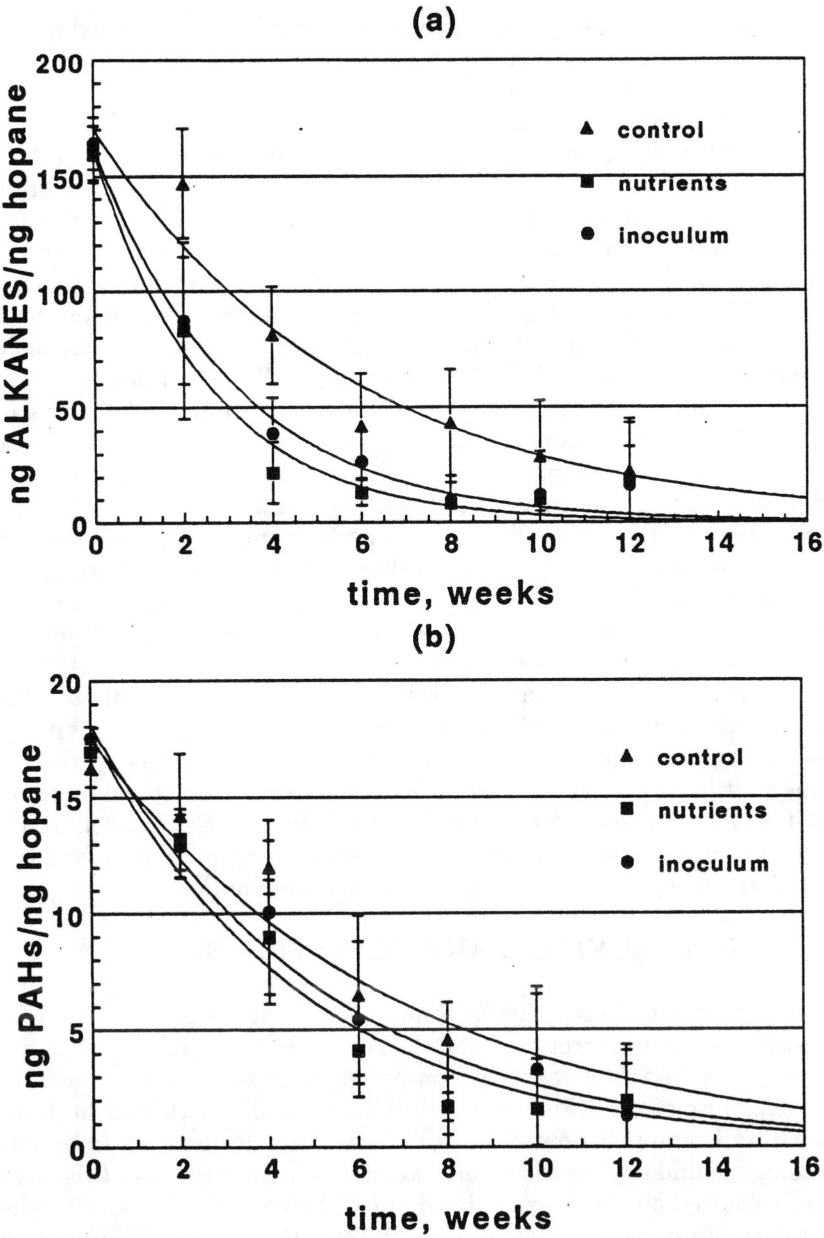

Figure 3. Decline in hopane-normalized total alkanes (a) and total aromatics (b) during the first 12 weeks of the study.

below the mean. Analyses of variance had only been completed for the first 8 weeks of the study at the time of this writing.

Statistically significant differences between treated and untreated plots were observed at weeks 2, 4, and 8 for the alkanes and week 8 for the PAHs. At no time were any significant differences observed between plots treated with nutrients alone and plots treated with nutrients and the indigenous inoculum. Notable from the data is the fact that substantial hydrocarbon biodegradation occurred in the untreated plots. This was attributed to the high levels of background nitrogen measured on the beach. Had the background nutrient levels been closer to those typical of Prince William Sound beaches (μg/L), the slopes of the control plots would likely have been much flatter, thus giving rise to highly significant differences between treated and control plots for both alkanes and PAHs perhaps as early as 2 weeks after the release. The latter conclusion is speculative, since such low background nutrient levels were not encountered in Delaware. Nonetheless, the goals of the project were met: bioremediation via nutrient enrichment was demonstrated unequivocally with statistically credible data; computed biodegradation rates were high, with an alkane half-life of approximately 2 weeks and a PAH half-life of approximately 4 weeks; and bioaugmentation even with indigenous organisms does not stimulate further declines in hydrocarbons beyond simple nutrient addition. Based on the fact that the nutrient content of Fowler Beach was measured at high enough levels to support significant biostimulative hydrocarbon decay, the recommendation would be not to encourage active bioremediation activities if an oil spill were to occur along the Delaware Bay shoreline. For coastlines having low natural input levels of nutrients, bioremediation should definitely be considered a major cleanup option.

SUMMARY AND CONCLUSIONS

The person considered the father of modern statistical experimental design, R. A. Fisher, wrote in 1930: "No one would now dream of testing the response to a treatment by comparing two plots, one treated and the other untreated" (Fisher and Wishart, 1930). Little did he know in all his wisdom how dominant pseudoreplicate designs would become in ecological and engineering field experiments. Emphasis in this chapter has focused on the overwhelming prevalence in the bioremediation literature of improperly designed or incorrectly analyzed field experiments. Experimental pseudoreplication is so widespread that information generated in support of bioremediation as a viable alternative oil spill countermeasure has led to skepticism among the user community, especially on the part of on-scene coordinators who must decide on appropriate cleanup procedures. We need to design field experiments that will provide the real world decision makers with the necessary information they need to implement the technology. Such has not been the case in this field.

One argument often put forth against use of independent replicate plots is the prohibitive cost of conducting research in this manner. Another is the requirement for a relatively large expanse of space needed to insure that plots are independent. With respect to the first point, I would argue that we cannot afford *not* to invest research funds in conducting field experiments properly. Think of all the studies that have been conducted so far (or just the ones mentioned in this chapter) that have led to our present state of knowledge. Have the funds in their support been well spent? Have the funding agencies been rewarded well for their investment? With regard to the second point, spatial needs are indeed substantial if true independence and high statistical power are desired. However, plots need not be huge to be representative, and a block of treatments need not be on the same experimental plane as others to be a valid replicate. Remember that a randomized complete block design by its very nature automatically accounts for block effects (i.e., differences in geomorphology between blocks of treatments), as long as the terrain *within* a block is approximately homogeneous for all treatments contained therein. In the *Exxon Valdez* project, Snug Harbor was long enough to accommodate four replicates of three treatments if the plots had measured 5 m × 10 m. Additional replicates could have been set up in different coves close by. The same is true for the Spitzbergen and Israeli experiments.

The reason why randomization is so important is to enable inferences on treatment effects over the whole experimental plane, not just the plots of interest. If a researcher has conducted a pseudoreplicate experiment and he assumes that no other effects except a treatment effect are present (see Pritchard *et al.,* 1991), then the inference will be limited to just the two plots of beach used in the experiment, rather than the whole beach area, since a random sample of the plots, not the whole beach, was used. Field scientists usually do not want to limit their conclusions to the experimental plots; rather, they want to be able to infer treatment effects over an entire beach. The only way to accomplish this objectively is to sample randomly over the whole beach (or a representative section).

Because they are so expensive, field experiments should be kept as simple as possible. This means limiting the number of treatments to as few as possible. Questions such as optimum concentration of oil, fertilizer, or inoculum or comparison of fertilizer types are best answered in the laboratory. A decision should be made up front, based on previously conducted laboratory experiments, what level and types of amendment(s) should be used in the field study. The field is not the place to answer basic questions such as these. The field should only be used to answer the simple questions "Does bioremediation work, and, if so, what is the rate enhancement factor?" All the energy and cost of a field study should be devoted to acquiring the highest quality data possible. This means fewer treatments and more replicates.

The Delaware study was designed to address the admonishments raised above. Based on the preliminary data presented, clear trends took shape early. Alkanes were biodegraded by almost 50% in the nutrient and inoculum plots (compared to only 10% in the no-nutrient control plots) after only 2 weeks of operation. The PAH biodegradation picture was somewhat less definitive with respect to treatment differences. The evidence suggests that sufficient nitrogen was naturally present on Fowler Beach to permit close to maximum hydrocarbon biodegradation without the need to apply additional fertilizer. This points out that bioremediation might not always be necessary if sufficient nutrients are naturally present at a spill site in high enough concentrations to effect natural cleanup. This is a very important finding. The point I wanted to make by presenting the experimental setup in such detail was to emphasize how important it is to expend one's energy in setting up the experiment correctly (such as plot layout as a function of elevation, treatment of all plots in identical fashion, staying with the original experimental design without making any changes in midstream) and keeping the experimental objective simple.

How can the "flood of statistical malpractice be stemmed?" (Hurlbert, 1984). Hurlbert suggested that scientific journal editors need only become slightly more educated in statistical design concepts to control the publication of pseudoreplicated experimental data. He eloquently opined: "When the coin of the realm is the published paper, nothing educates so well as an editorial rejection or request for major revision." We all need to become more aware of modern statistical concepts, especially the more recent spatial analysis ideas that could vastly advance the whole field of hazardous waste bioremediation, not just oil spills. Spatial analysis may be the answer to the heterogeneous contamination problems so characteristic of field situations. The plea I make is to keep all experiments simple, and replicate, replicate, replicate!

ACKNOWLEDGEMENT

I wish to express my appreciation to Dr. Dennis King of Statking Consulting for his valuable advice and review of the statistical concepts discussed in this chapter.

REFERENCES

Bragg, J.R., Prince, R.C., Harner, E.J. and Atlas, R.M. 1994. "Effectiveness of Bioremediation for the Exxon Valdez Oil Spill." *Nature,* 368:413–418.

Bushnell, L.D. and Haas, F.F. 1941. "The Utilization of Certain Hydrocarbons by Microorganisms." *J. Bacteriol.,* 41:653–673.

Cressie, N.A.C. 1991. *Statistics for Spatial Data.* John Wiley and Sons, Inc., New York.

Fisher, R.A. and Wishart, J. 1930. "The Arrangement of Field Experiments and the Statistical Reduction of the Results." *Imperial Bureau of Soil Science (London),* Technical Communications Number 10:1–23.

Fisher, W.S. and Foss, S.S. 1993. "A Simple Test for Toxicity of Number 2 Fuel Oil and Oil Dispersants to Embryos of Grass Shrimp, *Palaemonetes pugio.*" *Marine Pollution Bulletin,* 26:385–392.

Freund, R.J., Littell, R.C. and Spector, P.C. 1986. *SAS System for Linear Models.* SAS Institute, Inc. Cary, N.C.

Halmo, G. 1985. "Enhanced Biodegradation of Oil." In: *1985 International Oil Spill Conference,* American Petroleum Institute, Washington, D.C.

Hayes, M.O. Gundlach, E. and d'Ozouville, L. 1979. "Role of Dynamic Coastal Processes in the Impact and Dispersal of the 'Amoco Cadiz' Oil Spill (March, 1978) Brittany, France." In: *Proc. 1979 Oil Spill Conference,* Washington, D.C., American Petroleum Inst. Pub. No. 4308, pp. 193–198.

Hurlbert, S.H. 1984. "Pseudoreplication and the Design of Ecological Field Experiments." *Ecol. Monographs,* 54 (2):187–211.

Jobson, A.M., McLaughlin, M., Cook, F.D. and Westlake, D.W.S. 1974. "Effect of Amendments on the Microbial Utilization of Oil Applied to Soil." *Applied Microbiology,* 27:166–171.

Leahy, J.G. and Colwell, R.R. 1990. "Microbial Degradation of Hydrocarbons in the Environment." *Microbiol. Rev.,* 54:304–315.

Lee, K. and Levy, E.M. 1987. "Enhanced Biodegradation of a Light Crude Oil in Sandy Beaches." *Proc. 1987 International Oil Spill Conference,* American Petroleum Institute, Washington, D.C.

Lee, K. and Levy, E.M. 1989a. "Biodegradation of Petroleum in the Marine Environment and Its Enhancement." In: *Aquatic Toxicology and Water Quality Management,* J.A. Nriagu (ed.), John Wiley and Sons, New York.

Lee, K. and Levy, E.M. 1989b. "Enhancement of the Natural Biodegradation of Condensate and Crude Oil on Beaches of Atlantic Canada." *Proc. 1989 International Oil Spill Conference,* American Petroleum Institute, Washington, D.C.

Lee, K. and Levy, E.M. 1991. "Bioremediation: Waxy Crude Oils Stranded on Low-Energy Shorelines." *Proc. 1991 International Oil Spill Conference,* American Petroleum Institute, Washington, D.C.

Lee, K., Tremblay, G.H. and Levy, E.M. 1993. "Bioremediation: Application of Slow-Release Fertilizers on Low-Energy Shorelines." *Proc. 1993 International Oil Spill Conference,* American Petroleum Institute, Washington, D.C.

Montgomery, D.C. 1991. *Design and Analysis of Experiments.* 3rd ed., John Wiley and Sons, New York.

Office of Technology Assessment. 1991. *Bioremediation for Marine Oil Spills.* U.S. Government Printing Office, Washington, D.C.

Owen, D. 1991. "Bioremediation of Marine Oil Spills: Scientific Validity and Operational Constraints." *Proc. 14th Arctic and Marine Oil Spill Program,* Vancouver, British Columbia. Minister of Supply and Services, Canada EN 40-11/5-1991.

Owens, E.H., Harper, J.R., Robson, W. and Boehm, P.D. 1987. "Fate and Persistence of Crude Oil Stranded on a Sheltered Beach." *Arctic (J. Arctic Inst. of N. America),* G.W. Hodhson (ed.), 40 (suppl. 1), 109–123.

Owens, E.H. and Robson, W. 1987. "Experimental Design and the Retention of Oil on Arctic Test Beaches." *Arctic (J. Arctic Inst. of N. America),* G.W. Hodgson (ed.), 40 (supp. 1), 230–243.

Peters, K.E. and Moldowan, J.M. 1993. *The Biomarker Guide.* Prentice Hall, Englewood Cliffs, NJ.

Prince, Roger C. 1993. "Petroleum Spill Bioremediation in Marine Environments." *Critical Reviews in Microbiology,* 19(4):217–242.

Pritchard, P.H. and Costa, C.F. 1991. "EPA's Alaska Oil Spill Bioremediation Project." *Environmental Science and Technol.,* 25:372–379.

Pritchard, P.H., Costa, C.F. and Suit, L. 1991. *Alaska Oil Spill Bioremediation Project Science Advisory Board Draft Report.* EPA/600/9-91/046a,b. U.S. Environmental Protection Agency, Gulf Breeze, FL.

Pritchard, P.H., Mueller, J.G., Rogers, J.C., Kremer, F.V. and Glaser, J.A. 1992. "Oilspill Bioremediation: Experiences, Lessons, and Results from the Exxon Valdez Oil Spill in Alaska." *Biodegradation,* 3:315–335.

Rosenburg, E., Legmann, R., Kushmaro, A., Taube, R., Adler, R. and Ron, E.Z. 1992. "Petroleum Bioremediation—A Multiphase Problem." *Biodegradation,* 3:337–350.

Science Advisory Board. 1992. *SAB Review of the Alaskan Bioremediation Oil Spill Project.* EPA-SAB-EPEC-LTR-92-015, U.S. Environmental Protection Agency, Washington, D.C.

Sendstad, E. 1980. "Accelerated Biodegradation of Crude Oil on Arctic Shorelines." *Proc. 3rd Arctic and Marine Oil Spill Program,* Environment Canada.

Sveum, P. 1987. "Accidentally Spilled Gas-Oil in a Shoreline Sediment on Spitzbergen: Natural Fate and Enhancement of Biodegradation." *Proc. 10th Arctic and Marine Oilspill Program,* Environment Canada.

Sveum, P. and Ladousse, A. 1989. "Biodegradation of Oil in the Arctic: Enhancement by Oil-Soluble Fertilizer Application." In: *Proc. 1989 International Oil Spill Conference,* American Petroleum Institute, Washington, D.C.

Swannell, P.J., Lee, K., and McDonagh, M. 1996. "Field Evaluations of Marine Oil Spill Bioremediation." Microbiol. Rev. 60(2):342–365.

Technicon Industrial Systems. 1972. *Nitrate and Nitrite in Water and Seawater.* Industrial Method No. 158-71W. Technicon Industrial Systems, Tarrytown, NY.

Venosa, A.D., Haines, J.R. and Allen, D.M., 1992. "Efficacy of Commercial Products in Enhancing Biodegradation of Weathered Crude Oil Contaminating a Prince William Sound Beach." *Journal of Industrial Microbiology* 10:1–11.

Venosa, A.D., Haines, J.R., Suidan, M.T., Wrenn, B.A., Strohmeier, K.L., Eberhart, B.L., Holder, E.L. and Wang, X. 1994a. "Research Leading to the Bioremediation of Oil-Contaminated Beaches." In: EPA/600/R-94/075, *Symposium on Bioremediation of Hazardous Wastes: Research, Development, and Field Evaluations,* June 28–30, 1994, San Francisco, CA, pp. 103–108.

Venosa, A.D., Suidan, M.T., Wrenn, B.A., Haines, J.R., Strohmeier, K.L., Holder, E.L. and Eberhart, B.L. 1994b. "Nutrient Application Strategies for Oil Spill Bioremediation in the Field." In: EPA/600/R-94/011, *Twentieth Annual RREL Research Symposium,* March 15–17, 1994, Cincinnati, OH, pp. 139–143.

Westlake, D.W.S., Jobson, A.M. and Cook, F.D. 1978. "*In-Situ* Degradation of Oil in a Soil of the Boreal Region of the Northwest Territories." *Canadian J. Microbiology,* 24:245–260.

Experience with in situ Microbial Anaerobic Reductive Dechlorination

Ralph E. Beeman, Ronald J. Buchanan Jr., Alfred A. Biehle, Michael D. Lee and John T. Leethem*
DuPont Environmental Remediation Services
300 Bellevue Parkway, Suite 390
Wilmington, DE 19809-3722

INTRODUCTION

Summary

The contamination of groundwater by halogenated organic compounds is a significant, worldwide problem. Conventional remediation approaches, such as pump-and-treat systems, may effectively contain the contaminated groundwater but do little to restore the aquifer. In addition, pump-and-treat technologies are generally expensive and do not treat or eliminate the source of contamination. New, innovative technologies are needed that reduce the contaminant mass at the source yet are cost-effective.

Current research reveals that innovative, although unproved, technologies exist to degrade subsurface contaminants in situ. In situ treatment may be less costly and may eliminate the huge capital expense of large, aboveground pump-and-treat systems. This chapter highlights the DuPont experience in the development and application of in situ anaerobic reductive dechlorination. The following types of in situ biological technologies are currently being evaluated for the treatment or degradation of chlorinated organic compounds:

- intrinsic bioremediation
- enhanced bioremediation
- microbial-augmented bioremediation

Each of these in situ technologies has unique features. Intrinsic bioremediation relies on the natural attenuation and degradation of

* Present address: ERM Southwest, 16300 Katy Freeway, Suite 300, Houston, TX 77094.

organic compounds. In enhanced bioremediation, amendments are added to the subsurface to optimize conditions for microbial growth and the metabolism of various organic compounds. Microbial-augmented bioremediation involves injecting genetically engineered or specialized microbes into the subsurface to attack specific organic compounds.

Each of these technologies or systems shows promise in site-specific circumstances, and further developmental efforts are under way. DuPont has focused on enhanced bioremediation using anaerobic reductive dechlorination as a potential technology to remediate chlorinated organic compounds. These compounds include perchloroethylene (PCE), trichloroethylene (TCE), dichloroethylene (DCE), carbon tetrachloride (CT), chloroform (CF), methylene chloride (MC), tetrachloroethane (TeCA), trichloroethane (TCA), and dichloroethane (DCA). This approach, however, can produce undesirable microbial metabolites such as vinyl chloride (VC). For anaerobic bioremediation to function as a useful remediation technology, pathways that minimize or eliminate such unwanted by-products must be selected and executed in situ.

This need led DuPont to establish an internal bioremediation technology team, which is charged with

- evaluating the current state of technology as it relates to all phases of bioremediation
- expanding the boundaries of bioremediation knowledge
- developing technologies that could be applied in the field to accelerate remediation activities

The bioremediation technology team has made several important scientific contributions to the state of the art in bioremediation. In particular, team members developed and patented an in situ technique for microbial anaerobic reductive dechlorination, which can accelerate contaminated groundwater remediation projects.

Objectives

The objectives of this chapter are to present

- the DuPont developmental and application experience with anaerobic reductive dechlorination
- protocols developed to evaluate and implement the anaerobic reductive dechlorination technology
- relevant conclusions and recommendations for future work

Structure

To accomplish these objectives, this chapter is structured as follows:

- Section 1—Introduction
- Section 2—Background
- Section 3—Protocols and Field Experience
- Section 4—Applications and Results
- Section 5—Conclusions
- Section 6—Future Work

BACKGROUND

Under anaerobic conditions, chlorides are sequentially removed from the chlorinated aliphatic compounds. As a result, PCE is reduced to TCE, TCE is reduced to DCE (*cis*-1,2-DCE is the predominant form), DCE is reduced to VC, and VC is reduced to ethene (DiStefano et al., 1991). This process, known as reductive dehalogenation, requires fermentative, sulfate-reducing, or methanogenic conditions (Bouwer, 1994). Chlorinated aliphatic compounds may also serve as electron acceptors for some bacteria. Fermentative bacteria break down complex substrates such as yeast extract into simpler compounds, which are then broken down by sulfate reducers and methanogens (Atlas and Bartha, 1993). Under sulfate-reducing conditions, microbes convert sulfate to hydrogen sulfide, typically using a simple organic compound (e.g., acetate), other fatty acids, or molecular hydrogen as the electron donor. Methanogenic bacteria use a limited number of simple organic substrates (e.g., acetate, trimethylamine, and methanol or molecular hydrogen and either carbon dioxide or bicarbonate) to form methane. High sulfate levels tend to be inhibitory to methanogenic bacteria.

Microbes also require suitable environmental conditions (e.g., pH, temperature, and salinity) and electron acceptors, nitrogen, phosphorus, and trace minerals to support growth and activity (Lee et al., 1988). The environmental conditions and groundwater redox potential control the rate and type of microbial activity that predominates in the aquifer.

For anaerobic bioremediation to occur, microbes must have a substrate to serve both as a substrate for growth and an electron donor. Appropriate electron acceptors are also required (Beeman et al., 1994). The optimal choice for the substrate and electron acceptor is site-specific and related to the type of microbes in the formation and the environmental conditions (Bouwer, 1994). Simple substrates (e.g., acetate) will support a narrower population of sulfate reducers and methanogens, which may not dechlorinate as well as a larger population stimulated by more complex substrates (e.g., yeast extract).

PROTOCOLS AND FIELD EXPERIENCE

Based on laboratory and field experience, DuPont developed a specific test sequence to determine the preliminary applicability of the microbial anaerobic reductive dechlorination process. A flowchart showing the test sequencing and critical decision points is presented in Figure 1. This flow chart tracks the test sequencing, from precursor study, laboratory optimization study, hydrogeologic modeling, and tracer test to the in situ pilot study.

Several field options were evaluated to test anaerobic reductive dechlorination in situ. These options were used to reduce risk and/or cost in making a decision to proceed to pilot-scale implementation. The two primary methods used in these studies were the single-borehole bioreactor test and the multiple-well/-piezometer test. The single-borehole bioreactor tests were used at a Texas site and a New York site, while the multiple-well/-piezometer tests were evaluated at two sites in Texas. Either test regime can take several months to complete.

Precursor Study

A precursor study is conducted to determine whether and to what extent indigenous biological activity is degrading the chlorinated organic compounds and which, if any, factors might limit this activity. Groundwater samples are collected from monitoring wells and analyzed for biological indicator parameters and geochemical parameters. The hydrogeology of the site is also investigated during the precursor study.

Laboratory Optimization Study

The laboratory optimization study is conducted to determine the ambient and optimal conditions for anaerobic reductive dechlorination of chlorinated organic compounds. The study evaluates the combination of substrate, electron acceptors, and nutrients that would provide the maximum rate and extent of dechlorination.

Tracer Test and Hydrogeological Modeling

During tracer testing, pairs of piezometers and wells are used to determine groundwater flow rates and direction. In addition, ground-

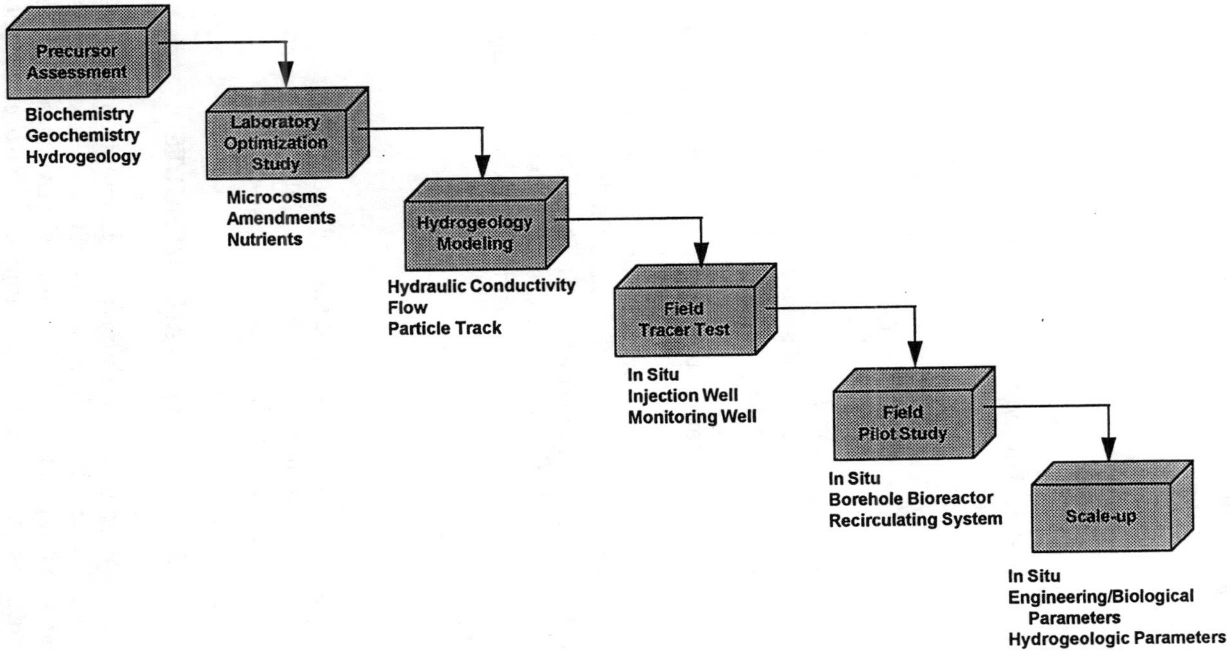

Figure 1. Flowchart of test sequence and decision points.

water modeling is used to predict preliminary flow rates and direction. After flow rates and directions are confirmed, in situ bioremediation experiments can commence.

Field Pilot

For the single-borehole bioreactor test, a well or series of wells are first identified as candidate locations based on hydrogeologic information and geochemical data. A conservative tracer (e.g., bromide) is added to the well, and the well is analyzed over time to determine the dilution rate and groundwater travel time. After the dilution rate and travel time are established, the amendment addition rate is calculated to achieve a predetermined concentration of nutrients and substrates in the well. Amendment addition then proceeds. The concentrations of tracer substrates, metabolic end products, environmental conditions, and chlorinated solvents are monitored over time.

In the multiple-piezometer well test, an amendment pilot skid is constructed to maintain and deliver amendments to the subsurface (see Figure 2 for a typical configuration). The skid consists of a substrate storage tank, nutrient reservoir, and ancillary pumps, controllers, valves, and piping and is connected to an injection well or wells, depending on the particular application. Downgradient extraction wells are installed and connected to the amendment skid, forming a closed-loop system.

Scaleup

Further engineering evaluations of the field test data are then conducted to determine system performance and viability. The results will eventually be used to design the scaled-up system.

APPLICATIONS AND RESULTS

In situ microbial anaerobic reductive dechlorination was tested at Victoria, Texas; Niagara, New York; and Gulf Coast, Texas. The studies were initiated to establish the viability and effectiveness of the technology in remediating groundwater contaminated with chlorinated aliphatic compounds such as chloroethenes, chloroethanes, and chloromethanes.

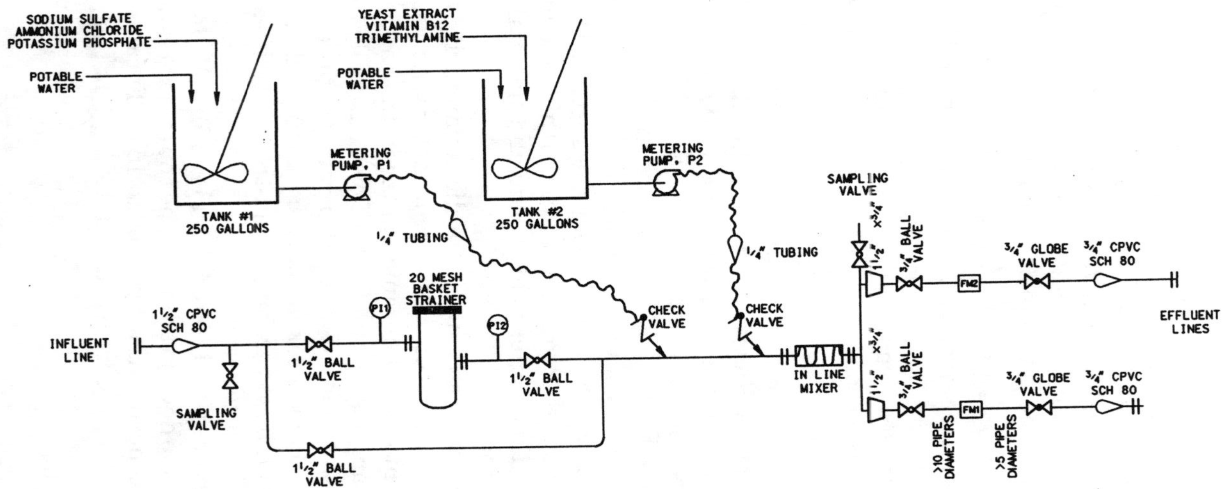

Figure 2. Typical pilot configuration (not to scale).

Victoria, Texas

The Victoria, Texas, study was motivated by literature reports (Suflita et al., 1988; Bagley and Gossett, 1990) and by the DuPont preliminary operating experience under sulfate-limited conditions, where PCE was reductively dechlorinated, but VC was produced. VC production possibly resulted from sulfate-limiting conditions experienced in situ; therefore, the current study was designed to determine whether

- Anaerobic reductive dechlorination could be applied to remediate effectively an aquifer underlying a landfill contaminated primarily with PCE.
- Sufficient sulfate levels could be maintained in situ to drive the biological reactions to completion (i.e., past VC to ethene or ethane).

Pilot tests were conducted at the site in an aquifer underlying the former West Landfill at the Victoria plant. The landfill comprised a series of trenches or cells 3 to 5 m (10 to 15 ft) deep, into which a variety of solid and liquid industrial wastes were placed over a 20-year period in the 1950s and 1960s (see Figure 3 for a cross section of the landfill). The test aquifer is a semiconfined sand unit (Zone B sand), the top of which is located approximately 18 m (60 ft) below the surface. The Zone B sand is a relatively continuous, 12 to 18 m (40 to 60 ft) thick, medium- to coarse-grained sand, overlain by interbedded sand and clay of the Pleistocene Age Beaumont Formation.

Aquifer test data indicate the hydraulic conductivity of the Zone B sand to be approximately 3.5×10^{-2} centimeters per second (cm/sec). The average content of total organic matter in the sand is 0.1% based on analyses of 22 samples. Groundwater flows naturally from northeast to southwest across the study area at a rate of approximately 0.3 m per day. Natural discharge takes place to a man-made canal along the southwest boundary; however, pump-and-treat activities contain such discharge within the study area.

PCE and benzene have migrated from landfill cells into the Zone B sand, forming a plume downgradient of the source that is approximately 170 × 500 m. PCE concentrations ranged to 15 milligrams per liter (mg/L) in the plume center, and between 1 and 2 mg/L in the pilot test area (see Figure 4).

TEST DESIGN

During Phase I, a 12.2 × 18.3 m (40 × 60 ft) test cell was established in the Zone B sand; the cell included extraction, monitor, and delivery (or

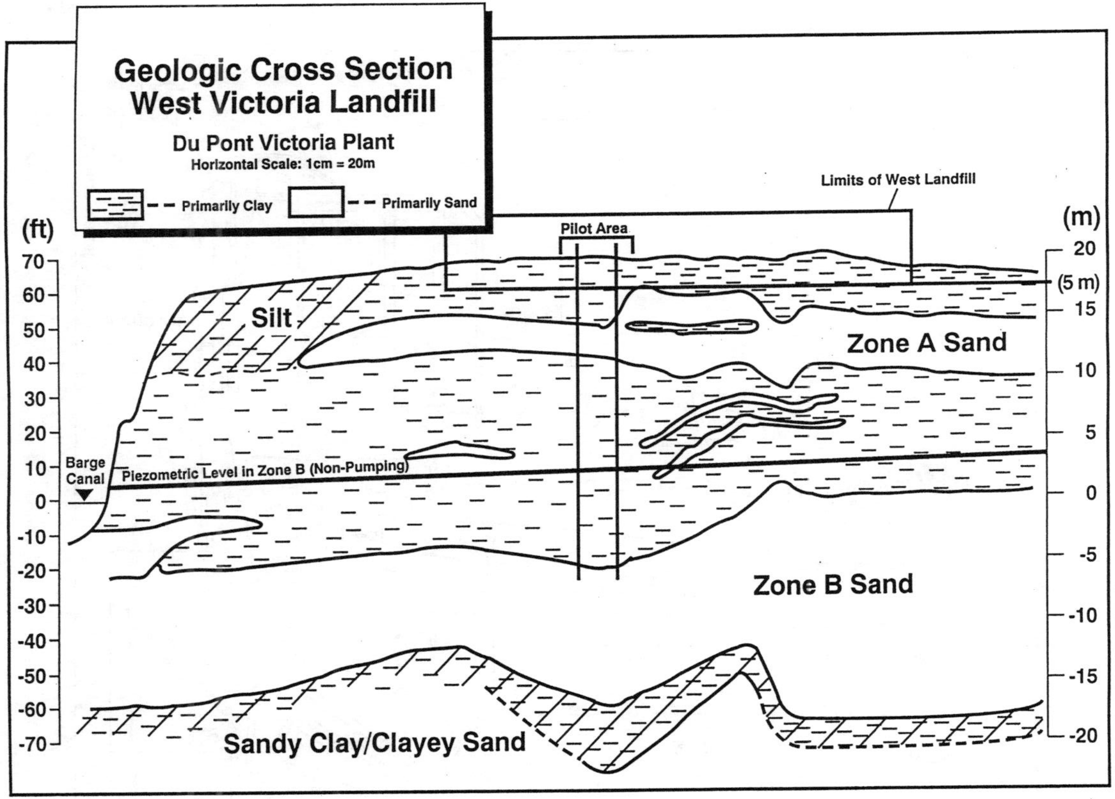

Figure 3. Victoria landfill cross section.

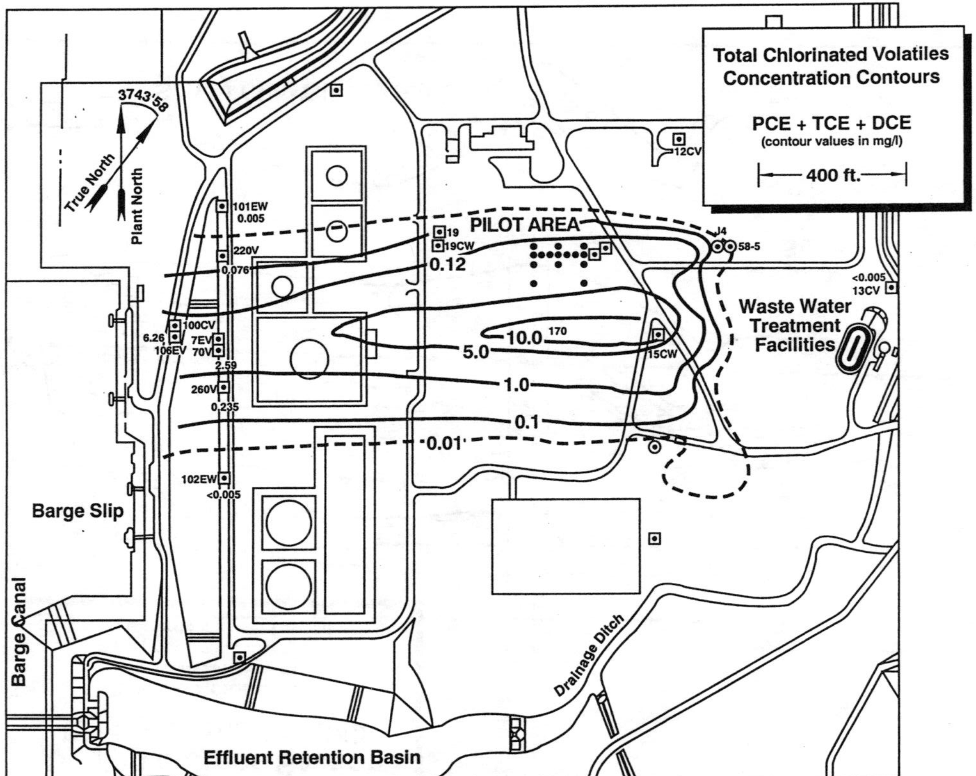

Figure 4. Total chlorinated volatile concentration contours.

Phase I: Demonstration of Perchloroethylene Biodegradation

Site Dimensions:
40 x 60 ft.

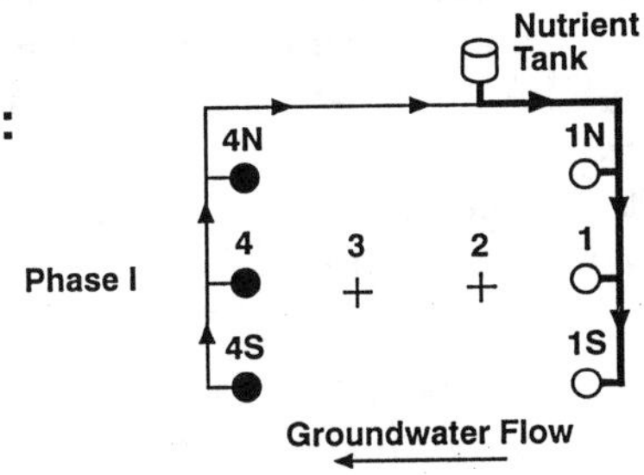

Nutrient Addition Rate: 1 gal/hr
Alternating Either:

- **Benzoate–38 ppm (0.3 mM)**
- **Mg SO4–45 ppm (0.5 mM)**
  Final Dilution Into the Well Bore

Typical Pumping Rates:
Wells 1S→4S–3 gpm
Wells 1→4–7 gpm
Wells 1N→4N–3 gpm

Figure 5. Layout and operating parameters for Phase I test cell.

recharge) wells (see Figure 5). All of the wells were screened in the top 4.6 m (15 ft) of Zone B and, therefore, did not fully penetrate the aquifer.

Wells 4N, 4, and 4S were extraction wells, while well 1N, 1, and 1S were delivery (or recharge) wells during Phase I (see Figure 5). Wells 2 and 3 were monitoring wells along the path of flow between the delivery and extraction wells. Pumping rates from wells 4N, 4, and 4S were 11.4, 26.6, and 11.4 liters per minute [L/min; or 3, 7, and 3 gallons per minute (gpm)], respectively. The water from the extraction wells was passed through a closed-loop mixing manifold where nutrient and substrate were added from the nutrient feed tank. Nutrient or substrate feed was delivered to the groundwater stream at a rate of 3.9 liters per hour (L/h). A final diluted concentration of either 0.3 millimolar [mM; 38 parts per million (ppm)] sodium benzoate (substrate) or 0.4 mM (56 ppm) magnesium sulfate was delivered to Zone B through the delivery wells. The extraction and delivery system was designed to minimize air stripping of the volatile organic compounds (VOCs). Groundwater flow rates in the test cell were estimated at 1.1 meters per day (m/day) during the pilot operations. The hydraulic retention time was approximately 16 days.

ANALYTICAL PROCEDURES

The samples were collected from wells and analyzed following EPA document SW-846 protocols. The monitoring wells were purged for three well volumes and then sampled using dedicated purge-and-gas bladder pumps. Gas chromatography/mass spectrometry (GC/MS) analysis using EPA document SW-846 method 8240 was conducted for halogenated volatile compounds and benzene. Gases were analyzed using GC with flame-ionization detection, following a modified EPA document SW-846 method 8015.

RESULTS

The results of the Victoria, Texas, study follow.

Aerobic Treatment

Phase I was divided into two periods: preliminary aerobic treatment lasting 203 days and followup anaerobic treatment lasting 133 additional days (to Day 336 of Phase I). Aerobic treatment consisted of sparging liquid oxygen into the delivered water stream in situ to raise dissolved oxygen levels above 4 mg/L. Efforts to degrade PCE in this manner generally were unsuccessful, although benzene concentrations were reduced in the pilot site to below the EPA document SW-846 method 8240 practical quantitation limit (PQL) of 0.005 mg/L after 18 to 22 weeks.

Figures 6, 7, and 8 show the variation in PCE, TCE, and 1,2-DCE groundwater concentrations, respectively, during Phase I in wells 2, 3, 4N, 4, and 4S. Concentrations of PCE, TCE, and DCE averaged roughly 10, 4, and 4 micromolar (μM; or 1.7, 0.5, and 0.4 mg/L), respectively, while the site was aerobic. VC was not detected throughout the period of aerobic operation (see Figure 9). In addition, the PCE, TCE, and DCE concentrations were stable during most of the aerobic treatment, indicating no significant biodegradation.

Anaerobic Treatment

Sodium benzoate was added on Day 203, when oxygen sparging was terminated. Figures 6 and 7 show that PCE and TCE concentrations began to decrease in monitoring wells and extraction wells within 35 days after the initiation of benzoate delivery. Eventually, PCE concentrations in monitoring wells 2 and 3 decreased to below the attainable PQL of 0.015 mg/L (0.1 μM). TCE levels also decreased with time through Phase I; however, the levels remained slightly above the PQL in wells 2 and 3.

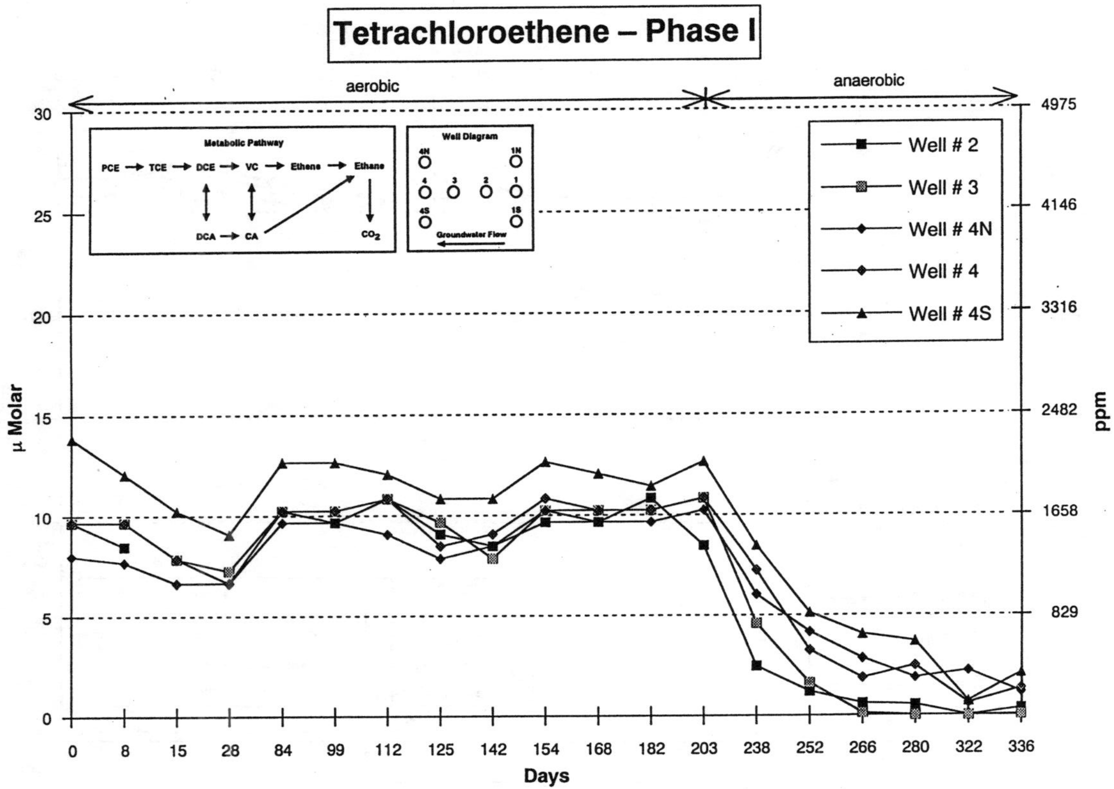

Figure 6. PCE concentration during Phase I.

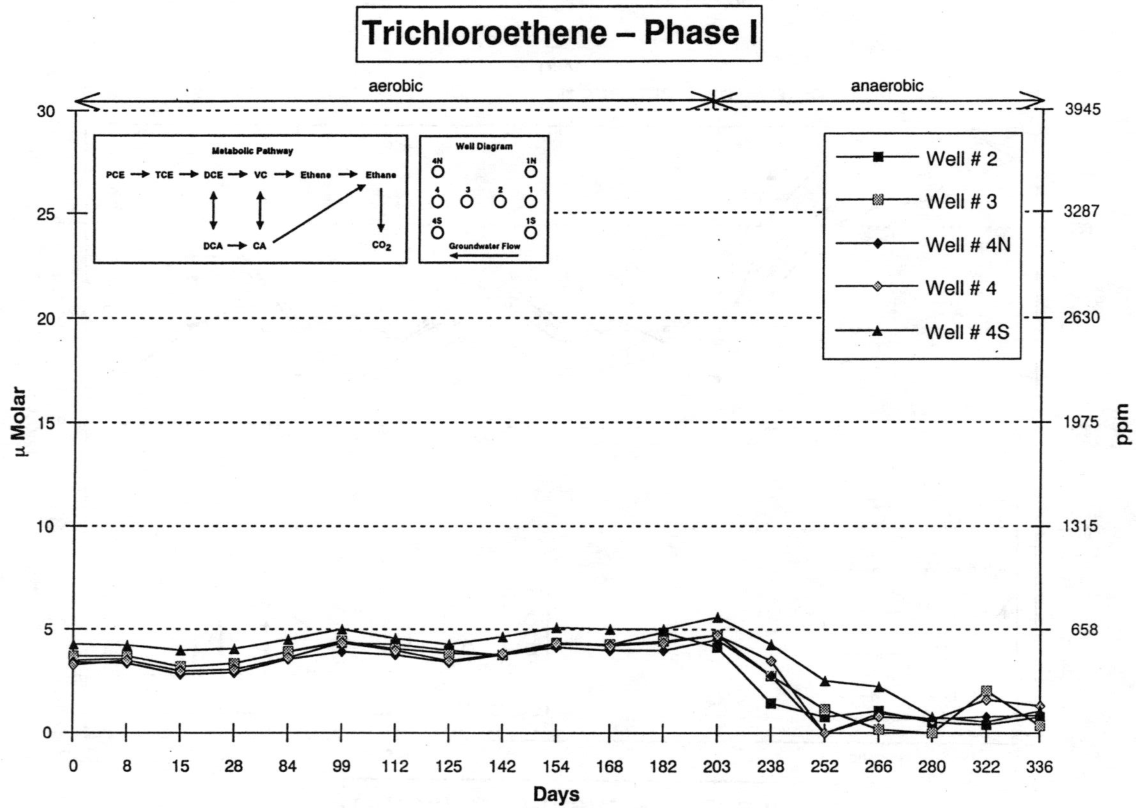

Figure 7. TCE concentration during Phase I.

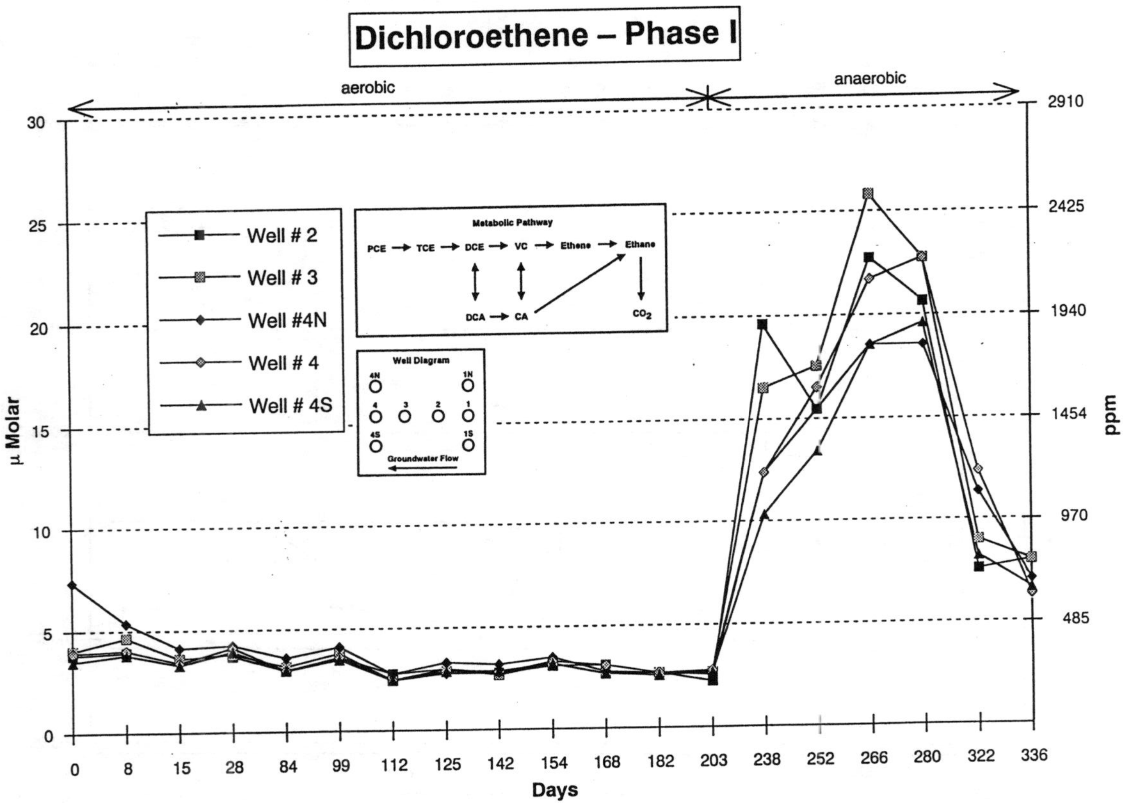

Figure 8. DCE concentration during Phase I.

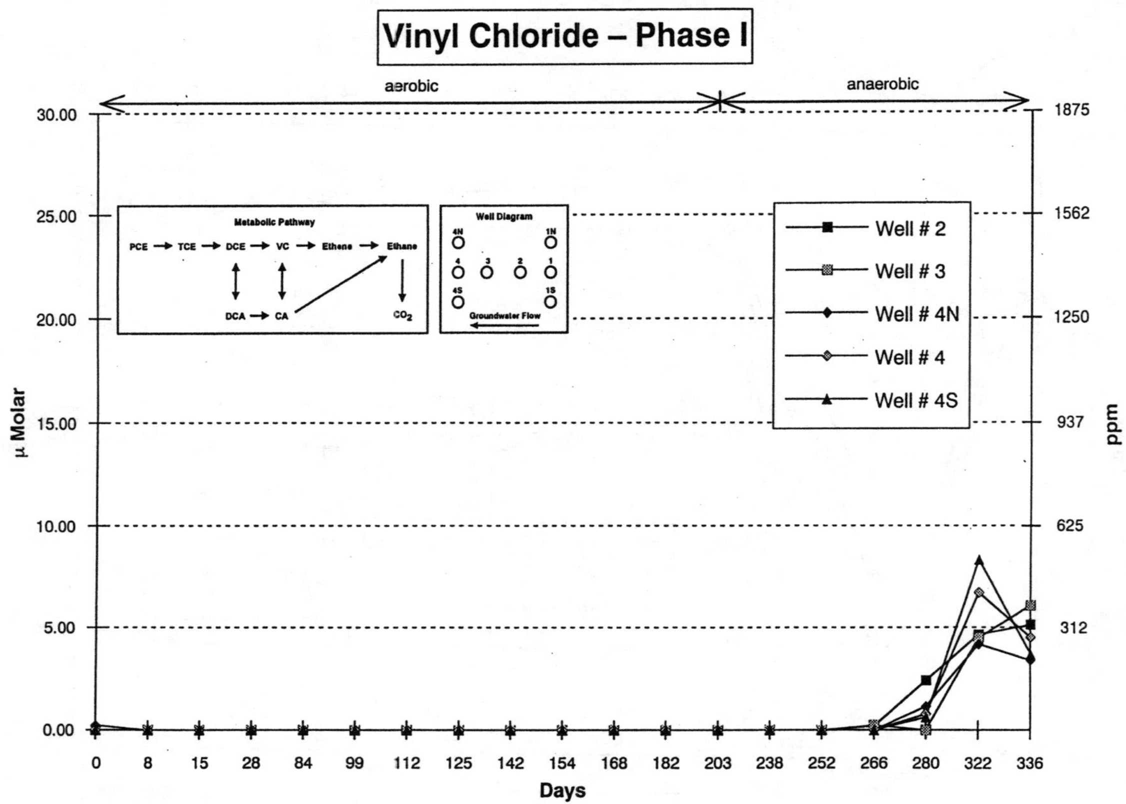

Figure 9. Vinyl chloride concentration during Phase I.

DCE concentrations increased in all wells—beginning with the anaerobic treatment at Day 203—until Day 280, when these began to decrease. DCE levels increased by a factor of seven- to eightfold, corresponding to near stoichiometric transformation of PCE and TCE to DCE by Day 280. PCE and TCE concentrations at Day 203 were approximately 12 and 5 μM, respectively, while DCE concentrations were approximately 4 μM at well 3 (see Figure 10). By Day 287, the DCE concentrations approached 26 μM, while both PCE and TCE concentrations decreased to near PQL at well 3. If all of the initial PCE and TCE concentrations were biotransformed to DCE, approximately 17 μM of DCE would have been produced. Added to the indigenous DCE (approximately 4 μM), the total DCE concentration should have approximated 22 μM, which is close to the 26 μM measured on Day 287 in well 3 (see Figure 10). Similar results were also observed at well 2 (see Figure 11).

The subsequent drop in DCE at Day 280 correlated with VC production. PCE and TCE concentrations fell below the PQL, and the lowest sulfate concentrations were observed at both wells 2 and 3 (see Figures 10 and 11). By Day 336, DCE concentrations decreased nearly 13 μM, while VC only increased approximately 5 μM, leaving approximately 8 μM of products unreconciled. The unreconciled products were possibly transformed into the gaseous end products ethane and ethylene; however, these gases were not monitored during Phase I. Ethylene production was confirmed in subsequent Phase II studies that were conducted at the same pilot site (Beeman et al., 1994).

At Day 280, sulfate concentrations became undetectable at well 3 (0.01 mM; 1 ppm), while the sulfate concentration at well 2 approached a low of 0.4 mM (50 ppm). The pilot site became limiting for sulfate, and the production of VC—an unwanted metabolic by product—commenced; therefore, the sulfate was increased in an attempt to control the further formation of VC.

Sulfate was added to the recharge water beginning on Day 285. The rebound in sulfate level at well 3 lagged somewhat behind well 2 because of the variable flow of water and chemical into the aquifer but is evident at Day 322, as shown on Figure 10. By Day 336, the sulfate level rebounded completely in both monitor wells. This rebound appears to be accompanied by the decreased DCE production and VC concentrations. Similarly, VC levels appeared to stabilize at the three extraction wells (i.e., wells 4N, 4, and 4S) by Day 336. The pilot site continued to operate under sulfate-reducing conditions to minimize the further production of VC.

To estimate the mass balance at the pilot site, individual halogenated concentrations in the feedstream were summed and compared to the

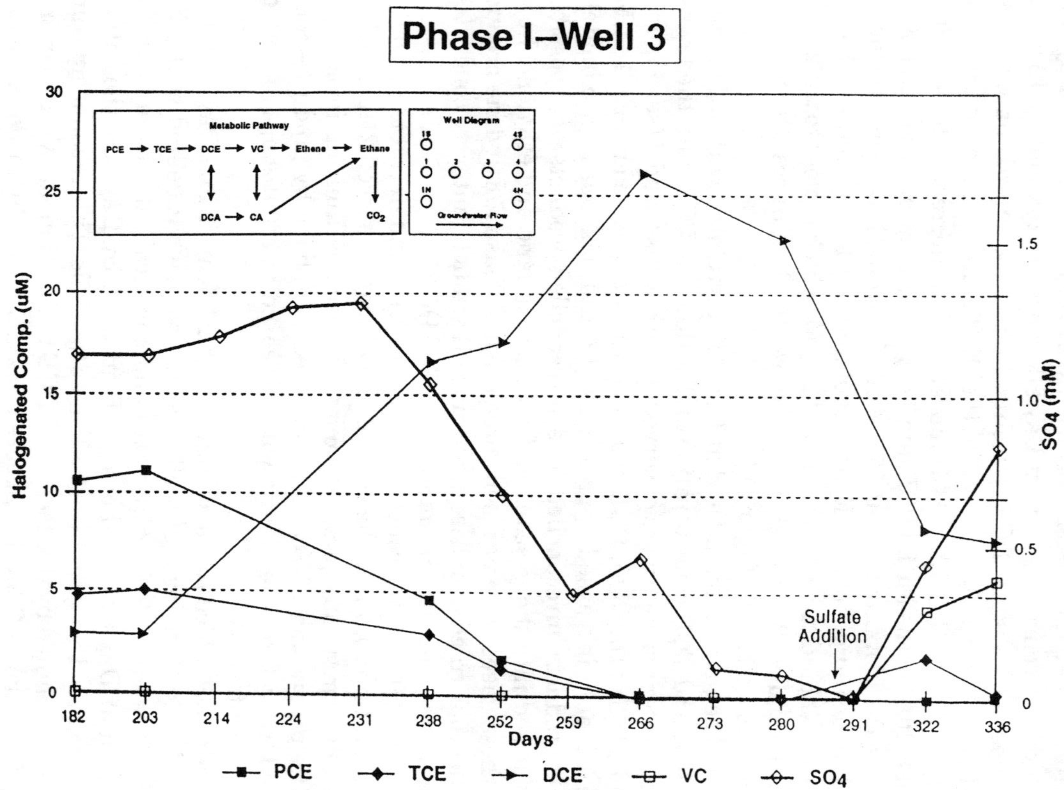

Figure 10. Well 3 data during Phase I.

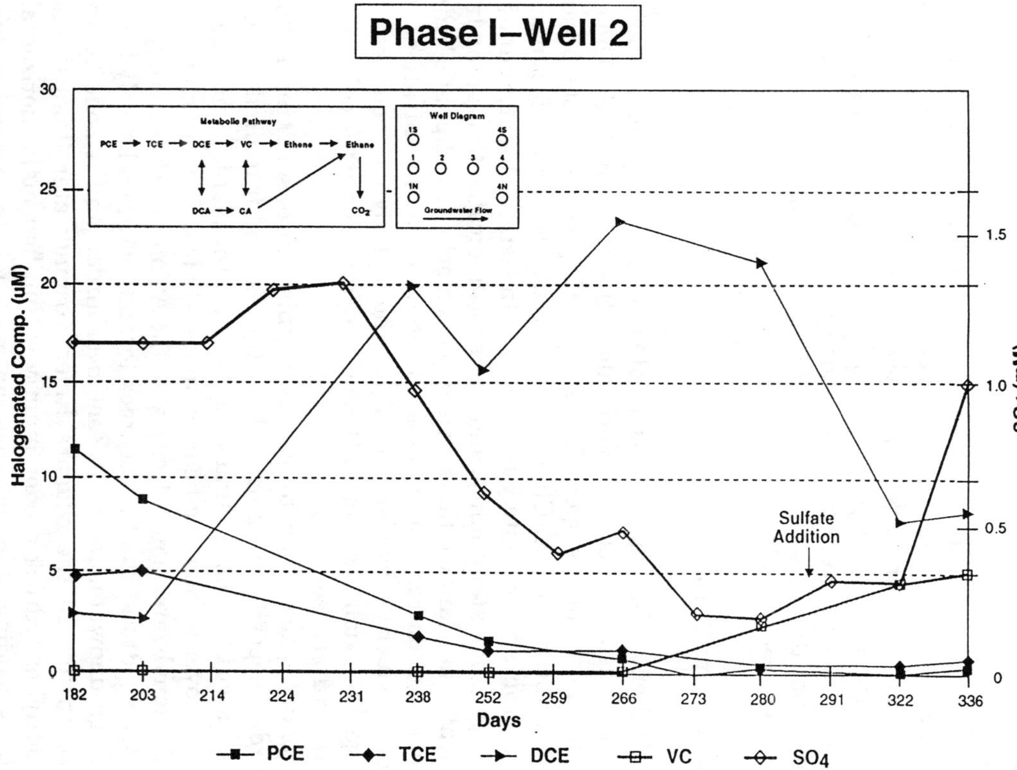

Figure 11. Well 2 data during Phase I.

products found in groundwater from well 6. Concentrations that were below the detection limit were ignored in these summations. The components in the feedstream totaled 4.58 μM, and the predominantly dechlorinated products in well 6 totaled 3.08 μM, the equivalent of a 67% recovery.

Product recovery at the pilot site was compared to the conservative tracer to account for dilution and dispersion with water beyond the site boundaries. A bromide anion tracer demonstrated that 74% of the bromide placed into the 1N-1S wells was recovered in the 7N-7S wells. The recovery of the end products from PCE degradation (67%) was compared to the recovery of the bromide tracer (74%). This comparison indicates that the major end products from PCE biodegradation in the aquifer were reconciled.

Niagara, New York

The DuPont Niagara plant is located in a heavily industrialized portion of Niagara Falls, New York, along the Niagara River. The site has been in continuous operation since 1898, producing both organic and inorganic chemicals. Chlorinated solvents were manufactured at the plant from approximately 1930 to 1975. Investigations of groundwater flow, groundwater quality, and transport indicated that groundwater beneath the plant contains a variety of chlorinated aliphatic hydrocarbons and other contaminants at a wide range of concentrations. An anaerobic reductive dechlorination pilot field study was conducted to investigate the viability of in situ biologic degradation of the chlorinated aliphatic hydrocarbons.

The plant is underlain by relatively thin, unconsolidated A Zone overburden deposits (primarily fill materials) and fractured dolomite of the Lockport Group. Groundwater beneath the plant is encountered in both the overburden and bedrock units. Hydraulic conductivity values and groundwater flow velocities for the A Zone overburden deposits and A Zone bedrock (the first bedrock formation) are relatively low. In general, groundwater in the A Zone overburden and A Zone bedrock flows south toward the Niagara River. Groundwater flows through the underlying bedrock predominantly within four major horizontally continuous bedding-plane fracture zones. Flow in the bedrock depends on the spacing, orientation, and interconnection of individual fractures but generally flows toward the north and northeast.

TEST DESIGN

The A Zone bedrock was selected for the Niagara bioremediation field pilot. The proposed location for testing was well MW-16A where the corresponding total VOC concentration ranges from 10 to 20 ppm total VOCs. Piezometers were installed approximately 2 weeks' travel time upgradient of well MW-16A. Tracer tests were conducted to verify the direction of groundwater flow from the newly installed piezometers to the existing monitoring well. The dilution factor between the injection point and the observation point was much greater than anticipated. As a result of the tracer tests at well cluster 16, the formation feeding task was altered from a dual-well design (injection at one point and observation at another) to a single-well injection test. Therefore, substrate was added and groundwater was sampled at well MW-16A. It was anticipated that the addition of yeast extract solution would drive the aquifer anaerobic and support anaerobic reductive dechlorination.

ANALYTICAL PROCEDURES

The wells were sampled weekly for a variety of analytical parameters to monitor microbial activity.

RESULTS

The initial sample at well MW-16A contained the following chlorinated solvents: Dichloromethane (DCM) at 0.059 mg/L; 1,1-DCA at 0.061 mg/L; *cis*-1,2-DCE at 0.64 mg/L; 1,1,1-TCA at 0.088 mg/L; TCE at 7.4 mg/L; and PCE at 0.11 mg/L. Groundwater pH was 11.4. The groundwater was anaerobic, with dissolved oxygen of 0.2 mg/L and a redox potential of –273 millivolts (mV). Prior to any additions, the groundwater contained 370 mg/L of sulfate, 0.08 mg/L of nitrite, less than 0.05 mg/L of nitrate, and 0.19 mg/L of total phosphate. Some organic substrates were present in the groundwater, with 59 mg/L total organic carbon (TOC) and 220 mg/L chemical oxygen demand (COD).

Well MW-16A was initially dosed with 650 g of yeast extract on Mondays and 1300 g of sodium sulfate on Tuesdays. After Day 19, the yeast extract additions were increased to 400 g on Mondays, Wednesdays, and Fridays. The sodium sulfate additions were maintained at 1300 grams on Tuesdays until Day 137, when additions ceased. Yeast extract additions were stopped completely on Day 187.

High levels of substrate were maintained in well MW-16A after the yeast extract additions were increased to 1200 g per week. During the first 19 days of limited yeast extract additions, the TOC and COD concentrations increased on the day of addition and then declined to background levels over the next 3 days (see Figure 12). After the quantity of yeast extract was increased on Day 19, the TOC levels were greater than 1000 mg/L after Day 37 and were generally greater than

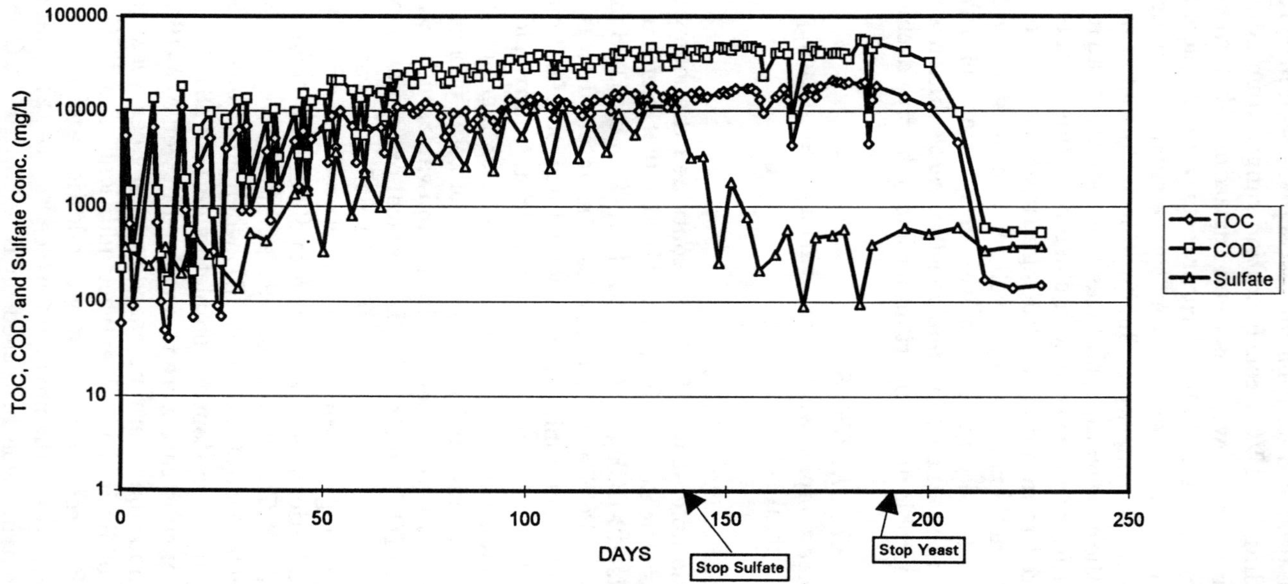

Figure 12. MW-16A TOC, COD, and sulfate concentrations.

10,000 mg/L after Day 68. The COD increased proportionately with the TOC. Both the TOC and COD fell to almost background levels within 27 days after yeast extract additions ceased.

Sulfate was the predominant electron acceptor in well MW-16A. The sulfate levels increased gradually and remained above the 1000 mg/L plateau after Day 64 and reached as high as 11,400 mg/L. After sulfate additions ceased on Day 137, the sulfate concentrations declined to background concentrations of approximately 300 mg/L within 6 days. The groundwater was anaerobic throughout the study, with redox potentials of −147 to −467 mV (see Figure 13). In contrast, dissolved oxygen (DO) levels were generally below 1.0 mg/L, but levels as high as 5.7 mg/L were measured. The high DO levels are thought to be a result of equipment problems or due to introduction of oxygen during sampling, because no oxygen would be expected under the highly reduced conditions shown by the redox potential.

High numbers of microbes were present, even during the low winter temperatures. The groundwater was turbid, had a strong anaerobic odor, and generated large quantities of gas (mostly carbon dioxide and some methane). Samples of the headspace above the well MW-16A groundwater that were collected on Day 163 contained 1.9 ppm volume/volume (v/v) methane using fixed-gases analyses, less than 0.2% v/v hydrogen, and 0.077% v/v carbon dioxide. VOCs were not detected (detection limits of 0.12 to 0.48 ppm) in the headspace above the well MW-16A groundwater, indicating that the contaminants were not simply being volatilized into the air by the microbial evolution of carbon dioxide and methane.

Nitrogen availability and winter temperatures may have limited the growth of the anaerobic microbes. Nitrate levels were less than 1.4 mg/L throughout the test, and very little nitrite was detected. Other forms of nitrogen, such as ammonia or total Kjeldahl nitrogen (TKN), were not measured. The phosphate increased substantially to greater than 100 mg/L after Day 7. The temperature declined from 15°C in November (Day 1) to a minimum of 9°C in early March (Day 107) and then gradually increased to approximately 13°C by the end of the test in July (Day 228; see Figure 14). The pH decreased from 11.4 on Day 1 to below 8.0 by Day 51. After the substrate additions ceased, the pH increased to between 9.7 and 10.8.

The chlorinated ethene concentrations decreased dramatically after the substrate levels stabilized above 1000 mg/L TOC on Day 50 (see Figure 15). The PCE concentrations decreased from a maximum of 0.81 mg/L to below the detection limit of 0.05 mg/L, a reduction of more than 93%. Concentrations of PCE, TCE, and *cis*-1,2 dichloroethene

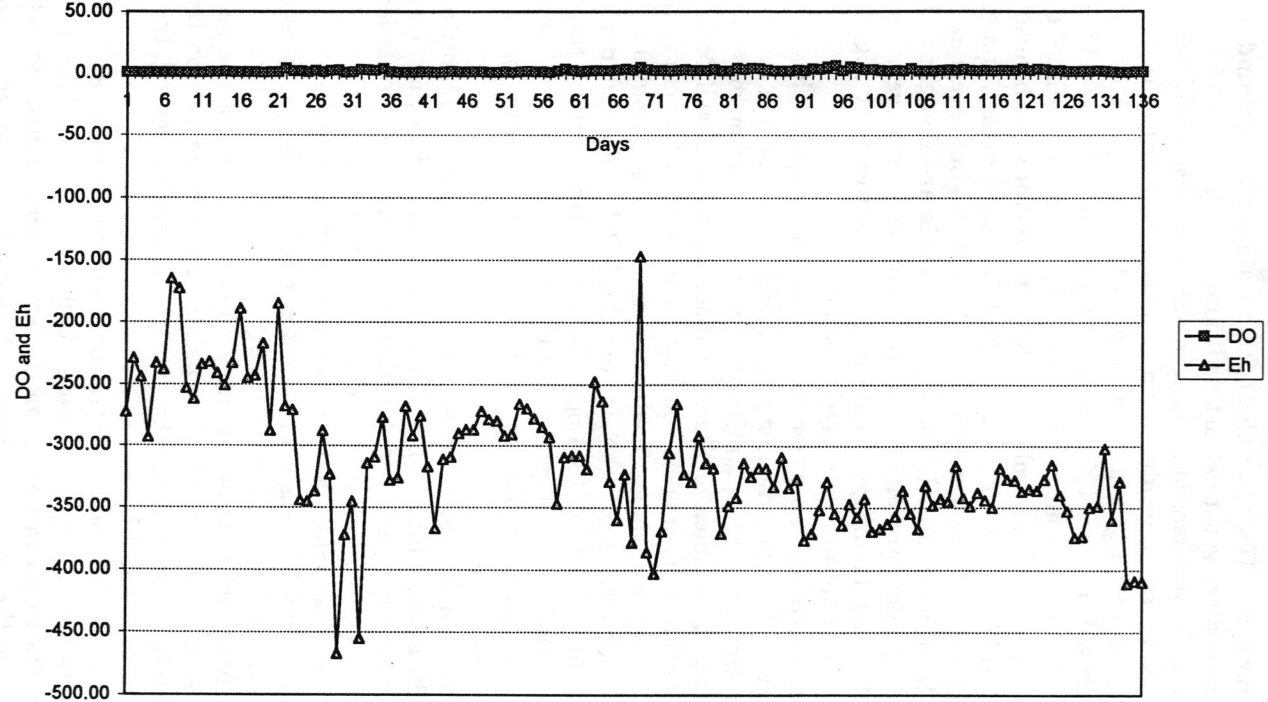

Figure 13. MW-16A DO and Eh.

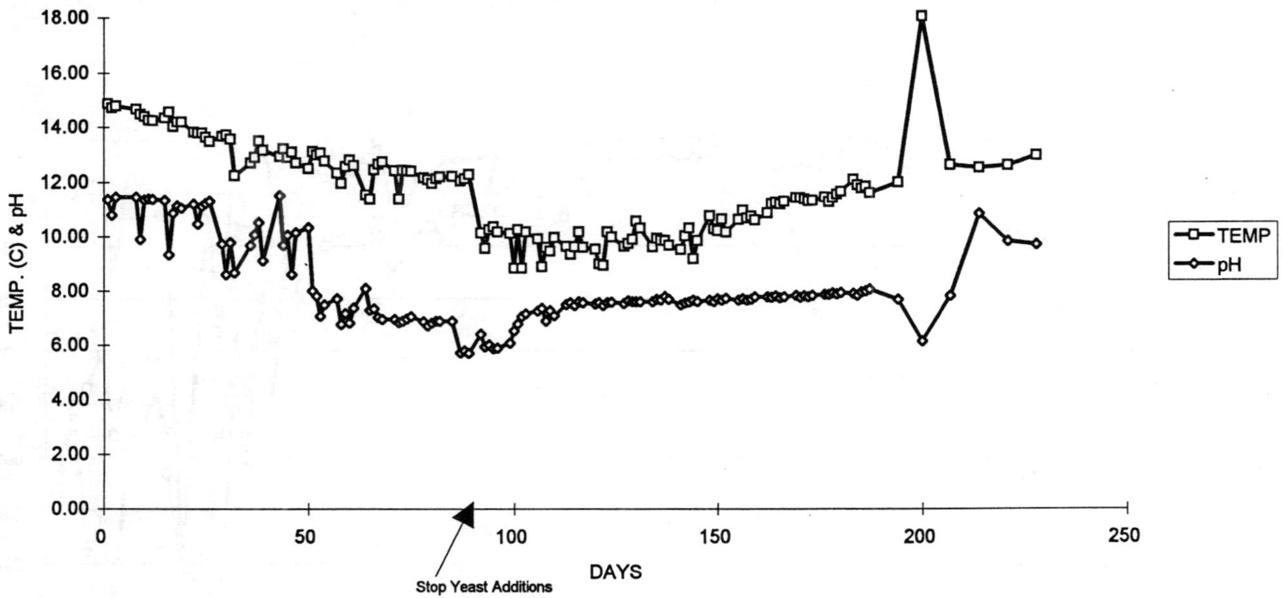

Figure 14. MW-16A temperature and pH.

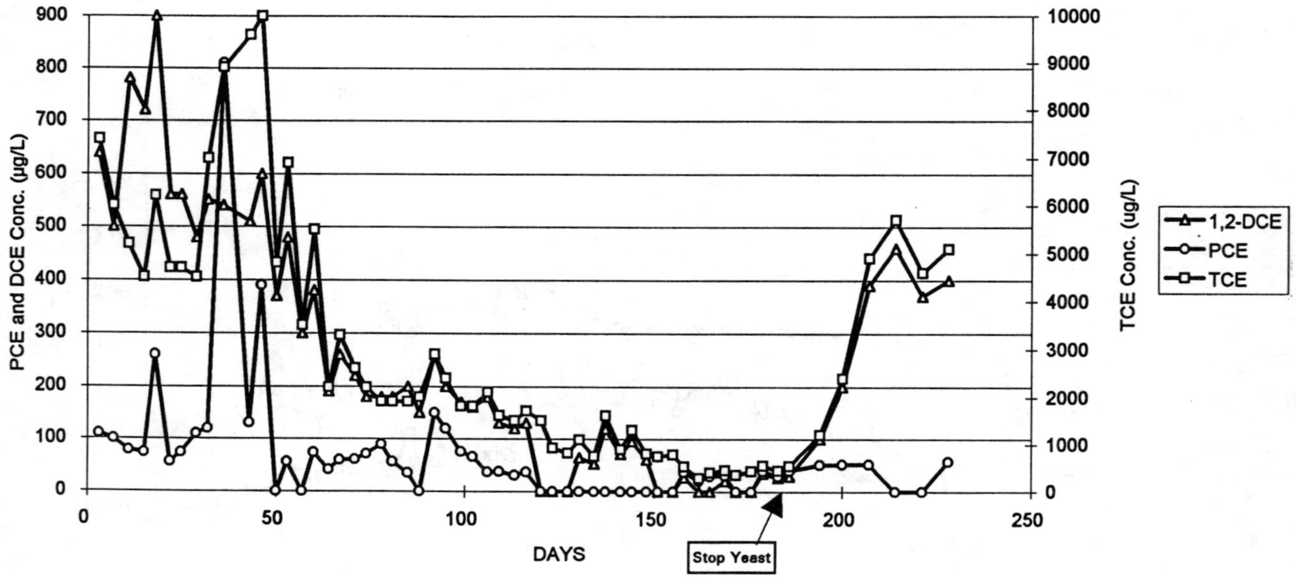

Figure 15. MW-16A chloroethene concentration.

(cDCE) were reduced by more than 97%. No increase in daughter products indicative of the typical reductive dechlorination pathway was observed. VC was not detected. After substrate additions ceased, PCE, TCE, and cDCE concentrations began to increase to almost pretreatment levels.

The chloroethanes were also biodegraded (see Figure 16). Concentrations of 1,1,1-TCA and 1,1-DCA were significantly reduced (removals greater than 91% and 73%, respectively). Chloroethane, the other compound in the degradation sequence, was not detected.

CF and DCM were the only two chloromethane series compounds detected in well 16A (see Figure 17) and behaved differently from the chloroethenes and chloroethanes. CF concentrations increased between Days 57 and 194 and then fell sharply after Day 228 from 0.31 to 0.093 mg/L, equivalent to a 70% removal. The DCM concentrations were fairly stable, ranging from less than 0.04 mg/L to 0.10 mg/L, with no discernible trends.

Gulf Coast, Texas, Site

Groundwater at a site on the Texas Gulf Coast is contaminated with several chlorinated organic compounds, including PCE, TCE, DCE, VC, CT, CF, and MC, which range from below detection limit to near solubility limits. Groundwater is contained on-site by an extensive groundwater pump-and-treat system. The time estimated to remediate the site to the groundwater protection goal is several decades; therefore, innovative technologies, such as in situ bioremediation, are being assessed as alternatives to the pump-and-treat operation.

Site geology consists of fluvial sand, silt, and interbedded clay. The lithology of the study area (in order of increasing depth) as interpreted from cone-penetrometer tests (CPT) and boring logs is as follows: upper clay layer (sandy to silty) between 3 and 13.7 m thick; sand layer consisting of a fining-upward sequence (fine-grained particles above coarse-grained particles) beneath the upper clay, ranging from 1.5 to 10.7 m thick; and a confining lower clay unit ranging from 1.5 to 15 m thick.

The uppermost aquifer is under confined conditions with two flow regimes. Groundwater in the lower section (regime) moves faster than in the upper section by virtue of the fining-upward sequence. The groundwater gradient in the area is controlled by an existing recovery well field located downgradient of the test site. Groundwater pumping rates for recovery wells 154 and 155 ranged from 3.8 L/min to 11.4 L/

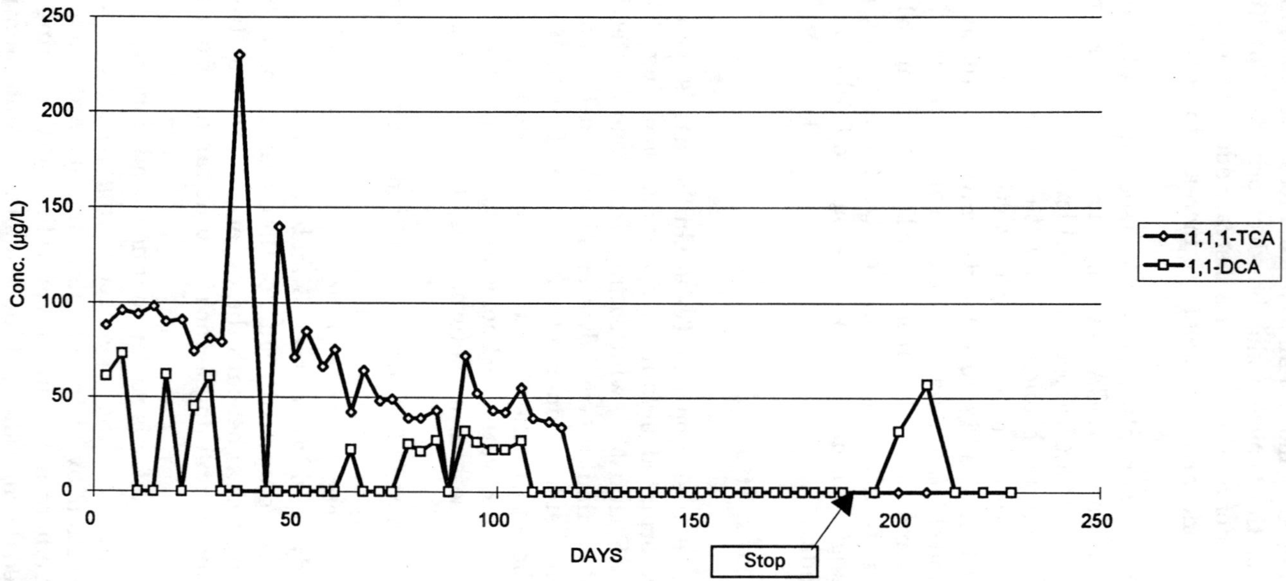

Figure 16. MW-16A chloroethane concentration.

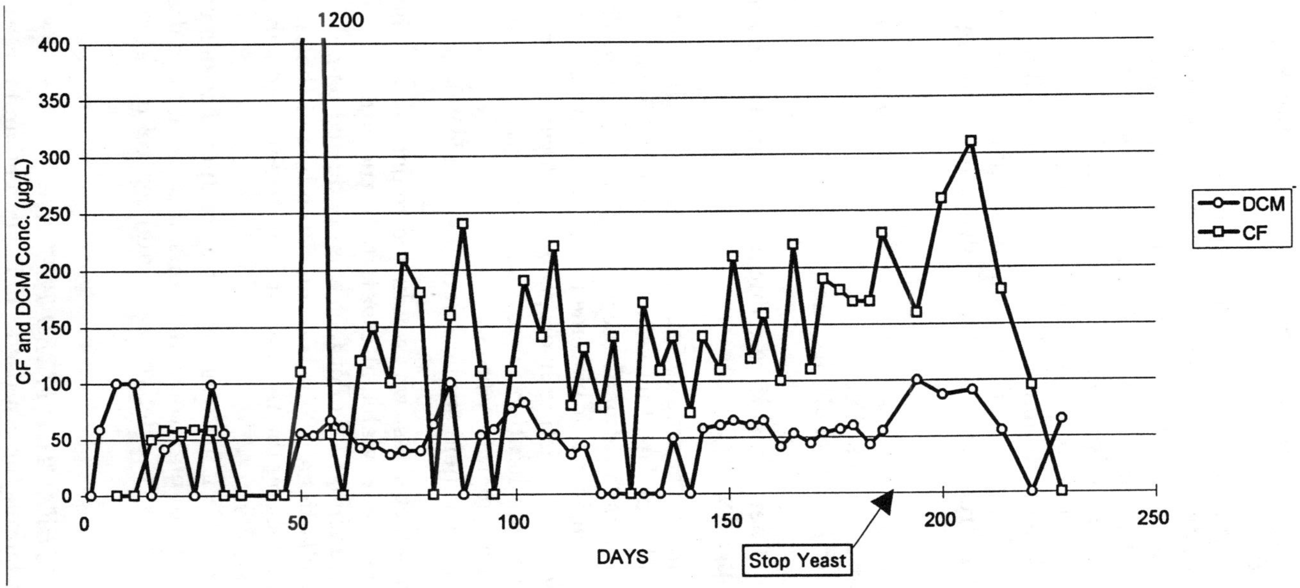

Figure 17. MW-16A chloroform and methylene chloride concentration.

min and from 30.4 to 45.6 L/min, respectively. Groundwater generally flows northward.

An initial assessment of the potential use of bioremediation at the site was conducted under aerobic conditions. The results were unfavorable oxidizing conditions, and the idea was abandoned. However, in situ bioremediation was reconsidered following the successful demonstration of anaerobic reductive dechlorination at the Victoria, Texas, site. The first step of the second assessment was to determine whether existing subsurface environmental conditions were conducive to anaerobic biological activity. A detailed examination of pertinent groundwater data in the area of interest revealed the following favorable conditions:

- biodegradation by-products of the contaminants, believed to be the result of in situ degradation
- the presence of sulfates/sulfites, indicating reduced anaerobic conditions
- the presence of aerobic, sulfate-reducing, and anaerobic reductive dechlorinating microorganisms

Previous groundwater monitoring data, together with soil chemical data, were compiled and evaluated for evidence of intrinsic bioremediation and to determine a rational microbiological approach. The results suggested that some intrinsic bioremediation may have been occurring. This information and existing hydrogeologic information was used to select an appropriate location for a field demonstration without conducting additional laboratory studies.

The following three field studies were then undertaken:

- A borehole bioreactor test was conducted to determine whether bioremediation was a viable alternative for the site.
- A substrate transmission pilot test was conducted using monitoring wells to gain a better understanding of the different mechanisms affecting the movement of nutrients and substrate through the aquifer.
- A more rigorous CPT investigation was conducted to demonstrate that the delivery of high concentrations of benzoate could inhibit microbial growth at the delivery well and control plugging of the aquifer.

BOREHOLE BIOREACTOR FIELD TESTING

The borehole bioreactor test used a delivery well and three downgradient monitoring wells (see Figure 18). The test area was selected

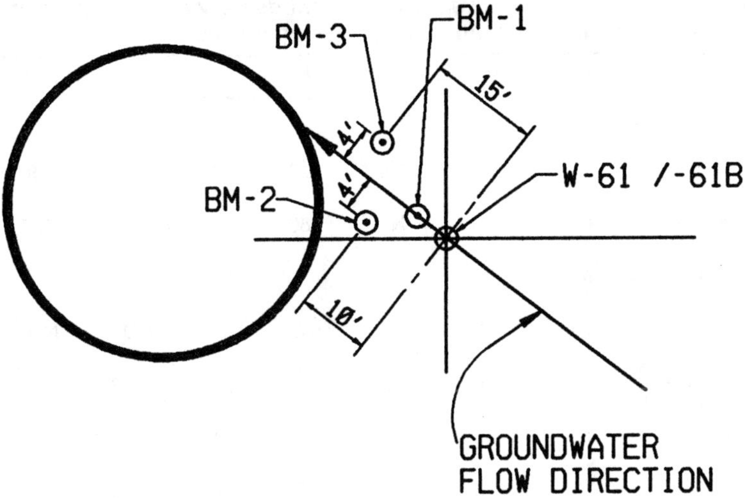

Figure 18. Bioreactor test location.

because of the low hydraulic gradient, the moderate contaminant concentration, and the favorable inorganic chemical concentrations.

The delivery well was screened from 7.3 to 11.8 m (24 to 39 ft) below the surface in an aquifer composed of fine- to medium-grained sand. Three downgradient monitoring wells were installed and constructed of 2-in., polyvinyl chloride (PVC), with screen lengths identical to those of the delivery well. The first well (BM-1) was placed 1.2 m (4 ft) from the delivery well in the path of expected groundwater flow. The second well (BM-2) was placed 45 degrees southwest of the expected direction of groundwater flow and 3 m (10 ft) downgradient. The final well (BM-3) was installed 45 degrees northeast of the expected direction of groundwater flow and 4.5 m (15 ft) downgradient (see Figure 18). The wells were placed in this formation to assure that the organic plumes and tracer would be intercepted. It was anticipated that groundwater would flow from the delivery well to well BM-1. The other wells (i.e., BM-2 and BM-3) were placed to ensure that the groundwater path was correctly identified.

Before the borehole test was initiated, a groundwater sample was collected from the delivery well and an upgradient well for volatile organic analysis (VOA) to establish background conditions (Table 1). Results of this sampling were used to compare and contrast with the results of the borehole bioreactor field test. Groundwater in the delivery and upgradient well contained detectable quantities of TCE, VC, CF,

Table 1. Volatile organic analysis from nutrient delivery and biomonitoring pre-yeast extract approximately 90 days after yeast extract (ppb).

Well/Date	PCE	TCE	DCE	VC	CT	CF	MC	CM
Pre-yeast								
Delivery well	ND	820	ND	610	ND	3900	210	ND
Upgradient well	910	1800	ND	380	ND	12000	2500	ND
30 days post yeast extract								
Delivery well	ND	ND	ND	ND	ND	ND	ND	ND
Upgradient well	ND	ND	ND	ND	ND	8	ND	ND
Biomonitor well 1	ND	ND	ND	ND	ND	ND	ND	ND
Biomonitor well 2	21	31	ND	ND	19	2400	93	ND
Biomonitor well 3	ND	ND	ND	ND	ND	8	ND	ND

ND = none detected (<5 ppb).

and MC prior to the addition of substrate and nutrients. PCE was detected in the upgradient well, but not in the delivery well.

The recipe of substrate and nutrients for the bioreactor test consisted of a dilute solution of sodium sulfate and yeast extract, with sodium bromide as a tracer. Water balance through the test area for the 45-day duration of the test was calculated to be 15,140 L (4000 gal). To obtain the appropriate concentration in the aquifer, sodium sulfate (500 mg/L), yeast extract substrate (1000 mg/L), and bromide (1 mg/L) were added to the bioreactor well. The solution was then dispersed into the aquifer by surging the well with a surge block (see Figure 19). The surge block was constructed of a block of wood with a diameter smaller than that of the well, connected with half-inch PVC pipe. Flexible rubber sheets cut to a diameter slightly larger than the inside diameter of the well were located above and below the block.

After nutrient was added, groundwater samples were collected from the bioreactor wells and analyzed for TOC, COD, pH, temperature, and DO concentration. Figure 20 demonstrates that the COD and TOC were high, immediately after substrate and nutrient addition, and that surging had moved substrate and nutrient into well BM-3. However, TOC and COD simultaneously declined on Day 19, indicating that the material had been used by the microorganisms. This decline is

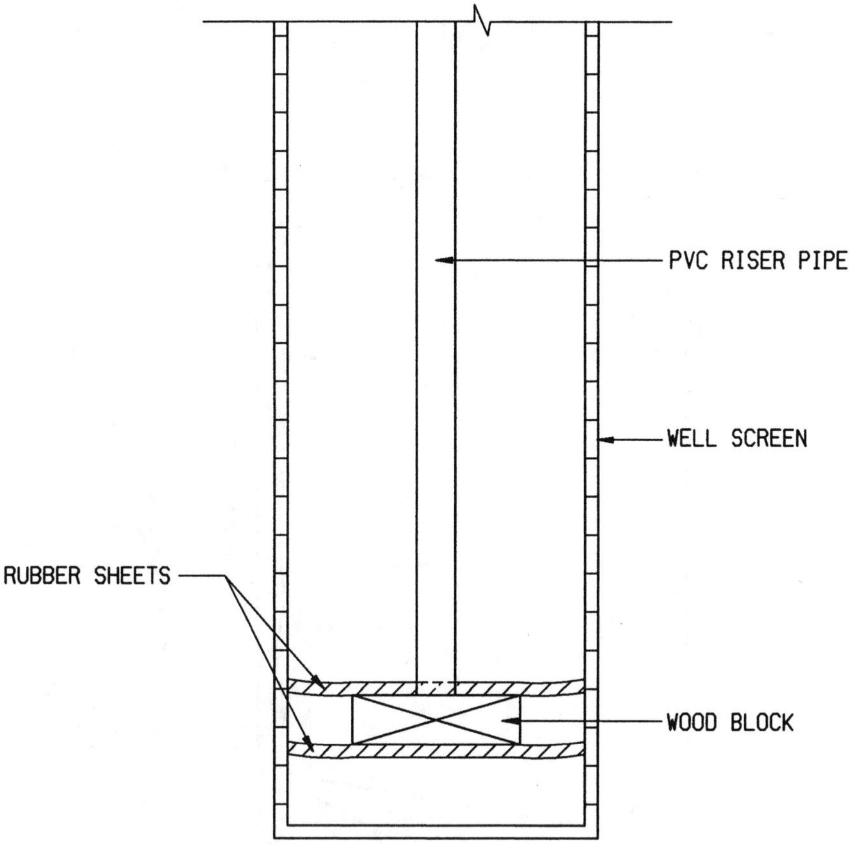

Figure 19. Surge block

consistent with microbial removal. Concomitantly, DO in the delivery well fell to zero in 1 week (data not shown). This suggests utilization by indigenous microbes and indicates that microorganisms are present to degrade yeast extract.

Samples were collected for VOA 1 month after the addition of yeast extract (see Table 1). The analytical results of these samples were compared to the results of the samples collected prior to addition of yeast extract. The comparison showed a complete disappearance of most of the chlorinated organics from the test area, with the exception of monitoring well BM-2. Samples taken post-treatment demonstrated a decrease of two to four orders of magnitude in chlorinated organic concentrations.

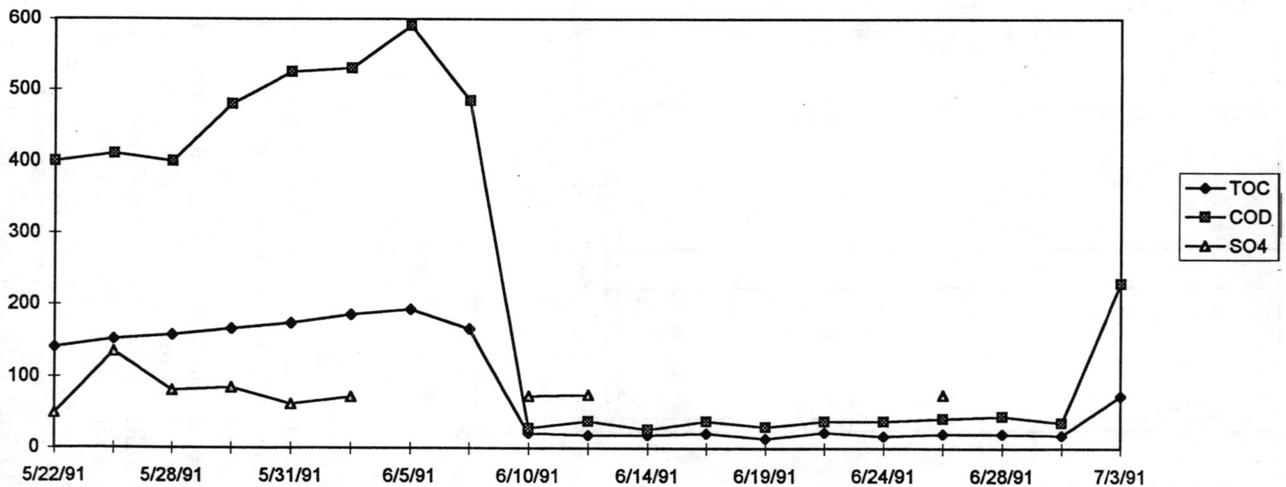

Figure 20. BM-1 COD/TOC (Gulf Coast bioreactor).

Historical groundwater quality data in the area was limited; however, soil samples from the area indicated that high levels of CT, CF, and PCE could be expected to be present in the groundwater. Chlorinated organic variability was large, but the variability alone could not explain the disappearance of these compounds (see Table 2) after the addition of yeast extract.

Anaerobic biological activity was apparently taking place at the site. The borehole bioreactor test had induced anaerobic conditions in the aquifer through the use of yeast extract. However, the disappearance of chlorinated organics, although intriguing, could not be directly attributed to biological activity, since no daughter products had been formed. Therefore, a more rigorous test was conducted to demonstrate anaerobic reductive dechlorination and daughter-product evolution.

SUBSTRATE TRANSMISSION STUDY

The previous test demonstrated that microbial reductive dechlorination was feasible at this site. Sulfate and low concentrations of sodium benzoate were successfully transmitted through this sandy aquifer to promote dechlorination. The injection of higher concentrations of substrate would be required to treat areas contaminated with higher levels of chlorinated compounds. Plugging of the delivery wells from microbial growth and sorption of substrate into the aquifer matrix becomes a concern at these higher substrate levels. One strategy to control microbial plugging of the delivery wells is to use an inhibitory concentration of substrate.

The objectives of the second test were to better understand the mechanisms affecting movement of substrate through the aquifer, to investigate whether delivery of high concentrations of substrate could inhibit microbial growth and subsequent plugging at the delivery well, and to effectively disperse substrate through the aquifer. During the second test, concentrated solutions of substrate (acetate and benzoate) were

Table 2. Comparison of the initial halogenated biotransformation products and the maximum value observed in the test.

Compound	Initial before Yeast Extract (ppb) (95% Std. Dev.)	Maximum Value (After Yeast Extract)
MC	ND	10
CM	ND	141
DCE	ND	72
VC	ND	141

ND = none detected (<5 ppb).

delivered to the aquifer. Movement of substrate and bromide tracer were monitored to determine their fate and transport. The substrate mixture used in this test is presented in Table 3.

The pilot system shown in Figure 21 consisted of three delivery wells (IW-1, IW-2, and IW-3), four monitoring wells (MW-1, MW-2, MW-3, and MW-4), and two recovery wells (RW-1 and RW-2). Wells MW-3 and MW-4 are located downgradient (8.2 and 22.6 m) from well IW-2 and are screened across the entire aquifer. At the predicted groundwater flow rate of 0.3 m/day, injectate would be expected to reach well MW-3 at 27 days and well MW-4 after 75 days. Well MW-2 is upgradient of the test site and is screened in the upper section of the sand. Well MW-1 is outside of the influence of the injectate and is also screened in the upper section of the aquifer. Wells RW-1 and RW-2 are approximately 39 m and 47 m downgradient from the delivery wells.

From May to October 1993 (131 days), 99,000 L of substrate were delivered by gravity feed, roughly at equal rates to the delivery wells. Sodium bromide (NaBr) was added at a concentration of 1000 mg/L (Br) in both the substrate and nutrients solutions to serve as a conservative tracer. Groundwater samples were collected from the monitoring and recovery wells over a period of 400 days.

Discrete groundwater samples were collected from wells MW-3 and MW-4 near the top of the aquifer (upper flow regime) and at a lower level (lower flow regime) starting at Day 63. Samples were collected from the upper flow regime of wells MW-1, MW-2, RW-1, and RW-2. The groundwater samples were analyzed for TOC, COD, acetate, benzoate, ammonia-nitrogen, bromide, sulfate, total phosphate, and methane.

Groundwater Monitoring Results

Concentrations of bromide, TOC, and COD in the study area are shown for the wells: MW-3 upper flow regime (see Figure 22), MW-3 lower flow regime (see Figure 23), MW-4 upper flow regime (see Figure 24), MW-4 lower flow regime (see Figure 25), RW-1 (see Figure 26), and RW-2 (see Figure 27).

Table 3. Injectate composition.

Chemical Components Substrate Mix	Weeks 0 to 6 (%)	Weeks 7 to 20 (%)
Sodium benzoate	2.5	1.0
Sodium acetate	2.5	1.0
Sodium bromide	0.1	0.1

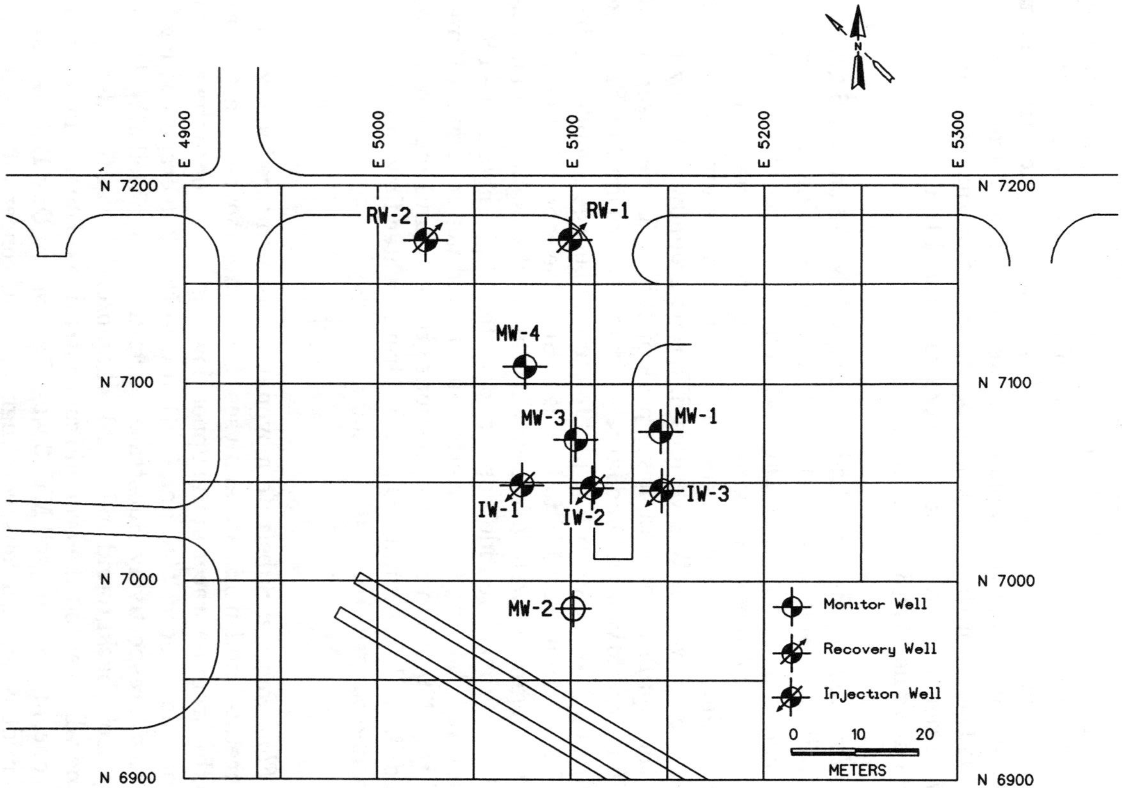

Figure 21. Pilot system well layout.

Bromide

Upper Flow Regime: Bromide was first detected in well MW-3 on Day 28 at a concentration of 2.0 mg/L. The highest concentration of bromide (455 mg/L) in well MW-3 was measured on Day 176. Concentrations declined to less than 10 mg/L after Day 204. Bromide was not detected consistently in well MW-4 until Day 121. Bromide concentrations in well MW-4 ranged between nondetectable and 163 mg/L from Day 176 to Day 421. Groundwater in the upper flow regime ranged from 0.05 to 0.07 m per day. The velocities were calculated from the arrival of bromide at wells MW-3 and MW-4.

Lower Flow Regime: Bromide had reached well MW-3 on Day 63, when the first deep sample was taken. A maximum bromide concentration of 670 mg/L was observed on Day 121, approximately 50 days before the peak in the upper flow regime. The difference in the arrival times for the maximum bromide concentration between the upper and lower flow regimes in well MW-3 verifies different groundwater flow rates within the aquifer. A second, lower peak of 405 mg/L was observed on Day 176 in well MW-3 and had largely dissipated by Day 239. Bromide reached well MW-4 on Day 121 and fluctuated between nondetectable and 290 mg/L from Day 176 to Day 421. Bromide concentration in the lower flow regime of well MW-4 was also generally higher than in the upper flow regime. Bromide was first detected at 0.6 mg/L in well RW-1 on Day 134, but the tracer concentrations did not go above 1.0 mg/L until Day 239. Bromide concentrations greater than 1 mg/L were not detected in well RW-2 until Day 421. Groundwater flow in the lower flow regime ranged from 0.08 to 0.09 m per day.

Substrate

Upper Flow Regime: Substrate movement was monitored using TOC, COD, acetate, and benzoate analyses. Substrate moved at a similar rate to the bromide tracer in the upper flow regime. Prior to the arrival of the bromide tracer, TOC and COD in well MW-3 averaged 21 mg/L and 54 mg/L, respectively. Substrate concentrations in well MW-3 were slightly above background on Day 28, when bromide was first detected, and generally tracked the bromide concentrations. The maximum substrate concentration in well MW-3 was observed on Day 176, the same day bromide concentrations were highest. TOC concentrations in well MW-3 declined to approximately 50 mg/L by Day 297. Benzoate moved slightly slower than acetate with higher benzoate concentrations than acetate detected from Day 204 to Day 239. Thereafter, the acetate concentrations were higher. Substrate levels increased in well MW-4

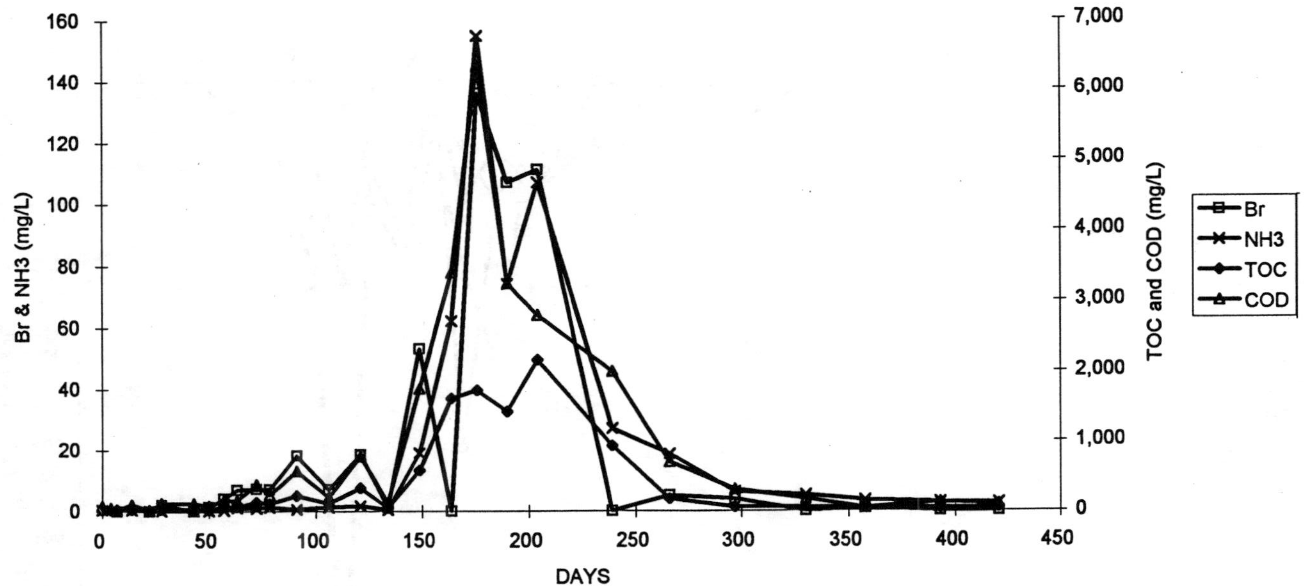

Figure 22. MW-3, upper flow regime.

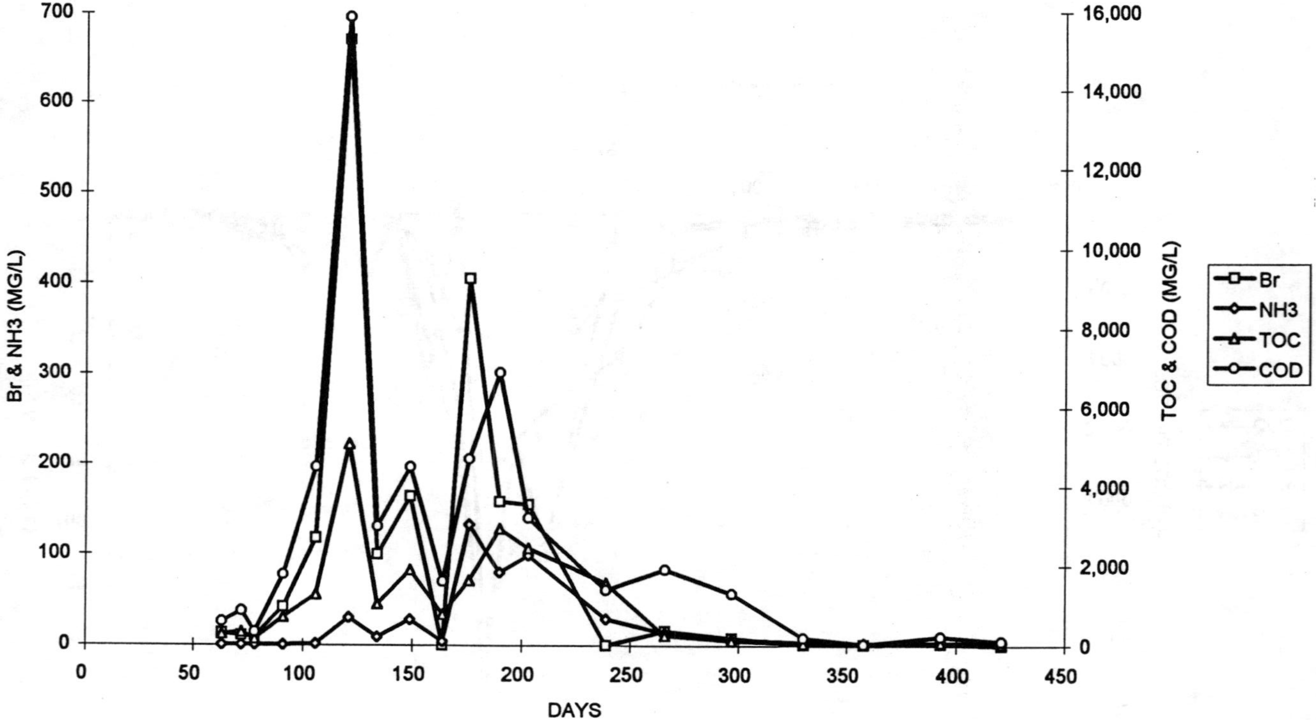

Figure 23. MW-3, lower flow regime.

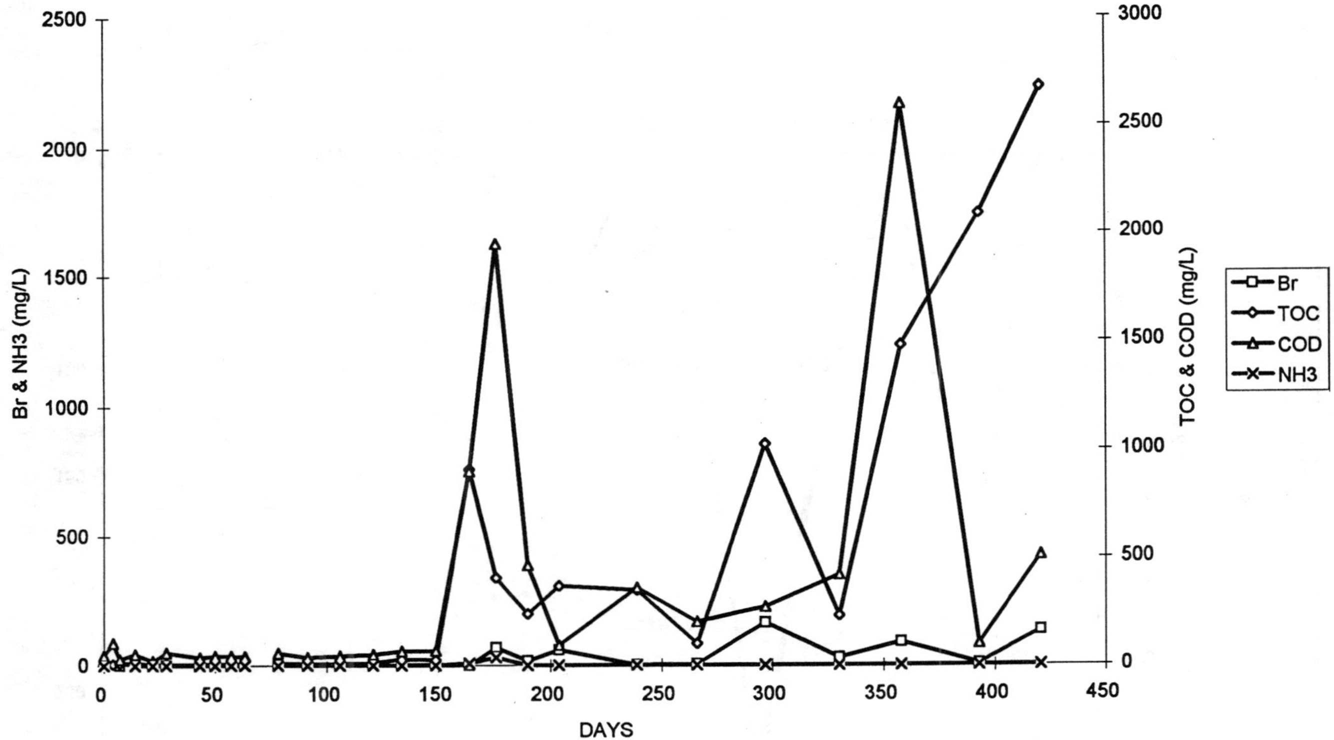

Figure 24. MW-4, upper flow regime.

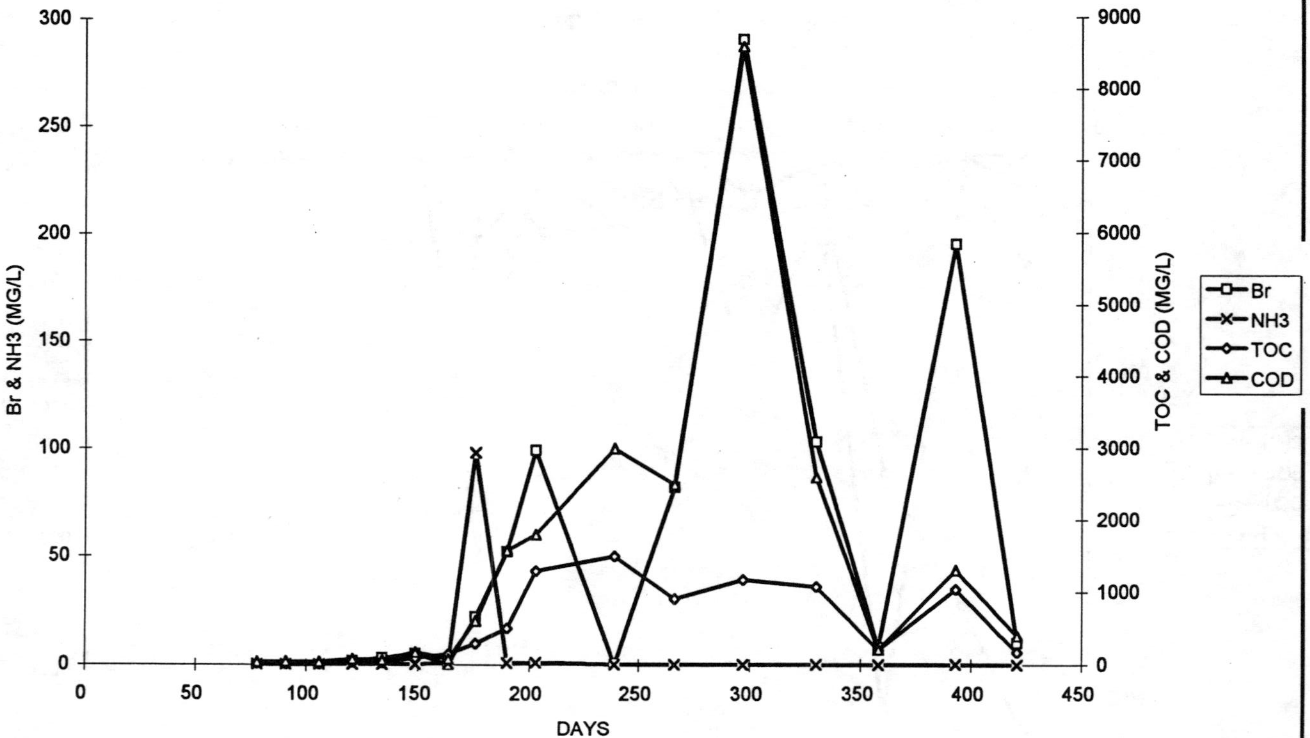

Figure 25. MW-4, lower flow regime.

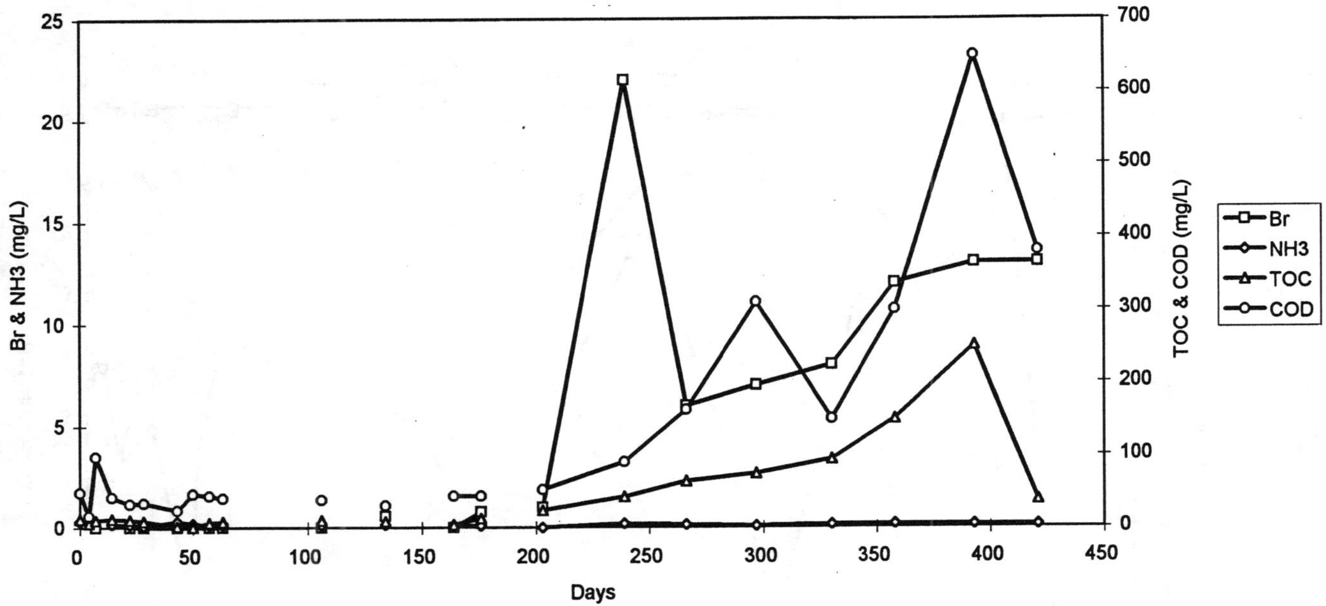

Figure 26. RW-1, upper flow regime.

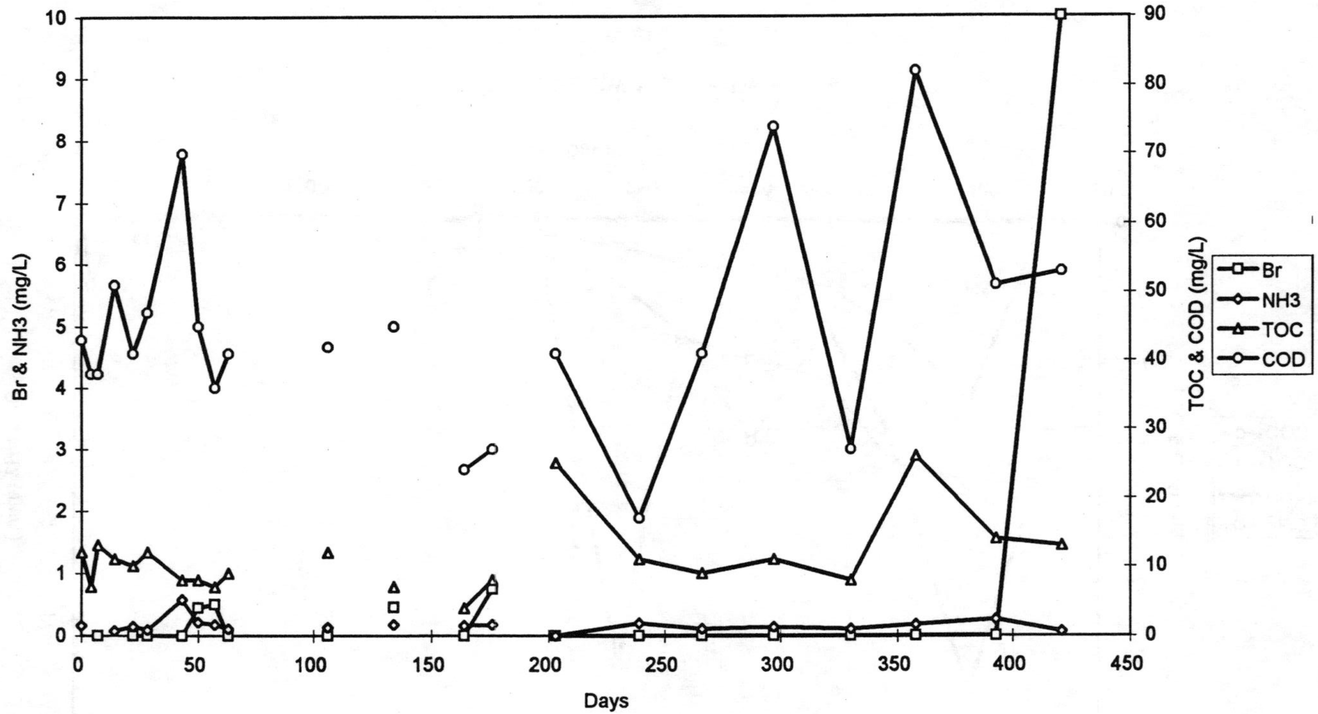

Figure 27. RW-2, upper flow regime.

beginning on Day 134, shortly after bromide was first detected in this well on Day 121. A second pulse of substrate reached well MW-4 about Day 300, and a third pulse was arriving when the study ended. The pulses were a result of the alternate injection of substrate and nutrients.

Lower Flow Regime: Substrate in the lower flow regime had reached well MW-3 when the first samples were collected. Maximum concentrations of TOC and COD in well MW-3 were measured on Day 121, when bromide concentrations were highest. A second pulse of substrate reached well MW-3 on Day 176. The substrate arrived at well MW-4 on Day 149, and concentrations peaked on Days 239 and 293. A slight increase in the substrate concentrations in well RW-1 was observed on Day 203, and the concentrations generally increased thereafter. Substrate concentrations were relatively constant in well RW-2.

CPT SURVEY

In addition to groundwater samples from the monitoring wells, a "snapshot" of the substrate plume was defined by collecting groundwater samples at 26 locations (from both flow regimes), 140 days after delivery started, using CPT technology. Logistical constraints (i.e., cultural features) limited the number of CPT samples that could be collected from the western-most substrate plume.

Bromide

Upper Flow Regime: Bromide distribution in the upper flow regime, as defined by samples from the CPT survey, is shown in Figure 28. Plumes emanating from each injection well are almost completely separated. The plume shape was variable in orientation, with the western-most plume oriented almost northwest and the eastern plume almost north. The maximum bromide concentration was 70 mg/L in CPT-15. The western-most plume associated with well IW-1 trended to the northwest and was 13.8 m by 5.8 m at the 100 mg/L contour. The center plume originating from well IW-2 moved north-northwesterly and was 7 m by 4.9 m at the 100 mg/L contour. The eastern-most plume moved due north and was 6.7 m by 3 m in size at the 100 mg/L contour.

Lower Flow Regime: The bromide concentration in the lower flow regime, as defined by samples from the CPT survey, is shown in Figure 29. The highest bromide concentration was found at CPT-3 (250 mg/L). Three plumes originating at the injection wells were evident. The bromide plume associated with well IW-2 moved 29 m. The plume associated with well IW-1 is 4.6 m by 21.3 m at the 100 mg/L contour.

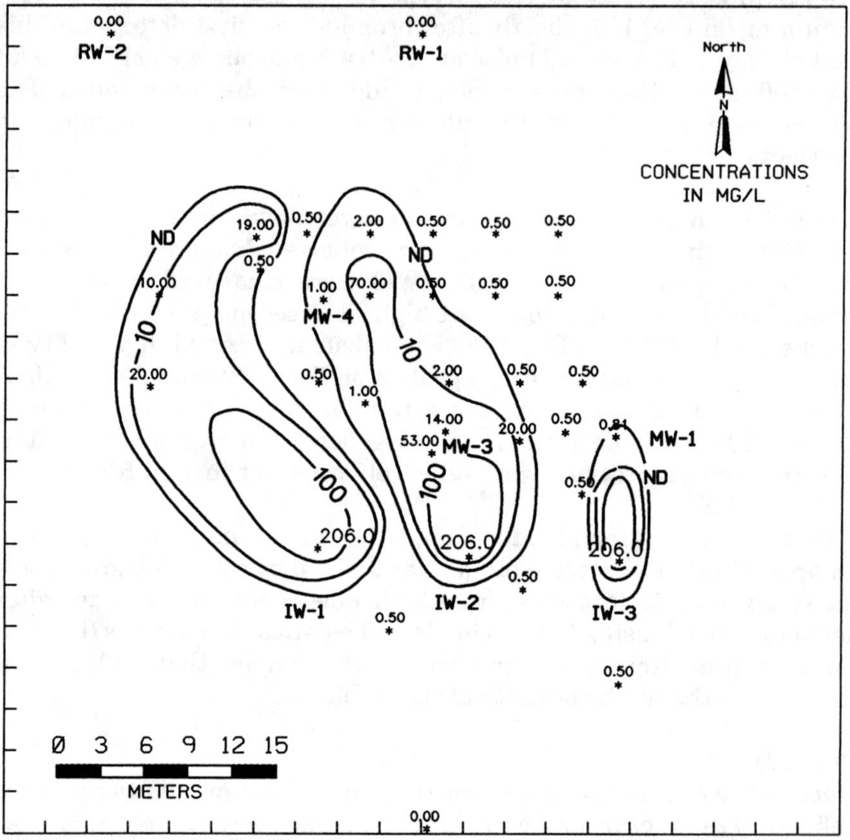

Figure 28. Bromide isoconcentration map, upper flow regime.

The plume associated with well IW-3 was estimated to be 7.6 m by 22.9 m. The entire plume was 29 m long and 33.5 m wide. There was a relatively slow rate of dispersion and diffusion compared to advection.

Bromide concentrations were approximately one order of magnitude higher in the lower flow regime than the upper flow regime. The greater areal extent of the lower bromide plume was due to higher permeability at the base and the density effects of the concentrated substrate solution.

Substrate

Upper Flow Regime: Figure 30 shows the substrate distribution in the upper flow regime during the CPT survey. TOC concentrations ranged from a high of 578 mg/L in well MW-3 to a low of between

1 and 47 mg/L on the plume edges. The highest COD concentrations were also found in well MW-3 (17,500 mg/L). Elevated acetate concentrations were generally associated with elevated TOC and COD levels, and the distribution of acetate was similar to TOC. However, benzoate was only detected around the center plume associated with well IW-2.

Lower Flow Regime: Substrate in the lower flow regime (see Figure 31) has a distribution similar to the bromide, except the plumes are separated and more distinct, and the highest concentrations of TOC extend only approximately half as far as the highest concentration of bromide. The highest concentrations of TOC were 2126 mg/L in CPT-3 and 2000

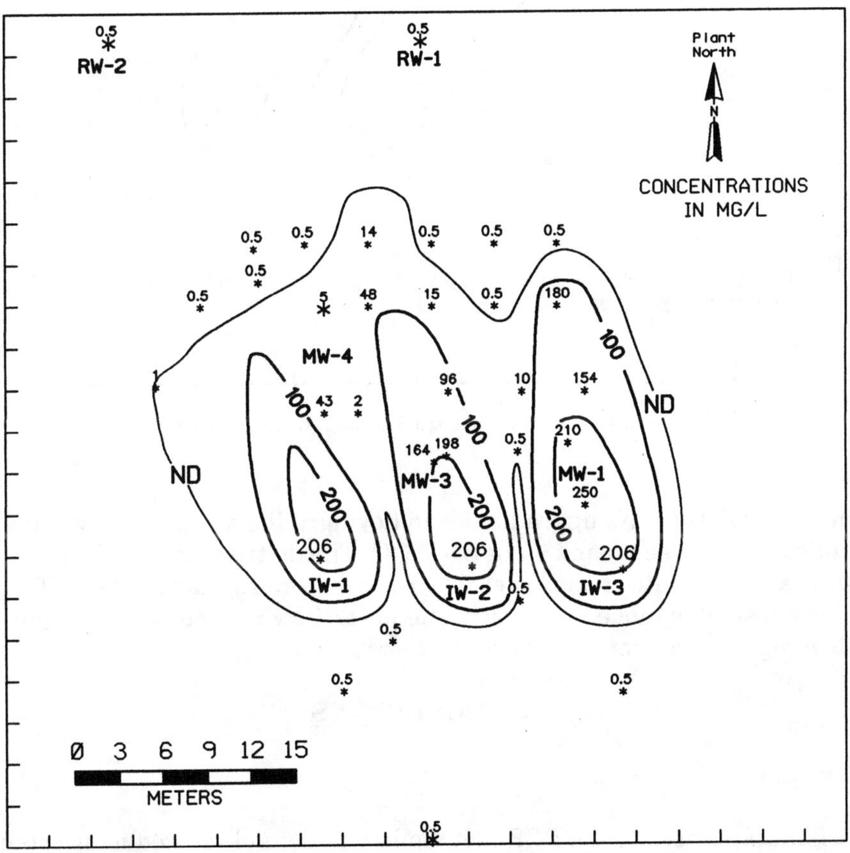

Figure 29. Bromide isoconcentration map, lower flow regime.

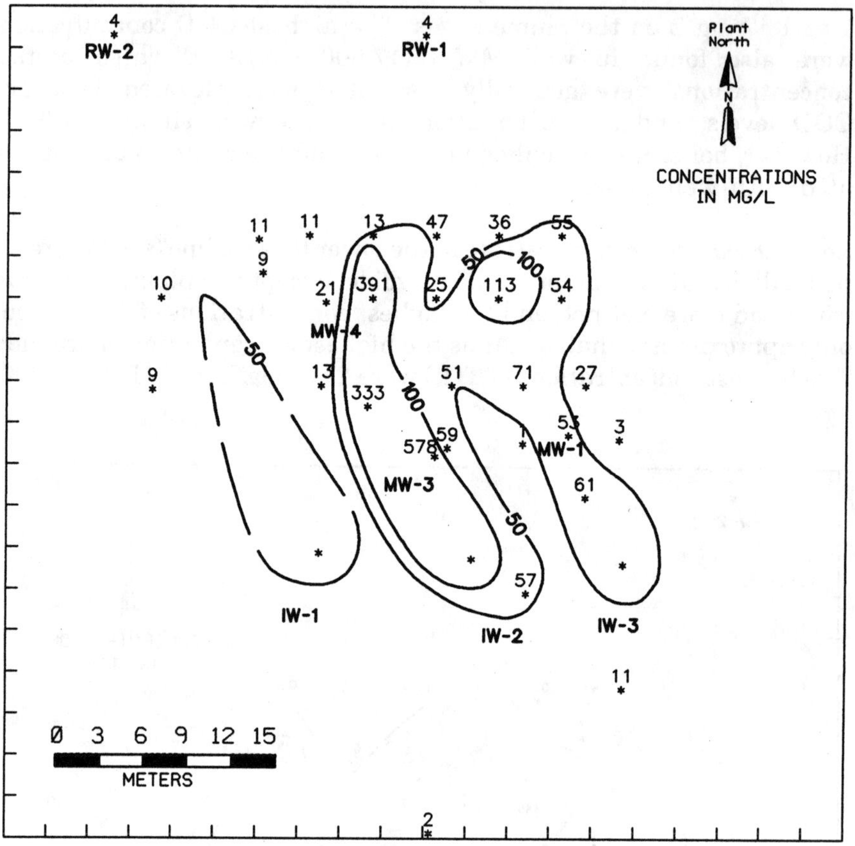

Figure 30. TOC isoconcentration map, upper flow regime.

mg/L in CPT-7. The upgradient and downgradient edges of the plume only have between 3 and 5 mg/L of TOC. The distributions of COD, acetate, and benzoate in the lower flow regime were similar to that of TOC. The substrates moved further in the lower flow regime than the upper flow regime and were more evenly distributed.

CONCLUSIONS

Victoria

Laboratory studies of PCE metabolism under sulfate-reducing conditions have shown TCE and DCE to be metabolic end products (Suflita et al., 1988; Bagley and Gossett, 1990). However, PCE metabolism under methanogenic conditions invariably produces VC, an unwanted microbial metabolic product (Suflita et al., 1988; Bagley and Gossett, 1990). If PCE could be biodegraded to DCE without the accumulation of VC, then anaerobic reductive dechlorination could become an acceptable remediation technology. These results, as well as the published literature, indicate that DCE could be produced in a controlled manner in selected aquifers by applying biological reductive dehalogenation under sulfate-reducing conditions.

The following three possible remediation scenarios can be envisioned to remove the remaining DCE:

- Complete reductive dehalogenation under sulfate-reducing

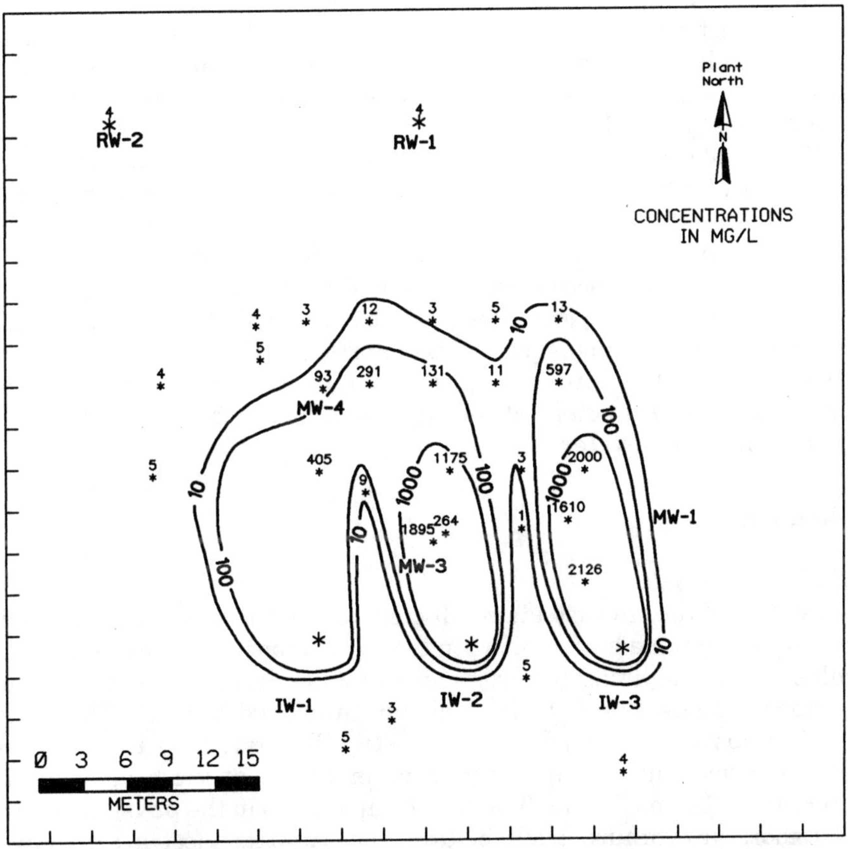

Figure 31. TOC isoconcentration map, lower flow regime.

conditions, such as demonstrated in Phase II studies at this site (Beeman et al., 1994).

- Allow the site to accumulate DCE and then convert the site conditions to either methanogenic or aerobic (methylotrophic).
- Recover and treat DCE-bearing groundwater ex situ (e.g., by air stripping or activated carbon adsorption).

If VC accumulates under methanogenic or acetogenic conditions, it could also be biotransformed under aerobic conditions or physically treated by stripping and vapor-phase removal.

Even partial dechlorination (i.e., PCE to DCE) can substantially reduce the remediation period for pump-and-treat programs. As an example, the average time to achieve a 90% reduction of PCE by a the standard pump-and-treat approach has been estimated to be about 50 pore volumes (Jackson and Patterson, 1989). However, if PCE could be biotransformed to DCE, then a 90% removal would require about 23 pore volumes by virtue of increased solubility and mobility. This would cut pump-and-treat time by more than half when compared to the removal of PCE.

The pilot program demonstrated that PCE, DCE, and VC can be anaerobically biodegraded to ethene and ethane. The concentrations of these chlorinated hydrocarbons were reduced to below analytical detection limits in a relatively short time. Microbial anaerobic reductive dechlorination was accomplished by adding a benzoate solution with sulfate to recirculated groundwater. The mass balance estimate, using bromide tracer in the test zone, was approximately 74% based on PCE. The results of this study suggest that microbial anaerobic reductive dechlorination is a potential remedial mechanism for chlorinated compounds in groundwater.

Niagara

TRACER TEST

Results of the tracer tests conducted during the initial phase of this anaerobic bioremediation pilot study have demonstrated the difficulties inherent to describing flow regimes in highly variable, relatively low-permeability overburden materials and fractured bedrock. The range in dilution rates observed in well MW-16A illustrate the wide variation in the permeability of the overburden materials at the Niagara plant and their effect on A Zone flow. The complexities in the physical nature of the A Zone overburden and bedrock zones and its effect on groundwater flow need to be better defined.

MW-16A

The hydrogeology of the area near well 16A was favorable to in situ biodegradation of chlorinated organic compounds. The substrate and sulfate levels increased in this well to levels that were probably in excess of that needed to promote dechlorination. The microbes responded to the yeast extract additions with a greater than 94% reduction from the peak concentrations of PCE, TCE, and cDCE. The percentage reductions from the peak concentrations of CF, 1,1,1-TCA, and 1,1-DCA concentrations also were substantial—94%, greater than 91%, and greater than 73%, respectively. The low levels of chlorinated solvents found in the period from Days 158 to 186 may be a result of movement of upgradient contaminated groundwater into the well. Typically under anaerobic reductive dechlorination conditions, an increase in the quantities of daughter product is observed. In this study, the concentrations of cDCE, TCE, and PCE were reduced concurrently, and there was little evidence for accumulation of the daughter products such as cDCE, vinyl chloride, or ethene. It appeared that either a biodegradation pathway other than reductive dehalogenation may have been occurring in this area or the kinetics were very rapid and the daughter products were not observed. Laboratory studies are planned to elucidate the biodegradation pathway. The concentrations of chlorinated solvents increased to pretest levels once substrate additions were ended.

The microbes were active even at the low wintertime temperatures, at the alkaline pH, and under potentially limiting nitrogen conditions. A very high microbial biomass was produced, which appeared to be largely fermentative as indicated by the relatively low levels of hydrogen sulfide and methane being generated and the high production of carbon dioxide. After the sulfate additions were ended on Day 137, there was a significant decrease in the TCE levels (1.6 mg/L to a minimum of 0.28 mg/L). This suggested that the high levels of sulfate may have been at least partially inhibitory to the dechlorination process. The microbes reduced the pH to near neutral conditions, but when the substrate additions were ended, the pH rebounded to the pretest levels.

Gulf Coast

An anaerobic in situ bioremediation field pilot test was operated at a Gulf Coast chemical manufacturing site. The objectives of the test were to determine if anaerobic reductive dehalogenation was a feasible technology and to gain a better understanding of mechanisms affecting the movement and distribution of substrate through the aquifer. An-

Table 4. Mass balance analysis of injectate plume.

Constituents	Upper Flow System (lb)	Lower Flow System (lb)	Injected Mass (lb)	Percent Change (%)
Bromide	161	170	332	0.0
COD	1655	6734	8862	–5.3
Acetate	1315	628	4097	–52.6
Benzoate	50	616	4765	–86.0
Ammonia-nitrogen	6	165	1210	–85.9
Phosphate	0	11	254	–95.6

Aquifer thickness calibrated such that bromide was conserved.

other objective was to investigate whether delivery of high concentrations of substrate would inhibit microbial growth at the delivery well and control plugging in the aquifer.

Groundwater concentrations of substrate were determined at 26 locations, 140 days after delivery started, using CPT technology. Little lateral dispersion was observed. CPT sampling confirmed that substrate delivery was accomplished and that delivery of high concentrations of substrate controlled plugging.

The strategy of injection at inhibitory levels was successful. Substrate was transported through the aquifer with little evidence of plugging in the injection wells or aquifer. The aquifer was heterogeneous, which impacted movement of injectate. The conservative tracer permitted a mass balance to substantiate injectate location. Mass balance calculations indicated that approximately 69% of the substrate was removed by biological activity or sorption (see Table 4). The loss of substrate suggests microbial utilization of benzoate and acetate.

FUTURE RESEARCH AND APPLICATIONS

Future research and application needs have been evaluated and include

- investigation of alternative microbial pathways and kinetics
- hydrogeologic model selection with bioparameter input
- enhanced understanding of fractured bedrock flow
- optimization of amendment command and control systems
- economic modeling of the technology

Filling these future needs will bring a greater understanding and increase the opportunity of success in applying and executing in situ anaerobic reductive dechlorination projects.

REFERENCES

Atlas, R. M. and R. Bartha. 1993. *Microbial Ecology Fundamentals and Applications.* The Benjamin/Cummings Publishing Company, Inc. Redwood City, California.

Bagley, D. M. and J. M. Gossett. 1990. "Tetrachloroethylene Transformation to Trichloroethylene and *cis*-1,2-Dichloroethylene by Sulfate-Reducing Cultures," in *Applied and Environ Microbiology,* 56:2511–2516.

Beeman, R.E., J.E. Howell, S.H. Shoemaker, E.A. Salazar and J.R. Buttram. 1994. "A Field Evaluation of in situ Microbial Reductive Dehalogenation by the Biotransformation of Chlorinated Ethylenes," in Hinchee, R. E., A. Leeson, L. Semprini and S.K. Ong (eds), *Bioremediation of Chlorinated and Polycyclic Aromatic Hydrocarbon Compounds.* Lewis Publishers, Boca Raton, Florida, pp. 14–27.

Bouwer, E. J. 1994. "Bioremediation of Chlorinated Solvents Using Alternate Electron Acceptors," in Norris, R. D., Hinchee, R. E., Brown, R., McCarty, P. L., Semprini, L., Wilson, J. T., Kampbell, D. H., Reinhard, M., Bouwer, E. J., Borden, R. C., Vogel, T. M., Thomas, J. M. and Ward, C. H., *Handbook of Bioremediation.* Lewis Publishers, Boca Raton, Florida, pp. 149–175.

DiStefano, T. D., J. M. Gossett and S. H. Zinder. 1991. "Reductive Dehalogenation of High Concentrations of Tetrachloroethene to Ethene by an Anaerobic Enrichment Culture in the Absence of Methanogenesis," in *Applied and Environmental Microbiology,* 57:2287–2292.

Freeman, D. L. and J. M. Gossett. 1989. "Biological Reductive Dechlorination of Tetrachlorethylene and Trichloroethylene to Ethylene Under Methanogenic Conditions," in *Applied and Environ. Microbiology,* 55:2144–2151.

Jackson, R. E. and R. J. Patterson. 1989. "A Remedial Investigation of an Organically Polluted Outwash," in *Ground-Water Monitoring Review,* 9:119–125.

Lee, M. D., J. M. Thomas, R. C. Borden, P. B. Bedient, J. T. Wilson and C. H. Ward. 1988. "Biorestoration of Aquifers Contaminated with Organic Chemicals," in *CRC Critical Reviews in Environmental Control,* 18:29–89.

Suflita, J. M., S. A. Gibson and R. E. Beeman. 1988. "Anaerobic Biotransformation of Pollutant Chemicals in Aquifers," in *Journal of Industrial Microbiology,* 3:179–194.

Acceleration of PCB Biodegradation *in situ:* A Field Study in the Hudson River

F. J. Mondello, M. R. Harkness, J. B. McDermott,
D. A. Abramowicz, J. J. Salvo, W. P. Flanagan, M. L. Stephens,
R. J. May, J. H. Lobos, K. M. Carroll, M. J. Brennan,
A. A. Bracco, K. M. Fish, G. L. Warner, P. R. Wilson,
D. K. Dietrich, D. T. Lin, C. B. Morgan and W. L. Gately

General Electric Company
Corporate Research and Development
K1, Room 3B38
P.O. Box 8
Schenectady, NY 12301-0008

INTRODUCTION

Polychlorinated biphenyls (PCBs) were produced and used as complex mixtures containing more than 60 different congeners. The chemical and physical properties of these mixtures (sold commercially as Aroclors) led to their widespread use as dielectric, heat transfer and hydraulic fluids, flame retardants, solvent extenders, and plasticizers. Within a 50-year period, approximately 1.4 billion pounds of PCBs were produced, and it is estimated that several hundred million pounds have entered the environment worldwide (Hutzinger and Veerkamp, 1981). The chemical and thermal stability of PCBs and the tendency of highly chlorinated congeners to bioaccumulate in fatty tissues have led to concerns regarding their potential human health effects.

Numerous studies conducted at the GE Corporate Research and Development Center (CRD) and elsewhere have demonstrated that PCBs, formerly believed to be nonbiodegradable, are biotransformed under a variety of conditions. There are two predominant and distinct types of PCB biotransformation activities: (1) reductive dechlorination (mediated by organisms that live in anaerobic environments such as river sediments) and (2) aerobic biodegradation (by bacteria that live in oxygenated environments such as the water columns and surficial sediment layers of rivers and lakes).

Aerobic biodegradation of PCBs has been observed and reported by many investigators since the early 1970s (e.g., Ahmed and Focht, 1973; Sundstrom and Hutzinger, 1976; Furukawa et al., 1978; Furukawa, 1982; Rochkind et al., 1986). The enzymatic pathway for this oxidative process is well understood for a number of bacterial systems. The predominant mechanism occurs via a 2,3-dioxygenase attack (Ahmed and Focht, 1973; Nadim et al., 1987), as shown in Figure 1. The hydroxylated product is then further metabolized, resulting in ring cleavage after attack via a second dioxygenase. The ultimate product of this enzymatic pathway is typically a chlorobenzoic acid. The chlorobenzoic acids are degradable by other aerobic bacteria and usually do not persist in the system (Focht and Hickey, 1990; Hernández et al., 1991). Therefore, even the transitory presence of these metabolites is strong evidence of aerobic PCB biodegradation.

Anaerobic bacteria are also capable of PCB biotransformations. These organisms, commonly found in anaerobic sediments, reductively remove chlorines from highly chlorinated PCBs, converting them to a less chlorinated form. In the Hudson River (NY), several sediment surveys have demonstrated that PCB dechlorination has been both widespread and extensive (Abramowicz *et al.,* 1995). This activity has resulted in a conversion of the original contaminant (Aroclor 1242) from a mixture predominantly comprised of tri- and tetra-chlorobiphenyls to one containing ~40–80% mono- and dichlorinated congeners (Figure 2). Similar PCB dechlorination has also been observed in other aquatic sediments [e.g., New Bedford Harbor (MA), St. Lawrence River (NY), and Sheboygan River (WI)]. The lightly chlorinated PCB congeners resulting from anaerobic dechlorination are readily degraded by a wide

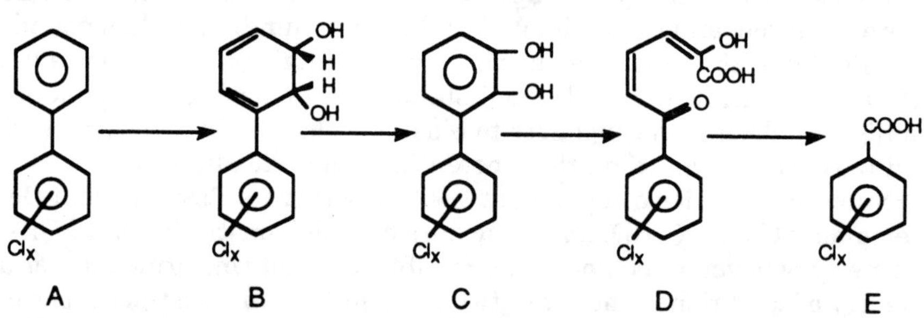

Figure 1. Degradation of polychlorinated biphenyls by the 2,3-dioxygenase pathway: (A) polychlorinated biphenyl (PCB); (B) chlorinated dihydrodiol; (C) chlorinated 2,3-dihydroxybiphenyl; (D) chlorinated 2-hydroxy-6-oxo-6-phenylhexa-2,3-dienoic acid; (E) chlorinated benzoic acid. The chlorobenzoates (E) are further biodegraded to CO_2, H_2O, Cl^-, and biomass by a variety of indigenous aerobic microorganisms present in Hudson River sediments.

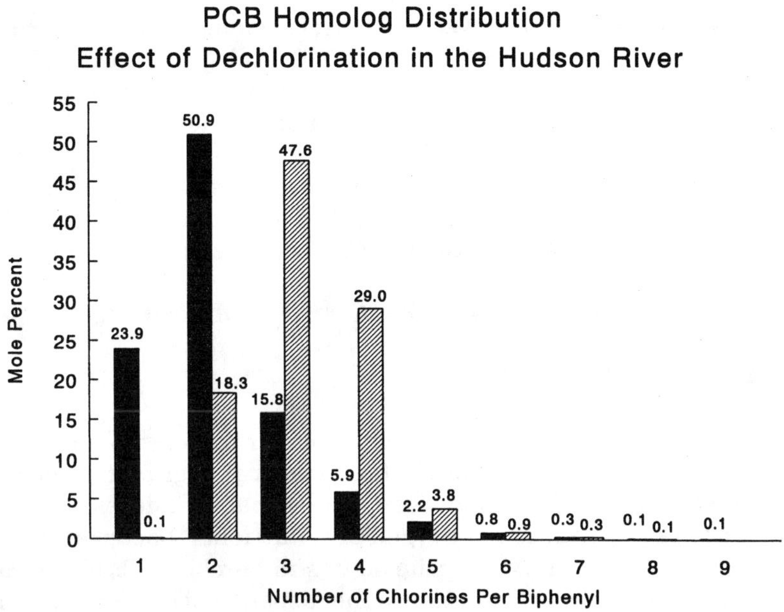

Figure 2. Histogram comparing the homolog distribution of PCBs in a typical H7 Hudson River sample (black bars) with that of unaltered Aroclor 1242 (shaded bars). The increased proportions of mono- and dichlorobiphenyls reflect extensive *in situ* dechlorination activity.

range of aerobic organisms (Bedard et al., 1986). The complementary nature of the anaerobic and aerobic systems provides the potential to completely biodegrade PCBs to carbon dioxide, water, and chloride ions.

A significant amount of information had been gathered regarding the oxidative degradation of PCBs by aerobic bacteria in laboratory experiments (reviewed in Abramowicz, 1990; Bedard, 1990). However, very little was known about this process in the field. Since the PCBs in upper Hudson River sediments had undergone extensive dechlorination by indigenous anaerobic bacteria, the site was well suited for a field study of aerobic biodegradation. Furthermore, a wide variety of aerobic PCB-degrading bacteria have been isolated from the upper Hudson River region. Included among these is *Alcaligenes eutrophus* strain H850, one of the most competent PCB-degrading strains identified to date (Bedard et al., 1986).

To our knowledge, the 1991 *in situ* Hudson River Research Study was the first PCB bioremediation study in a river sediment environment. The objectives of this study were to (1) demonstrate that aerobic PCB biodegradation could occur under field-like conditions, as suggested by prior laboratory investigations; (2) identify the key variables that influence the rate and extent of PCB biodegradation in Hudson River sediments; and (3) investigate the potential for aerobic biodegradation to occur in these sediments via indigenous PCB-degrading microorganisms.

Some of the information in this report was previously published by Harkness *et al.* (1993) in *Science,* Vol. 22. Copies of this article and the final report from the study are available upon request.

EXPERIMENTAL DESIGN AND METHODS

Experimental Design

The Hudson River Research Study was performed to evaluate the parameters that potentially affect the biological degradation of PCBs in sediments under field conditions. The variables addressed in this study include the addition of oxygen and nutrients, bacterial inoculation, and the effect of mixing mode on *in situ* aerobic PCB biodegradation. Mixing is necessary to enhance the dispersal of oxygen and nutrients into the sediment. A high-mix condition was created to suspend the lighter fractions of the sediment into the water column. A low-mix condition was created using a plow-like rake rotating at 3 rpm to turn the sediment over periodically without resuspending large amounts of dense material. In both cases, only the top 6–8″ of sediment were mixed, because this is where the PCB levels were highest.

The bacterial strain *Alcaligenes eutrophus* H850 was isolated in 1983 from PCB-containing dredge spoils several hundred yards upstream from the study site (Bedard et al., 1984). The degradative competence of this organism has been well characterized using a variety of PCB congeners in the laboratory. *A. eutrophus* H850 is able to degrade an unusually broad spectrum of PCB congeners, including some tetra-, penta-, and hexachlorobiphenyls (Bedard et al., 1987). During this study, H850 was added to two of the caissons to establish if bacterial inoculation would enhance PCB biodegradation in a competitive environment.

The experimental design for the river study is summarized in Table 1. Two of the caissons, R101 and R104, were used as controls for the high- and low-mix agitation systems, respectively. No nutrients, oxygen, or organisms were added to these reactors in order to limit aerobic biological activity. It was difficult in practice to exclude all oxygen from these systems; thus, each caisson was aerobic to some degree. This was not a significant problem for the low-mix caisson R104, but a leaky

Table 1. Experimental protocol.

R101: Physical Control for the Turbine Agitated (High-Mix) System
- Turbine agitated system
- Physical mixing only
- Anaerobic environment
- No nutrients, oxygen, or organisms added

R104: Physical Control for the Rake Agitated (Low-Mix) System
- Rake agitated system
- Physical mixing only
- Anaerobic environment
- No nutrients, oxygen, or organisms added

R102: Effect of Added Organisms under High-Mix Conditions
- Turbine agitated system
- Nitrogen, phosphate, and biphenyl added
- Oxygen on demand via hydrogen peroxide
- H850 addition

R103: Effect of Added Organisms under Low-Mix Conditions
- Rake agitated system
- Nitrogen, phosphate, and biphenyl added
- Oxygen on demand via hydrogen peroxide
- H850 addition

R105: Stimulation of the Indigenous Population under Low-Mix Conditions
- Rake agitated system
- Nitrogen, phosphate, and biphenyl added
- Oxygen on demand via hydrogen peroxide
- No addition of organisms

R106: Stimulation of the Indigenous Population under Low-Mix Conditions
- Same as R105

seal in R101 combined with the rapid mixing resulted in elevated oxygen levels in the water column. This was particularly pronounced when the reactor was opened for sampling. This phenomenon was minimized by pouring several liters of liquid nitrogen into the caisson to purge the headspace of oxygen prior to closing the caisson and restarting the mixers after sampling.

R102, R103, R105, and R106 were experimental caissons. Each received identical additions of nitrogen, phosphate, and biphenyl at the start and at intervals throughout the study. The actual addition schedule is shown in Table 2. Nitrogen and phosphate were added in the form of reagent grade diammonium phosphate, ammonium sulfate, and potassium phosphate to maintain the predetermined minimum nutrient levels in the caissons. Nitrogen and phosphate analyses were performed on a biweekly basis to monitor these levels. Biphenyl was added on a fixed schedule at 10- and 20-day intervals throughout the experiment. The addition of biphenyl served two purposes: (1) it is a growth substrate for nearly all PCB-degrading bacteria identified thus far, and (2) it is an inducer of the PCB degradative pathway and thus promotes maximum levels of activity (Rochkind et al., 1986; Nadim et al., 1987). The temperature and pH in all reactors were monitored but not controlled.

The experimental caissons differed from each other by agitation mode and bacterial inoculation with H850. R102 contained a high-mix, turbine agitation system, while the other experimental caissons contained low-mix, rake agitators. R102 and R103 were inoculated with H850 at various points during the study. R105 and R106 were duplicate caissons in which only the indigenous bacterial populations were stimulated. The duplication was used to determine the inherent variability in the experiment. The effect of mixing may be assessed by a direct comparison of caissons R102 and R103. The impact of H850 inoculation may be determined by comparing R103 with R105 and R106.

Site Characterization and Platform Design

The Hudson River Research Station (HRRS) was located immediately offshore from the west bank of the Hudson River, 193.5 river miles upstream from the Battery in New York City. The site was in the town of Moreau in Saratoga County, New York, approximately 2 miles downstream of Fort Edward, with access provided from West River Road. The area, which has been designated Hot Spot #5 by the New York State Department of Environmental Conservation, will be called H7 in this report (GE designation). This site was selected for evaluation based upon earlier laboratory work using this sediment. Samples taken from this location over the past several years have demonstrated extensive anaerobic dechlorination activity. Also, aerobic PCB-degrading bacteria such as *A. eutrophus* H850 (used in this study) had been isolated from the banks of the Hudson River near this site.

Physical Plant Description

The HRRS platform was a 25 ft × 37.5 ft rectangular steel structure erected approximately 10 ft offshore. The longer side (37.5 ft) ran parallel to the shoreline. The river platform supported six caisson reactors

Table 2. HRRS—nutrient and organism addition log.

Date	Reactor	$(NH_4)_2PO_4$	$(NH_4)_2SO_4$	KH_2PO_4	Biphenyl	H850
8-9-91	R102	1.0	2.0	—	1.0	0.19
8-9-91	R103	1.0	2.0	—	1.0	0.19
8-9-91	R105	1.0	2.0	—	1.0	—
8-9-91	R106	1.0	2.0	—	1.0	—
8-22-91	R102	0.5	—	—	—	—
8-22-91	R103	0.5	—	—	—	0.19
8-22-91	R105	0.5	—	—	—	—
8-22-91	R106	0.5	—	—	—	—
8-27-91	R102	—	—	—	—	0.21
8-30-91	R102	0.5	—	—	0.25	—
8-30-91	R103	0.5	—	—	0.25	—
8-30-91	R105	0.5	—	—	0.25	—
8-30-91	R106	0.5	—	—	0.25	—
9-9-91	R102	—	—	—	0.25	—
9-9-91	R103	—	—	—	0.25	—
9-9-91	R105	—	—	—	0.25	—
9-9-91	R106	—	—	—	0.25	—
9-12-91	R102	—	—	0.5	—	—
9-12-91	R103	—	—	0.5	—	—
9-12-91	R105	—	—	0.5	—	—
9-12-91	R106	—	—	0.5	—	—
9-20-91	R102	—	—	—	0.25	—
9-20-91	R013	—	—	—	0.25	—
9-20-91	R105	—	—	—	0.25	—
9-20-91	R106	—	—	—	0.25	—
9-30-91	R102	—	—	—	1.00	—
9-30-91	R103	—	—	—	1.00	0.21
9-30-91	R105	—	—	—	1.00	—
9-30-91	R106	—	—	—	1.00	—
10-24-91	R103	—	—	—	—	2.10

All notations in kilograms.

plus a gas handling system and local monitoring equipment. The majority of the support functions, such as hydrogen peroxide and nitrogen supply, set point control, data acquisition, and motor control were land-based. The detailed design, construction, and installation of the HRRS was subcontracted by GE to OHM Corporation (Findlay, OH). An aerial view of the platform is depicted in Figure 3.

Construction of the platform was done by a barge mounted crane. A silt curtain was installed around the platform to contain any silt disturbed during construction. The 6-ft diameter caissons were driven with a vibratory hammer an average of 8 ft into the sediment. This caisson depth, coupled with the very low sediment permeability, provided a virtual seal at the reactor bottom. The water depth in the caissons ranged from 2.5 to 4 ft. The footprint of the six caisson reactors on the river bottom was approximately 25 ft × 15.5 ft. The criteria for placement of the reactors were (1) average PCB concentration in excess of 50 ppm and (2) moderate variation in PCB concentration from point to point.

The results of a site survey conducted by O'Brien & Gere Engineers indicated a high degree of variation in PCB concentration throughout the area. The difference between adjacent points was often large, despite the fact that samples were taken only 6 ft apart. This is a common feature of heterogeneous environmental samples that complicates any field experiment or operation. This spatial variability would manifest itself when the survey results were compared to the PCB concentrations measured in the caissons at the beginning of the experiment.

The controllers and monitors for the process variables were housed in a motor control trailer. The variables continuously monitored included the following:

1. Temperature
2. pH
3. Dissolved oxygen (DO)
4. Water level
5. Motor torque
6. Gas flows (through resin traps)
7. Hydrogen peroxide flow

The gas flows and hydrogen peroxide flows were integrated with totalizers. The process variable information was displayed with LED readouts and recorded for later analysis with strip chart recorders and a personal computer with data acquisition software. The process information was also monitored on a personal computer at CRD in Schenectady via modem.

Caisson Details

The caissons were constructed from 6-ft internal diameter spiral welded carbon steel pipe with a wall thickness of 1/2 in. The caisson tops were 3/8-in. thick flat steel plates bolted onto a flange welded to the top of the caisson walls. The gasket material sealing the caisson top was 1/2 in. Gore-tex™. Access to the caisson internals was through the caisson top. Four rectangular sampling ports located 90° apart allowed sediment samples to be taken at four angular positions and at varying radial positions. An 18-in. manhole provided additional access. All monitoring probes entered through ports on the caisson lid. All seals were maintained with 1/8-in. viton gaskets.

One objective of the experimental design was to assess the impact of agitation on the rate and extent of PCB biodegradation. Therefore, two caisson configurations were employed for the study, each utilizing a different mode of mixing the sediment. A high-mix mode of agitation was created in two reactors using turbine agitators manufactured by Lightnin (Rochester, NY). The intent was to suspend the lighter fractions of the sediment in the aqueous phase. A 5-hp variable speed AC motor (0 to 80 rpm) was used to turn a four-blade impeller through the water column at a constant speed of 40 rpm. At this speed, the 3-ft diameter turbine blade afforded large pumping volumes (22,250 gpm at maximum rpm) at relatively low tip speeds.

The remaining four reactors utilized a low-mix mode of agitation, created with a rake mechanism designed to slowly plow through the sediment. The rake agitators were manufactured by EIMCO (Salt Lake City, UT). The assembly was a drive mechanism for a 12-ft diameter thickener or clarifier. The rake design consisted of four arms at 90°, each with four plows. The plows were 1 ft long and had three angled plates welded onto a 3 in. × 3 in. right angle. The plows were arranged in a staggered configuration so that the plow faces would not ride one behind the other. This prevented the formation of mounds of unmixed sediment in between the plow furrows. The rake was driven with a 5-hp motor and set to plow the top 6–8 in. of sediment at a constant 3 rpm. Caissons with this rake agitator also had small water column agitators to induce slow circulation and to eliminate any gradients in the water column.

Hydrogen Peroxide Delivery System

The hydrogen peroxide used as an oxygen source in the experiment was obtained as a 10% solution in water from Degussa Corporation

Figure 3. Aerial photograph of the Hudson River Research Station platform during construction. Next page: Close-up photograph of a high-mix reactor.

SERIES
DANGER
HIGH VOLTAGE

(Richfield, NJ). It was delivered in an ISO container mounted on a trailer along with a feed pump system. Oxygen was delivered to the experimental caissons on demand via controllers linked to dissolved oxygen probes. The dissolved oxygen (DO) probes and monitoring system (5340D and 671D, respectively, Great Lakes Instr. Co.) provided a continuous readout of DO in ppm and a system output of 4–20 mA to a loop controller. The controller (MYCRO 352, Moore Prod. Co.) metered hydrogen peroxide flow to the caissons when the dissolved oxygen levels fell below a set point. Total peroxide usage was tracked for each experimental caisson.

Growth of *Alcaligenes eutrophus* Strain H850

The *Alcaligenes eutrophus* strain H850 used to inoculate caissons R102 and R103 was grown aerobically at 30°C in a phosphate-buffered mineral salts medium (PAS) containing biphenyl as a carbon source and supplemented with 0.005% yeast extract (Bedard et al., 1984). The bacteria were grown in 15-L and 300-L fermenters with vigorous mixing and aeration. Large-scale cultures were inoculated to an optical density (OD_{615}) of 0.1 with stored frozen H850 cells (e.g., 20 g H850 wet weight per 300 L volume). Growth occurred over a 24-h period to a final OD_{615} of 2.5. Cells were harvested by concentration with a Millipore tangential flow filtration apparatus (40 ft^2 of 0.22 micron Duropore membrane), followed by high-speed centrifugation. The cell paste was transferred to plastic bags, quickly frozen as a thin film in liquid nitrogen and then stored at −80°C until needed.

PCB-degradative competence of the cells was assessed at various times during the growth phase of the culture and after harvesting. This was done by conducting a resting-cell assay using a defined PCB congener mix (Bedard et al., 1986). No significant change in the PCB-degradative competence was observed before or after cell harvesting. The viability of the cells after harvesting was determined through plate counts on PAS/biphenyl plates.

To inoculate a caisson, the frozen H850 cells were diluted at the study site using phosphate buffer to resuspend the paste. In general, the caissons were inoculated with about 0.2 kg of H850 to give an OD_{615} of 0.1 (approximately 1.0×10^8 cfu/ml).

Sampling Procedures

Sediment cores for PCB analysis were taken from the caissons at the beginning and end of the experiment, and at 1-week intervals throughout the study. In an attempt to reduce PCB concentration gradients, the sediments were premixed for 24 h using the rake or turbine agitators and then allowed an additional 24 h to settle before the initial samples were taken (T_0). Twelve cores were taken from each caisson at the initial and final time points. This large number of samples was deemed necessary to achieve representative averages, given the anticipated variability of the PCB distributions in the sediment. Four cores were taken from each caisson at the intermediate time points. While conclusions should not be drawn based on data from any single intermediate time point due to the small sample size, these values gave an indication of the rate at which PCB concentrations changed over time.

The initial sampling took 2 days. Cores were taken from the inner, middle, and outer position of each port in an attempt to achieve a radial distribution of samples. A similar protocol was followed for the final caisson sampling (T_f), without the initial 24-h premix step.

In addition to samples for PCB analysis, samples for bacterial counts, chlorobenzoates, and biphenyl were also taken throughout the study. A minimum of one aqueous sample and two sediment samples per caisson per week were obtained to track both aerobic heterotrophic and biphenyl/PCB-degrading bacterial populations. Additional samples were taken during the first several weeks of the experiment and after each H850 addition to the caissons. Samples for biological analysis were obtained by subsampling sediment cores taken directly from the caissons. Aqueous-phase samples were obtained by pouring water from the top of the core tubes into a sterile 50-ml sample tube. The sediment samples for biological analysis were taken by removing 3–5 g of material from the top 1–2 in. of sediment remaining in the core. All samples were stored on ice during transfer to CRD for analysis.

The caissons were sampled for chlorobenzoates daily during the first week of the experiment, three times per week from the second through the fourth weeks, and weekly thereafter. In general, one aqueous sample and two sediment samples were collected from each caisson on each sampling day. On sampling days when sediment cores were taken for PCB analysis, four, rather than two, sediment samples were collected from each caisson. Sediment and aqueous samples were analyzed for biphenyl concentration on a weekly basis. Aqueous samples for chlorobenzoate analysis were obtained either by lowering a dip bucket through a caisson sample port or by collecting the aqueous supernatant from sediment cores taken for PCB analysis. In either case, 350- to 400-ml aqueous samples were collected in glass jars and capped with Teflon®-lined lids. Sodium hydroxide (10 ml of 0.75 N NaOH) was added to each of the aqueous samples, which were then transported to CRD on ice and stored at 4°C. The combination of pH adjustment and cold storage was used to inhibit further microbiological activity in the samples.

All the caissons were vented to the atmosphere through vapor traps containing XAD-2 resin to retain any PCBs that may have volatilized from these systems. The vapor traps were removed and the resin analyzed for PCBs after the first week and at 3-week intervals throughout the duration of the study.

PCB Analysis

PCB analysis for this study was done by Northeast Analytical Laboratory, an independent, commercial environmental laboratory in Schenectady, New York. Samples containing large amounts of liquid were centrifuged in their containers to separate the solid and liquid phases. In some cases, alum was used to facilitate this process. The clear aqueous phase was removed, and the liquid samples were combined for each caisson and extracted with methylene chloride.

The sediment samples were partially air-dried, homogenized, weighed, and subsampled for PCB analysis. The subsampled material was completely air-dried in a low flow fume hood at room temperature. No substantial loss of lower-chlorinated PCB congeners was observed during drying (data not shown). The dried sediment was then ground into a fine powder. Foreign objects and unusually large stones that would artificially skew the analysis were removed at this point. The sediment was reweighed into a tared soxhlet thimble and extracted with hexane/acetone for 16 h.

XAD-2 resin samples from the vapor traps were weighed into tared soxhlet thimbles and extracted in bulk with hexane/acetone for 16 h. The extract cleanup procedure for all samples included the use of elemental mercury to remove sulfur, sulfuric acid to remove readily oxidizable hydrocarbons and colored biogenic compounds, and florisil to remove polar acidic compounds from the sediment extract. This procedure eliminates compounds that interfere with PCB analysis on the gas chromatograph (GC). PCB analyses were performed on two Varian 3400 GCs, using DB-1 (J&W Scientific), bonded polydimethylsilicone, fused silica capillary columns (30 m × 0.25 mm inner diameter), and electron capture detectors. One hundred eighteen PCB peaks are resolved using this column system. A mixture of Aroclors 1242 and 1260, spiked with known amounts of mono- and dichlorobiphenyls, was used as the primary calibration standard.

Chlorobenzoate and Biphenyl Analysis

Chlorobenzoate (CBA) concentrations were monitored throughout this study using a congener-specific method developed by Flanagan and May (1993).

Bacterial Plate Counts

The bacterial plate counts done as part of this experiment were designed to track the relative numbers of PCB/biphenyl-metabolizing and total aerobic heterotrophic bacteria present in the caissons. The colony-forming units (cfu) reported do not represent the total number of bacteria present in the samples, since exhaustive efforts were not employed to remove all the bacteria from the sediment particles, and all bacteria do not grow on the same agar medium.

The colony-forming units of indigenous biphenyl-metabolizing bacteria were measured by direct plate counts with biphenyl as the sole source of carbon and energy. Supplemental additions of nutrients such as yeast extract could not be used in the PAS-biphenyl plates without obscuring the analysis due to increased background growth of nonbiphenyl-degrading microbes. Some biphenyl-degrading organisms previously isolated from the Hudson River sediment grew very poorly without the addition of trace amounts of yeast extract. Therefore, the numbers counted in this study are conservative estimates of the total number of biphenyl-degrading organisms in the caissons. Since virtually all biphenyl-degrading bacteria examined are able to degrade PCBs to some extent, the number of these bacteria was used to estimate the PCB-degrading population. Enumeration of *Alcaligenes eutrophus* H850 was on the basis of its recognizable colony morphology, which was distinguishable from that of indigenous biphenyl-degrading organisms.

Toxicity Testing

Sediment samples taken before and after the study were subjected to whole-sediment toxicity testing at the Battelle Great Lakes Environmental Center (Traverse City, MI). Whole-sediment, acute toxicity tests were performed using *Ceriodaphnia dubia,* an EPA-approved toxicity indicator organism. In accordance with Battelle's standard operating procedures, the test species was exposed to whole sediment and the mortality of the organisms monitored over a 48-h period.

Statistical Analysis

Two-sample *t*-tests were performed using initial and final sampling measurements of PCB concentration, total core PCB levels, and PCB concentrations normalized to total organic carbon (TOC), and PCB peak 61. In addition, a Wilcoxon rank sum test, a nonparametric counterpart to the *t*-test was also performed on the data. This test is valid under less restrictive assumptions and only requires that the samples come from continuous populations. The results from this test closely duplicated the *t*-test results for each of the caissons at the 5% significance level using both the original and transformed values of the variables.

RESULTS

The Hudson River Research Study began on August 9, 1991, with the addition of nutrients and bacterial strain H850 to the experimental caissons. The study was conducted for 10.5 weeks until October 21, and the final sampling took place October 22–24.

PCB Biodegradation

Demonstrating *in situ* biodegradation in field studies is complicated by difficulties in establishing mass balance and separating biotic from abiotic loss processes. To minimize these difficulties, several methods were used to examine PCB changes within the reactors: (1) direct measurements of PCB concentration, (2) PCB concentration measurements normalized to that of a recalcitrant "internal standard" reference (peak 61; 3,4,3′,4′-CB/2,3,6,3′,4′-CB), and (3) PCB concentration measurements normalized to the TOH content of each sample.

A strong indicator of biological activity against PCBs is the preferential loss of certain congeners. Aerobic biodegradation is known to be a congener-specific process, with lightly chlorinated PCBs being degraded more readily than others (reviewed in Abramowicz, 1990; Bedard, 1990). It was therefore expected that mono- and dichlorobiphenyls would be preferentially degraded relative to the more highly chlorinated forms. Such results would also be consistent with laboratory studies using these sediments (Harkness *et al.,* 1992).

The initial PCB concentrations in each caisson varied substantially and ranged from approximately 6 ppm in R101 to 50 ppm in R105 (all PCB concentrations determined on a sediment dry weight basis). All of the experimental reactors showed statistically significant reductions in PCB levels regardless of the quantitation method used (Table 3).

Table 3. PCB concentration changes.

Caisson	Treatment	Starting PCB Conc. (mg/kg)	Percent Change: Direct PCB Conc.	Percent Change: TOC Normalized	Percent Change: Peak 61 Normalized
R101	High mix, control	6.0 ± 1.9	+8.7[a]	−30.7[a]	−14.4
R102	High mix, H850	20.0 ± 11.0	−41.0	−44.7	−42.4
R103	Low mix, H850	30.0 ± 10.6	−36.8	−55.5	−37.8
R104	Low mix, control	39.9 ± 15.6	−41.8	+8.4[a]	−4.3[a]
R105	Low mix, indigenous	49.7 ± 27.8	−72.6	−53.1	−40.5
R106	Low mix, indigenous	39.1 ± 17.5	−68.5	−46.0	−38.7

[a]Indicates changes were not statistically significant at a 95% confidence limit.

The most consistent results were obtained using the two normalization techniques, which indicated losses between 38 and 55% over the course of the study. Bar charts showing the average PCB congener distributions for T_0 and T_f sediment samples for each caisson are shown in Figures 4a through 4f. Several of the important PCB congener peaks are identified for reference. These results are normalized using a PCB peak (peak 61), recalcitrant to biodegradation as an internal standard.

Significant congener-specific changes were evident in all of the experimental caissons. Reductions in concentration are especially evident for 2-, 2,2′-/2,6-, 2,3-/2,4′-, and 2,3,6-/2,6,3′-chlorobiphenyls (peaks 2, 5, 8, and 16, respectively), while congeners like 2,6,2′-chlorobiphenyl (peak 10), and some of the tri- and tetrachlorobiphenyls (peak 30 and above) changed much less.

Interpreting the data from the two control reactors (R101 and R104) is complicated by the fact that the statistical significance of the results depends upon the method of quantitation. For example, no significant loss of PCB is observed in R104 when using either normalization method. However, an examination of direct PCB concentration shows a statistically significant reduction of 41.8%. This inconsistency can be attributed to a sampling phenomena observed in the slow mix reactors. During the course of the experiment, movement of the rakes through the sediment increased the fraction of PCB-rich organic particles that did not rapidly settle prior to T_f sampling. The influence of this phenomenon on PCB recovery is shown by comparing the recovery of PCBs from R104 samples taken before and after 1 week of settling (48.9 ± 4.8 μg/g after settling, 23.1 ± 9.9 before settling). These data strongly suggest that redistribution and not biodegradation is responsible for

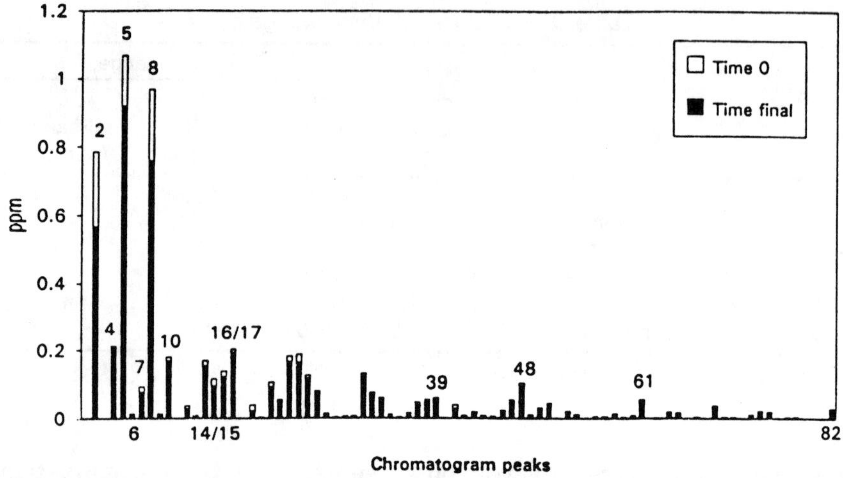

Figure 4a. Histogram showing peak-specific PCB concentrations in core samples from caisson R101 at T_0 and T_f. Histograms represent averages of all the cores taken at that time point, normalized to peak 61 to emphasize congener-specific changes. Selected PCB peak identifications are: peak 2 (2-CB); peak 5 (2,2′-/2,6-CB); peak 8 (2,3-/2,4′-CB); peak 10 (2,6,2′-CB); peak 16 (2,3,6-/2,6,3′-CB); peak 17 (2,3,2′/2,6,4′-CB); peak 39 (2,3,6,4′-/2,6,3′,4′-CB); peak 48 (2,3,5,6,2′-/2,3,6,2′,5′-CB); peak 61 (3,4,3′,4′-/2,3,6,3′,4′-CB); peak 82 (2,3,4,2′,4′,5′-/2,3,5,6,3′,4′-CB).

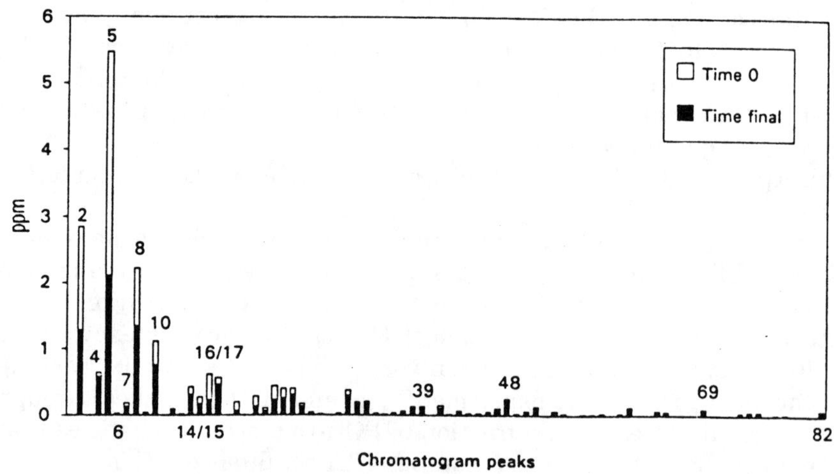

Figure 4b. Histogram showing peak-specific PCB concentrations in core samples from caisson R102 at T_0 and T_f. Histograms represent averages of all the cores taken at that time point, normalized to peak 61 to emphasize congener-specific changes. Selective congener depletion is evident. Peak identifications are as indicated in Figure 4a.

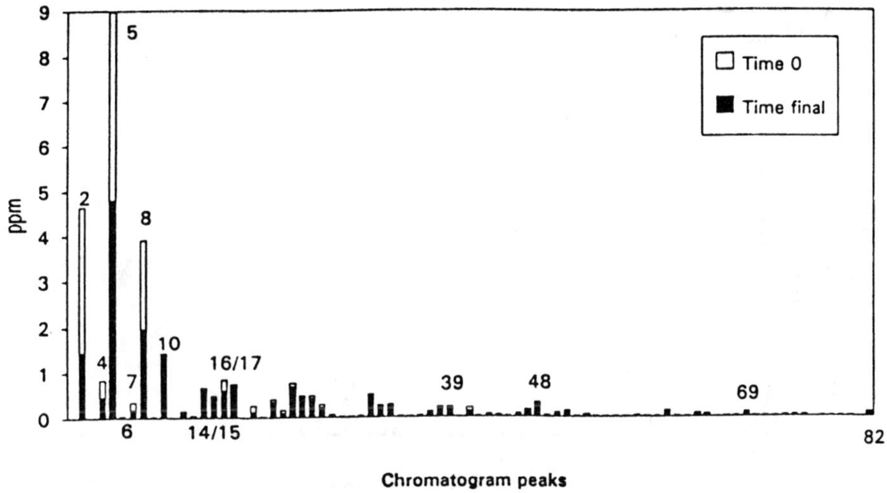

Figure 4c. Histogram showing peak-specific PCB concentrations in core samples from caisson R103 at T_0 and T_f. Histograms represent averages of all the cores taken at that time point, normalized to peak 61 to emphasize congener-specific changes. Selective congener depletion is evident. Peak identifications are as indicated in Figure 4a.

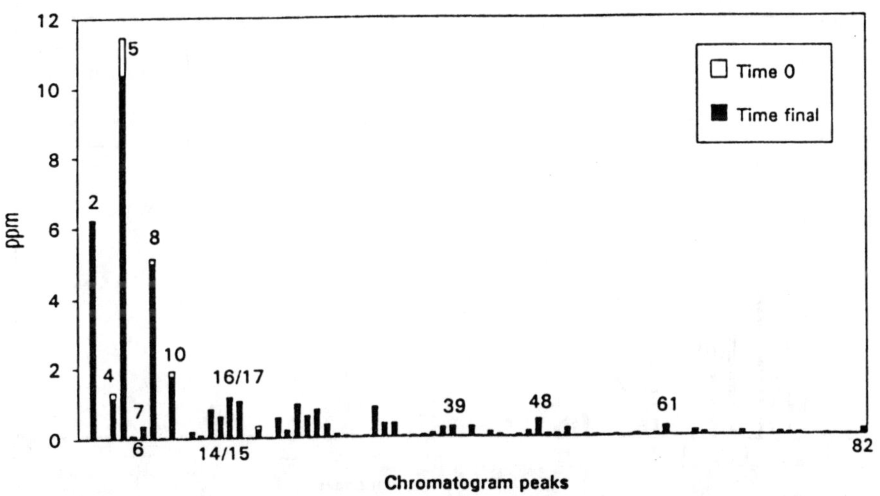

Figure 4d. Histogram showing peak-specific PCB concentrations in core samples from caisson R104 at T_0 and T_f. Histograms represent averages of all the cores taken at that time point, normalized to peak 61 to emphasize congener-specific changes. Little selective congener depletion is evident. Peak identifications are as indicated in Figure 4a.

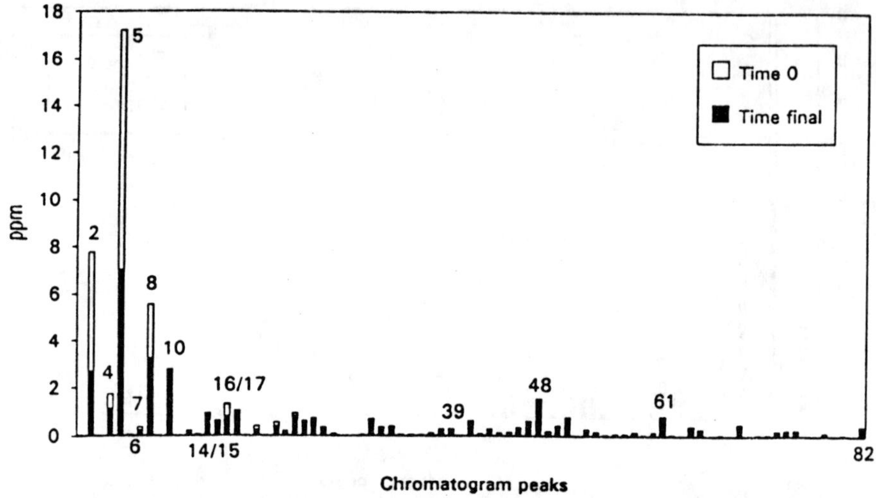

Figure 4e. Histogram showing peak-specific PCB concentrations in core samples from R105 at T_0 and T_f. Histograms represent averages of all the cores taken at that time point, normalized to peak 61 to emphasize congener-specific changes. Selective congener depletion is evident. Peak identifications are as indicated in Figure 4a.

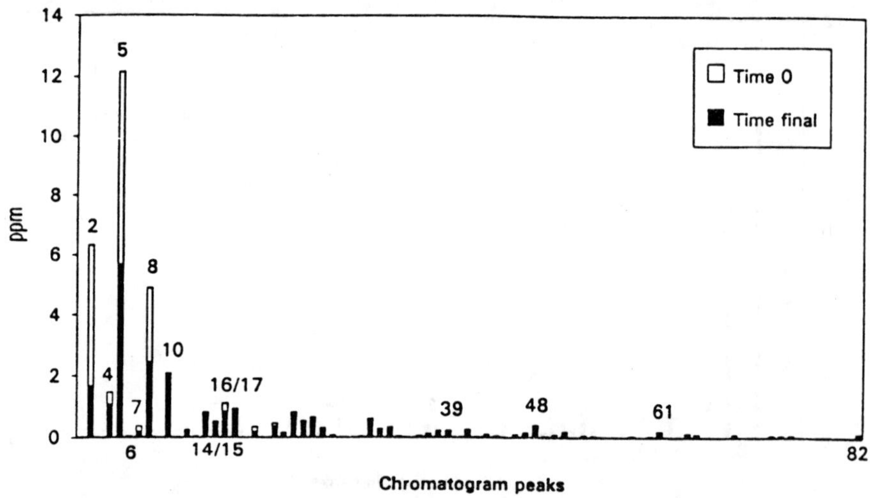

Figure 4f. Histogram showing peak-specific PCB concentrations in core samples from caisson R106 at T_0 and T_f. Histograms represent averages of all the cores taken at that time point, normalized to peak 61 to emphasize congener-specific changes. Selective congener depletion is evident. Peak identifications are as indicated in Figure 4a.

the observed reduction in direct PCB concentration in this reactor. The absence of congener specific PCB changes in R104 and the absence of significant physical losses (monitored by vapor traps and swipe sample of vessel walls) also support the conclusion that no significant PCB losses occurred in this caisson.

Both normalization procedures indicate a loss of PCBs from control caisson R101; however, the statistical significance of this depletion is uncertain (Table 3). As stated earlier, a problem with the reactor seal prevented exclusion of oxygen from this reactor, resulting in somewhat aerobic conditions. An examination of histogram Figure 4a reveals a pattern of congener depletion similar to that of the other high-mix reactor R102, particularly in the largest congener peaks (2-, 2,2′-/2,6-, and 2,3-/2,4′-chlorobiphenyl). Although this depletion is much less extensive than in the experimental reactors, these results suggest that some or all of the PCB depletion in R101 results from aerobic biodegradation.

No toxicity was evident in any of the sediment samples taken from the caissons. Control non-PCB-contaminated river sediments also exhibited no toxicity. This indicated that the aerobic biodegradation of PCBs and the addition of nutrients used in this study posed no acute toxicity hazard to the test species.

Chlorobenzoate Analysis

Aerobic biodegradation of PCBs proceeds through a series of degradation intermediates, as shown in Figure 1 (reviewed in Abramowicz, 1990; Bedard, 1990). The detection of one or more of these metabolic compounds during *in situ* biodegradation would establish a clear link between the observed loss of PCB and the induced microbiological activity. An analytical method was therefore developed to detect low concentrations of chlorinated benzoates (Flanagan and May, 1993) resulting from PCB biodegradation. Aqueous and sediment samples were collected from each caisson throughout the experiment and analyzed for chlorobenzoates. Attention focused primarily on the detection of mono- and dichlorobenzoates, since most of the PCBs in the H7 region were mono- and dichlorobiphenyls.

During the experiment, chlorinated benzoates were detected as transient intermediates in the aqueous phase of all of the experimental low-mix caissons R103, R105, and R106. In R105 and R106, which contained only indigenous microflora, the monochlorobenzoates appeared earlier (10–20 days) than the dichlorobenzoates (15–35 days), presumably because the biodegradation of monochlorinated biphenyls was faster. Figure 5 shows the chlorobenzoate results from R105, which are representative of those obtained in the other slow-mix reactors.

Increases in CBA concentration corresponded with increasing populations of indigenous biphenyl-metabolizing bacteria and depletion of biphenyl and monochlorinated biphenyls in those caissons during this time period. Increased chlorobenzoate levels were not detected in the corresponding low-mix control, R104. These results provide strong evidence linking the loss of PCBs in these experimental caissons with the stimulated aerobic microbial biodegradation.

Since chlorobenzoates are readily biodegraded to CO_2, H_2O, and Cl^- by many naturally occurring microorganisms (Hartmann et al., 1979; van den Tweel et al., 1987; Focht and Hickey, 1990; Hernandez et al., 1991), the rapid disappearance of these metabolites from the reactors was expected. In addition, a variety of aerobic chlorobenzoate-degrading microorganisms have been isolated from sediments near the field study site (data not shown).

Caissons R102 and R103 were dosed several times with amounts of *A. eutrophus* H850 calculated to yield a concentration of about 10^8 cells/ml. The survival of *A. eutrophus* H850 in aqueous samples

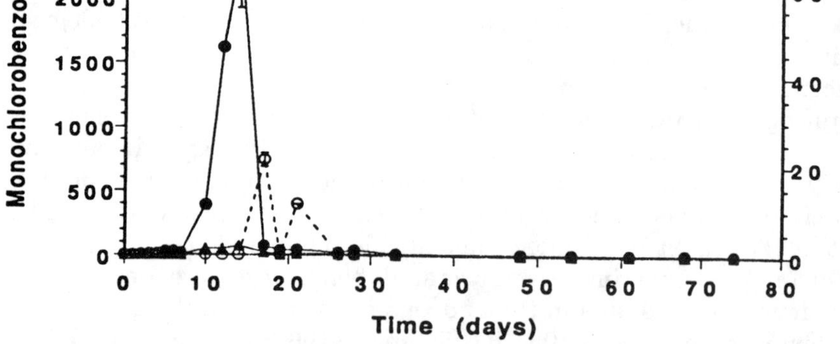

Figure 5. Congener-specific, aqueous-phase mono- and dichlorobenzoates in R105. Monochlorobenzoates appear earlier than dichlorobenzoates, consistent with a preferential attack on more lightly chlorinated congeners. Points represent average values for two analyses with ranges shown.

from the two reactors after each dosing is shown in Figure 6. After the initial dose, the concentration of *A. eutrophus* H850 dropped below 10^5 cfu/ml after ten days in both caissons. A second dose to R103 resulted in a similar decline in the *A. eutrophus* H850 population after 6 days, decreasing from 2×10^7 to 10^4 cfu/ml. A third dose of H850 to R103 at day 54 appeared to survive slightly better than the previous two, but still declined below 10^5 cfu/ml after 2 weeks. The cfu of *A. eutrophus* H850 in the aqueous samples taken from R103 within 24 h of dosing were reasonably close to the calculated value of 10^8 cells/ml. The second dose of *A. eutrophus* H850 to R102 resulted in less than 10^5 cfu/ml 1 day after addition and no further doses were attempted. It should be noted that the pH in R102 had dropped to about 5 at this time (Figure 7).

The profiles for the growth and decline of biphenyl-degrading bacteria were very similar in the three low-mix experimental caissons. The population of these organisms reached maxima of approximately 10^8 cfu/ml in 3 weeks, followed by a period of decline that leveled off at about 10^4 and 10^5 cfu/ml. The slight increase observed near experiment day 54 may have been in response to the 1.0 kg of biphenyl that was added to each of the four experimental caissons on day 50 (0.25 kg was added at 10-day intervals from day 20 to day 40). Very few indigenous biphenyl-degrading bacteria were detected in the samples from the controls, R101 and R104.

The decline in the biphenyl/PCB-degrading population observed in the intermediate weeks of the study may have been due to substrate limitation once the initial addition of biphenyl was depleted after the first 2 weeks (Harkness *et al.*, 1992). This resulted in a gradual decline in the number of biphenyl-utilizers to a concentration (10^4 to 10^5 cfu/ml) that could be maintained by the amount of available biphenyl. The slight increase in the biphenyl-metabolizing population from days 54–60 after the 1.0 kg addition supports this hypothesis, as does the absence of biphenyl in the aqueous phase over the period of decline (Harkness *et al.*, 1992). This suggests the biphenyl-utilizing population will not persist long at high levels in the environment once biphenyl is no longer available.

Resting-cell assays were performed on some of the predominant PCB-degrading bacteria isolated from the experimental caissons. As shown in Table 4, these organisms were capable of degrading many of the lightly chlorinated PCBs typically found in Hudson River sediment. Although the indigenous strains did not possess the broad congener specificity of *A. eutrophus* H850, some were capable of degrading a variety of tetra- and pentachlorobiphenyls.

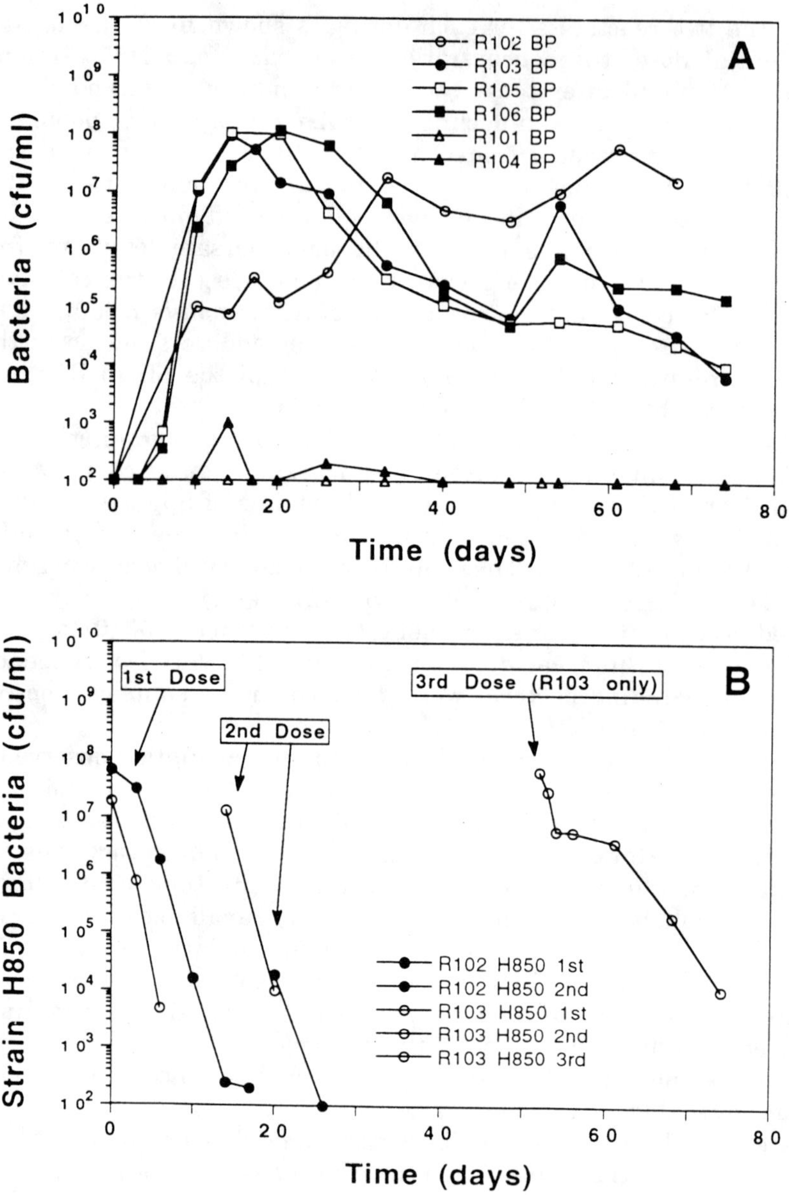

Figure 6. (a) Colony-forming units (cfu) of indigenous biphenyl-degrading bacteria in sediment samples versus time. (b) Colony-forming units of *A. eutrophus* H850 versus time in caissons R102 and R103. Plate counts were made using PAS agar medium with biphenyl.

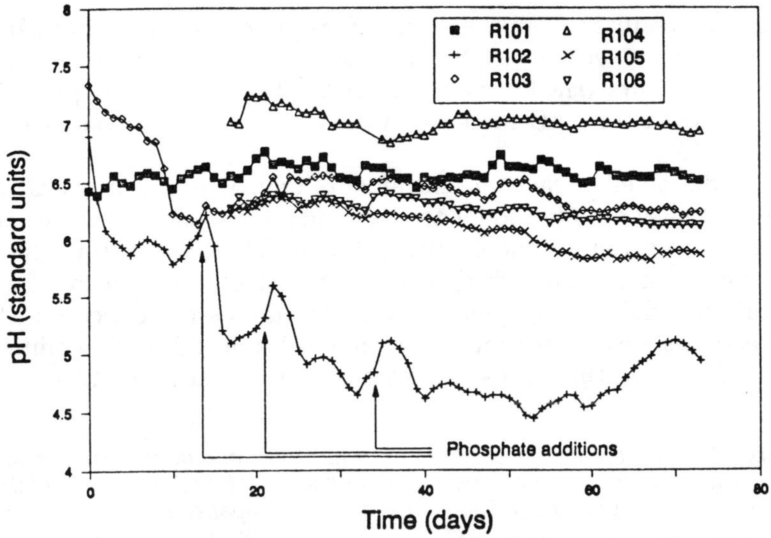

Figure 7. pH readings in the caissons over time. Phosphate additions are indicated by arrows. Technical problems prevented accurate pH readings in caissons R104-R106 during the first 2 weeks of the study.

Reactor Operation

Analysis of samples taken after 10.5 weeks (T_f) reflected the impact of both nutrient addition and enhanced biological activity in the experimental caissons. Increased levels of sulfate and ammonia-N in the aqueous phase of the experimental caissons were a direct result of nutrient addition, although the elevated sulfate level in control reactor R104 is perplexing. The increase in nitrate and nitrite in the experimental caissons probably resulted from the activity of nitrifying bacteria. This was especially pronounced in R102. Available phosphate levels remained low in all the caissons, despite repeated phosphate amendments to the experimental reactors. An increase in alkalinity was evident in the low-mix caissons.

The daily pH measurements are shown in Figure 7. The pH of the control caissons was stable throughout the experiment at 6.5 and 7.0 for R101 and R104, respectively. The pH in the experimental rake caissons declined slowly from 7.0 to 6.0 or less by week 11. pH data for the first 2 weeks in caissons R104–R106 are not shown because electrical interferences disrupted pH measurements in those caissons during that period. The pH in R102 changed the most, falling rapidly and remaining below 5.0 for most of the experiment. This pH drop apparently resulted from the strong nitrifying activity in that caisson and was not unprecedented, having also been observed in some laboratory experiments (Harkness *et al.* 1992). The addition of phosphate caused a short-term pH increase in the caissons, which was most evident in R102.

Daily dissolved oxygen measurements appear in Figure 8. Control setpoint values are plotted for the experimental caissons, while actual readings are given for the control caissons. The DO control setpoints were increased gradually during the first week of the experiment while the peroxide delivery system was monitored. Dissolved oxygen levels were established at less than saturation (8.5 ppm at 23°C) to minimize oxygen losses to the headspace so that an oxygen balance could be

Table 4. PCB degradative ability of bacteria isolated from the experimental caissons. The strain number indicates the caisson of origin. The degadative profile of H850 is included for comparison.

PCB CONGENER	BACTERIAL CULTURES															
	2A	2B	2C	2D	3A	3B	3C	5A	5B	5C	5D	6A	6B	6C	6D	H850
2-	⬤	⬤	⬤	⬤	⬤	⬤	⬤	⬤	⬤	⬤	⬤	⬤	⬤	⬤	⬤	⬤
4-	●	nd	⬤	●	●	●	nd	●	nd	⬤	nd	nd	⬤	⬤	nd	⬤
2,3-	⬤	⬤	⬤	⬤	⬤	⬤	⬤	⬤	⬤	⬤	⬤	⬤	⬤	⬤	⬤	⬤
2,6-					·	·		·		•						
2,2'-	●		⬤	●		•		·		·				⬤	•	⬤
2,3'-	⬤	nd	⬤	⬤	⬤	⬤	nd	⬤	nd	⬤	nd	nd	●	⬤	nd	⬤
2,4'-	⬤	⬤	⬤	⬤	⬤	⬤	⬤	⬤	⬤	⬤	⬤	⬤	⬤	⬤	⬤	⬤
4,4'-	⬤	⬤	⬤	⬤	⬤	⬤	⬤	⬤	⬤	⬤	●	⬤	⬤	⬤	●	•
2,3,2'-	·	nd	•			·	nd		nd	·	nd	nd		·	nd	⬤
2,4,2'-		nd					nd		nd		nd	nd			nd	⬤
2,5,3'-	⬤	nd	⬤	⬤			nd		nd	·	nd	nd		⬤	nd	⬤
2,6,2'-																·
2,6,3'-										•						⬤
2,6,4'-		nd					nd		nd		nd	nd			nd	·
2,4,4'-	⬤	⬤	⬤	⬤	⬤	⬤	⬤	⬤	⬤	⬤	•	⬤	⬤	●	•	•
2,5,4'-	●	·	⬤	●	·	·	·	•	·	⬤	•	·	•	⬤	·	⬤
2,5,2'-	·		⬤	●	·	·	·	·			●		·	⬤	•	⬤
2,5,2',5'-																⬤
2,3,2',5'-	●		⬤	•							·			•	•	⬤
2,3,2',3'-			·		●	⬤	•	⬤	●	⬤	·		⬤			⬤
2,4,2',4'-																·
2,4,3',4'-	·		•			·		·		•			·			
2,5,3',4'-			●	·										●		●

PERCENT PCB BIODEGRADED AFTER 24 HOUR RESTING-CELL ASSAY:

⬤ = 80-100% ● = 60-80% • = 40-60% · = 20-40% < 20%

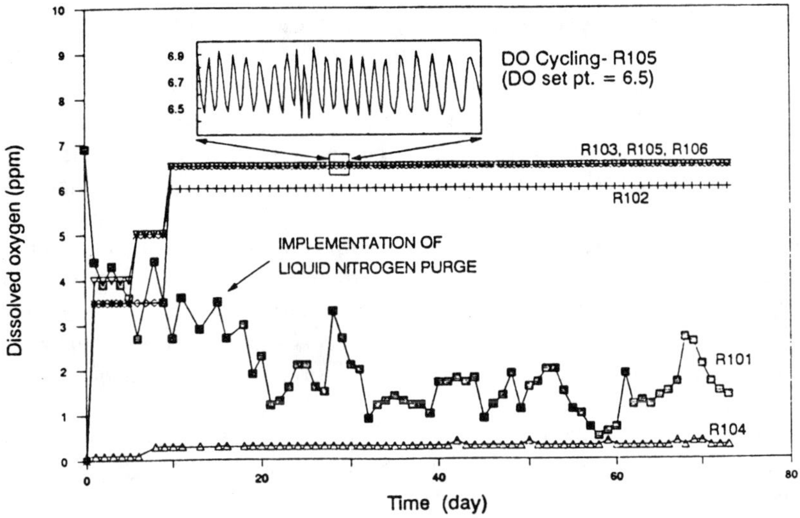

Figure 8. Daily dissolved oxygen measurements in individual caissons. DO setpoint values are given for caissons R102, R103, R105, and R106. Representative cycling of dissolved oxygen about the setpoint in these caissons is illustrated in the inset.

established. The setpoint in R102 was slightly lower than the others (6.0 versus 6.5) because that caisson ran at a higher temperature and, hence, had a lower oxygen solubility at saturation. The actual DO cycling around the control setpoint is illustrated by the inset in Figure 8. The liquid nitrogen purge protocol was implemented in caisson R101 on day 14 of the experiment. It was generally successful in reducing the DO level in that caisson to below 2 ppm. DO levels below 2 ppm could not be reliably measured with the probes used in this study.

DISCUSSION

In the past, claims of biodegradation in the field were often based solely upon apparent reductions in contaminant concentrations over time. It is now recognized that real-world difficulties (sampling problems, heterogeneous contaminant concentrations, limited mixing, and abiotic losses) seriously reduce the value of such studies. It is no longer acceptable to infer that biodegradation has occurred or that microorganisms are the causative agents with such limited evidence. In this study several convergent lines of evidence support the conclusion of stimulated PCB biodegradation in Hudson River sediment.

(1) PCB losses were congener specific and consistent with known selectivity patterns observed in laboratory biodegradation studies. Two independent normalization methods were used to correct for variations in sampling and extraction. Both methods demonstrated that losses of 38–55% occurred in the experimental caissons, while small or insignificant losses occurred in the control caissons. In the first method, PCB values were expressed on a total organic carbon (TOC) basis for each core. As previously discussed, hydrophobic contaminants such as PCBs typically reside in1 the organic fraction of the sediment. The results of expressing the initial and final PCB concentrations on a TOC basis showed that PCB losses ranged from 44.7–55.5% in the experimental caissons, while statistically significant decreases were not observed in either control caisson.

A second method for quantifying PCB biodegradation levels involved normalization of the data using a "nondegradable" PCB peak as an internal standard. This method is limited by the fact that no PCB congeners present at the site have been demonstrated to be completely nondegradable by aerobic PCB-degrading microorganisms. Therefore, this normalization technique results in a *conservative measure of actual losses.* This bias can be minimized by selecting a less biodegradable congener as the normalization standard. PCB losses using peak 61 (3,4,3′,4′-/2,3,6,3′,4′-CB) as an internal standard ranged from 37.8–42.4% in the experimental caissons. Again, only small changes were observed in the control caissons. These conservative estimates are consistent with the slightly larger losses of 44.7–55.5% determined using the TOC technique.

(2) The detection of metabolic intermediates that were consistent with the PCB congeners degraded. The large proportion of 2-CBA observed is consistent with the known fate of 2-chlorobiphenyl, 2,2′-chlorobiphenyl, 2,3-chlorobiphenyl, and 2,4′-chlorobiphenyl, all proceeding through the 2-CBA intermediate. The low levels of di-CBAs observed (2,5- and 2,4-CBA) are consistent with the low initial levels of PCB congeners that proceed through these intermediates (2,5-, 2,5,2′-, 2,4-, 2,4,2′-chlorobiphenyl). The detection of these PCB metabolites corresponding with the depletion of the appropriate congeners offers the *strongest* evidence that *in situ* biodegradation is responsible for the observed PCB losses.

(3) The number of biphenyl/PCB-degrading bacteria in the reactors increased during the course of the study. This was determined by plate counts of biphenyl-degrading organisms in the aqueous phase of the caissons. Dramatic increases (five to six orders of magnitude) in the number of active biphenyl/PCB-degrading microorganisms were ob-

served only in the experimental caissons indicating that oxygen, nutrient, and biphenyl additions directly stimulated this microbial population. Similar increases were also observed in sediment samples from the experimental reactors (Harkness *et al.,* 1992). These data again support the conclusion that *in situ* PCB biodegradation occurred in the experiment.

(4) Co-reactants utilized in biodegradation (oxygen, biphenyl, and other nutrients), were consumed during the study. Decreases in biphenyl levels were observed and expected since biphenyl is often used as a growth substrate by PCB-degrading bacteria. Oxygen was also utilized in the caissons, as shown by the continuous oxygen demand in the experimental reactors.

(5) Biological activity was readily distinguishable from abiotic losses in control reactors. This was possible even though a small, but measurable, level of oxygen was present in control reactor R101 (allowing for some biodegradation), and significant redistribution of PCBs was observed in R104. Other steps to assess the impact of abiotic processes included monitoring vapor traps to measure volatilization, swiping the sides of the caissons to evaluate adherence of PCBs to the vessel walls, and investigating if Fenton chemistry contributed to PCB depletion (Sediak and Andren, 1991). No significant PCB losses occurred via these abiotic routes.

The results of this field study demonstrated no notable difference in the extent of PCB biodegradation in the high-mix caisson (R102) compared to the low-mix caissons (R103, R105, R106). Also, the extent of PCB biodegradation was not obviously affected by inoculation with PCB-degrading strain H850.

The extent of PCB biodegradation in both laboratory and field experiments is limited by the bioavailability of PCBs in the sediment. This phenomenon is commonly observed to varying degrees whenever the biodegradation of hydrophobic contaminants (like oil or PCBs) is carried out in the presence of soils or sediments. At PCB concentrations typical for this Hudson River site (20–40 ppm), only about 50% of the target compounds are readily available for the bacteria to degrade over the relatively short time course of the experiment (Carroll et al., 1994). It was this available PCB fraction that was degraded during the field study. The remaining resistant fraction was not available to the PCB-degrading bacteria and is unlikely to be available to other organisms as well (Larsson et al., 1992).

Although the rate of PCB biodegradation was enhanced by oxygen and nutrient levels, limited biodegradation appeared to occur even in control caisson R101 (no added nutrients or biphenyl). Low levels of oxygen (<2 ppm) were present in that caisson, the result of a leaky seal and equilibration of headspace oxygen with the water column. This suggests that *in situ* PCB biodegradation occurred even under “unfavorable” conditions (minimal mixing, no microbial inoculation, low oxygen levels, and no added nutrients or biphenyl) during this study. Therefore, it is possible that aerobic PCB biodegradation is occurring naturally in the environment, a hypothesis further supported by the detection of chlorobenzoic acids in undisturbed Hudson River sediments (see below).

Chlorobenzoate Detection in Undisturbed Hudson River Sediments

The presence of PCB-dechlorinating and -degrading bacteria in the environment strongly suggests that these processes are ongoing at PCB-contaminated sites. The first conclusive demonstration of naturally occurring *in situ* PCB biodegradation was provided by Flanagan and May (1993) using Hudson River sediment. Core samples were collected from a variety of undisturbed regions of the Hudson River including the H7 site, ~60 ft upstream of the experimental platform. The cores were frozen and sliced into 1-in. sections, which were thawed, homogenized, and analyzed for PCBs and CBAs via gas chromatography/mass spectroscopy (GC/MS).

Chlorobenzoates were detected in all cores of PCB-contaminated sediment, and the CBA concentration versus depth generally paralleled that for PCBs. In most cases, however, the concentration of CBAs was approximately 1000-fold less than the PCB concentration. Duplicate analyses demonstrated excellent reproducibility, given the low level of analyte. The majority of the detected CBAs were monochlorinated benzoates consistent with those expected from the aerobic degradation of 2- and 2,2′-dichlorobiphenyl. Traces of dichlorinated benzoates (2,5- and 2,4-) were observed at low, nonquantifiable levels. CBAs were not detected in sediments collected from a section of the Hudson River (Spier Falls) that did not contain PCBs (Flanagan and May, 1993). These data provided the first qualitative evidence of aerobic PCB biodegradation in the environment.

CONCLUSIONS

Results from this field study in Hudson River sediments demonstrate that aerobic PCB biodegradation can occur *in situ*. This conclusion is supported by several independent lines of evidence, as well as by a comprehensive set of PCB analyses, including PCB data normalized to total organic carbon levels and to an "internal standard" reference peak. The major conclusions that can be drawn from this field study are

1. Indigenous PCB-degrading microorganisms are present in these sediments and can biodegrade the lightly chlorinated PCBs typically present in upper Hudson River sediments.
2. *In situ* aerobic PCB biodegradation can be accelerated in Hudson River sediments by the addition of nutrients and supplemental oxygen.
3. Inoculation of extensively dechlorinated sediments with large numbers of a purified PCB-degrading organism resulted in little or no benefit over the degradation achieved via stimulating the growth of the indigenous microflora alone.
4. Rapid biodegradation of chlorobenzoates indicates that PCBs can be fully metabolized through these intermediates by indigenous microflora present in Hudson River sediments.
5. Evidence from this field study, together with chlorobenzoate analyses in undisturbed river sediments, suggests that aerobic PCB biodegradation occurs naturally in the Hudson River. This process complements the extensive anaerobic dechlorination previously documented.

POST HRRS DEVELOPMENTS

Laboratory Models of Undisturbed River Sediments

Although evidence of both aerobic and anaerobic PCB biotransformations has been obtained from Hudson River sediment, it is difficult to determine the rates at which these processes occur in undisturbed environments. One approach to predicting such rates is the use of microcosms to partially simulate river conditions. Recently, a test tube microcosm system has been developed to model the fate of PCBs in the Hudson River (Fish and Principe, 1994). The advantages of this system are homogeneity, reproducibility, convenience, and cost-effectiveness. In addition, the microcosms contain both aerobic (within 2 mm of the surface) and anaerobic zones (beneath 2 mm), making it possible to study both processes simultaneously.

The microcosms are prepared using sieved Hudson River sediment amended with PCBs and placed into Pyrex tubes (25 by 100 mm). Racks of these tubes are then submerged in tanks containing Milli-Q purified or Hudson River water and equipped with circulating pumps. At appropriate time points, the sediment cores are removed from the tubes, sectioned, and PCB analysis performed using GC and GC/MS.

The usefulness of this system depends upon how closely results from the microcosms mimic those observed in the environment. Recently, the ability of the microcosms to predict the fate of Aroclor 1242 released into the Hudson River has been evaluated (Fish and Principe, 1994). In the anaerobic zone, extensive PCB dechlorination resulted in congener distribution patterns frequently observed in the Hudson River. In near surface sediments, PCBs were transformed by a combination of aerobic biodegradation and reductive dechlorination. This resulted in a significant reduction in PCB concentration and produced a congener distribution corresponding to what has been designated "pattern Y" in Hudson River sediment (Fish and Principe, 1994). These data suggest that the microcosm system is a valid model for predicting PCB fate in the Hudson River.

As a result of being conducted at laboratory temperatures (~24°C), PCB biotransformations in the microcosms occur very rapidly (extensive degradation and dechlorination within 70 days) and are probably not representative of environmental rates. Conducting these studies at river temperatures should provide reliable estimates of environmental biotransformation rates.

Metabolite Analysis

Since the HRRS study, refinements have been made in the detection and identification of PCB metabolites from environmental samples. In addition to chlorobenzoates, the presence of chlorinated hydroxybiphenyls in river sediment can now be used as evidence of naturally occurring aerobic PCB biodegradation (Flanagan and May, 1993). Unlike CBAs, chlorohydroxybiphenyls retain the biphenyl ring, virtually eliminating any possibility of source ambiguity. Among the hydroxylated PCB metabolites identified in Hudson River sediment are a variety of chlorinated 2,3-dihydrodiol derivatives and chlorinated 2,3-dihydroxybiphenyls (Figure 1). These metabolites were detected only in PCB-contaminated samples and were absent from sediment uncontaminated by PCBs (Flanagan and May, 1993).

A method to simplify the identification of PCB metabolites using GC/MS in selective ion monitoring mode has also been developed. One problem has been that the number of potential metabolites from each congener is large, and only a few are commercially available for use as standards. This has been addressed by using mutant PCB-degrading strains, which accumulate the desired metabolites when incubated with selected congeners. The material obtained from these strains has been used to construct a database containing relative retention time data for a wide variety of PCB metabolites.

Since the measurement of retention times is critical to metabolite identifications, a relative retention time index based on a set of fatty acid methyl esters (FAME) is utilized. The advantages include (1) increased precision of retention time assignments; (2) more robust measurements, which are not influenced by small changes in conditions; and (3) the assignments being "portable," enabling different laboratories to achieve the same results given different samples.

The database was created using the computer program Excel (version 4.0, Microsoft Inc.) and thus far contains degradative intermediates derived from 22 individual PCB congeners. This database should simplify metabolite identification by making it possible to search by particular index values or chlorination number (May and Mondello, in preparation).

ACKNOWLEDGEMENTS

We thank OHM Corporation (Findlay, Ohio), for their assistance in the design, construction, and installation of the research facility.

REFERENCES

Abramowicz, D.A. 1990. Aerobic and anaerobic biodegradation of PCBs: a review. In: *CRC Critical Reviews in Biotechnology,* G. G. Steward and I. Russell, (eds.), Boca Raton, FL. CRC Press, Inc., Vol. 10 3:241–251.

Abramowicz, D.A., J.F. Brown, M.R. Harkness and Michael K, O'Donnell, "*In situ* Anaerobic PCB Dechlorination and Aerobic PCB Biodegradation in Hudson River Sediments", in *The Implementation of Biotechnology in Industrial Waste Treatment and Bioremediation,* ed. Dr. Robert F. Hickey, Lewis Publishers, pp. 57–92 (1995).

Ahmed, M. and D.D. Focht. 1973. Degradation of polychlorinated biphenyls by two species of *Achromobacter. Can. J. Microbiol.,* 19:47–52.

Bedard, D.L. 1990. In: *Biotechnology and Biodegradation,* D. Kamely, A. Chakrabarty and G. S. Omenn (eds.), *Adv. Appl. Biotechnol. Series, Vol. 4,* Portfolio Publishing Co., the Woodlands, TX pp. 369–388.

Bedard, D.L., M.J. Brennan and R. Unterman. 1984. *In* G. Addis and R. Komai (ed.), *Proceedings of the 1983 PCB Seminar,* Electrical Power Research Institute, Palo Alto, Calif.

Bedard, D.L., R. Unterman, L.H. Bopp, M.J. Brennan, M.L. Haberl and C. Johnson. 1986. Rapid assay for screening and characterizing microorganisms for the ability to degrade polychlorinated biphenyls. *Appl. Environ. Microbiol.,* 51:761–768.

Bedard, D.L., Wagner, R.E., Brennan, M.J., Haberl, M.L. and Brown, J.R., Jr. 1987. Extensive degradation of Aroclors and environmentally transformed polychlorinated biphenyls by *Alcaligenes eutrophus* H850. *Appl. Environ. Microbiol.,* 53:1094–1102.

Carroll, K.M., Harkness, M.R., Bracco, A.A. and R.R. Balcarcel. 1994. Application of a permeant/polymer diffusional model to the desorption of polychlorinated biphenyls from Hudson River sediments. *Environ. Sci. Technol.,* 28:253–258.

Fish, K.M. and J.M. Principe. 1994. Biotransformation of Aroclor 1242 in Hudson River test tube microcosms. *Appl. Environ. Microbiol.,* 60:4289–4296.

Flanagan, W.P. and R.J. May. 1993. Metabolite detection as evidence for naturally occurring aerobic PCB biodegradation in Hudson River sediments. *Environ. Sci. Technol.,* 27:2207–2212.

Focht, D.D. and W.J. Hickey. 1990. Degradation of mono-, di-, and trihalogenated benzoic acids by *Pseudomonas aeruginosa* JB2. *Appl. Environ. Microbiol.,* 56:3842–3850.

Furukawa, K., F. Matsmura and K. Tonomura. 1978. *Alcaligenes* and *Acinetobacter* strains capable of degrading polychlorinated biphenyls. *Agric. Biol. Chem.,* 42:543–548.

Furukawa, K. 1982. Microbial degradation of polychlorinated biphenyls. In: *Biodegradation and Detoxification of Environmental Pollutants,* A.M. Chakrabarty (ed.), Boca Raton, FL: CRC Press, Inc., pp. 33–57.

Harkness, M.R., J.B. McDermott, D.A. Abramowicz, J.J. Salvo, W.P. Flanagan, M.L Stephens, F.J. Mondello, R.J. May, J.H. Lobos, K.M. Carroll, M.J. Brennan, A.A. Bracco, K.M. Fish, G.L. Warner, P.R. Wilson, D.K. Dietrich, D.T. Lin, C.B. Morgan and W.L. Gately. 1992. *1991 In situ Hudson River research study: A field study on biodegradation of PCBs in Hudson River sediments.* Final project report. GE Corporate Research and Development, pp. 1–265.

Harkness, M.R., J.B. McDermott, D.A. Abramowicz, J.J. Salvo, W.P. Flanagan, M.L. Stephens, F.J. Mondello, R.J. May, J.H. Lobos, K.M. Carroll, M.J. Brennan, A.A. Bracco, K.M. Fish, G.L. Warner, P.R. Wilson, D.K. Dietrich, D.T. Lin, C.B. Morgan and W.L. Gately. 1993. *In situ* stimulation of aerobic PCB biodegradation in Hudson River sediments. *Science,* 259:503–507.

Hartman, J., W. Reinecke and H.-J. Knackmuss. 1979. Metabolism of 3-chloro-, 4-chloro- and 3,5-dichlorobenzoate by a *Pseudomonad. Appl. Environ. Microbiol., 37:421–428.*

Hernandez, B.S., F.K. Higson, R. Kondrat, D.D. Focht. 1991. Metabolism of and inhibition by chlorobenzoates in *Pseudomonas putida* P111. *Appl. Environ. Microbiol.,* 57:3361–3366.

Hutzinger, O. and W. Veerkamp. 1981. Xenobiotic chemicals with pollution potential. In: *Microbial Degradation of Xenobiotics and Recalcitrant Compounds,* Leisinger, T., Hutter, R., Cook, A.M. and Nuesch, J. eds., Academic Press, New York, pp. 3–45.

Hutzinger, O., S. Safe and V. Zitko. 1974. *The Chemistry of PCBs,* CRC Press, Cleveland.

Larsson, P., L. Colivin, L. Okla and G. Meyer. 1992. *Environ. Sci. Technol.,* 26:346.

Nadim, L., M.J. Schocken, F.J. Higson, D.T. Gibson, D.L. Bedard, L.H. Bopp and F.J. Mondello. 1987. In: *Proceedings of the 13th Annual Research Symposium on Land Disposal, Remedial Action, Incineration, and Treatment of Hazardous Waste.* EPA/600/9-87/015. U.S. Environmental Protection Agency, Cincinnati, Ohio.

Rochkind, M.L., J.W. Blackburn and G.S Sayler. 1986. In: *Microbial decomposition of chlorinated aromatic compounds.* EPA/600/2-86/090. U.S. Environmental Protection Agency.

Sediak, D.L. and A.W. Andren. 1991. Aqueous-phase oxidation of polychlorinated biphenyls by hydroxy radicals. *Environ. Sci. Tech.,* 25:1419–1427.

Sundstrom, G. and O. Hutzinger. 1976. The metabolism of PCBs. *Chemosphere,* 5:267–298.

van den Tweel, W.J.J., Kok, J.B. de Bont, J. A.M. 1987. Reductive dechlorination of 2,4-dichlorobenzoate to 4-chlorobenzoate and hydrolytic dehalogenation of 4-chloro-, 4-bromo-, and 4-iodobenzoate by *Alcaligenes denitrificans* NTB-1. *Appl. Environ. Microbiol.,* 53:810–815.

The Lasagna™ Process for *in situ* Bioremediation of Low-Permeability Soils

S. V. Ho, C. J. Athmer, J. M. Brackin, M. A. Heitkamp,
P. W. Sheridan, D. Weber and P. H. Brodsky
Monsanto Company, U4E
800 N. Lindbergh Blvd.
St. Louis, MO 63167

INTRODUCTION

Bioremediation is conceptually a very attractive *in situ* technology for cleanup of contaminated soils. However, in actual field applications bioremediation has enjoyed very limited success, primarily due to the physical difficulty of bringing together the necessary components (e.g., microbes, nutrients, contaminants) in the subsurface soil for degradation to occur. This tough distribution problem is further magnified when dealing with low-permeability soils such as clayey, silty soils, which are both very resistant to liquid flow and prone to channeling.

We are currently developing a novel, *in situ* technology aimed at cleaning up contamination in low-permeability soils. Our approach involves coupling electroosmosis with treatment zones that are installed directly in the contaminated soils. Used since the 1930s for dewatering clays, silts, and fine sands (Casagrande, 1952), electroosmosis has recently received increasing attention as an *in situ* method for soil remediation (Shapiro et al., 1989; Hamed et al., 1991; Bruell and Segall, 1992; Acar et al., 1992; Shapiro and Probstein, 1993). Under electroosmosis, water is moved through the soil from the anode to the cathode, carrying with it the contaminants, and then to the surface for further treatment or disposal. Advantages with electroosmosis include uniform water flow through heterogeneous soil, high degree of control of the flow direction, and very low power consumption. There are, however, several major drawbacks associated with electroosmosis for remedial applications, such as low speed (liquid flow induced by electroosmosis typically moves about 1″ per day for clay soils), additional aboveground treatment and unstable long-term operation resulting from soil drying and cracking, steep pH gradient in the soil bed, and precipitation of metals and minerals near the cathode (Bruell and Segall, 1992; Segall and Bruell, 1992; Acar et al., 1992; Shapiro and Probstein, 1993). The pH changes primarily result from water electrolysis as the predominant electrode reactions, generating acid at the anode and base at the cathode according to the following half-cell reactions:

$$\text{Anode: } 2H_2O - 4e^- \rightarrow O_2 + 4H^+ \qquad E_o = -1.23 \text{ volts}$$

$$\text{Cathode: } 2H_2O + 2e^- \rightarrow H_2 + 2OH^- \qquad E_o = -0.83 \text{ volts}$$

where E_o is the standard electrochemical reduction potential. Recent efforts to mitigate the pH problem primarily involve conditioning the anode and cathode solutions through external recirculating loops (Lageman, 1993; Acar et al., 1993). In this chapter, we will describe the concept of coupling electroosmosis with *in situ* treatment zones and show how such an approach can effectively address the difficult delivery problems required for successful *in situ* bioremediation.

INTEGRATED *IN SITU* REMEDIATION TECHNOLOGY (LASAGNA™)

General Lasagna Concept

The general concept is to utilize electroosmosis to move contaminants from the soils into "treatment zones" where the contaminants are removed from the moving groundwater by adsorption, immobilization, or degradation. One possible configuration of the process has the following components:

1. Create permeable zones in close proximity sectioned through the contaminated soil region and turn them into treatment zones by introducing appropriate materials (sorbents, catalytic agents, microbes, oxidants, buffers, etc.). Hydraulic fracturing and related technologies may provide an effective and low-cost means for creating such zones horizontally in the subsurface soil (Murdoch, 1992). The degradation zones can also be vertical, which can be constructed using sheet piling, trench, slurry wall, etc. Instead of being discrete, the treatment zone can also be continuous using *in situ* soil mixing devices.
2. Utilize electroosmosis as a liquid pump for flushing contaminants from the soil into the treatment zones for degradation. Since these

zones are deliberately located close to one another, the time taken for the liquid to move by electroosmosis from one zone to the adjacent one can be short. In the horizontal configuration, the zones above and below the contaminated soil area can be injected with graphite particles during the hydrofracturing process to form in-place electrodes. For highly nonpolar contaminants, surfactants can be introduced into the water or incorporated into the treatment zones to solubilize the organics. For mixtures of organics and metals, the treatment zones can contain sorbents for binding the metals and microbes or catalysts for degrading the organics.

3. Liquid flow can be periodically reversed, if needed, simply by switching the electrical polarity. This mode would enable multiple passes of the contaminants through the treatment zones for complete sorption/destruction. The polarity reversal also serves to minimize complications associated with long-term applications of one-directional electroosmotic processes as discussed above. Optionally, the cathode effluent (high pH) can be recycled directly back to the anode side (low pH), which provides a convenient means for pH neutralization, as well as simplifies water management.

Electrodes and treatment zones can be of any orientation depending upon the emplacement technology used and the site/contaminant characteristics. Schematic diagrams of two typical layered configurations, horizontal and vertical, of electrodes and treatment zones are shown in Figure 1. Monsanto Company has coined the term "LASAGNA"[1] to identify its products and services based on this integrated *in situ* remediation technology.

The Lasagna Process for *in situ* Bioremediation

The need for uniform delivery of nutrients, microbes, and so on to the subsurface soil, a great challenge in bioremediation applications, is effectively handled in the Lasagna process through the operations of electroosmosis (or, more generally, electrokinetics for both water and ion migration) and permeable zone formation. Both biostimulation (enhancing indigenous microbial activities) and bioaugmentation (introducing new or enhanced microbial capability) approaches can be utilized for accomplishing *in situ* microbial degradation. For biostimulation the addition will be primarily chemicals, whereas both chemicals and microorganism additions will be involved in bioaugmentation. The

[1] "LASAGNA" is a trademark of Monsanto Company.

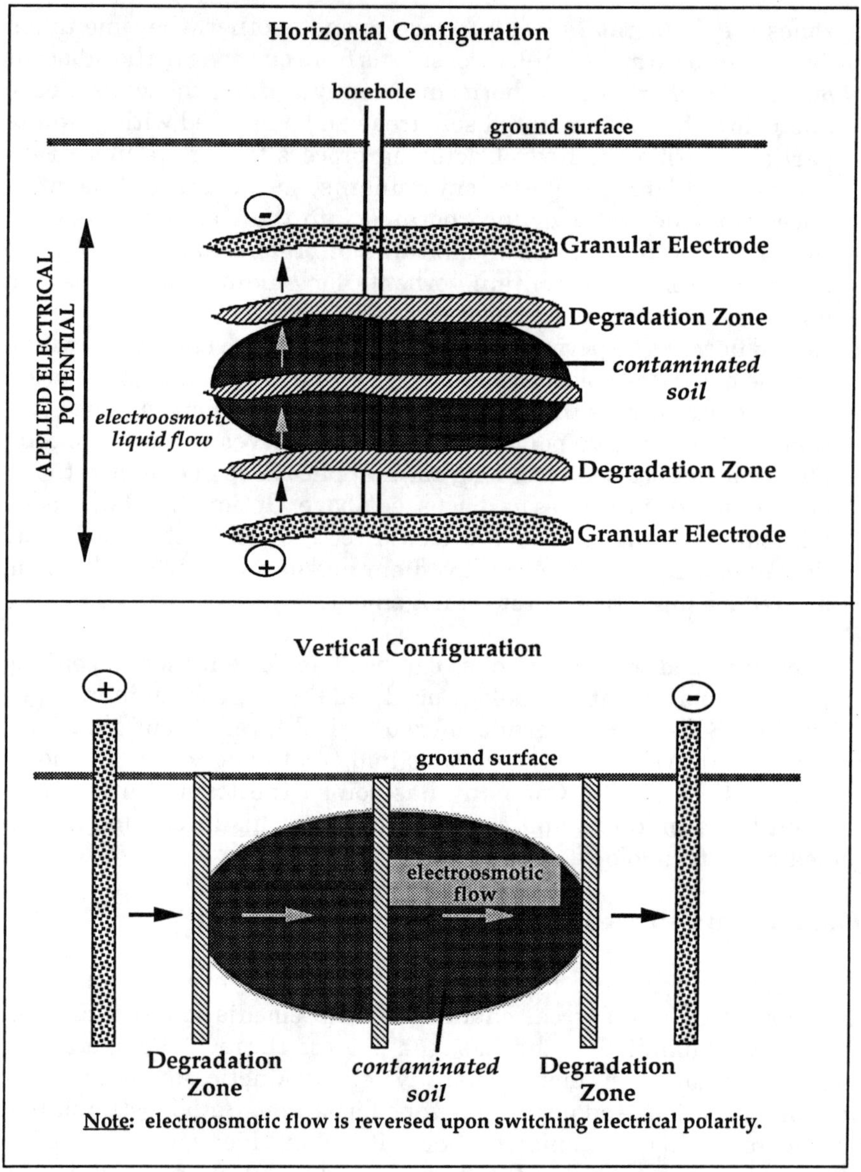

Figure 1. Schematic diagram of the integrated *in situ* remediation process—horizontal and vertical configurations.

Lasagna process offers two ways for the "additives" to be introduced into and distributed throughout the subsurface soil region of interest: (1) incorporate the additives into treatment zones during or subsequent to zone formation or (2) dissolve the additives in the water circulated through the soil by electroosmosis.

EXPERIMENTAL METHODS

Setup for Electroosmotic Removal of PNP Coupled with *in situ* Adsorption

The electroosmosis cell design shown in Figure 2 was used to simulate the configuration encountered in using hydrofracturing to install electrodes and treatment zones in subsurface soils. A cylindrical tube (4″ inside diameter and 9-1/4″ long) was used to house the electrodes and the packed soil. Each electrode consisted of about 50 g granular activated carbon packed into a 1/2″ layer in direct contact with the soil. Connection to each electrode was made with a carbon rod (3/16″ in diameter and about 4″ long) through the side of the tube wall. Water inlet and outlet were conducted through the packed carbon electrodes. A layer of clean clay was packed in the back of each electrode both to

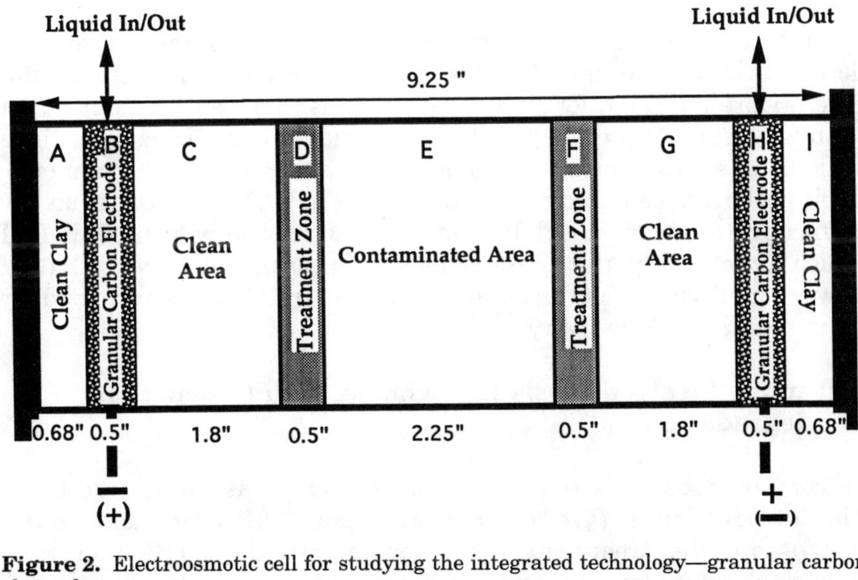

Figure 2. Electroosmotic cell for studying the integrated technology—granular carbon electrodes.

keep the carbon particles in place and to prevent water leakage. Ports were installed along the tube wall to serve as posts for measuring the voltage drops across the various soil sections between the electrodes. Georgia kaolinite clay (air-dried, air-floated kaolinite obtained from Thiele Kaolinite Co., Wrens, Georgia) was packed in the midsection of the cell. The characteristics of this clay are well documented in the literature (Hamed et al., 1991), with hydraulic conductivity on the order of 10^{-8} cm/sec.

P-nitrophenol (PNP) was used as the model organic compound for preparing contaminated clay soil. Typically, about 500 g dry kaolinite clay was mixed with 300 g aqueous solution containing 1050 mg PNP/L, which resulted in a clay paste of 37.5 wt % moisture and a loading of 390 μg PNP/gram wet clay. This PNP-contaminated clay section was bracketed at each end with a 1/2″ layer of sand and carbon particles (160 g sea sand and 4 g activated carbon or 2.4% carbon by weight), occupying about 100 cm^3 in volume. The activated carbon used was obtained from Calgon and exhibited high adsorption capacity for PNP (0.5 g PNP/g carbon). The sand-carbon layers thus represented permeable adsorption zones. Uncontaminated kaolinite clay (about 38 wt % moisture) was packed next to each sand-carbon layer to simulate clean soil. Glass fiber filters were placed between different soil layers to facilitate soil sampling. The well water used in the experiments to simulate groundwater had a pH of about 8 and conductivity of about 600 μS/cm.

At the end of each experiment, several small clay samples at different locations in the contaminated clay section, as well the whole clay section itself, were analyzed for PNP. The analysis involved extracting PNP from clay samples with 0.1N NaOH solution and measuring the level of PNP in solution by spectrophotometric absorption at 400 nm or by high-performance liquid chromatography (HPLC). One extraction was sufficient to remove all PNP from clay. For carbon, which binds PNP much more tightly, the extraction solution contained 0.1N NaOH and 2 wt % methylene chloride, and repeated extractions were carried out to maximize PNP recovery.

Setup for Electroosmosis Experiment with Concurrent Biodegradation

The electroosmosis cell setup was the same as above, except for the introduction of *Pseudomonas* sp. strain PNP1 (Heitkamp et al., 1990) into the treatment zones for in situ degradation of PNP. Bacterial solutions used for inoculation into the electroosmosis unit were typically prepared by growing *Pseudomonas* sp. PNP1 in L-salts media that contained about 100 mg PNP/L. A known volume of this bacterial solution was injected into the treatment zones prior to the electroosmosis experiments. Microbial supports packed into the treatment zones included sand, powdered and granular activated carbon, and sawdust. Oxygen was supplied for biodegradation by air sparging through tubes connected to the bottom of the treatment zones.

Instead of air, a solid slow-release oxygen (SO) material was also tested as an oxygen source for the microbial degradation. This material was provided to us from EPA/University of Cincinnati (Vesper et al., 1994). It contained sodium percarbonate ($Na_2CO_3 \cdot 1.5H_2O_2$), which released hydrogen peroxide upon contact with water, followed by further decomposition to oxygen. The solid percarbonate powder was encapsulated in poly(vinylidene chloride) to slow down the release rate. The SO material has an oxygen content of about 7 wt %. Pine sawdust was selected as the packing material of choice in the biotreatment zones for studying the use of the solid oxidant. One gram of anhydrous ethanol-washed SO was loaded into each sawdust zone. Previous shake flask studies in our laboratories have indicated that some unknown inhibitory components that leached from the granular SO could be removed by washing the SO in anhydrous ethanol followed with overnight drying.

For microbial analysis, the unit was disassembled at the end of each run, and a total of 20 samples were taken from the anode zone, the contaminated clay middle zone, and the cathode zone. Six clay samples of approximately 1 g each were taken from the zones next to the anode and the cathode, and eight clay samples of approximately 1 g each were taken from the contaminated clay zone. Each clay sample was place in a preweighed 20-ml sterile glass tube, and the exact weight of the tube with the clay was measured. Each clay sample was vigorously mixed with 10 ml of L-salts and used to prepare tenfold serial dilutions in L-salts. The 10^{-3}–10^{-6} dilutions were plated onto two types of media: an L-salts media containing PNP (50 mg/L), yeast extract (5 mg/L), and Difco agar (1.5%) and a second standard plate count agar containing tryptone (5 g/L), yeast extract (2.5 g/L), dextrose (1 g/L), and Difco agar (1.5%). The plates were incubated 2–4 days, after which the colonies on each plate were enumerated. The tubes containing the clay were dried overnight at 100°C and then reweighed to determine the dry weight of the clay sample. The concentration of bacteria in each clay sample was reported as colony forming units (cfu)/gram dry weight of clay.

Biodegradation Batch Studies under Anaerobic Environment

The ability of the slow-release oxygen material to support the aerobic biodegradation of PNP was studied in batch experiments under strict anaerobic environment. All the sterilized tubes contained 39 ml of a 100 mg/L PNP solution in 40-ml EPA tubes with a screw cap containing a septum. The tubes were labeled, passed into an anaerobic hood, and incubated in a nitrogen atmosphere. Also passed into the hood were small glass vials containing varying amounts of SO and a centrifuge bottle containing resuspended cells of *Pseudomonas* sp. PNP1. A 1-ml aliquot of the active bacterial suspension was inoculated into each tube. This level of inoculum would normally degrade all of the PNP within 24 hours of incubation under fully aerobic conditions (Heitkamp et al., 1990). The tubes received varying amounts of SO and were capped tightly.

Sampling involved removing 1 ml from each tube using a 23-gauge/1-inch needle and a 3-ml syringe. Samples were centrifuged or filtered to remove cells, and PNP level was analyzed with a spectrophotometer or by HPLC. The experimental design included a positive control (incubated aerobically outside the anaerobic hood) and a negative control (incubated anaerobically without SO addition). Most studies included transferring the tubes from the anaerobic hood to an aerobic environment at the end to demonstrate rapid and complete degradation under aerobic conditions.

Studies of Microbial Movement under Electric Field

The ability to efficiently track the movement in clay of chemical-degrading pseudomonas during electroosmosis was facilitated by inserting the *E. coli lacZ* and *lacY* reporter genes into the chromosome of the bacterial isolates (Barry, 1986). The introduction of the *lacZY* marker genes into selected pseudomonas enables them to utilize lactose as a sole carbon source, a metabolic trait not found in most fluorescent pseudomonas.

The genetically marked strains can be isolated on media containing the chromogenic substrate 5-chloro-4-bromo-3-indolyl-beta-D-galactopyranoside (X-gal). Transformed microbes carrying the *lacZY* genes produce beta-galactosidase, which cleaves X-gal and gives rise to bacterial colonies that are blue when X-gal is incorporated into the solid media. Addition of the rifampicin resistance phenotype to the *lacZY*-expressing cells substantially enhances selectivity without compromising chemical-degradation abilities.

PROCEDURE FOR INSERTION OF THE lacZY REPORTER GENES INTO SELECTED PSEUDOMONAS STRAINS

The bacterial isolate transformed with the *lacZY* marker genes was *Pseudomonas putida* PNP1. The parental strain had been made resistant to rifampicin at 50 mg/L. [*E. coli* isolates carrying the broad host-range plasmid HB101/pRK2013 (helper plasmid) or *E. coli* HB101/pMON 7197 (Tn7-*lac* delivery plasmid) were obtained from G. Barry, Monsanto Co.]

The introduction and subsequent transposition on the Tn7-*lac* elements into the genome of the parental pseudomonas strains was carried out via a triparental mating between the parental pseudomonas, *E. coli* HB101/pRK2013 and *E. coli* HB101/pMON 7197. The three cultures were grown in 1 ml of Luria-Bertania broth containing the appropriate antibiotic for each strain for 18 h at 30°C. The helper and delivery plasmids were sequentially pelleted and resuspended in 200 μL of LB broth, and this mix was added to the cell pellet of *Pseudomonas* PNP1. The mating mixture was spotted onto a LB plate, incubated for 48 h, and then a serial dilution of the mating mixture was made in L-salts. Transformants of the pseudomonad parental strain were isolated on *Pseudomonas* F agar (Difco Co.) containing rifampicin (50 mg/L) and X-gal (0.006%). Bacterial cells that contained chromosomally inserted *lacZY* genes were detected by the appearance of blue colonies on the solid X-gal media. Several blue colonies of the transformed pseudomonad strain were grown on L-salts media containing 1% lactose to confirm stable expression of the *lacZY* genes.

EXPERIMENTAL SETUP AND OPERATION

The setup shown in Figure 3 was used for studying the microbial movement during electroosmosis. The unit is similar to the one shown in Figure 2 but with only one treatment zone (biozone) in the middle of the EO cell. This middle zone was in contact with an 8-cm thick PNP-contaminated clay zone on the anode side and an 8-cm thick clean clay zone on the cathode side. Typically, the contaminated clay zone contained about 1000 g of clay that was mixed with a 1000 mg/L PNP solution to obtain a final concentration of 350 to 400 μ PNP/gram wet clay. The biozone had a volume of 100 cm^3 and contained various materials such as coarse sand, fine sand, granular activated carbon, or sawdust, either as a single component or in various combinations. The bacterial inoculum was an overnight suspension of *Pseudomonas* sp. PNP1-*lacZY*. The isolate was grown in L-salts with 1% glucose and yeast extract (5 mg/L) for 18 h, pelleted by centrifugation, and resuspended in 50 ml of quarter-strength L-salts. The biozone was

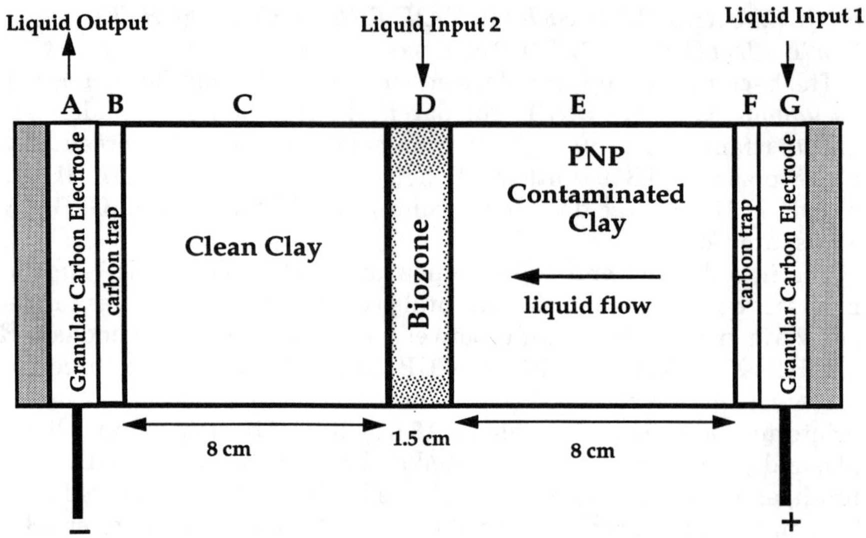

Figure 3. Electroosmotic cell for studying the movement of *Pseudomonas* sp. PNP1 *lacZY* under electric field.

inoculated by injecting the microbial solution into the bottom of the zone through a port of the EO unit. The final concentration of cells in the biozone was about 5×10^{10} cell/100 ml of biozone material. The biozone was aerated by attaching a 1/16-inch air line to the bottom port leading into the biozone, and air was slowly bubbled through the zone. A second mariotti bottle was also attached to the bottom port as a source of water to keep the biozone from drying out due to constant aeration.

Upon completion of each electroosmosis study, 1-g samples of clay were taken from both the clean and PNP-contaminated EO zones to analyze for *Pseudomonas* sp. PNP1-*lacZY*. Four of the clay samples were taken from the top of each clay zone, and four samples were taken from the bottom portion of each clay zone. A 1-g sample of the biozone material from the top, middle, and bottom of the biozone were also analyzed. A serial dilution in L-salts buffer was performed on each sample, and the diluted microbes were plated on the selective solid media. After a 3-day incubation, the levels of *Pseudomonas* sp. PNP1-*lacZY* present in clay, sand, carbon, or sawdust were measured and reported as cfu/ml. Microbial strain *Pseudomonas* sp. PNP1-*lacZY* was quantified in clay and solid biozone support matrices by isolation on the selective solid media *Pseudomonas* F-agar, which was supplemented with rifampicin (50 mg/L) and 0.006% X-gal. Due to the high selectivity of the plating media, only the *Pseudomonas* sp. PNP1-*lacZY* isolate was able to grow on the media, as demonstrated by the appearance of blue, rifampicin-resistant colonies.

Electrokinetic Parameters

For each run, the electroosmotic permeability can be calculated using the following equation:

$$Q = k_e\, i_e\, A \tag{1}$$

where

Q: volumetric flow rate by electroosmosis (cm^3/sec)
k_e: coefficient of electro-osmotic permeability (cm^2/volt-sec)
i_e: voltage gradient applied across the soil mass (volt/cm)
A: cross-sectional area perpendicular to flow (cm^2)

Note that the electroosmotic flow is proportional to the applied voltage gradient and that the electroosmotic conductivity has the units of velocity over field strength (cm/sec over volt/cm). The cross-sectional area of the cell was about 81 cm^2.

RESULTS AND DISCUSSION

Coupled Electroosmosis with *in situ* Adsorption

Prior to biodegradation studies, the coupling of electroosmosis with *in situ* adsorption was investigated to establish the ability of the process to clean up the PNP-contaminated clay soil. In this experiment, electroosmosis was carried out in the direction of A to G for 6 days (passing an amount of water equivalent to 1 pore volume of each contaminated clay section) and then reversed (G to A) for an extended period (14 days, 2.8 pore volumes). The cathode effluent (at alkaline pH) was brought back to the anode zone (at acidic pH), thereby accomplishing both total water recycle and pH neutralization. pH probes were inserted in the sand/carbon zones for periodic measurements of pH (with the power turned off).

Figure 4 shows the profiles of pH and conductivity of the system during one run. The pH's of both sand/carbon zones remained fairly stable between 6 and 6.5. The pH of the cathode effluent went up from 8.5 to 9.5 during the first pass, but settled down to 8–8.5 throughout the second pass. The pH's of the clay samples measured at the end of the run did not show a pattern between the electrode but was lowest

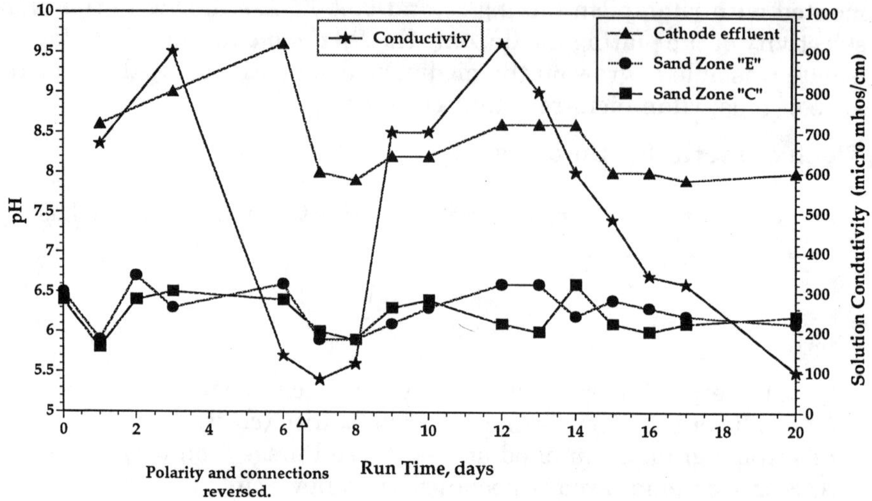

Figure 4. Conductivity and pH profiles for electroosmosis with total water recycle.

near the anode (3.8) and highest (5.2) at the cathode. In between these extremes, the values varied from 3.8 to 4.9 with no readily discernible pattern. The fairly moderate pH values obtained in our electroosmosis system versus more extreme pH's observed by others (Bruell and Segall, 1992; Acar et al., 1992; Shapiro and Probstein, 1993) are believed to be a consequence of many factors. The major ones are water recycling and the use of activated carbon, which was found to be a good solid buffer of pH, as electrodes and in the treatment zones. Conductivity of the cathode effluent seemed to cycle with the polarity reversal, probably reflecting precipitation and redissolution of minerals in the water as a function of pH.

More than 99% of the PNP was removed from the contaminated zone, with about 21% of the initial load captured in sand/carbon zone C and 68% in sand/carbon zone E (Figure 5). About 90% of the initial PNP was accounted for at the end of the run. The average electroosmotic conductivity was 1.3×10^{-5} cm^2/V-sec. The total power usage was about 28 kwh/m^3 of contaminated soil (average of 7.2 kwh/m^3 soil per pore volume of water transported through the contaminated soil section). Thus, the cost of electricity at this small scale is less than $3 per cubic meter of soil treated, based on $0.1 per kwh. For larger scales where the treatment zone spacing is about 2 feet or larger, depending upon the power applied and the length of treatment, the electricity cost is probably in the range of 10 to 20 dollars per cubic meter of contaminated soil.

The above study clearly demonstrates that electroosmosis could flush an organic contaminant uniformly from clay soil into adjacent permeable layers where the contaminant is removed from the pore fluid by a solid adsorbent. Also, clay soil containing permeable layers appears to behave similarly to homogeneous clay soil in terms of electroosmotic conductivity and power consumption. Finally, the electrical polarity and flow reversal does seem to stabilize the electroosmosis process, a beneficial feature for long-term operation.

Coupled Electroosmosis with Biodegradation

For the coupling of electroosmosis with *in situ* biodegradation of PNP, we used *Pseudomonas* sp. strain PNP1, a bacterium reported to mineralize high concentrations of PNP as a sole source of carbon and energy (Heitkamp et al., 1990). Previous radiometric studies with this bacterium showed that over 70% of PNP was degraded to carbon diox-

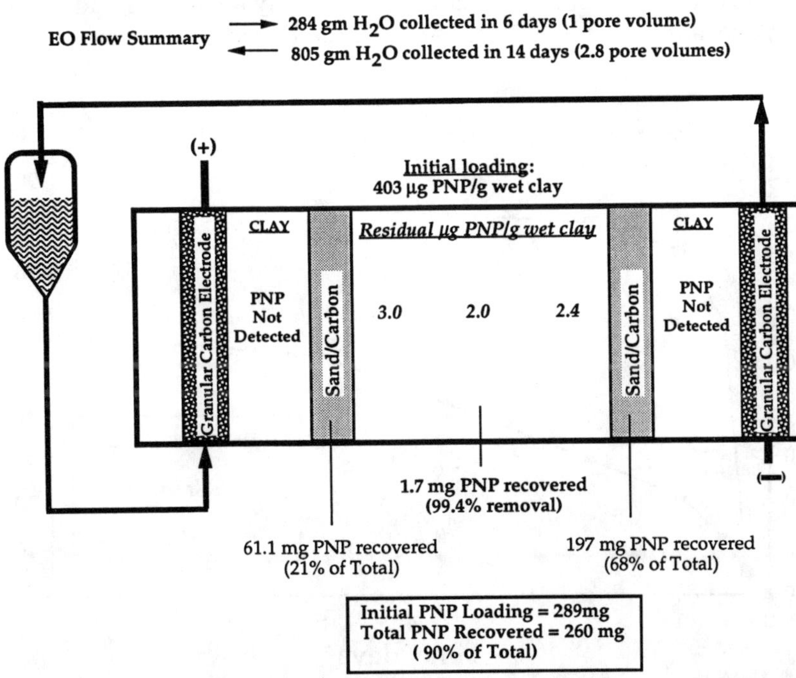

Figure 5. PNP mass balance for electroosmosis with total water recycle.

ide, and only 3.7 to 6.7% remained as highly polar, water-soluble metabolites. These studies also showed that *Pseudomonas* sp. strain PNP1 incorporated 8.4 to 11.8% of PNP into cell residues, lost <1% as volatile metabolites, and reduced levels of undegraded PNP in L-salts (Leadbetter and Foster, 1958) media to less than 50 μg/L.

Shake flask experiments were conducted first to study the effects of microbial supports on PNP biodegradation characteristics, followed by soil column experiments.

EFFECTS OF MICROBIAL SUPPORTS ON PNP DEGRADATION IN SHAKE FLASK STUDIES

Depending upon its pore size and binding affinity for PNP, the microbial support could strongly affect the bioavailability of PNP. Figure 6 shows degradation rate comparison for PNP free in solution or adsorbed on activated carbon of two different particle sizes (powder and granular). These were batch experiments carried out aerobically in shake flasks with the solution mixtures well-agitated. PNP in solution was found to be completely degraded within 1 day, which was typical for the conditions used. However, it took 2 days to degrade about 80% of the PNP adsorbed on the powdered carbon, with only about five additional percent degraded on day 8. For PNP adsorbed on granular carbon, the

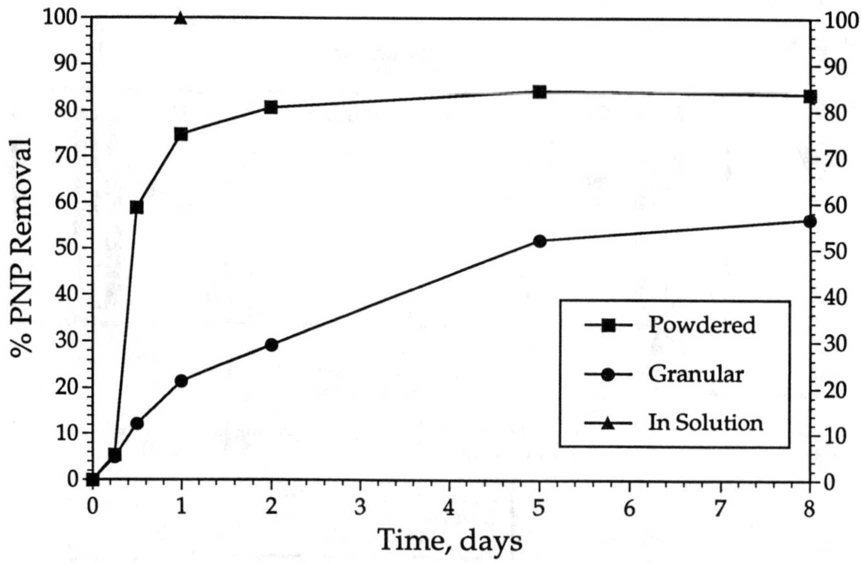

Figure 6. Effects of supports on PNP biodegradation in shake flasks.

amount of PNP degradation climbed slowly, reaching less than 60% after 8 days. It is clear that both the adsorption and diffusion (particle size effect) significantly reduce the degradation rate of PNP.

ELECTROOSMOSIS WITH IN SITU BIODEGRADATION USING AIR SPARGING

The coupling of electroosmotic removal of PNP from clay soil with its *in situ* biodegradation in the treatment zones was studied. The experimental setup was similar to the one used for *in situ* PNP adsorption (Figure 2), except for the presence of PNP-degrading microorganisms in the treatment zones. Support materials used for microbial attachment in the treatment zones include activated carbon, sand, and sawdust. Air was sparged into the treatment zones to provide oxygen to the system. Electroosmotic characteristics of these experiments in terms of flow rate, electroosmotic permeability, and power consumption were similar to those reported above for the adsorption studies. Results are summarized below for each type of support.

Granular Activated Carbon

The treatment zones consisted of sand and granular activated carbon and were inoculated with a cell suspension of the PNP-degrading microorganisms just before starting electroosmosis. PNP loading was 375 μg/g wet clay. The cell was operated for 9 days, 3 days in one direction (0.75 pore volume) and then 6 days in the opposite direction (1.74 pore volume). Moisture content of the three clay zones ranged from 36 to 38% at the end of the run, essentially the same as the initial 37.5%. The pH's of the clay soils were between 4.5 and 5.5, but those of the two treatment zones were 6.7 and 6.8. About 98% of the initial PNP was found to be uniformly removed from the contaminated zone. However, total PNP recovered from all areas was about 90%, most of which bound to the activated carbon in the treatment zones. Thus, very little PNP biodegradation occurred. This lack of significant PNP degradation was attributed to the inaccessibility of PNP inside the pores of the carbon particles to the microbes, as indicated by the shake flask results presented above.

Powdered Activated Carbon

Powdered activated carbon was used in the treatment zones to minimize diffusional limitation to PNP degradation that was observed with granular carbon. Also, the bacteria were cultivated on the powdered carbon prior to packing into the electroosmosis unit to maximize microbial attachment to the carbon surface. PNP loading was the same as with the granular carbon run reported above. The cell was operated for 8 days, 4 days in one direction (1.1 pore volumes) and then reversed for 4 days (0.74 pore volume). Analysis of the soil samples showed that 94% of the initial PNP load was removed from the contaminated clay zone. However, only 77% of the total PNP was recovered from all the zones, which shows that some degradation (23% of total) of PNP did occur (average degradation rate = 8 mg PNP/day). Microbial analysis of the soils in the electroosmotic cell after the run showed very little movement of the PNP-degrading microorganisms from the treatment zones into the surrounding clay soils.

Sand Alone

Only sand was used in the treatment zones (no activated carbon) to eliminate the interference of carbon adsorption to PNP bioavailability. PNP loading was also reduced from 375 to 86 μg/g wet clay to enhance the percentage of PNP loss due to biodegradation. Bacterial solutions were injected into the sand zones prior to starting up electroosmosis. The system was operated for 10 days, 3 days in one direction (0.75 pore volumes), reversed for 5 days (0.99 pore volume), and then reversed again for 2 days (0.24 pore volume). Analysis shows that about 90% of the initial PNP was removed from the contaminated clay, but only 13% of the initial PNP load was recovered from all areas (Figure 7). Thus, it is possible that up to 87% of the loaded PNP could be lost due to biodegradation (degradation rate ≅5.1 mg PNP/day). Microbial analysis showed good levels of colonization in both sand zones. Soil samples taken from both the center clay area and the areas near the electrodes showed excellent colonization of the PNP-degrading bacteria throughout all the soil sections (Figure 8), indicating that the microbes were able to move away from the inoculation point and penetrate clay soil under the electric field.

Sawdust

We also investigated the use of natural cellulosic materials such as sawdust instead of sand or activated carbon in the treatment zones. It is well documented in the biofiltration literature that these natural fibers provide excellent surfaces for microbial attachment and growth (Leson and Winer, 1991). Bacterial solution was mixed with sawdust during packing of the electroosmosis unit. PNP loading was 87 μg PNP/g wet clay. The system was operated for 4 days (0.56 pore volume) and then reversed for 8 days (1.1 pore volumes). Analyses showed that 90% PNP was removed from the contaminated zone, 13% recovered in the two sawdust zones, none in the clean zones, resulting in a possible 77%

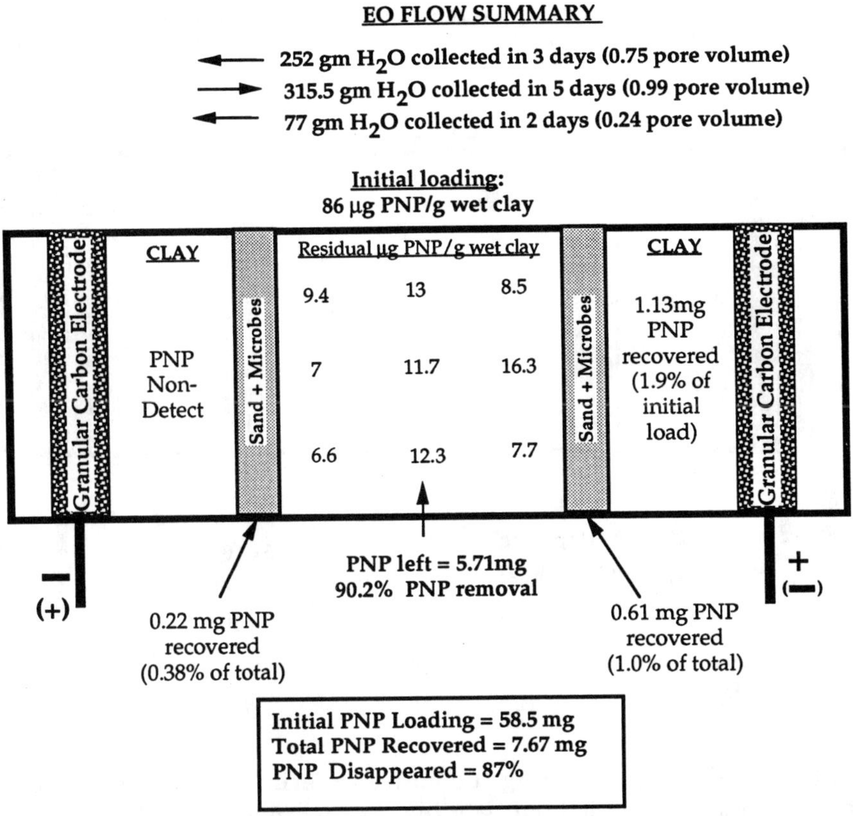

Figure 7. Coupled electroosmosis with biodegradation of PNP with sand in treatment zones.

of the loaded PNP gone due to biodegradation (degradation rate ≅ 4 mg PNP/day). It was observed that sawdust was able to retain water much better than sand, a property beneficial to both microbial activity and the electroosmotic process. These results indicate that sawdust, like activated carbon, is a very good support for microbial attachment but without the tight adsorption that adversely affects the degradation rate.

Interestingly, while the sawdust areas were found to contain inoculated microbes at high levels (3.6×10^7 to 9.8×10^8 cfu/g dry wt of sawdust), no microorganisms were detected in any of the 14 samples taken from the center clay area or from the areas between the treatment zones and the electrodes. The bacteria were thus retained completely

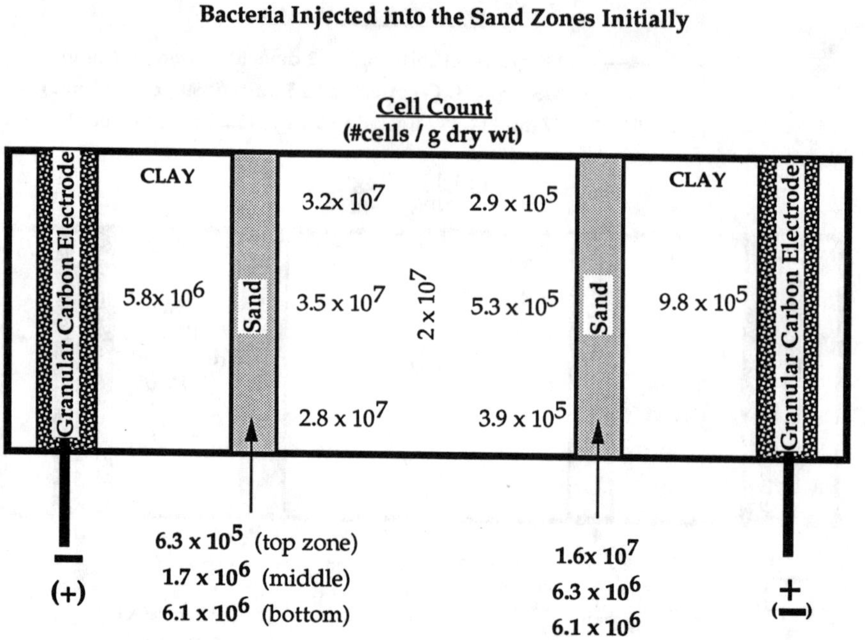

Figure 8. Microbial migration in clay under electroosmosis with sand in treatment zones.

in the sawdust areas during the entire experiment. Another sawdust experiment was carried out at a higher PNP loading of 409 μg/g wet clay. The system was operated for 10 days with polarity reversal (1.5 pore volumes total). Analyses showed that 75% of initial PNP was removed from the contaminated zone, and a total of 42% of the initial PNP was gone, resulting in an average biodegradation rate of 12 mg PNP/day. It should be noted that the initial PNP concentration in the interstitial solution in the contaminated area was over 1000 mg/L, a normally toxic level to microbes in solution. Similarly, the *Pseudomonas* sp. PNP1 bacteria were found only in the sawdust regions (1.1×10^7 to 4.3×10^8 cfu/g dry wt of sawdust); no microorganisms were detected in any of the clay areas.

The results obtained in this section demonstrate that electroosmosis can be coupled with *in situ* biodegradation for remediating clay soils contaminated with organic compounds. Also interestingly, it appears that the types of microbial supports used may influence whether the injected microbes would remain in the treatment zone or migrate throughout the clay soil matrix. This particular aspect will be investigated more closely later with a genetically marked PNP degrader in a simpler experimental setup.

ELECTROOSMOSIS WITH IN SITU BIODEGRADATION USING A SLOW-RELEASE OXYGEN (SO) MATERIAL

The encapsulated sodium percarbonate obtained from EPA/University of Cincinnati with an equivalent oxygen content of 7 wt % (Vesper et al., 1994) was used to provide the required oxygen for the electroosmosis experiments. Qualitative investigation of the rate of oxygen release from SO using a Gilson Oxygraph (model OXY-5, Gilson Medical Electronics, Middleton, Wisconsin) showed relatively rapid oxygen release. Batch degradation studies showed that the SO material could support aerobic microbial degradation of PNP in the absence of oxygen from the air. The rapid rate of oxygen release caused a significant increase in pressure in the tubes containing high dosages of SO. In addition, testing with one batch of SO showed some toxicity or inhibition to microbial degradation at SO levels greater than 100 mg SO per 40 ml solution (Figure 9). We found that this adverse effect could be removed by washing the SO material with ethanol. However, recent studies in our laboratory using a more recently formulated SO did not show these adverse effects or microbial inhibition.

Sawdust was used as the microbial support and mixed with SO in EO experiments. A number of experiments was carried out with two PNP loading levels (30 and 130 μg PNP/g wet clay) and three levels of SO (0.3, 1, and 2 g). PNP degradation was found to range from 40 to 50% of the initial loading. However, in a control experiment with the treatment zones containing noninoculated sawdust, but not sterilized, almost 30% of the initial PNP was unaccounted for. So it is not conclusive whether significant biodegradation of PNP by the inoculated microbes occurred in these preliminary EO experiments with the solid oxidant.

It was observed in the above EO experiments that oxygen was released very quickly, causing visible expansion of the treatment zone in the first few days. In one experiment, oxygen probes were inserted into the two treatment zones, which contained sawdust, SO, and bacteria, and periodic readings were taken to determine the rate of oxygen release and consumption over time. The results shown in Figure 10 show that most of the oxygen was released within the first two days of the experiment. This may explain the relatively low degree of biodegradation obtained since a longer time is required to move PNP from the contaminated soil into the treatment zones. There is concern, based on these preliminary results, whether the SO material used in this

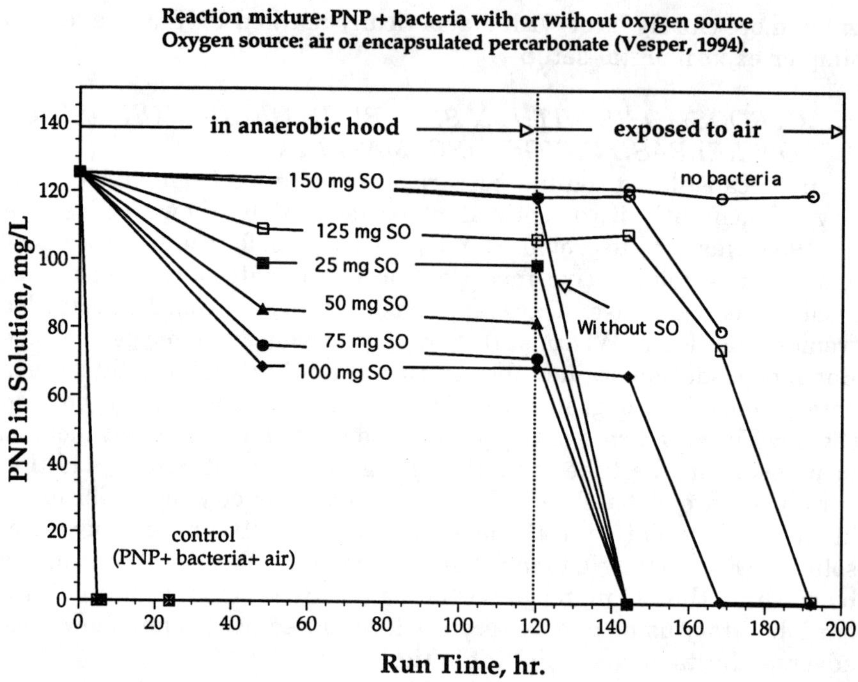

Figure 9. PNP biodegradation in shake flasks using a solid oxidant.

study has the slow release rates needed to support biodegradation in the integrated process.

MICROBIAL MOVEMENT UNDER THE ELECTRIC FIELD

The *Pseudomonas* sp. PNP1 *LacZY* bacterium was used to study microbial movement in the clay soil under the electric field. The setup in Figure 3, which had only one biozone, was used without polarity reversal in order to simplify interpretation of the experimental results. A control experiment was run using the coarse grade sand and 2% GAC in the biozone, but without PNP1 inoculation to prevent degradation. The unit was run for 7 days, resulting in over 2 pore volumes of water passing through the biozone. This was much longer than normal in order to allow all of the PNP to move into the biozone. Analyses showed that most of the PNP load was recovered from the biozone (83% of total) with a total mass balance of 92% for PNP.

The various materials packed in the biozone for studying their effects on microbial movement are listed in Table 1. Since the focus of these

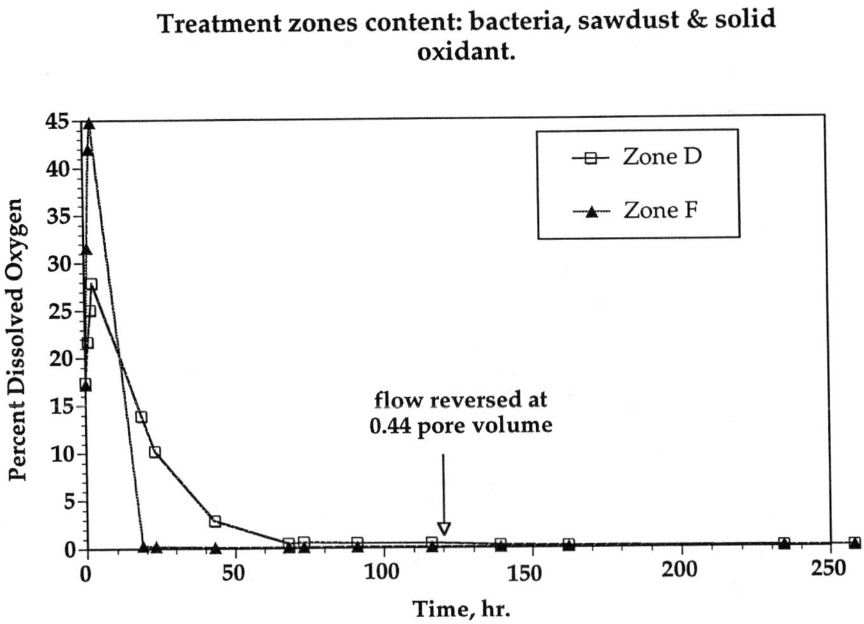

Figure 10. Time release profiles of a solid oxidant in treatment zones during electroosmosis.

experiments was on microbial movement rather than achieving high biodegradation efficiency, each run was operated in one direction to move PNP from the contaminated clay through the biozone, passing from 0.8 to 1.5 pore volumes. Inlet solution was a quarter-strength L-salt with a conductivity of 400–500 μS/cm. Electrical current was held at 3 mA and voltage ranged from 10 to 20 volts (voltage gradients from 0.5 to 1 V/cm). PNP loading was from 320 to 460 μg/g wet soil. Typically, from 40 to 60% of the initial PNP loaded was gone after each run, which in comparison with the 92% mass balance obtained in the control experiment supports the occurrence of biodegradation of PNP in the electroosmosis experiments.

Results for microbial movement during electroosmosis with the various supports are, however, somewhat less conclusive. As shown in Table 1, some degree of microbial movement from the biozone into the clean clay zone was observed for all runs. There may be less movement with GAC support and better microbial viability when the microbes were cultivated on the carbon prior to packing into the EO cell. There was no sign of microbes moving from the biozone back into the contaminated clay toward the anode, which would be expected if electromigra-

Table 1. Analysis of microbial movement and PNP degradation during electroosmosis.

Packing Materials in Biozone	Pore Volume of Liquid Passed	Loading (μg PNP/g wet clay)	% PNP Removal from Contaminated Clay Section	Total % PNP Recovered	Microbe Movement into Clean Clay Zone	Microbial Mass Balance
2% GAC/coarse sand	2.22	420	96	95[a]	Control, no bacteria	
100% coarse sand	0.78	337	42.7	92[b]	1 cm top, 2 cm bottom	2.5 log increase
100% fine sand	0.90	452	99.9	59	3 cm bottom	2 log decrease
100% GAC	0.88	461	94.5	46	1 cm top, 1 cm bottom	1 log decrease
2% GAC/coarse sand	0.81	317	90.2	52	2 cm top, 1 cm bottom	2.5 log increase
2% GAC/coarse sand	1.53	351	98	62.9	1 cm top, 3 cm bottom	1.5 log decrease
100% pine sawdust	0.73	420	71.1	59	1 cm top, 2 cm bottom	2 log decrease
Sawdust/2% carbon	0.85	407	75.5	63.3	1 cm top, 6 cm bottom	2.3 log decrease
50/50 sand/carbon[c]	0.95	407	77.8	43	1 cm top, 1 cm bottom	no change

[a]Mass balance control study, no bacteria injected.
[b]Large breakthrough of PNP into clean clay zone.
[c]Cells were pregrown on carbon prior to pacing into the biozone.

tion was operating since the microbial cell's surface is negatively charged. There are two possible reasons for this lack of movement towards the anode. First, the electrical potential applied is probably too low for significant movement of the microbes by electromigration. Second, the level of PNP loading on the clay for these experiments is equivalent to a PNP concentration of about 1000 ppm in the pore water, which is toxic to the microbes and thus might have prevented them from moving. These results are from preliminary studies, and further investigation is needed for definitive understanding of the complicated interactions between microbial characteristics, the nature of the solid matrices (soil and microbial support), the electroosmotic process, and the electric field with regard to microbial movement in the integrated *in situ* remediation process.

SUMMARY AND CONCLUSIONS

An innovative integrated *in situ* remedial concept has been developed in which electroosmosis is coupled with treatment zones for cleanup of contamination in low-permeability soils. Some important conclusions that can be made from the experimental data are

- Electroosmosis could operate with clayey soil mass containing embedded permeable (sand, granular carbon, saw dust) layers.
- The process allows total recycling of the cathode effluent back to the anode region. This mode of operation not only has the beneficial effect of moderating the system's pH and conductivity, but also significantly simplifies water management in field operation.
- The process was demonstrated experimentally utilizing electroosmosis to remove a water-soluble organic contaminant (PNP) from kaolin clay and deliver it to an adjacent permeable region where it is either adsorbed by activated carbon or aerobically biodegraded.
- The slow-release oxygen material used, sodium percarbonate encapsulated in poly(vinylidene chloride), was shown to support the aerobic biodegradation of PNP in an anaerobic environment. However, further development appears necessary to utilize SO to support long-term aerobic biodegradation in the integrated *in situ* remediation process.
- The nature of the materials used for microbial supports in the treatment zones may affect the microbial migration during electroosmosis. This is, however, only one of many variables,

which include strength of the electric field (electromigration), chemical loading (chemotaxis), soil type, the type of microorganisms, and whether they are inoculated as solution into the treatment zones or pregrown on the support.

- The approach of genetically tagging common soil microorganisms with a unique gene for aiding in its detection was successfully implemented. Transformed microbes carrying the *lacZY* genes were phenotypically distinct. This provides a very simple and selective method for tracking the movement and viability of microorganisms inoculated into soils and other nonsterilized solid matrices.

We are extending the study to include widespread contaminants such as chlorinated solvents (e.g., dichloroethane, trichloroethylene, tetrachloroethylene), different soil types, and other modes of degradation (anaerobic, chemical, electrochemical), as well as scaling up the technology. A consortium of industry (DuPont, GE, and Monsanto), in collaboration with two federal agencies (EPA and DOE), has also been formed to combine expertise and resources for accelerating the development of this technology for large-scale implementation.

ACKNOWLEDGEMENTS

The authors would like to express their appreciation to Jay Wendling for the HPLC analysis of PNP and to Gerald Barry for providing helper plasmid, Tn7-lac delivery plasmid and technical consulting to facilitate the genetic marking of *Pseudomonas* sp. PNP1.

REFERENCES

Acar, Y.B., Li, H. and Gale, R.J. 1992. *ASCE,* Vol. 118, No. 11, 1837–1852.

Acar, Y.B. and Alshawabkeh, A. N. 1993. "Principle of Electrokinetic Remediation." *Environ. Sci. Technol.,* Vol. 27, No. 13, 2638–47.

Barry, G. 1986. "Permanent Insertion of Foreign Genes into the Chromosomes of Soil Bacteria." *Bio/Technology,* Vol. 4, May, 446–449.

Bruell, C.J. and Segall, B.A. 1992. *J. Environ. Eng.,* Vol. 118, No. 1, Jan/Feb, 68–83.

Casagrande, L. 1952. *J. BSCE,* 39:51–83.

Hamed, J., Acar, Y.B. and Gale, R.J. 1991. *ASCE,* Vol. 112, February, 241–271.

Heitkamp, M.A., Camel, V., Reuter, T.J. and Adams, W.J. 1990. *Appl. Environ. Microbiol.* 56:2967–2973.

Lageman, R. 1993. "Electroreclamation—Applications in The Netherlands." *Environ. Sci. Technol.,* Vol. 27, No. 13, 2648–50.

Leadbetter, E.R. and Foster, J.W. 1958. *Arch. Furmikrobiol.,* 30:91–118.

Leson, G. and Winer, A.M. 1991. *J. Air & Waste Manage. Assoc.,* 41:1045–1054.

Murdoch, L. 1992. "Some Recent Development in Delivery and Recovery: Hydraulic Fracturing and Directional Drilling" to appear in the *Proceedings of ETEX'92—The 2nd Annual Environmental Technology Exposition and Conference,* Washington, DC, USA, April 7–9.

Shapiro, A.P., Renauld, P. and Probstein, R. 1989. "Preliminary Studies on the Removal of Chemical Species from Saturated Porous Media by Electroosmosis." *Physicochemical Hydrodynamics,* Vol. 11, No. 5/6, 785–802.

Shapiro, A.P. and Probstein, R.F. 1993. *Environ. Sci. Technol.,* 27:283–291.

Segall, B.A. and Bruell, C.J. 1992. *J. Environ. Eng.,* Vol. 118, No. 1, Jan/Feb, 84–100.

Vesper S.J., Murdoch, L.C., Hayes S. and Davis-Hoover, W. 1994. "Solid Oxygen Source for Bioremediation in Subsurface Soils." *J. Hazardous Materials,* 36:265–274.

Vertical Circulation Flows for in situ Bioremediation in Aquifers

J. STAMM
Institute of Hydromechanics
University of Karlsruhe
Kaiserstrasse 12
D-76128 Karlsruhe, Germany

INTRODUCTION

In many countries, especially in the highly industrialized ones, contaminations of soil and groundwater are the consequence of the intensive use of hazardous substances. Because of the resulting risk for man and environment, a strong effort has been made in the last decade to find innovative in situ remediation techniques and/or to improve existing methods to remove the contaminants. In situ techniques are, for the most part, gentle to the environment and are cost-effective systems. Furthermore, alternatives to pump-and-treat techniques, which have been proven inadequate in many cases, are needed (e.g., US EPA 1989a, 1989b, 1990). Experiences with pump-and-treat remediation installations show that the expected cleanup time was exceeded in many cases, caused mostly by the heterogeneity of the aquifer and the adsorption of the contamination onto the soil matrix (Kinzelbach et al., 1991).

Various existing methods are convenient to handle either the saturated or the unsaturated underground. In the unsaturated zone, soil air is used as a solvent and transport medium for contaminants, and analogous free groundwater is used in the saturated zone. Some treatment methods are based only on physical cleaning, while others are supported by biological degradation processes (Hinchee, 1994). Based on an extensive set of experiments, Wilson et al. (1990) concluded that attempts to mobilize immiscible liquid contaminant ganglia in a porous media by pumping processes alone would require unrealistically high hydraulic gradients.

In this section several techniques are presented that are based on a vertical circulation flow and that enable physical and biological remediation of the saturated and the unsaturated subsoil. Vertical circulation flows around wells with at least two screen sections are universally applicable remediation tools. They are employed in several techniques, e.g., the so-called Groundwater Circulation Well (German: *G*rundwasser-*Z*irkulations-*B*runnen, abbr.: GZB), the Vacuum Vaporizer Well (German: UVB), and the Groundwater Flushing Circulation Well (German: SZB). These techniques can be combined with any appropriate in-well or on-site method. Their hydraulic flow regime can be used to support biological degradation in the aquifer. Moreover, an addition of nutrients and/or electron acceptors for stimulating bioreclamation processes is possible when the groundwater passes the well casing. Within the well itself the water moves vertically. The contaminated groundwater enters the well at the bottom and stripped or treated water leaves at the top or vice versa (Figure 1). In the vicinity of the well, an area of vertical flow circulation is created. One well should be used to remediate only one aquifer (confined or phreatic) at a time and should not connect different aquifers.

The vertical circulation flow and transport regime around UVB and GZB has been the subject of continuing numerical and field investigations at the University of Karlsruhe in Germany. This section discusses the different remediation techniques, treatable contaminants, and numerical results concerning the influence of hydrogeological conditions and well parameters on the sphere of influence.

VERTICAL GROUNDWATER CIRCULATION FLOW SYSTEMS

At numerous sites in Germany and more recently in the United States, vertical circulation wells have been used for in situ groundwater remediation where the underground is contaminated by strippable substances, e.g., volatile chlorinated hydrocarbons, BTEX (Herrling et al., 1992, 1993, 1994). As an alternative to conventional hydraulic redevelopment measures (pumping, off-site cleaning, and reinfiltration), the contaminated groundwater is stripped by air in a below-atmospheric pressure field in the UVB. Gvirtzman and Gorelick (1992) give a detailed overview of the concept of in situ vapor stripping for removing VOCs from groundwater. A significant advantage of the installation is the fact that no water is pumped to the ground surface. So, no additional wastewater treatment is necessary. Further, from the regulatory point of view, no reinjection occurs.

The vertical circulation flow applications differentiate the three types of wells. The UVB eliminates strippable substances in the vacuum zone of the upper part of the casing. Often, these wells are used for vapor extraction at the same time to support the remediation of the vadose zone (Section 2.1). The GZB does not have this negative pressure field within the well casing, but delivers the same hydraulic flow field in the aquifer (Section 2.2). This technique enables an in-well treatment, as well as a combination with an on-site treatment. Both types, UVB and GZB, are employed in the saturated groundwater zone. In contrast to these, the SZB is used in the unsaturated zone and delivers a multiphase air–water flow (Section 2.3), which is especially suitable for eliminating biologically degradable contaminants from the unsaturated zone.

The basic vertical circulation flow operating mode is explained using the example of an upward directed, standard circulating well.

The UVB Technique

The UVB technology is used to transport dissolved aqueous contaminants from the aquifer into the stripping zone of the well where they are transferred into the gaseous phase. Furthermore, by improving the environmental conditions needed by the naturally occurring microorganisms in the aquifer, an effective reduction of contaminants is focused within the sphere of influence of the well.

UVB—STANDARD CIRCULATION

When the aquifer is contaminated by a dense nonaqueous phase liquid (DNAPL, heavier than water), an upward operating vertical circulation well is used (Figure 1). The upper, closed part of the well is maintained at subatmospheric pressure by a ventilator. This lifts the water level within the well casing. The fresh air for the upper part of the well casing is introduced through a fresh air pipe. The upper end is open to the atmosphere, and the lower end terminates in a pinhole plate. The height of the pinhole plate is adjusted such that the water pressure is lower there than the atmospheric pressure. Therefore, ambient fresh air is drawn into the system. The reach between the pinhole plate and the water surface in the well casing is the stripping zone, in which an air bubble flow develops. The well casing is divided by a separating plate in a lower and upper part. A pump transports the water from the lower section to the stripping area. This causes a potential gradient between the two screen sections, which induces the circulation flow in the aquifer. The upward-streaming, stripped groundwater leaves the well casing through the upper screen section in the reach of the groundwater surface, which is lifted in a phreatic aquifer by the previously explained pump processes and the below-atmospheric pressure (Figure 1a). The circulation cell that is created transports dissolved oxygen into the aquifer and supports aerobic biological degradation. Additionally, soil air is drawn from the surrounding contaminated unsaturated zone at many sites (Figure 1a). Stripped air and possibly soil air are transported through the ventilator and across activated carbon, where the contamination is adsorbed. Thus, only clean air escapes into the atmosphere.

The cleaning effect of the well is based on reduced pressure, which reinforces the escape of volatile contamination out of the water, and is a result of the intermixing of the air, of the considerable surface area of the air bubbles, and of the high concentration gradient between water and clean air. In this sense, the permanent vibration caused by the air bubbles is beneficial to the escape process of the contamination.

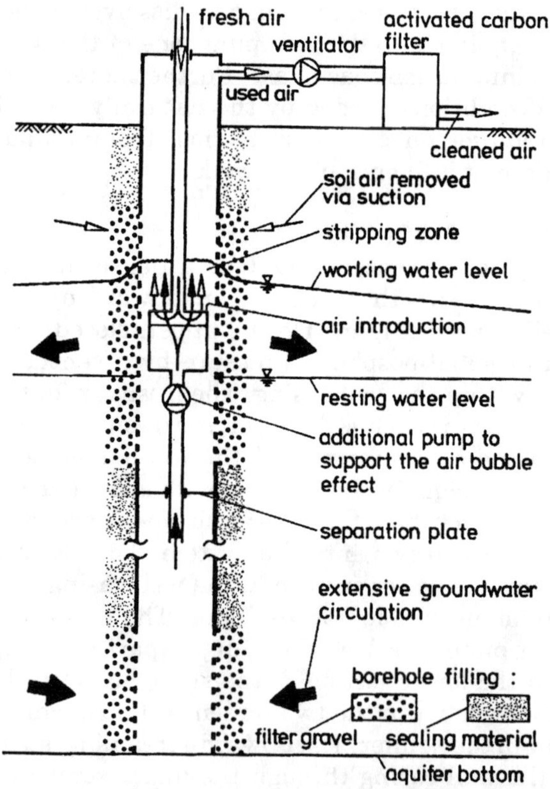

Figure 1a. Vacuum vaporizer well (UVB) with standard circulation.

This vibration is transmitted as compression and shear waves into sediment and fluid. The influence of the vibrations on the mobility of the contaminants outside the circulation well is the matter of recent investigations in Karlsruhe, Germany. Laboratory investigations by Reddi and Challa (1994) have shown that vibrations enhance the mobilization of immiscible liquid ganglia in sands.

In the well design represented in Figure 1a, precipitation of iron, manganese, and calcium may occur, depending on the water quality at a remediation site. In general, the water entering the well contains very little oxygen. When it leaves the well after flowing through the stripping zone, it is fully saturated with oxygen. Supersaturation with carbon dioxide at the well influx is decreased in the stripping zone by the exchange with the air. The pH value increases, and the above-mentioned precipitations can occur. To reduce these effects, which may eventually seal the upper screen section, the closed on-site air circuit is introduced to keep pCO_2 high and pH low: the cleaned air is recycled as fresh air input (Figure 1b). Only those pollutants adsorbed by activated carbon are removed from the circulating air. The stripped substances not adsorbed remain in the groundwater and will not contaminate the atmosphere. However, special cleaning methods incorporated into the air circulation are conceivable.

When using the closed air circuit, the solid well shaft reachs down under the groundwater table, and so any contaminated soil air must be separately removed via suction. Efficient technologies exist for that. Furthermore, the upper well casing and the pipe system must be sealed so that no fresh air can enter the closed system of air circulation.

UVB—REVERSE CIRCULATION

Groundwater pollutions with light nonaqueous phase liquids (LNAPL) are mainly situated in the reach of the groundwater table. In that case, the hydraulic flow field has to be reversed, which means that the cleaned groundwater leaves the well through the lower screen section and returns in an extensive circulation to the upper screen. In this way, the aquifer material surrounding the well is also remediated (Figure 2). The downward operating mode of the vertical circulation well—based on the flow direction within the well—prevents an unwanted spreading of LNAPLs through the depth of the aquifer. The phreatic groundwater surface is slightly lowered by the pumping process. The amount is small, however, caused by the below-atmospheric pressure, which has an opposite effect. The artificial groundwater circulation determines the sphere of influence of a well and is superimposed with the natural groundwater flow.

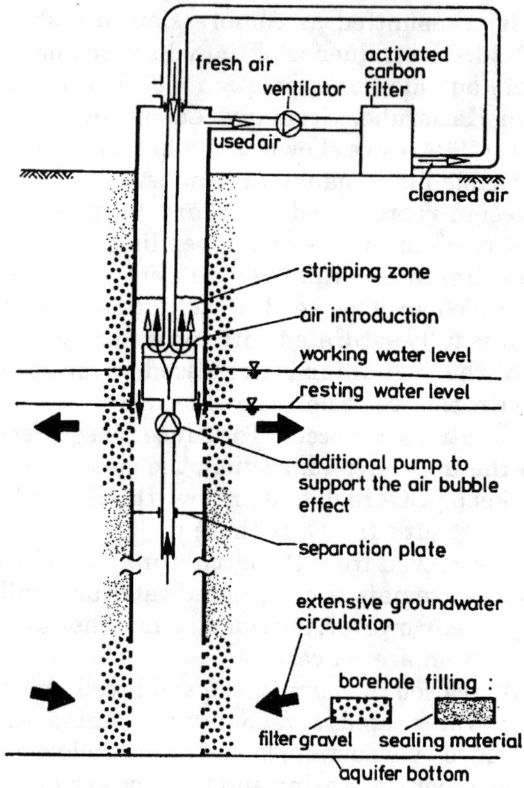

Figure 1b. UVB standard circulation with closed on-site air circuit.

UVB FOR ENHANCED BIOREMEDIATION

The cleaning effect within the circulation cell surrounding a UVB is based on two processes: (1) physical dissolution of contaminants into the aqueous phase and transport to the well and (2) elimination of contaminants by biological degradation processes supported by an improved nutrient and/or electron acceptor supply. When the groundwater is contaminated by substances that are suitable for bioreclamation, the addition of nutrients within the remediation well can enhance the microbial activity in the aquifer. Any appropriate gaseous or liquid substance, soluble in water, can be added in measured quantities while the groundwater passes the well casing. In this case the aquifer itself is used as a bioreactor. The two-phase flow of gas and water in the stripping zone ensures that the gaseous or liquid addition is mixed with the groundwater in an optimal way.

For in situ bioremediation of, e.g., BTEX or phenols, essentially oxygen is needed. The in situ stripping process with air of a UVB provides oxygen saturation. While the groundwater quantity captured by a well remains relatively constant, the quantity of circulating water around the UVB can be enhanced by a stronger pump in the well casing, which results in a faster rotation velocity of the circulation flow. Thus the amount of oxygen supplied for in situ bioremediation can be considerably increased. Carbon dioxide (and where denitrification is occurring, nitrogen, N_2) by-products are removed from the groundwater by the same stripping process, so biogas clogging in the aquifer will be avoided (Herrling et al., 1994). Since the hydrocarbons mentioned are generally lighter than water, a downward operating UVB as shown in Figure 2 is used. The contaminated air is cleaned by using, e.g., biofilters. A UVB

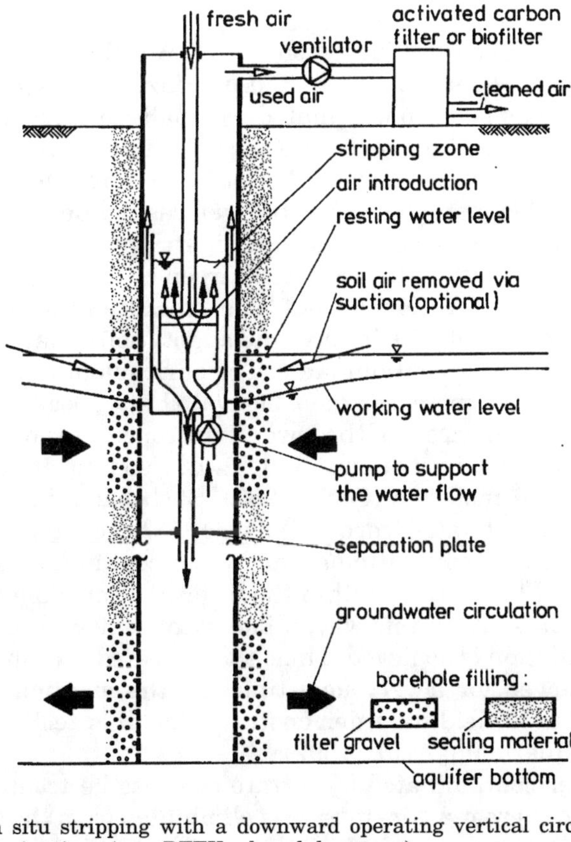

Figure 2. In situ stripping with a downward operating vertical circulation well for LNAPL contaminations (e.g., BTEX, phenol, kerosene).

technique that combines vapor extraction with the stripping process is generally used for those sites; thus, oxygen is supplied to the vapor zone and capillary fringe for biodegrading processes. Bae et al. (1995) tested the feasibility of a down-well gas-transfer apparatus for adding O_2, or other gases as well (such as methane for chlorinated aliphatic hydrocarbon cometabolism), directly into the recirculating groundwater within a downward operating groundwater circulation well and demonstrated the suitability of this well system for the addition of gas.

UVB WITH IN-WELL REACTOR

Generally, every vertical circulation can be combined with an in-well treatment unit between the lower and upper screen, e.g., a special bioreactor or other technologies for adsorbable contaminants (Figure 3). These in situ techniques can be directly combined with the in situ stripping for eliminating the produced biogas (e.g., CO_2, N_2) or for introducing oxygen.

A specially designed well based on the UVB technology of Figure 3 was utilized for in situ bioremediation of triazine. At a polluted site the subsurface has been contaminated by small leakages at a storage facility over a period of many years. The contaminated water, entering the well through the lower screen, is pumped through an activated carbon filter that is installed within the well casing. Before leaving the vertical circulation well, the groundwater is stripped and supplied with oxygen. The pesticides on the activated carbon are removed by adsorption when the well operation started. The exposure to triazines over a relatively long period of time induces the adaptation of natural groundwater bacteria to these contaminants. The enrichment of triazine onto the activated carbon increases their bioavailability, because the accumulation leads to an excess of the lower limit concentration for biological degradation. Certain strains of bacteria develop the ability to break down the triazine molecule (Cook, 1987). Such bacteria are also adsorbed onto the activated carbon particles. The contaminant flows through the filter by circulating groundwater containing oxygen. The bacteria proceed to multiply within the filter and consume the oxygen and triazines present. In this way, a biological remediation of the triazine contamination is achieved. Thus, the activated carbon unit needs not to be renewed for a very long time. At the mentioned polluted site, the measured field data demonstrate that this technique works successfully (Buermann et al., 1992).

Groundwater contaminated by nitrate can also be treated with an in-well unit. In these reactors nitrate is eliminated from the circulating groundwater in a rapid and efficient process that involves catalytic

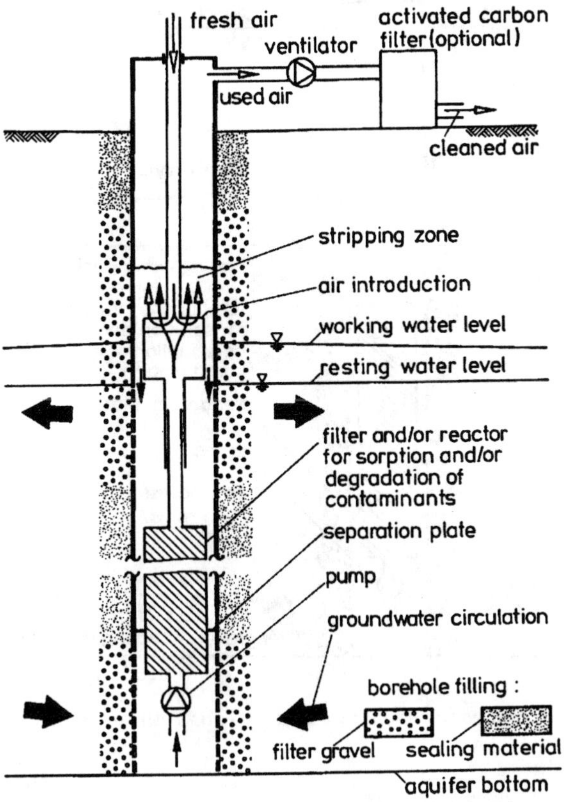

Figure 3. Vacuum vaporizer well (UVB) with in-well treatment reactor.

reduction by immobilized enzymes. The use of oxygen-independent reductases is necessary (no inhibition in present of oxygen). The reduction is driven by an electrical current and results in complete conversion of nitrate to N_2 without residues; the dissolved N_2 is finally removed via in situ stripping using the UVB technique. The development of the technique has been completed on a laboratory scale and is in preparation for field applications (Mellor et al., 1992; Herrling et al., 1994).

The GZB Technique

The plain vertical groundwater circulation well system is represented in the GZB technique (Figures 4a and 4b). In contrast to the above explained systems, there is no vacuum air stripping within the well

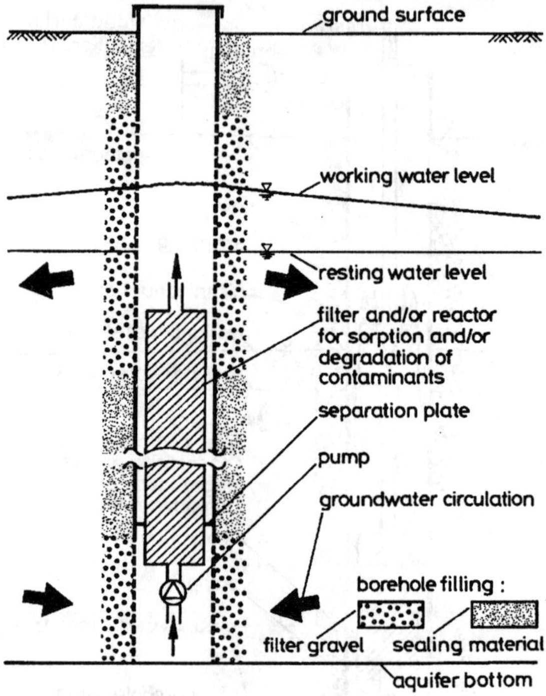

Figure 4a. Groundwater circulation well (GZB) with in-well treatment unit.

casing. Nevertheless, the hydraulic flow system, with respect to the extension and rotation velocity of the circulation cell, is the same. Figures 4a and 4b show that the vertical groundwater circulation can be combined with any appropriate in-well and/or on-site technique for groundwater remediation, e.g., the use of a catalyst for nitrate reduction (Figure 4a) or the elimination of dissolved heavy metals in an on-site reactor and groundwater reinfiltration in the same well (Figure 4b).

The SZB Technique

In the unsaturated zone bioremediation is affected by the soil moisture content. For most biological degradation processes, the optimal water content is in a range between 50% and 80%. When the vadose zone is contaminated with light nonaqueous phase liquids (LNAPLs, e.g., diesel, BTEX, mineral oil) (Schwille, 1984), the groundwater flushing circulation well (German abbr.: SZB) makes it possible to establish an adjustable, vertical, unsaturated multiphase flow in the contaminated region, demonstrated in Figure 5 (Stamm, 1995). The flushing water is pumped from the upper saturated groundwater zone and/or the capillary fringe to a reactor that is situated at the upper end of the well casing near the ground surface. The water enters the otherwise closed casing through a local screen at that height. At the reactor, the water is (optionally) stripped with air in a vacuum. Before the stripping, nutrients can be added for bioremediation processes or surfactants for an enhanced flushing process. The stripping with air leads to oxygen saturation of the water.

Before the water leaves the reactor zone, it flows through a zone with a special granulated material separating the LNAPL from the water; LNAPL remains on top of the granulated material and can be pumped off as free product from time to time. The water, optionally provided with nutrients or tensides, then leaves the upper well casing through a special infiltration zone near the ground surface. The vacuum

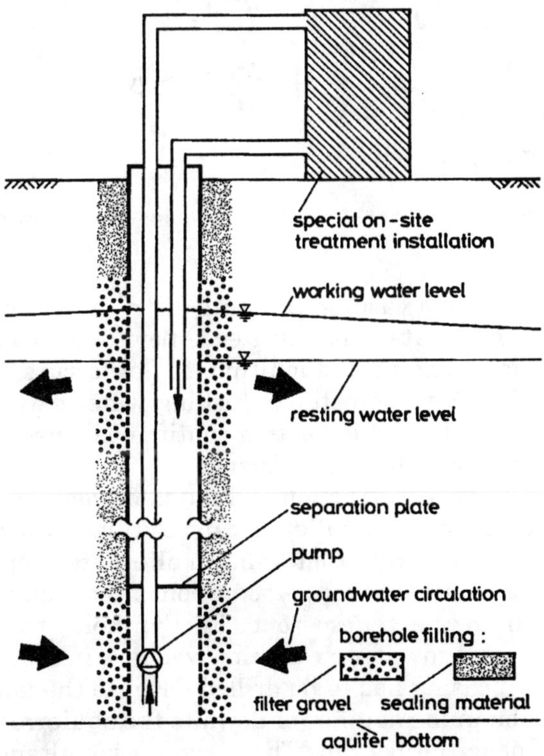

Figure 4b. Vertical circulation well (GZB) combined with on-site treatment installation.

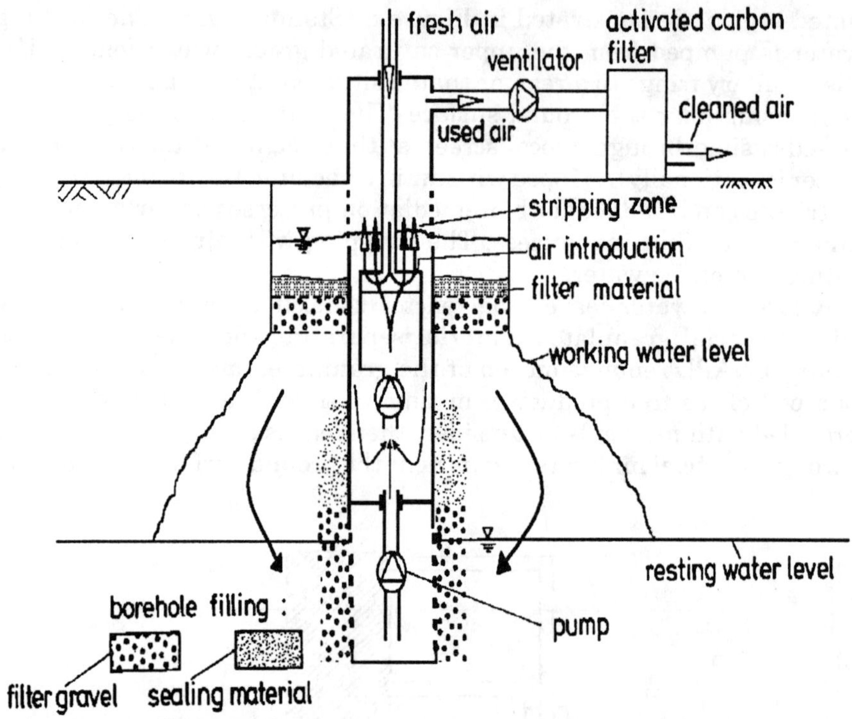

Figure 5. Multiphase (air–water) vertical circulation flow in the vadose zone.

in the reactor leads to a vacuum pressure distribution under the filter material, and this dictates unsaturated inflow conditions. Therefore, a vertical, circular, unsaturated multiphase flow occurs. The so-called dynamic water hill with adjustable flushing water saturation is produced around the well, and leads to a continuous supply of nutrients and oxygen within its sphere of influence.

Using the same well construction and the same vacuum pump (ventilator), the above explained installation can be utilized after some simple modifications as an air extraction well. An alternate operation may be advantageous for an accelerated remediation of the mentioned LNAPL contamination from the vadose zone. Furthermore, the SZB can be combined with a groundwater circulation well flow in the lower aquifer. For this, the well casing has to be drilled down to the lower datum of the contamination in the aquifer (or down to the aquitard) with a third screen section near the bottom. Then, even a simultaneous upward groundwater circulation operation (reverse mode as shown in Figure

2) is possible. The evaluation of the multiphase flow regime of the groundwater flushing well is a matter for future numerical investigations.

Special Arrangements

IN SITU TREATMENT WALL GENERATED BY VERTICAL CIRCULATION FLOWS

When a UVB or a GZB is situated in a contamination plume, the polluted upstream groundwater of quantity Q_0 is captured by the well. Here it is mixed with the circulating water influx to the total well discharge Q and treated by suitable in situ or on-site measures. Due to mass conservation, the same quantity Q_0 is released by the well flowing downgradient while the difference of cleaned water enters the circulation cell in the ratio $1 - Q_0/Q$. If the width of a plume is larger than the capture zone of a single well, several vertical circulation wells are arranged in one line normal to the natural groundwater flow (Figure 6). The distance between each system is determined in the way that no water can pass the well line without being caught and treated. The natural groundwater flow field is only locally influenced because no groundwater is extracted. Thus, a line of several wells represents a treatment wall, and no contamination coming from upstream can pass the line of wells, assuming the treatment efficiency is approximately 100%.

The aquifer zones that are involved in the remediation can be differentiated into the active and passive treated regions. Within the sphere of influence of the vertical circulation flow field, an active remediation takes place. The downgradient flowing outflux of the remediation well comprises the passive treated zone (e.g., oxygen enrichment). Depending on the contamination expansion, different well location scenarios are suitable (Figure 7).

MULTIPLE CIRCULATION CELLS IN ONE AQUIFER

The sphere of influence of a vertical circulation well depends on the distance of the two screen sections or, in the case of a fully penetrating well, on the aquifer thickness, H, as shown later. The larger the circulation cell is, the larger the travel time is for the circulation. Density-driven DNAPL migration through thick aquifers can result in more or less tall contaminant distribution over depth, first of all in aquifers with low natural gradient (Schwille, 1984). So, for thick aquifers ($H >$ 30m) it may be beneficial to install several circulation cells, one upon another. This produces smaller circulation cells and shortens circula-

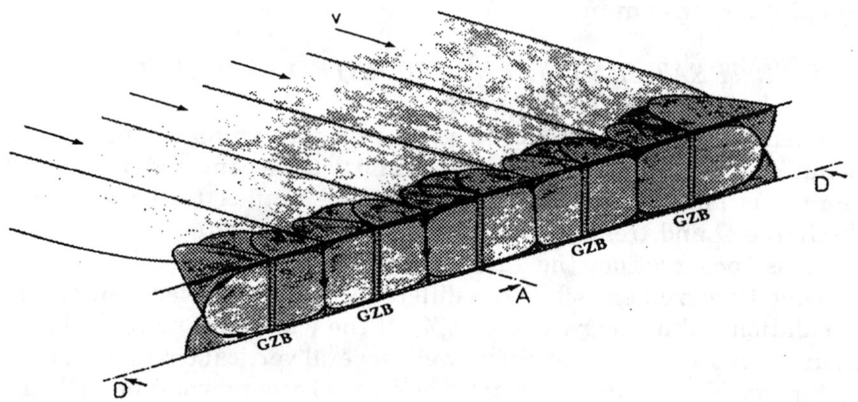

Figure 6. Vertical cross section through an in situ treatment wall with five vertical circulation wells.

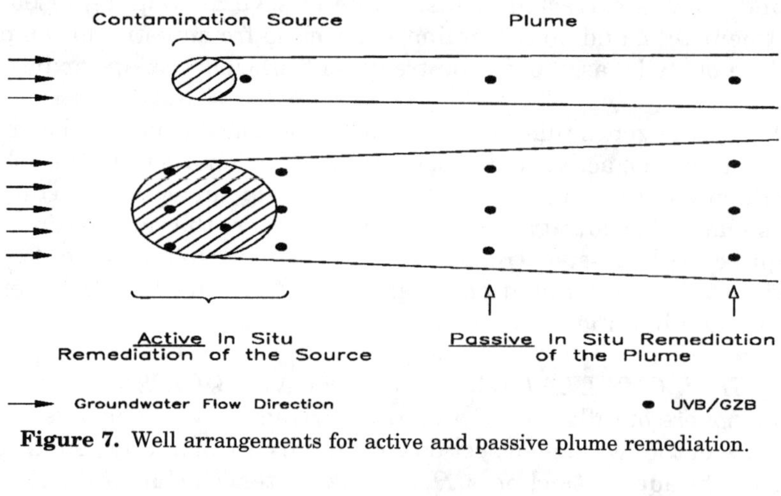

Figure 7. Well arrangements for active and passive plume remediation.

tion flow time. Figure 8 shows a system sketch with two different operation modes. In both demonstrated modes, the circulation flows are counterrotating. The hydraulic flow system can be combined with in situ stripping, an in-well treatment reactor, or any appropriate on-site technology.

The use of multiple circulation cells in one deep aquifer may also be advantageous when the hydrochemical properties of the soil and groundwater are drastically dependent on the aquifer depth. In that case, mixing groundwaters over a huge vertical range should be avoided.

PARTIAL EXTRACTION/INFILTRATION

The remediation of the contamination source should always have the highest priority in the cleanup strategy. When the pollution is present as free product, perhaps in the hot spot or on top of a local low-permeable region, the product recovery should be a high priority. The hydraulic gradient of the circulation flow field supports the movement of NAPL to the well. The NAPL influx into the remediation well can be pumped on-site with a separate pump. This leads to a partial extraction of the circulation well (Figure 9a).

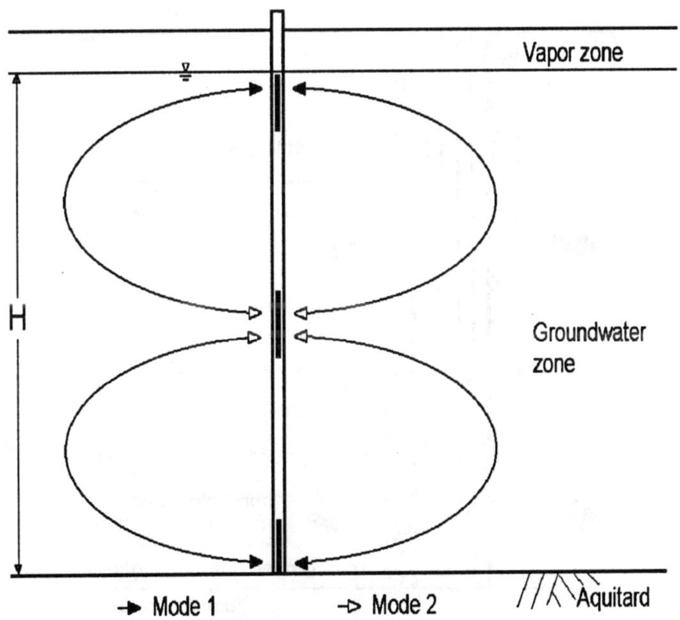

Figure 8. Multiple circulation well. Two different operation modes are sketched.

Furthermore, the vertical circulation well can be used as a pump or infiltration well for standard on-site remediation, too. In such a case, partial withdrawal or infiltration is taken from (Q_E) or added to (Q_I) the total volume of groundwater flowing through the well casing (Figure 9b). It is possible to extract or to infiltrate water without changing the groundwater head at the well top for a specific ratio of the extracted or infiltrated flow to the vertical circulation around the well (see Section 3.4). This has been numerically investigated for unconfined aquifer conditions by Herrling and Stamm (1992). At some distance from the well, the groundwater head at the aquifer top slightly deviates from the resting position. If the well capacity is low, pumping is continuous at a much higher rate, and the infiltration can be realized for a much greater quantity, even for a shallow distance between the surface and the groundwater table. While pumping, the turbulent mixing of air

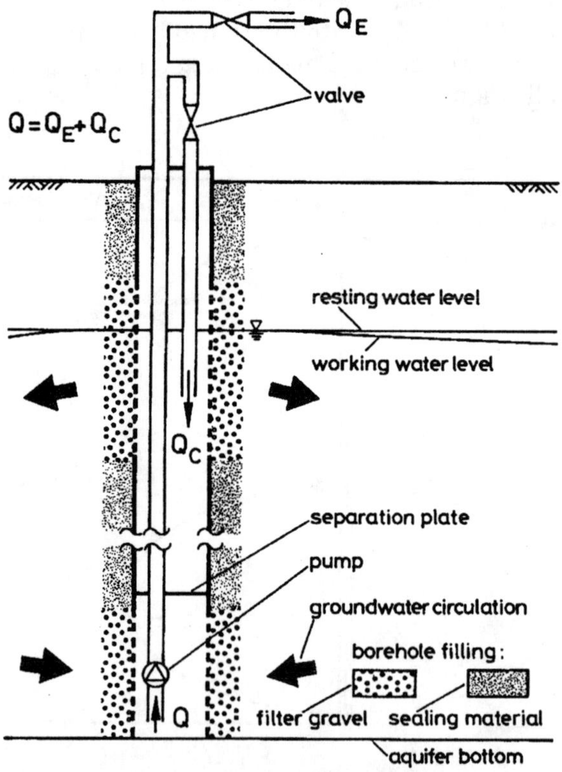

Figure 9a. Groundwater circulation well with partial extraction system.

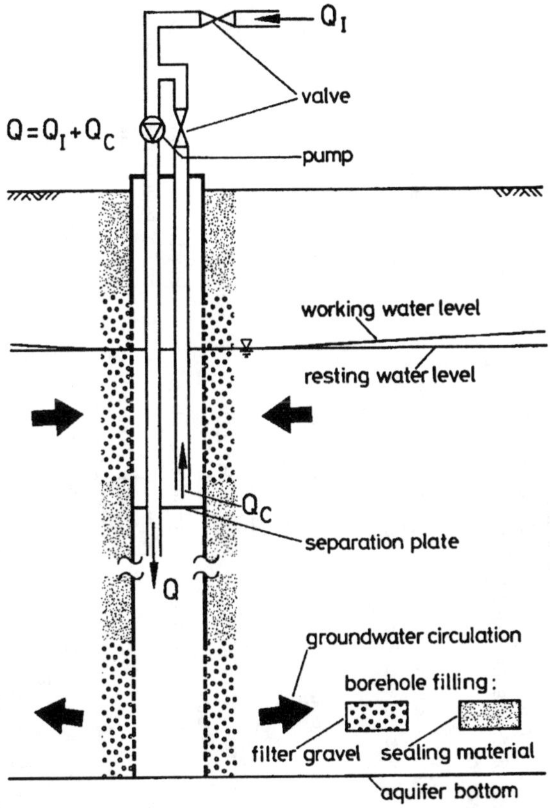

Figure 9b. Partial infiltration system.

and water in the artificial sand pack of the well construction, which often causes unwanted precipitation reactions, can be avoided.

COMPUTATION OF THE HYDRAULIC CIRCULATION FLOW FIELD

The radius of influence of the hydraulic circulation flow field depends on a large number of hydrogeological and remediation well parameters, e.g., horizontal and vertical hydraulic conductivity (K_H, K_V), aquifer thickness (H), natural flow gradient (v), total well discharge (Q), and screen section length (a). For practical realization, the user needs to have knowledge of the active circulation cell diameter and extent of passive reached aquifer zones (Figure 7), as well as information about

Table 1. Overview of vertical circulation flow applications.

Contaminants	Technique	Method	Special Features	Zone of Applicability
1. VOCs: DNAPL (e.g., PCT, TCE, VC)	upward operating UVB[a]	Physical in situ stripping plus aeration/biodegrad.	Vapor extraction soil–air treatment	Groundwater and vadose zone
LNAPL (e.g., BTEX, kerosene, phenol)	downward op. UVB	Physical in situ stripping plus aeration/biodegrad.	Vapor extraction soil–air treatment	Groundwater and vadose zone
2. Nonvolatiles: NAPL	UVB/GZB[b]	Aeration/biodegration	Addition of nutrients (gaseous or liquid)	Support of biological degradation within the aquifer
DNAPL (e.g., creosote, PCP)	UVB/GZB	Biodegradation with in situ bioreactor	Second pump system	Free product recovery
	SZB[c]	Air–water flushing in the vadose zone	Addition of nutrients	Biostimulation and degradation of residual contaminants
Other biodegradable substances	SZB		Combination with UVB possible	treatment of water saturated zone
Pesticides, Nitrate	UVB/GZB	Bioreactor: e.g., activated carbon, enzymatic biodegradation	In-well bioreactors	Water saturated zone anaerobic degradation

[a]UVB: vacuum vaporizer well.
[b]GZB: groundwater circulation well.
[c]SZB: groundwater flushing well.

the system's flow time characteristics. Numerical methods allow the evaluation of all these details. However, the computational effort can be minimized by recognizing through prior dimensional analysis the parameter combination that controls the system behavior. Furthermore, the dimensional analysis supports the development of universal diagrams for dimensioning vertical groundwater circulation wells. Nevertheless, the great scope of investigated variations necessitates a fast algorithm for the evaluation of the complex three-dimensional flow field (Herrling et al., 1990; Herrling and Stamm, 1992). This requires the use of some suitable simplifications and assumptions.

The most essential simplification is that only confined aquifer conditions have been investigated. This approximation results in an error on the order of 1% concerning the radius of influence (MacDonald and Kitanidis, 1993). For confined conditions, the circulation flow around each well can be calculated for the special case of radial symmetry, and the total field of several, mutually influenced wells together with the natural horizontal groundwater flow field can be calculated by superposition of the simple single systems. However, each individual flow system must have the same aquifer thickness, and local hydraulic properties must be radial symmetric. Anisotropy and horizontal layers can exist. The calculations are carried out for steady-state conditions, and only convective transport is considered for estimating the capture zone. Furthermore, for the computation of the flow field around a vertical circulation well, it is assumed that the local subatmospheric pressure field and density effects can be neglected.

The radial symmetric calculation is based on the equation

$$\frac{\partial}{\partial r}\left(2\pi r k_r \frac{\partial h}{\partial r}\right) + \frac{\partial}{\partial z}\left(2\pi r k_z \frac{\partial h}{\partial z}\right) = 0 \tag{1}$$

formulated in cylindrical coordinates r and z (Bear, 1972). h defines the piezometric head, and k_r and k_z the anisotropic hydraulic conductivities. Dirichlet and Neumann boundary conditions are used

$$h - \bar{h} = 0 \text{ and } v_i n_i + \bar{v} = 0 \tag{2}$$

The bar denotes prescribed values, v_i the velocity vector and n_i the unit vector normal to the boundary. For the extraction or infiltration well, the boundary condition at the outer borderline is described by a flux condition

$$\bar{v} = \lambda(\bar{h} - h) \tag{3}$$

[see Equation (2)]. This artificially enlarges the model area to reduce the influence of the boundary on the results. λ defines the quotient of the horizontal hydraulic conductivity and the distance, L, between the required and the real boundary (K_H/L). $\bar{h}$ defines the groundwater head at the required boundary (resting water level).

The numerical computation of the radial symmetric flow has been performed using a Galerkin finite element method with linear shape functions and triangular elements (Zienkiewicz, 1977; Pinder and Gray, 1977). The above-mentioned superposition of the different flow fields has been realized on a rectangular three-dimensional discretization with variable grid distances by interpolating and adding the respective velocity vectors. Thus, complex 3D flow fields, e.g., defined on 200,000 grid points, can be computed simply, quickly, and accurately. A particle tracking method has been used to calculate the streamlines.

Current measurements of the active (circulation cell) and the passive treated zones around a vertical circulation well at the demonstration site in Karlsruhe-Knielingen, Germany, confirmed the numerical results.

Sphere of Influence and Capture Zone of a UVB/GZB

The general character of a standard vertical circulaton flow is demonstrated in Figure 10. The streamlines represent the flow regime in a longitudinal section parallel to the natural groundwater flow. It is evident that with decreasing natural flow gradient, the region of circulation increases. In absence of natural groundwater flow, the theoretical sphere of influence is infinite. Therefore, for practical use a radius R is defined within which a certain amount (Q_R) of the well discharge (Q) circulates. In the presence of a natural gradient, the circulation flow field is restricted by a stagnation point [see Figure 10(b)]. Figure 10(d) shows an in-line arrangement of several vertical circulation wells. The strong vertical circulation flow between the two wells, which is extremely beneficial in a highly polluted area near the contamination source, is evident. Two wells aligned in this manner can be used to generate an anaerobic biodegradation zone around the first circulation well (e.g., GZB without air stripping and, if necessary, with a carbon source added within the well casing) and an aerobic zone around the second well (e.g., created by a UVB). With this arrangement, an adaptation of the oxidic conditions of the contaminant-dependent suitable biodegradation processes can be achieved. Sequential anaerobic/aerobic biological treatment can mineralize halogenated organic compounds that are difficult to degrade under a single set of environmental conditions.

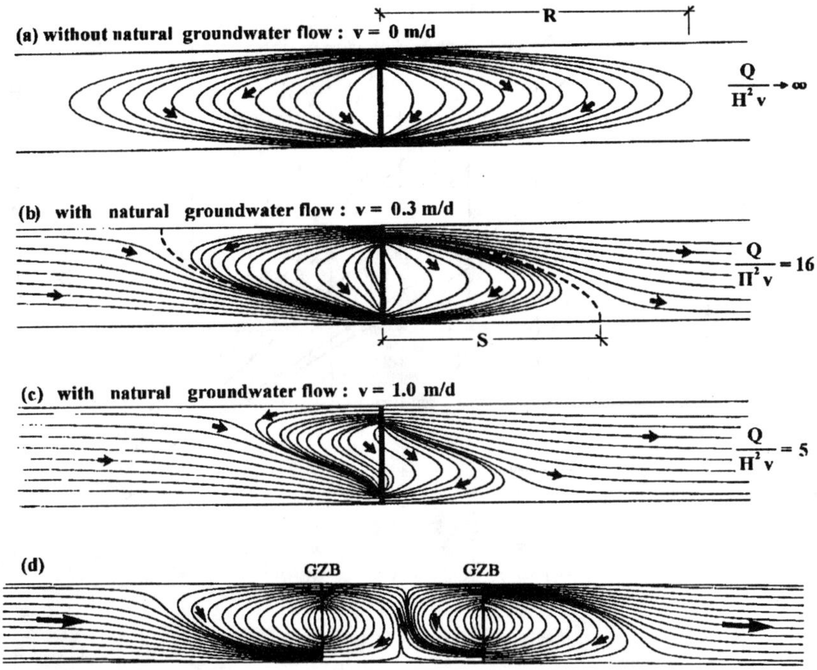

Figure 10. Streamlines in a longitudinal section for different gradients: (a) 0.0 m/d, (b) 0.3 m/d, and (c) 1.0 m/d; (d) two GZB in line. K_H/K_V = 10.

In Figure 11(a), results for the radius of influence (R) are presented for the ratios Q_R/Q = 0.98 and 0.8, in a dimensionless diagram. The radius of influence is independent of the discharge through the well, but dependent on the anisotropy K_H/K_V (here shown for anisotropies up to 10). Under normal circumstances, the length of the screened sections has only a small influence on the radius of influence. Nevertheless, a totally screened well casing should be avoided because of the possibility of hydraulic short-circuiting.

Figure 11(b) presents a dimensionless diagram that describes the differences (Δh) between the hydraulic heads at the top and bottom of a vertical circulation well through which a discharge (Q) is pumped. Δh, or the dimensionless parameter $\Delta h/H$, is dependent on the parameter $Q/(H^2 K_H)$ and the ratios K_H/K_V and a/H. For screens arranged symmetrically, the rise of the hydraulic head at the top of the well amounts to $\Delta h/2$, and the decrease is $-\Delta h/2$ at the bottom (both referring to the position of rest). When K_H is known (e.g., by pump test)—along with H, Q, and a—Figure 11(b) and the measured Δh allow an estimate of

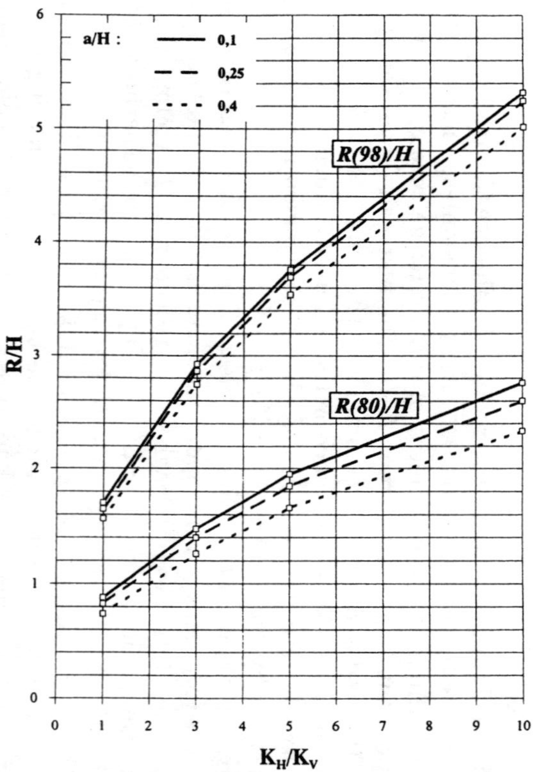

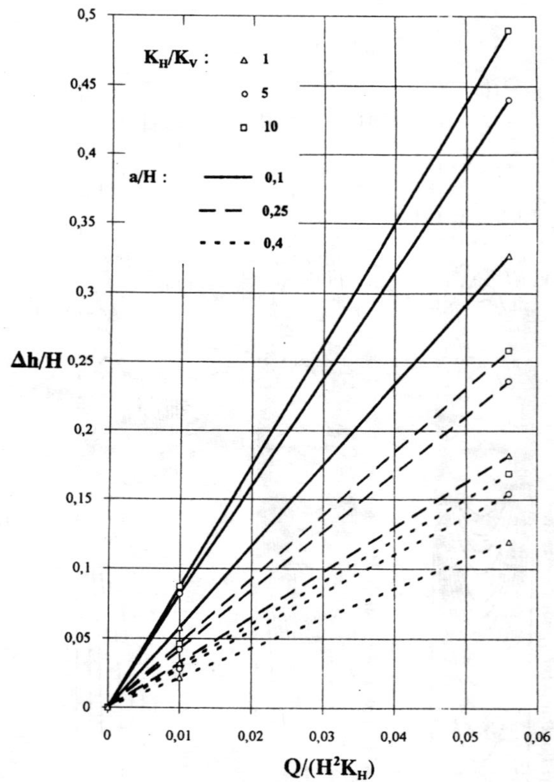

Figure 11. (a) Sphere of influence (R) for different ratios Q_R/Q, and (b) differences (Δh) of the hydraulic heads between upper and lower screen section.

the anisotropy at a specific site. An alternative analytical approach for the determination of the anisotropy using these data is shown in Kabala (1993). In the literature the values of anisotropy vary largely. Bear (1972) declares the spectrum of anisotropy from 1 to 42, Marsily (1986) from 1 to 100, and Busch et al. (1993) from 2 to 10.

Figure 12 exhibits a spatial view of the separating stream surfaces of all nine water bodies in connection with the flow around three vertical circulation wells in a line perpendicular to the natural gradient. The natural groundwater flow comes from the left side. In Figure 12(b) the water bodies are separated for clarification. At the left front side, the separating stream surfaces of the contaminated groundwater captured by the remediation wells can be seen. In the center the water bodies are shown, which consists of treated groundwater and demonstrates the circulation flow region around each GZB unit. At the right side of Figure 12, the separating stream surfaces of the cleaned groundwater flowing downstream are displayed.

The characteristic shape of the capture zone in an upstream cross section of a vertical circulation well in standard mode and its notations are shown in Figure 13. Using these notations, Figure 14 enables the dimensioning of a remediation installation. In Figure 14(a), the horizontal distance (S) of the stagnation point from the well axis is described. The ratio S/H is dependent on the parameters $Q/(H^2v)$, K_H/K_V, and a/H. The location of the stagnation point is highly sensitive to the anisotropy of the aquifer. The length of the screen section is of small importance within usual proportions. The knowledge of the distance (S) to the stagnation point can be used to determine the positioning of the measuring equipment. The operation of a GZB can also be monitored

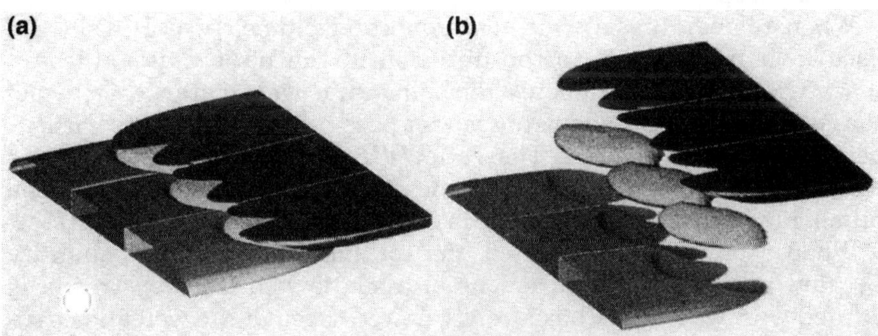

Figure 12. Separating stream surfaces of the different water bodies around three UVB/GZB units: captured, circulating, and flowing downstream water in (a) a real situation, and (b) separated for clarification.

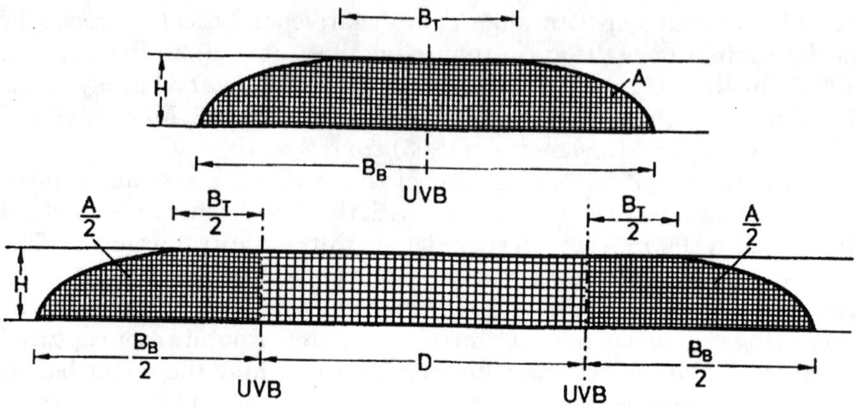

Figure 13. Notations in an upstream cross section through the capture zone for one and two GZB installations.

using depth-dependent measurements between the stagnation point and the well.

The results of Figure 14(b)–(d) have been calculated for an upstream distance of $5H$ from the well and for a constant ratio of $a/H = 0.25$.The results are discussed for wells that pump upward. The widths B_T and B_B of the upstream zone, measured at the aquifer top and bottom, are shown in Figure 14(b). The ratios B_T/H and B_B/H are again dependent on the ratios $Q/(H^2v)$, K_H/K_V, and a/H. For small values of $Q/(H^2v)$, the upper part of the capture zone does not reach the top of the aquifer. This implies that, for remediating a plume, a minimum well discharge (Q) is required. Again, the results are quite sensitive to the degree of the anisotropy.

When remediating a wide contamination plume, several wells are used in a line normal to the direction of the natural groundwater flow. The length (D) denotes the maximum well distance at which the contaminated groundwater cannot pass between the wells without being cleaned or treated. The ratio D/H is dependent on the same parameters as before. When a plume of width W is to be cleaned, the number (n) of well installations can be estimated by $n = (W - B_T)/D + 1$.

When a plume is remediated, the contaminated water of quantity Q_0, flowing into the capture zone of a single well from upstream, is diluted with water that has already flowed through the well and circulates around it. Thus, the contaminant concentration of the water within the well casing will be lower than in the upstream plume; near a contamination source the situation is reversed. Figure 14(d) illus-

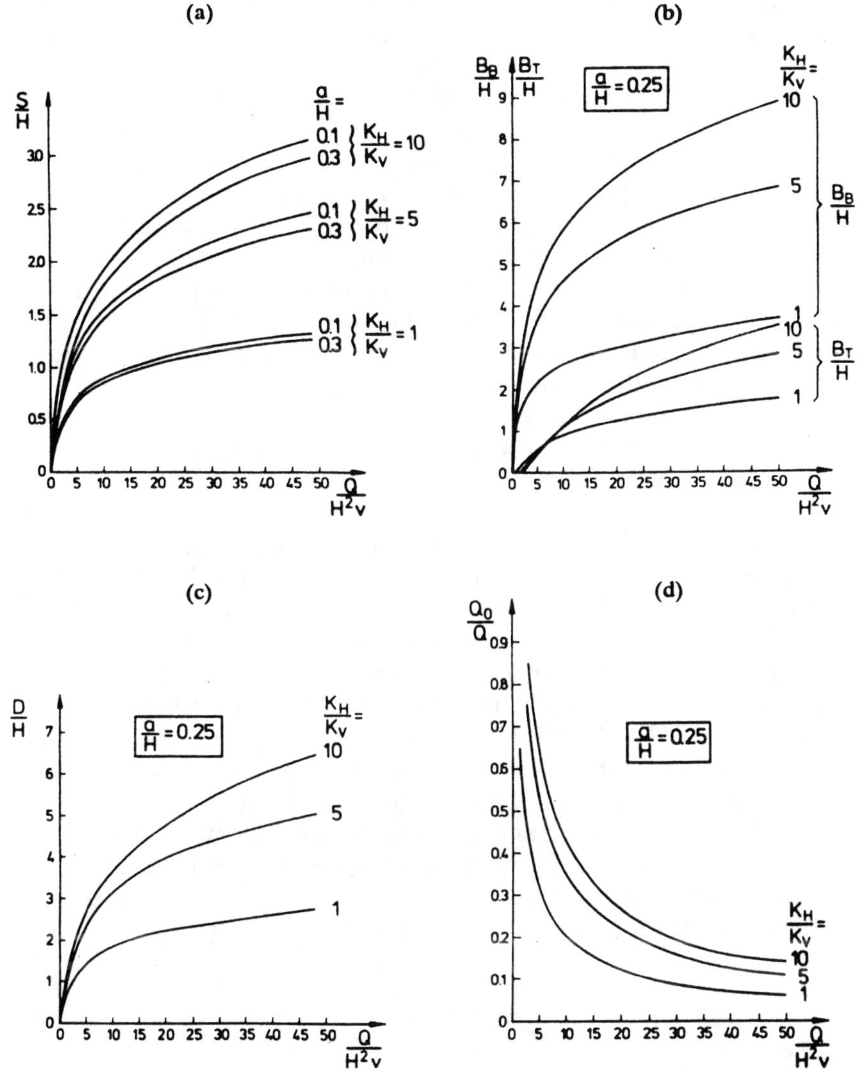

Figure 14. (a) Distances (S) of the stagnation point from the well axis; (b) widths (B_T and B_B) of the upstream capture zone at the aquifer top and bottom; (c) maximum well distance (D) at which the contaminated groundwater cannot pass between the wells without being treated; (d) upstream discharge (Q_0) in the capture zone, which is diluted with the circulating water to the total well discharge (Q).

trates the part Q_0 of the total well discharge Q. The ratio Q_0/Q is again dependent on the same parameters as the widths of the upstream capture zone. Figure 14(d) can be used to estimate the expected contaminant concentration in water within the well casing for the dimensioning of a well installation. It may help to evaluate the progress of remediation at a site when concentration data of the upstream plume and the water within the well are determined.

The sphere of influence of the circulation cell around a GZB at sites with natural groundwater flow is of special interest. In the direction of the natural gradient, the influence has a maximum expansion of S to the upstream and downstream sides, as can be seen in Figures 10 and 14(a). Normal to this direction, the maximum radius of the sphere of circulation is approximated by $(B_B + B_T)/4$ [Figure 14(b)], and in the case of several wells in one line by $D/2$ [Figure 14(c)].

Residence Time Characteristics in the Vicinity of the Recirculation Well

A knowledge of the velocity distribution within the sphere of influence of the circulation system has primary significance for the estimation of the groundwater flushing intensity at any distance from the circulation center. Consequently, it allows a first estimate of the nutrient and/or electron acceptor supply situation in the aquifer as a result of an in-well nutrient addition or the oxygenation of the treated water.

The flow time T_{Zirk} that a fluid element needs to make one circulation (from the outflow to the inflow screen of the remediation well) depends strongly on the streamline followed and maximum distance (r) from the well that the element reaches. Furthermore, the total well discharge Q, the aquifer thickness H, the anisotropy K_H/K_V, and the porosity of the soil matrix n have important influences on T_{Zirk}. The functional relationship can be expressed in a dimensionless form as shown in Equation (4):

$$\frac{T_{Zirk} \cdot Q \cdot n}{H^3} = f\left(\frac{r}{H}, \frac{K_H}{K_V}, \frac{a}{H}\right) \tag{4}$$

Figure 15 presents the diagrams for the time dimensioning for a fixed screen length ratio $a/H = 0.25$, applicable for the case that there is no natural groundwater gradient. For any given circulation radius r, aquifer thickness H, well discharge Q, and porosity n, the required travel time T_{Zirk} can be calculated [Figure 15(a)]. It is obvious that, with increasing well distance r, the travel time increases logarithmically, too. From this diagram, the number of circulations N for any radius

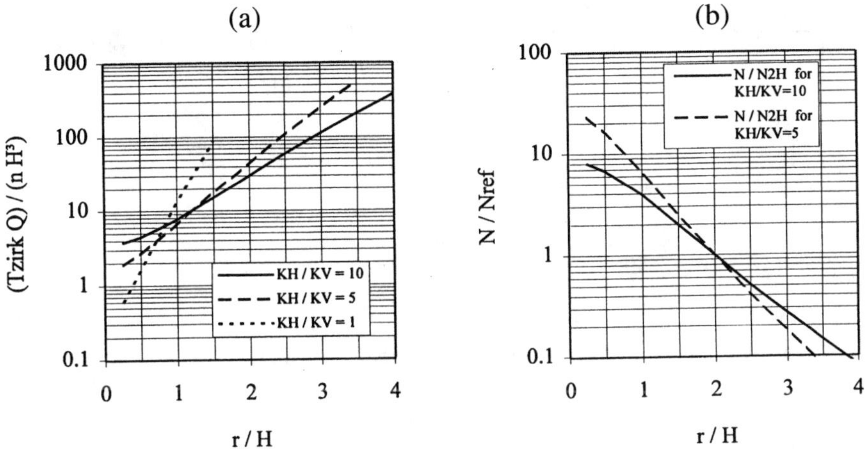

Figure 15. (a) Dimensionless circulation time versus the circulation radius for the anisotropies 5 and 10 and (b) number of circulations N in relation to one reference circulation N_{ref} in the distance $2H$.

with respect to a reference circulation N_{ref} in a given reference radius can be easily derived [Figure 15(b)]. Here, the reference radius of $2H$ (two times the aquifer thickness) was used. For example, for an anisotropy of 10, the groundwater circulates in the well distance $r = 1H$ approximately five times faster than in a distance of $2H$. It is possible to calculate the time for a single reference circulation, and afterwards any other circulation time can be simply determined using Figure 15(b).

In the presence of natural groundwater flow, a stagnation point occurs and limits the maximum circulation extension. Even in this case, the diagrams can be used under the restriction that no values for r/H greater than S/H should be considered. For the practical usage of these diagrams for sites with natural groundwater flow, the introduced additional uncertainty as a consequence of this simplification is negligible relative to the uncertainty concerning the assumption of homogeneous conditions for natural aquifers (Schäfer and Kinzelbach, 1991). Figure 16 demonstrates the characteristic shape of isochrones (equidistant lines in time) in a longitudinal section through the upstream captured water body and the circulation flow of a GZB. The impedance of the groundwater flow close to the stagnation point is visualized by the accumulation of isochrones in that region. The closer the water is to the lower influx screen, the faster it flows and the larger is the distance of isochrones.

The associated isochrones in a cross section through the capture zone in an upstream well distance of $5H$ are presented in Figure 17. The

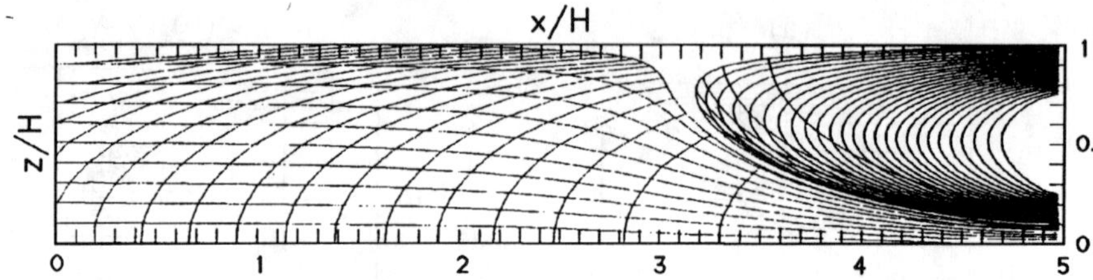

Figure 16. Dimensionless isochrones in a longitudinal section through a vertical circulation well with $\Delta t = 0.864\ H^3/Q$ (for $a/H = 0.25$; $Q/H^2v = 25$; $K_H/K_V = 5$).

evaluated case study considers a natural groundwater flow velocity of $v = 0.173$ m/d (Darcy), which yields a genuine flow time of 43.4 days for 50 m ($H = 10$ m) and a porosity of $n = 0.15$. Due to the vertical circulation flow ($Q = 18\ m^3/h$), the required flow time in the level of the lower influx screen decreases, while it increases in the horizon of the upper screen section, because here the water passes the stagnation point.

At this point, it should be mentioned that the evaluation of these diagrams take only conservative transport processes into account. Due to biodegradation processes that are accompanied by nutrient consumption, this diagram overestimates the remediation progress. The introduction of a retardation factor RE enables the consideration of nutrient consumption in a very simple way. RE reflects a linear delay in progress of the biodegradation and coupled nutrient front (McCaulou et al., 1995). The retardation factor doesn't affect the shape of the isochrones, but the isochrone values increase. The faster the biodegradation process is, the faster the contaminant consumption is and the lower is the value of the isochrone. The retardation factor itself is a function of the total contamination mass. The consideration of diffusion-limited exchange between the mobile pore water and immobile biomass, as well as the kinetic rated transfer between the adsorbed and dissolved organic car-

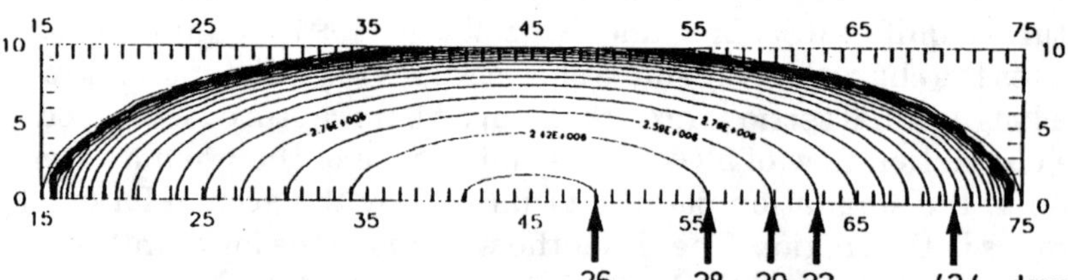

Figure 17. Isochrones in a cross section through the capture zone. Case study for: $K_H/K_V = 5$, $a/H = 0.25$, $n = 0.15$, $Q = 18\ m^3/h$, $H = 10$ m, $v = 0.173$ m/d, well distance $5H$.

bon within the aquifer, would lead to a further delay in remediation progress (Baveye and Valocchi, 1989; Kinzelbach et al., 1991; Lensing et al., 1994).

Vertical Circulation Flows in Stratified Aquifers

The diagrams in Figures 11 and 14 for the dimensioning of the vertical circulation remediation technique are based on the assumption that homogeneous aquifer conditions occur. However, aquifers with a more or less distinct stratification are widespread. In the following, the influence of a vertical variation of the hydraulic conductivity will be discussed.

For the investigations, a numerical model test aquifer with seven equidistant horizontal layers was employed. The layers are denoted in order of increasing distance from the middle of the aquifer (see Figure 18). The center layer is denoted position 0; the neighboring layers, position 1; and the outermost, position 3. Each layer has a thickness d that is equivalent to a seventh of the aquifer thickness H and the layer anisotropy, K_H/K_V, is equal for every layer. The model was used to determine the influence of a single layer in various positions and with various permeabilities. The parameter α is introduced to define the ratio between the hydraulic permeability of the layer and of the entire aquifer.

Figure 19 demonstrates the effect of layer position and ratio of permeability α on the effective anisotropy of the whole system. In absence of natural groundwater flow, the sphere of influence of a GZB with a centric layer (d = H/7) that is 100 times less permeable than the ambient aquifer (α = 1/100, and the layer anisotropy is five) is identical to the radius of influence in a homogeneous aquifer with an anisotropy of $K_H/$

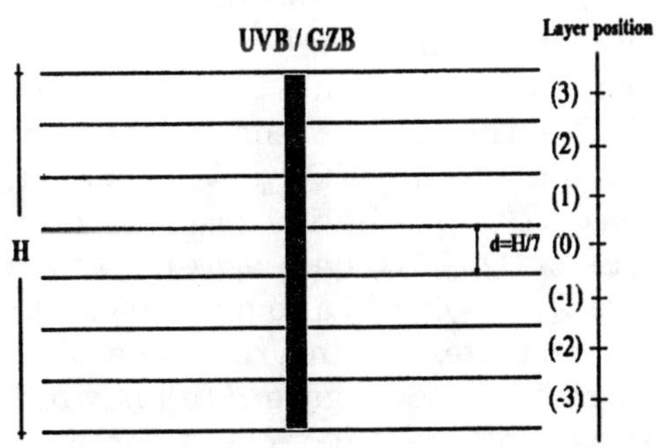

Figure 18. Structure of the heterogeneous model aquifer with the used notations.

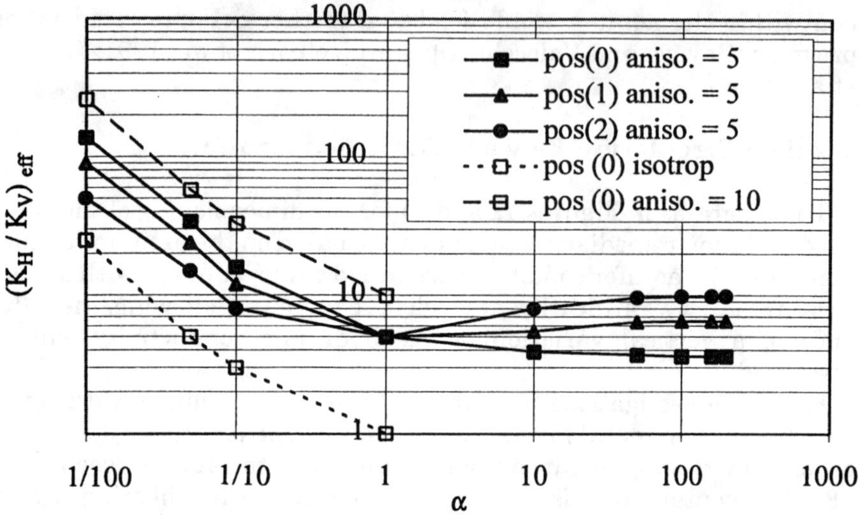

Figure 19. Effective anisotropy of an aquifer in dependency of the position and permeability ratio of an intermediate layer.

K_V = 140. And as the anisotropy increases, the radius of influence of a GZB increases, too. $\alpha < 1$ leads to an extreme enlarging of effective anisotropy, while for $\alpha > 1$ changes are negligible. Furthermore, it is obvious, the closer the layer is in the middle of the aquifer (position 0), the higher is the influence on the circulation flow regime. The lower short dashed line demonstrates the influence of a centric layer to the effective anisotropy for the case that every single layer is isotropic. The upper dashed line is valid for a layer anisotropy of ten. The results show that the effective anisotropy is proportional to the layer anisotropy.

A consequence of increasing effective anisotropy, accompanied by enlarging circulation cell diameters, is a decrease in circulation velocity. Figure 20(a) shows in a dimensionless form the dependence of the relative circulation quantity within a certain travel time (Q_{Tzirk}/Q) on the anisotropy (K_H/K_V). This diagram enables the percentage of circulating water within a definite time (Q_{Tzirk}/Q) of the total well discharge Q to be determined. The qualitative result is that with increasing anisotropy the ratio of (Q_{Tzirk}/Q) decreases for a fixed time.

The knowledge of Q_{Tzirk}/Q, or in a common form Q_r/Q (Q_r defines the portion of Q, which circulates within a radius r around the well), allows the corresponding radius r [see Figure 20(b)] in which this quantity of Q, circulates during the time T_{zirk} to be determined. The calculated results for high anisotropies (dashed and dotted lines) would be more accurate for a more discrete numerical grid. The importance of these data is clear, because now it is possible to approximate the supply of nutrients and/or electron acceptors over time and location within the circulation cell.

For the case where natural groundwater flow is present, there are also diagrams available, but their presentation would exceed this contribution. Nevertheless, Figures 11 and 13 enable the extrapolation of the dimensioning parameters, using the effective anisotropy of Figure 19.

Dimensioning of the Partial Extraction/Infiltration System

When the nonaqueous phase liquid (NAPL) contamination in the subsurface exists not only in the dissolved form, but also as NAPL pool, e.g., at the hot spot or on top of a local, low-permeable stratigraphic unit (Waterloo Centre for Groundwater Research, 1989), the NAPL recovery should be the first remediation step. For this, the vertical circulation well can be combined with a partial extraction.

Presently, numerical results are only available for the case of absence of natural groundwater flow and for confined aquifer conditions. When a GZB (Figure 9a) is used to split the vertical well discharge *(Q)* into a quantity (Q_E) that is extracted, and an amount (Q_C) that is reinfiltrated through the upper screen section and that generates a circulation flow,

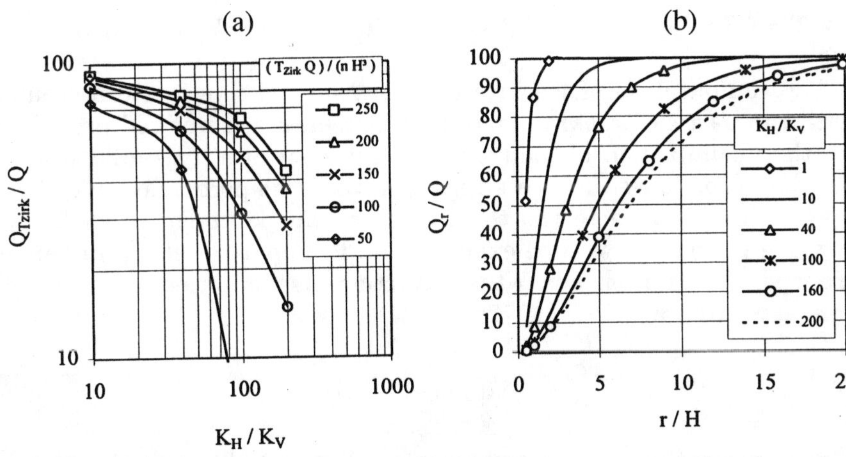

Figure 20. (a) Ratio of circulation quantity within a definite time (Q_{Tzirk}/Q) vs. effective anisotropy (K_H/K_V) and (b) ratio of circulation quantity within a definite radius (Q_r/Q) vs. relative well distance (r/H) for several K_H/K_V.

the opportunity exists for avoiding any change of heads at the well top for a special ratio of Q_E/Q_C. This ratio is only dependent on K_H/K_V and a/H. Figure 21 demonstrates the results for $a/H = 0.25$. The value Q_E can be about two to three times as large as Q_C, depending on the aquifer anisotropy. For this special ratio of Q_E/Q_C, the decrease of the head at the well bottom (no change of the heads at the well top) is Δh [Figure 21(b)]. In a dimensionless description, $\Delta h/H$ is dependent on the parameters $Q_E/(H^2K_H)$, K_H/K_V, and a/H.

In case of infiltration of a quantity (Q_I) in a GZB (Figure 9b), the ratio Q_I/Q_C can be taken from the diagram of Figure 21(a) as well, when no change of heads are demanded at the well top. The increase of head Δh at the well bottom can be estimated from Figure 21(b).

At some distance *(r)* from the well axis, a maximum head deviation *(f)* from the resting water level will result at the aquifer top. The dimensionless parameter r/H is dependent on K_H/K_V and a/H, and f/H (negative for groundwater extraction and positive for infiltration) depends on $Q_E/(H^2K_H)$, K_H/K_V and a/H. Figures 21(c) and 21(d) represent results for $a/H = 0.25$.

When a groundwater extraction Q_E (or Q_I) is prescribed, the resulting changes of heads at the bottom and top of a GZB can be calculated for any Q by using the diagrams of Figure 11(b), 21(a), and 21(b) as individual results can be superimposed. Another case is to prescribe Q_E (or Q_I) and a maximum change of head at the bottom or top of a GZB and to estimate Q. In both cases the values of H, K_H, and K_V must be known.

THE TRANSPORT CHARACTERISTICS OF A UVB/GZB

The computation of isochrones, based on the flow pattern created with the particle tracking method, represents a simplified procedure for the visualization of the transport behavior of a hydraulic system. The particles move along a streamline with the pore velocity v_e. This approach only characterizes the advective transport term. In the following, their influence will be examined and compared with a numerical transport simulation using the advection-dispersion equation, which can be written as (Bear, 1972):

$$\frac{\partial c_T}{\partial t} = -\nabla(\vec{v}_e \cdot c_T) + \nabla(D \cdot \nabla c_T) \qquad (5)$$

c_T denotes the tracer concentration, ∇ the Nabla-operator, D the tensor of hydrodynamic dispersion, and t the simulation time. The exemplary simulation of a continuous conservative tracer addition

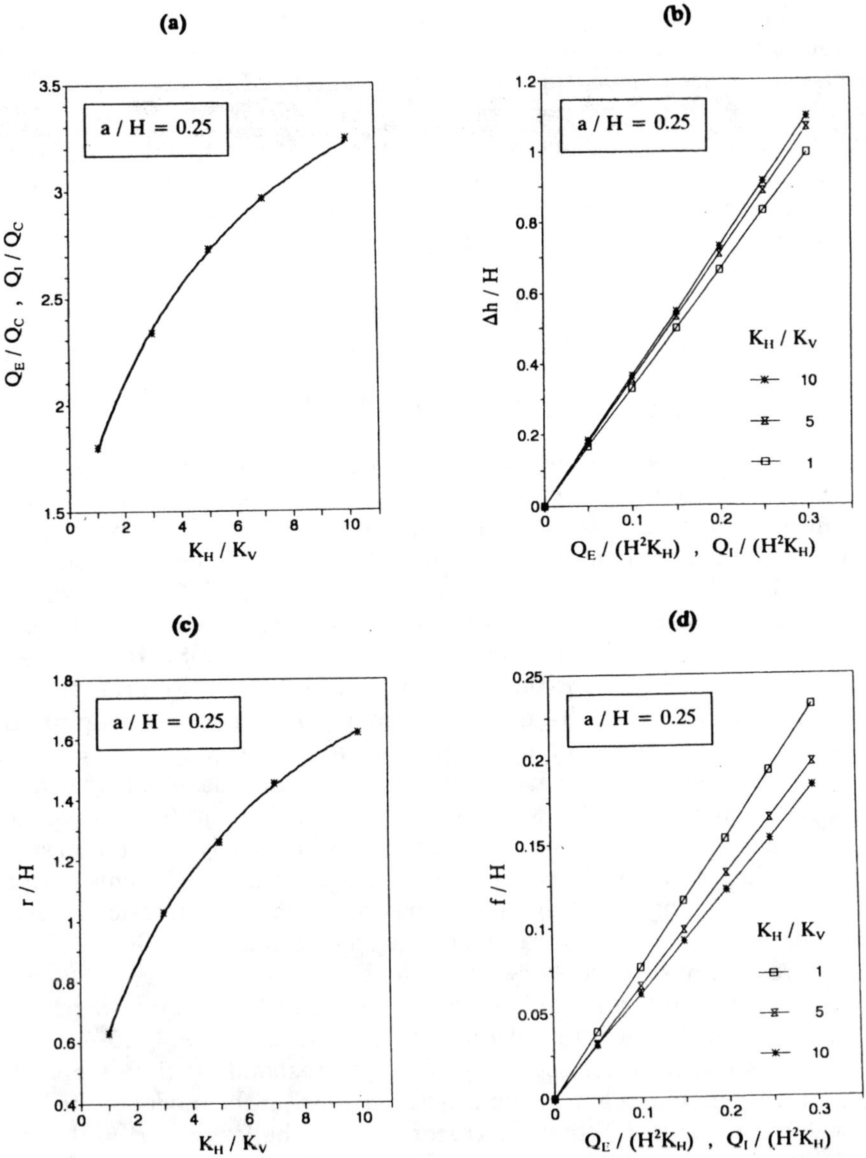

Figure 21. (a) Ratio Q_E/Q_C and Q_I/Q_C for no head change at the well top, (b) head changes Δh at the well bottom, (c) distance r from the well axis with maximum head deviation f, and (d) maximum head deviation f.

within a standard vertical circulation well was performed assuming a radial symmetric, resting aquifer. For this, the computational effort can be reduced dramatically in comparison to a three-dimensional simulation. Within the upper well casing, the tracer input is realized using a Dirichlet boundary condition (Γ_1). The restriction that no mass flux should occur at the closed model borders (Γ_2) is achieved employing a Neumann B.C. Further, the initial concentration c_0 has to be prescribed onto the whole calculation domain Σ.

$$
\begin{aligned}
c\ (r,z,t) &= c_\Gamma(r,z,t) && \text{onto } \Gamma_1 \text{ for } t>t_0 \\
\frac{\mathrm{P}c}{\mathrm{P}n} &= 0 && \text{onto } \Gamma_2 \text{ for } t>t_0 \\
c\ (r,z,t) &= c_0(r,z,t_0) && \text{onto } \Sigma
\end{aligned}
\tag{6}
$$

The simulated aquifer has a length of 83 m and a depth of 6.6 m. The hydraulic conductivities are $K_H = 5.6*10^{-4}$ m/s in the horizontal and $K_V = 5.6*10^{-5}$m/s in the vertical direction, and the porosity is n = 0.18. The screen length is 1 m both at the top and bottom of the well. The flow field is simulated for a well discharge of Q = 4 m^3/h. The prescribed constant influx tracer concentration in the upper well casing is $c_T(t)=c(\Gamma_1) = 1$ and the initial condition $c_T(t_0) = c_0 = 0$. So, the presented isolines mark concentration differences of 0.1 each. Figures 22(a)–(c) show a sequence of the tracer transport progress for a longitudinal dispersivity of α_L = 3 m and a transversal dispersivity of α_T = 0.3 m.

In contrast to the horizontal flow from a fully penetrating conventional infiltration well, which affects the entire aquifer, the vertical circulation well only has a local influence. At the early simulation stage, the concentration isolines are rather perpendicular to the flow pattern. With increasing time, this configuration changes dramatically. After 24 hours [Figure 22(b)] the tracer concentration in the lower part of the well is approximately $c_T = 0.25$, and the shape of the $c_T = 0.2$ tracer isoline close to the lower screen is more parallel to the flow pattern. In this simulation interval the tracer isolines for c_T = 0.5 are quite congruent with the isochrones of the corresponding time step. This affinity remains as long as no tracer "outflow" of the model in the lower well casing occurs. When the tracer reaches the lower part of the well, driven by the circulation flow in the aquifer, the total mass influx per time unit decreases. From that point in time, the isochrones and isolines of concentration are no longer compatible.

The values of dispersivity α_L and α_T determine the length of the tracer front. The larger the dispersivity, the longer is the concentration gradient front. After 11 weeks the $c_T = 0.1$ isoline reaches a maximum

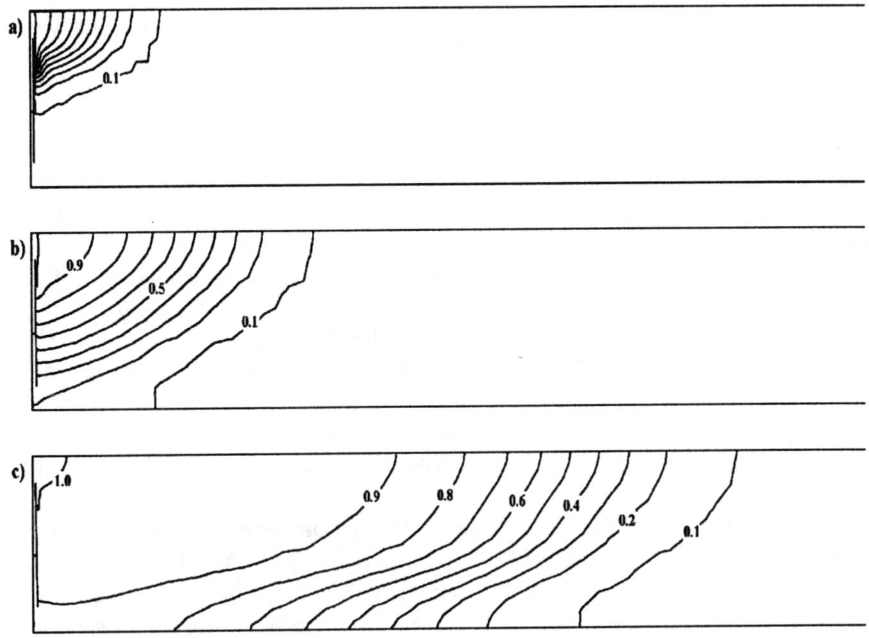

Figure 22. Sequence of radially symmetric tracer concentration for different time steps: (a) t = 90 min, (b) t = 24 h, and (c) t = 11 weeks.

extension at the aquifer top, which is similar to the value R_{98}/H of Figure 11(a). The effect of dispersion onto the tracer breakthrough curve at the lower screen section is minor, because the variability of travel times for particles on neighboring streamlines is much greater than the variability of travel times for individual particles onto a single streamline as a consequence of longitudinal dispersivity.

SITE APPLICATIONS

The use of circulation flows for in situ remediation has become widespread, as mentioned in the introduction. At currently more than 100 sites in the United States and Europe, the technique is used for physical and biological groundwater and subsoil remediation. Table 2 gives an overview above the broad spectrum of applicability of circulation flows with respect to hydrogeology and contamination types that were carried out as research and development wells within the U.S. SITE program.

The University of Karlsruhe, Germany, operates a groundwater circulation well test site in Karlsruhe-Knielingen, where the groundwater is contaminated with chlorinated hydrocarbons (PCE and TCE) up to concentrations of 200 μg/L. The aquifer lies in the Rhine River valley and consists of fluvial sediments (sandy gravel with fine to medium sand layers) with a hydraulic conductivity of $5.6*10^{-4}$ m/s and a specific discharge of 8 cm/d. The contamination plume was detected in the upper part of the aquifer, down to a continuous horizontal silt layer, which is 10 m under ground surface. The upper screen section of the UVB 400 (400 mm annulus, 600 mm borehole) is between 3 m and 5 m under ground surface (a_T = 2m), the lower screen section (a_B = 1m) between 8.4 m and 9.4 m. The groundwater enters the well through the lower screen. The test site is equipped with 13 monitoring wells with 42 sampling points, in order to investigate the flow and transport regime of the vertical circulation well (Figures 23 and 24) (Stamm et al., 1995).

The measured hydraulic heads in the upper and lower filter agree very well with the numerically computed values as shown Fig. 11b (Table 3). The expansion of the circulation body is validated by the breakthrough curves of concentration (Figure 25), here shown for a choice of monitoring wells, namely in the upgradient E5, downgradient E12, side gradient E7, and the UVB itself.

The upstream distance between the UVB and E5 is approx. 6.0 m. E7 is located 7.2 m beside the UVB, perpendicular to the natural groundwater flow direction. After the remediation start on October 19, 1994, the PCE concentrations were depleted in the vicinity of the well, excluding in E5 (deep), which represents the upstream contaminated influx body [compare Fig. 10(b), left side under circulation zone]. The concentrations measured in E7 suggest that the circulation cell encapsulates this monitoring well, too. The stripping process leads to very low outflux concentrations of the UVB (shallow) that after ten days became close to the lower detection limit. The influx into the lower UVB screen depletes more slowly, because that water is mixed with the contaminated upstream influx. In the last week of October, the total discharge was reduced from Q = 4 m^3/h to 2 m^3/h, which results in a higher ratio of Q_0/Q (Q_0= contaminated upstream influx), and therefore in higher PCE influx concentrations [compare Figure 14(d)].

Using an effective porosity of n_e=0.15, the transport velocity is approximately. 0.5 m/d. The breakthrough curves for E13, which is located 25 m downgradient in the symmetry axis, agree well with the duration of transport velocity. Moreover, that diagram demonstrates that the cleaned water downstream approaches the horizon of the outlet screen

Table 2. Groundwater circulation well sites in the United States.

1. Location	Troutman, NC	Pt. Hueneme, CA	Rochester, NY	Gainesville, FL	Charlotte, NC	March AFB, CA
2. Method	UVB 400, air lift	Three UVB 200 biocurtain plus single UVB 400	UVB 400 in situ bioreactor + off gas bioreactor	SZB & UVB 600 in situ bioreactor inoculated with EPA 505	UVB 400 3-Screen, dual circulations	UVB 400
3. Main contamination	Gasoline (BTEX)	BTEX/MTBE (gasoline)	CHC	Creosote LNAPL/DNAPL	CHC	CHC
4. Aquifer type: lithology	Saprolite clayey silt with sand	Fine to medium sand	Glacial till, fine-grained sand and silt	Fine to medium sand	Saprolite silty sand with clay	Alluvial fan, silty sand
5. Hyd. conduct.	$1.9*10^{-4}$ cm/s	$1.0*10^{-3}$ cm/s	$1.0*10^{-3}$ cm/s	$1.0*10^{-3}$ cm/s	$1.8*10^{-3}$ cm/s	$7.5*10^{-3}$ cm/s
6. Sat. thickness	5.9 m	5.0 m	5.8 m	8.5 m	44 m	13.7 m
7. Nat. velocity	0.0078 m/day	0.033 m/day	0.004 m/day	0.003 m/d	0.044 m/day	0.17 m/d
8. Meas. sphere of influence	20 m	38 m (UVB 400) diameter	26 m upgradient diameter	45 m diameter	min. 20 m side-gradient (no other mon. wells available)	12.2 m (tracer) 26 m (deconta.)
9. Detection method (conc. Top 8)	Decontamination	Trace dyes rhodamine WT, eosin, fluorescein	Trace dyes rhodamine WT, pyranine	Trace dyes rhodamine WT, eosin, fluorescein	Decontamination and DO increase	Tracer and decontamination

Table 3. Measured and computed piezometric head differences, Δh, between the two screen sections of the circulation well in Karlsruhe, for varying total discharges, Q.

Discharge Q [m^3/h]	1.0	2.0	3.0	3.2
$\Delta h_{measured}$ (m)	0.32	0.71	0.84	1.04
$\Delta h_{calculated}$ (m)	0.34	0.68	1.02	1.09
Difference (%)	−5.9	4.4	−17.6	−4.5

faster (here: at the top of the aquifer) than at the opposite screen (Figure 26). The uppermost sampling point in E13 is located in a less permeable fine sand layer, which results in a retarded breakthrough.

CONCLUSIONS

The broad applicability of vertical circulation flow systems around groundwater remediation wells (UVB, GZB, SZB) has been demonstrated for various in situ and on-site remediation techniques. The hydraulically vertical circulation flow regime can be used to support biological degradation processes of contaminants:

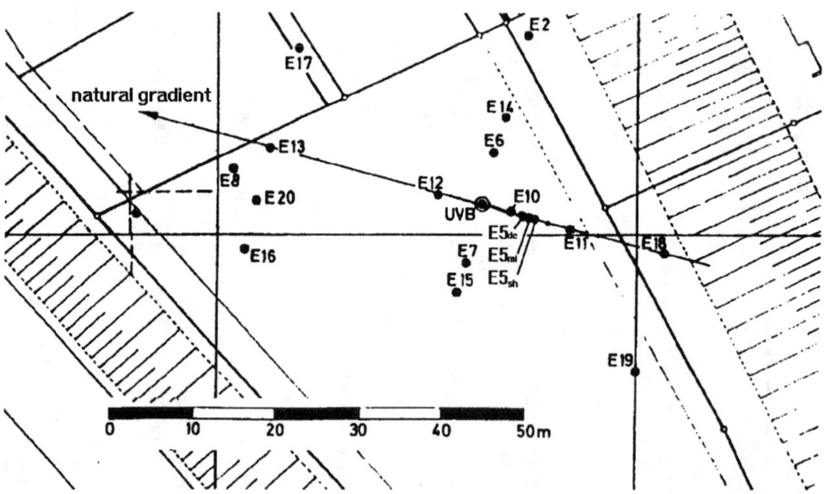

Figure 23. Map of the test site in Karlsruhe-Knielingen, Germany.

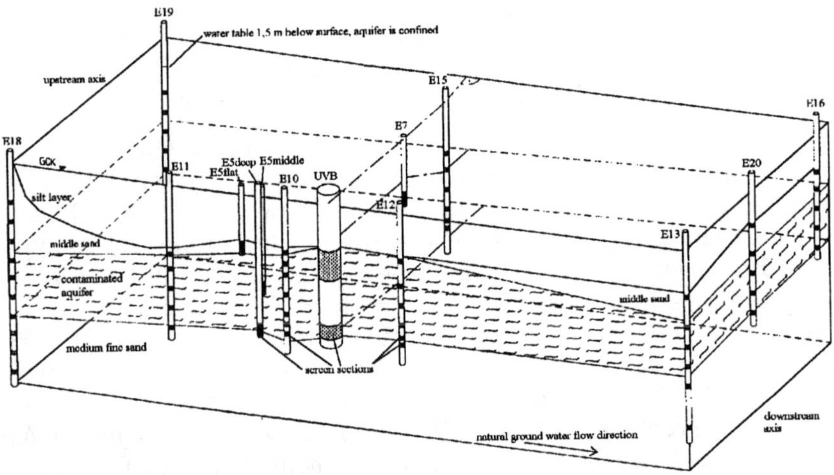

Figure 24. Spatial view of the southern part of the multileveled test site.

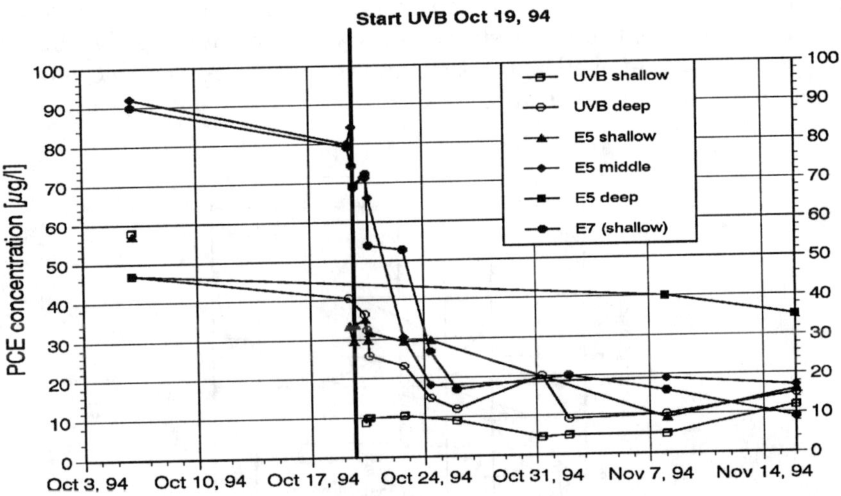

Figure 25. Tetrachlorethene (PCE) concentrations as a function of time in the UVB and in the ambient aquifer.

- in the aquifer, using it as a bioreactor in combination with in situ stripping
- within the well casing, employing special in-well treatment units (bioreactor, biofilter) and using the stripping process for its oxygen supply in the aquifer or for removing biogas
- in the unsaturated zone, in combination with soil air removal

On the basis of confined aquifer conditions, the resulting 3D flow system for several GZB units have been discussed (in terms of sphere of influence, capture zone, head changes, flow times), as well as its transport characteristics. By means of extended numerical computations, dimensionless diagrams have been created that allow the dimensioning of a vertical circulation flow remediation system.

In general, no groundwater extraction is necessary. But in the presence of a subsurface NAPL pool, the recovery of NAPL in its fluid phase should be the first remediation step. For this, the hydraulic gradient regime of the vertical circulation flow field can be used to support NAPL flow to the well, where a suitable pumpoff measure for the free product can be realized. In this case the well works in a partial extraction mode. Even at low well capacity a continuous remediation operation can take place. Furthermore, this system can be used to enlarge the aquifer's groundwater abundance in comparison to a conventional groundwater extraction. One of the surprising results with considerable significance to practice is that well extraction or infiltration is possible without head changes at the well top.

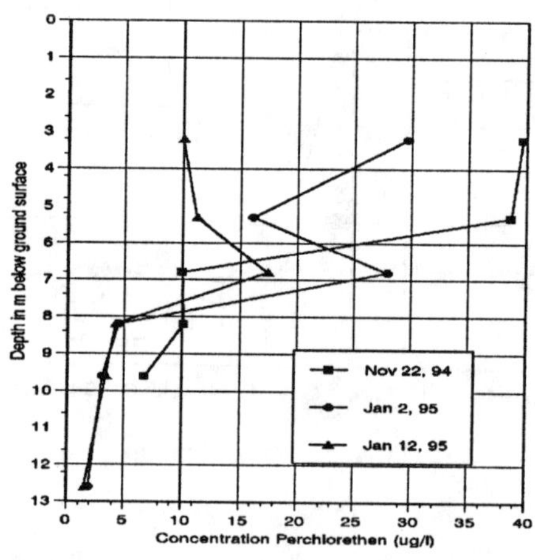

Figure 26. Concentrations over depth for different points of time in the downgradient monitoring well E13.

When the aquifer has a very large depth and the remediation well does not reach the aquitard, the sphere of influence can be approximated using the well length (from the upper to the lower screen section) as *H*. This extends the degree of utilization of the diagrams even to partially penetrating wells (Herrling et al., 1991).

The investigation regarding the influence of heterogeneity on the extent of the active and passive treated aquifer zones shows that layering can be expressed in the effective parameter for the anisotropy (K_H/K_V). The effective anisotropy increases with increasing differences in the hydraulic conductivity of the layers. That means stratification enlarges the sphere of influence of the vertical circulation region. Nevertheless, a vertical circulation well should not connect two different aquifers that are separated by an aquitard.

The presentation of isochrones in a dimensionless form enables the evaluation of treatment time requirements and provides a basis for the approximation of nutrient and/or electron acceptor supply in the aquifer. The radially symmetric transport simulations that have been carried out support the utility of the isochrone concept for practical design.

ACKNOWLEDGEMENTS

The present work originated from a longstanding collaboration with Dr. Bruno Herrling at the University of Karlsruhe, who regrettably passed away in October 1994. The presented work resulted from research projects financed by IEG mbH, D-72770 Reutlingen (Germany). I would like to thank B. Bernhardt of IEG who was helpful in supporting the research effort and, together with IEG, is also patent holder on the technologies described in Section 2 (see references). Finally, I would like to acknowledge the field data that has been supplied by various circulation well operators and the many helpful discussions and contributions by my colleagues H. J. Lensing and M. Scholz.

REFERENCES

Bae, J., L. Semprini and P.L. McCarty. 1995. Apparatus for down-well oxygen transfer into contaminated aquifers, *J. Environmental Engineering, ASCE,* 121(8):565–570.

Baveye, P. and A. Valocchi. 1989. An evaluation of mathematical models of the transport of biologically reacting solutes in saturated soils and aquifers, *Water Resour. Res.,* 26 (6):1413–1421.

Bear, J., ed. 1972. *Dynamics of Fluids in Porous Media.* Elsevier, New York.

Buermann, W., G. Bott-Breuning and R. Krug. 1992. "Groundwater remediation using the vacuum-vaporizer-well," *Proceedings of Envirotech,* Vienna, Austria, 1992, W. Pillmann (Ed.), Industrial Waste Management, pp. 723–732.

Busch, K.F., L. Luckner and K. Tiemer, ed. 1993. *Geohydraulik. Lehrbuch der Hydrogeologie Band 3,* Gebrüder Borntraeger, Berlin.

Cook, A. 1987. Biodegradation of *s*-triazine xenobiotics, *FEMS Microb. Rev.,* 46:93–116.

Gvirtzman, H. and S.M. Gorelick. 1992. The concept of in-situ vapor stripping for removing VOCs from groundwater, *Transport in Porous Media 8,* Kluwer Academic Publishers, Netherlands, pp. 71–92.

Herrling, B., W. Buermann and J. Stamm. 1992. "In-situ remediation of volatile contaminants in groundwater by a new system of 'Vacuum-Vaporizer-Wells,'" *Proceedings of the International Conference on Subsurface Contamination by Immiscible Fluids,* 18–20 April, Calgary, Canada, pp. 351–359.

Herrling, B., W. Buermann, J. Stamm and M. Schoen. 1990. "UVB technique for in-situ groundwater remediation of strippable contaminants: Operation and dimensioning of wells," *Proceedings of Envirotech,* Vienna, Austria, 1990, W. Pillmann (Ed.), Hazardous Waste Management, pp. 631–640.

Herrling, B. and J. Stamm. 1992. "Numerical results of calculated 3D vertical circulation flows around wells with two screen sections for in situ aquifer remediation," *Proceedings of Computational Methods in Water Resources IX, Vol. 1: Numerical Methods in Water Resources,* Elsevier Applied Science, London, pp. 483–492.

Herrling, B., J. Stamm and E.J. Alesi. 1993. "Physical and biological in situ aquifer remediation using vertical circulation flow (UVB technique)," *Proceedings of the Environmental Remediation Conference ER'93*, Oct. 24–28, Augusta, Georgia, USA, pp. 1019–1025.

Herrling, B., J. Stamm, E.J. Alesi, G. Bott-Breuning and S. Diekmann. 1994. "In situ bioremediation of groundwater containing hydrocarbons, pesticides, or nitrate using vertical circulation flows (UVB/GZB technique)," *Proceedings of the Second International Symposium on In Situ and On-Site Bioreclamation,* April 1993, San Diego, CA, USA. *Air Sparging for Site Remediation,* pp. 56–80.

Herrling, B., J. Stamm and W. Buermann. 1991. "Hydraulic circulation system for in situ bioreclamation and/or in situ remediation of strippable contamination," *Proceedings of the First International Symposium on in situ and On-Site Bioreclamation,* April 1991, San Diego, CA, USA. *In Situ Bioreclamation,* pp. 173–195.

Hinchee R.E. 1994. "Air sparging state of the art," *Proceedings of the Second International Symposium on In Situ and On-Site Bioreclamation,* April 1993, San Diego, CA, USA. *Air Sparging for Site Remediation,* pp. 1–13.

Kabala, Z.J. 1993. The dipole flow test: a new single-borehole test for aquifer characterization, *Water Resour. Res.,* 29 (1):99–107.

Kinzelbach, W., W. Schöfer and J. Herzer. 1991. Numerical modeling of natural and enhanced denitrification processes in aquifers, *Water Resour. Res.,* 27 (6):1149–1159.

Lensing, H.J., M. Vogt and B. Herrling. 1994. Modeling of biologically mediated redox processes in the subsurface, *Journal of Hydrology* 159 (1994), 125–143.

MacDonald, T.R. and P.K. Kitanidis. 1993. Modeling the free surface of an unconfined aquifer near a recirculation well, *Ground Water,* Vol. 31, No. 5:774–780.

Marsily, G. de., ed. 1986. *Quantitative Hydrogeology.* Academic Press, Inc., London.

McCaulou, D.R., W.T. Weinig and G.R. Walter. 1995. "Evaluation of vertical circulation wells for enhanced bioremediation of contaminated aquifers," *Proceedings of the Third International Symposium on in situ and On-Site Bioreclamation,* April 1995, San Diego, CA, USA, 3(2):pp. 495–501.

Mellor, R.B., J. Ronnenberg, W.H. Campbell and S. Diekmann. 1992. Reduction of nitrate and nitrite in water by immobilized enzymes, *Nature,* 355:717–719.

Pinder, G.F. and W.G. Gray, ed. 1977. *Finite Element Simulation in Subsurface Hydrology.* Academic Press, New York.

Reddi, L.N. and S. Challa. 1994. Vibratory mobilization of immiscible liquid ganglia in sands, *J. Environmental Engineering, ASCE,* 120(5):1170–1190.

Schäfer, W. and W. Kinzelbach. 1991. "Numerical investigation into the effects of aquifer heterogeneity on in situ bioremediation," *Proceedings of the First International Symposium on in situ and On-Site Bioreclamation,* April 1991, San Diego, CA, USA. *In situ Bioreclamation,* pp. 196–226.

Schwille, F. 1984. Migration of organic fluids immiscible with water in the unsaturated zone. *Pollutants in Porous Media.* Ecol. Stud. 47, Springer Verlag, Berlin.

Schwille, F., ed. 1988. *Dense Chlorinated Solvents in Porous and Fractured Media.* Lewis Publ. Inc., New York.

Stamm, J. 1995. "Vertical circulation flows for vadose and groundwater zone in situ (bio-)remediation," *Proceedings of the Third International Symposium on in situ and On-Site Bioreclamation,* April 1995, San Diego, CA, USA, 3(2):483–493.

Stamm, J., M. Scholz and M. Löseke. 1995. "3D Vertical circulation flows around groundwater circulation wells (GZB) for aquifer remediation: numerical calculations and field experiments" *Proceedings of "Contaminated Soil '95,"* ed. by W.J. van den Brink et al., Kluwer Academic Publishers, Netherlands, pp. 171–181.

US EPA. 1989a. *Evaluation of Ground Water Extraction Remedies. Vol. 1, Summary Report,* EPA/540/2-89/054, Cincinnati, Ohio.

US EPA. 1989b. *Evaluation of Ground Water Extraction Remedies. Vol. 2, Case Studies 1–19,* EPA/540/2-89/054a, Cincinnati, Ohio.

US EPA. 1990. *Handbook on in situ Treatment of Hazardous Waste-Contaminated Soils.* EPA/540/2-90/002, Washington, D.C.

U.S. Patent Office, Bernhardt, Patent No. 4,892,688, Prio. 07/08/1988; Bernhardt, Patent No. 4,943,305, Prio. 06/23/1988; Bernhardt et al., Patent No. 4,950,394, Prio. 02/19/1988; Bernhardt, Patent No. 5,082,053, Prio. 09/16/1989; Bernhardt, Patent No. 5,095,975, Prio. 09/16/1989; Bernhardt, Patent No. 5,143,606, Prio. 11/22/1990; Bernhardt, Patent No. 5,147,535, Prio. 12/07/1989; Bernhardt, Patent No. 5,171,103, Prio. 04/05/1989; Bernhardt, Patent No. 5,220,958, Prio. 08/29/1990; Bernhardt, Patent No. 5,318,698, Prio. 11/22/1991; Bernhardt, Patent No. 5,281,333, Prio. 02/06/1992; Bernhardt, Patent No. 5,403,476, Prio. 05/29/1992; Bernhardt, Patent No. 5,380,126, Prio. 06/03/1992; Bernhardt, Patent No. 5,345,655, Prio. 02/19/1992.

Waterloo Centre for Groundwater Research (WCGR). 1989. University of Waterloo Short Course, "Dense Immiscible Phase Liquid Contaminants in Porous and Fractured Media," Nov. 6–9, 1989, Kitchener, Ontario, Canada.

Wilson, J.L., S.H. Conrad, W.R. Mason and W. Peplinski. 1990. *Laboratory Investigation of Residual Liquid Organics from Spills, Leaks and the Disposal of Hazardous Wastes in Groundwater,* EPA/600/6-90/004, Washington, D.C.

Zienkiewicz, O.C., ed. 1977. *The Finite Elements Method.* McGraw-Hill, New York.

Slurry Phase Bioremediation of Hazardous Materials: Bioslurry Reactor Application

SALIL PRADHAN, BILL Y. LIU, AND VIPUL SRIVASTAVA
Institute of Gas Technology
1700 S. Mt. Prospect Road
Des Plaines, IL 60018

INTRODUCTION

Bioremediation refers to the breakdown of organic contaminants by microorganisms. Bioremediation does not lead to confinement or cross media transfer of the contaminants, but to the destruction of the pollutants using natural processes to environmentally safe treatment end products, such as carbon dioxide and water. In addition, bioremediation is generally more economical than energy-intensive processes, such as incineration, that require high temperature and energy.

The success of bioremediation is controlled by several factors: the solubility of the contaminants in water, soil characteristics, and microbial activity in the soil. Generally, the contaminants are hydrophobic and are attached to the soil particles and, hence, are not easily available to the microorganisms. Slurry phase bioremediation overcomes these drawbacks and improves the bioavailability of the contaminants by enhancing the desorption of the contaminants from the soil through proper control of various operational parameters.

SLURRY PHASE BIOREMEDIATION TECHNOLOGY

Slurry phase biological treatment is a relatively new development for the remediation of hazardous wastes. The slurry phase bioremediation process has been highly effective for a variety of wastes, such as pentachlorophenol, herbicides, wood-preserving waste (Ross, 1990–91), and polynuclear aromatic hydrocarbons (PAHs) (Liu *et al.*, 1994). Rate of degradation of contaminants is 10 times faster than other bioremediation approaches currently in use (Ross, 1990–91; Srivastava *et al.*, 1991).

Slurry phase bioremediation involves the formation of 10–40% (by weight) solids slurry in a bioslurry reactor, where optimal conditions are provided for microbial growth leading to biodegradation. The bioslurry reactor can be operated either in a batch or a semicontinuous mode. Following the bioslurry treatment, the treated solids are dewatered.

In the bioslurry reactor, the bioavailability of the contaminants is promoted by physical, chemical, and biological actions. Mechanical agitation is provided to ensure a homogenous mixture in the bioreactor. Nutrients in the form of nitrogen and phosphorus sources are added, and air is introduced for aeration through a diffuser system. In some cases, foreign microorganisms acclimated to the contaminants of concern are added to enhance the biodegradation caused by indigenous microbes.

Different designs of the bioslurry reactors have been used for soil and sludge remediation. Praxair Inc. uses a two-stage slurry process called Linde's Mixflo System (VISITT, 1994). In the first stage, the slurry in the reactor is pumped and pressurized up to 2 to 4 atmospheres. Pure oxygen is then injected into the slurry, and the resulting mixture is passed through a pipeline contactor, where 60% of the injected oxygen dissolves. In the second stage, the oxygen-slurry mixture is re-injected into the reactor through a liquid-liquid eductor, which ingests nonoxygenated slurry and mixes it with oxygenated slurry.

The EIMCO Equipment Company uses Biolift reactors that utilize airlift mechanism for recirculation of the solids in the slurry (Brox, 1989; Brox and Hanify, 1991). The settled solids are raked to a central airlift or multiple airlifts for recirculation. In addition to the airlift, flexible membrane tube diffusers are provided on the rake arms to effect the aeration of the slurry. The biolift reactor is gas sealed, and volatile emissions are recirculated. Other bioslurry reactor configurations that have been developed and tested include the completely mixed design by OHM (Woodhull, 1993).

Quality of Air Emissions from Slurry Phase Bioremediation

Another major advantage of slurry phase bioremediation is that the air emissions, if any, associated with this treatment can be easily trapped and treated with an adsorption unit, such as granular activated carbon (GAC) column. Results of a bioslurry study, conducted by the EPA (Lauch et al., 1992), for the removal of PAHs from creosote-contaminated soils, showed that low concentrations of benzene, toluene, ethylbenzene, and xylene (BTEX) and low concentrations of 2- and 3-ring PAHs were detected during the first few days of the treatment. All the contaminants diminished to concentrations that were below the detection limits after 5 days of operation. Another EPA study, testing the emissions of organics from bioslurry reactors treating soil contaminated with wood-preserving waste, concluded that major emissions are observed during the loading of the reactor, and a small amount may continue for the first few days (Lewis et al., 1992). However, the problem of air emissions observed during the loading of a reactor can be similar or even greater in loading incinerators or other nonbiological treatment processes.

Treatment Costs Associated with Slurry Phase Bioremediation

The most significant factors that affect the treatment cost are the target contaminant, its concentration, and the quantity of waste to be remediated. Other factors that affect the cost include waste preprocessing, amount of debris, residual waste characteristics, and depth of contamination. Some other factors that also influence treatment cost are labor costs, site preparation, and utility rates. Full-scale slurry phase bioremediation systems, operated by more than 10 companies, are documented in the EPA literature (VISITT, 1994). Based on the available information from full-scale projects, estimated costs are summarized in Table 1. Estimated price ranges do not include all indirect costs associated with the treatment, such as excavation and permits.

Table 1. Estimated treatment costs for full-scale bioslurry systems.

Company	Waste Facility	Media	Contaminant	Estimated Cost per Cubic Yard
OHM	Wood treatment	Soil and sludge	PAHs, PCBs, BTEX	\$75–\$250
RETEC	Petroleum refining	Sludge	PAHs, BTEX	\$30–\$600
EIMCO (Biolift Reactor)	NA	Soil and Sludge	PAHs, BTEX	\$100–\$180
Praxair, Inc. (Mixflo System)	Petroleum refining and reuse	Soil and sludge	PAHs, PCBs	\$100–\$300
Bogart Environmental	Wood treatment	Sludge	PAHs, herbicides, pesticides	\$40–\$100

An economic analysis of slurry phase bioremediation is presented later in this chapter.

CASE STUDY—INTEGRATED CHEMICAL BIOLOGICAL TREATMENT

This case study describes bench- and field-scale evaluation of the slurry phase bioremediation of a former manufactured gas plant (MGP) site in New Jersey.

Background

In 1950, natural gas replaced manufactured gas as the major gaseous fuel; hence, the production of manufactured gas came to an end. Operations and residuals management practices of many former manufactured gas plants resulted in the contamination of their site soils with hazardous organic compounds. One of the main contaminants of concern is polynuclear aromatic hydrocarbons (PAHs). PAHs consist of two or more fused benzene rings in linear, angular, or cluster arrangements. Some of these PAHs are known carcinogens and are an environmental hazard. As a result, these former MGP sites require cleanup to ensure environmental safety. Over 1500 MGP facilities have been identified to date in the United States.

The 2-, 3-, and some of the 4-ring PAHs are biodegradable and can support the growth of bacteria. The remaining persistent PAHs can be biodegraded by secondary utilization, that is, the fortuitous biodegradation of compounds that function as cosubstrates. However, the biodegradation of many recalcitrant contaminants such as the higher ring (4- to 6-ring) PAHs is often slow, incomplete, and/or insufficient (Srivastava et al., 1991).

To enhance the rate and extent of biodegradation, the Institute of Gas Technology has developed an integrated chemical biological treatment for remediation of contaminated waste matrices.

Integrated Chemical Biological Treatment (MGP-REM Process)

As the name suggests, the integrated chemical biological treatment, also referred to as the MGP-REM Process, combines a mild, but powerful, chemical treatment with the biological treatment. The chemical treatment uses Fenton's reagent to chemically transform the organic compounds in the waste matrix. The Fenton's reaction, the interaction of hydrogen peroxide (H_2O_2) and ferrous ions (Fe^{+2}), generates highly reactive free hydroxyl radicals that oxidize the organic contaminants into more water-soluble and, hence, readily biodegradable compounds, which can be further degraded in the subsequent biological treatment (Ravikumar and Gurol, 1992).

The biological and chemical treatment steps of the MGP-REM Process can be applied in different sequences and varying number of treatment cycles, depending upon the nature and degree of contamination and upon regulatory treatment endpoints. The MGP-REM Process has two distinct advantages over the conventional bioremediation. Firstly, it accelerates the transformation of organic pollutants into more readily biodegradable forms and thus enhances the rate of biodegradation. Secondly and more importantly, it improves the extent of biodegradation by rendering the recalcitrant organics more bioavailable for further biodegradation.

Methods

This section describes the field-scale bioslurry remediation of soil from an MGP site in New Jersey. The study began with soil characterization and bench-scale evaluation of soil from the site. The results of the bench-scale study showed that the treatment had good potential for field-scale evaluation.

BENCH-SCALE STUDY

The bench-scale slurry reactor study was performed in 10-L stainless steel reactors. Each reactor was equipped with an air sparger and mechanical mixer to ensure complete mixing of the slurry contents. The soil slurry experiments employed a mixture of contaminated soil and water with a solids content of about 15%. The treatment scheme evaluated was the MGP-REM Process (integrated chemical/biological treatment), which started with 3 weeks of biological treatment, followed by a chemical treatment with 2.5% hydrogen peroxide and then a biological treatment for final polishing. The total operation time was 34 days. Following the encouraging results obtained at the bench-scale level (as described in the results section), the field-scale evaluation was initiated.

FIELD-SCALE STUDY

The schematic of the field-scale bioslurry system is shown in Figure 1. The slurry tank was divided into three independent, but interconnected, reactor units, designated as the slurry reactor, the bioreactor,

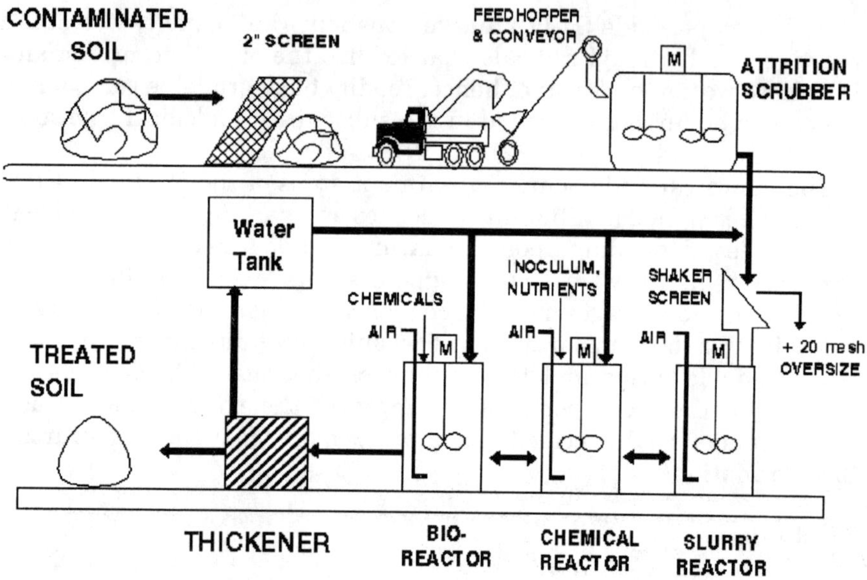

Figure 1. Schematic of the slurry phase bioremediation system.

and the chemical reactor. Once the treatment in the reactor units was complete, the slurry would be pumped to the thickener for dewatering purposes. The excess water or the overflow of the thickener was recycled into the process via the water management tank. Off-gas from the reactor system was passed through two granular activated carbon (GAC) canisters operated in series to adsorb any VOCs released during the process to eliminate the possibility of air pollution.

OPERATING PROTOCOL

As shown in Figure 1, excavated soil was screened to remove objects greater than 2 inches. The screened soil was then transported to an attrition scrubber where soil was mixed with water to make a 50% soil slurry. Slurry from the attrition scrubber was sent to the 20-mesh vibrating screen to remove particles greater than 1 mm, which were back-filled into the source location.

The slurry reactor, chemical reactor, and biological reactor had identical capacities of 2100 gallons. Slurry passing through the screen was diluted to 25–35% solids in the slurry reactor as required for the following chemical/biological treatment. During the biological treatment, oxygen was provided thorough aeration to maintain dissolved oxygen concentration (DO) in the soil slurry between 2 and 8 mg/L. Nutrients (nitrogen and phosphorus) were added to ensure optimal conditions for microbial growth. The pH of the soil slurry was maintained between 6.5 and 7.5, and the temperature ranged between 65 and 85°F. Prior to the chemical treatment, the slurry was allowed to settle for 4 hours, and the supernatant (about half of the total volume) was transferred to an inoculum reactor, where it was aerated and mixed using a submersible pump. This procedure served as a way to oxidize only the PAHs in the soil while preserving the microbial population in the supernatant under ideal biological conditions in the inoculum tank.

During the chemical treatment, 50% solution of hydrogen peroxide (H_2O_2) was added to the chemical reactor to achieve the designed volume H_2O_2/volume slurry ratio. Commercial-grade $FeSO_4 \cdot 7H_2O$ was added prior to the addition of H_2O_2 to achieve a predetermined Fe^{+2} concentration in the slurry. After the chemical oxidation was completed, the supernatant with active microbial population was added back to the soil slurry.

The slurry reactor system was operated in a batch mode. The treatment conditions used in the reactor system are shown in Table 2.

ANALYSES

Slurry samples were taken at the beginning, midpoint, and end of each treatment scenario for PAH analyses. PAHs in the samples were determined using a modified USEPA (United States Environmental Protection Agency) SW-846 Method 8270B analysis of soxhlet extracts. Soxhlet extraction was performed by EPA SW-846 Method 3540 using a 1 : 1 mixture of acetone and hexane as the extraction solvent. The modifications of Method 8270B involved a 30-meter long XTI-5 (Restek) column (0.25-mm ID and 0.5-μm film) installed in a Hewlett Packard 5890 II gas chromatograph/5971 mass selective detector. The injector temperature was 270°C, and the following temperature program was used: 40°C for 4 min followed by an increase to 300°C at a rate of 10°C per min: isothermal at 300°C for 20 min. The PAH analysis was done by Analab, Edison, New Jersey.

Table 2. Operating conditions in the reactor system during the field-scale study.

Parameter	Treatment Range
Slurry Volume, gals.	1800–2100
Total Solids, %	25–35
Slurry Residence Time, days	10–20
Air Flow Rate, SCFM	0–25
Hydrogen Peroxide, % (vol./vol.)	0.5–2.0

Table 3. Air monitoring procedures for the field study.

Analytical Parameter	USEPA Method	Air Flow Rate	Cartridge
BTEX	TO-10	25 cc/min	Carbotrap-300
PAHs	TO-1	1000 cc/min	XAD-2 Resin

The Microtox Toxicity Test was used to determine the toxicity of the soil samples. The Microtox reagent contains bioluminescent bacteria that are grown under optimal conditions. This reagent is exposed to the test sample to measure the toxic effect of the sample on the bacteria. The test system measures the light output of the luminescent bacteria after they have been challenged by a sample of unknown toxicity and compares it to the light output of a control (blank). The degree of light loss indicates the degree of toxicity of the sample.

AIR MONITORING SYSTEM

In order to prevent organic emissions, off-gas from the reactor units was passed through two granular activated carbon (GAC) canisters operated in series. Air samples were collected before and after the GAC units and were analyzed by Aqua Air Analytical, Raritan, New Jersey, for benzene, toluene, ethyl benzene, and xylene (BTEX) and PAHs. The procedures used for collecting and analyzing the air samples are shown in Table 3.

Results and Discussion

BENCH-SCALE STUDY RESULTS

Initial characterization of the soil used for the bench-scale study is shown in Table 4. Results of the bench-scale batch study are summa-

Table 4. Soil characterization for bench-scale study.

Parameter	Value
pH	5.4
Organic matter, %	5
Phosphorus, ppm	25
Potassium, ppm	70
Nitrate-N, mg/kg	5.5
Sand, %	81
Fines, %	19

Table 5. PAH degradation data for bench-scale study.

PAHs, mg/kg	Day 0	Day 7	Day 34	Reduction, %
Total	1013.7	204.4	44.8	95.5
Noncarcinogenic	782.5	93.3	30.1	96.2
Carcinogenic	231.2	111.1	14.7	93.6

rized in Table 5. The slurry treatment reduced the total PAHs from over 1000 mg/kg to less than 50 mg/kg in 34 days. More importantly, the carcinogenic PAHs were reduced to less than 15 mg/kg from their original concentration of about 230 mg/kg (more than 93% reduction). These results indicate that the combination of chemical/biological treatment in the slurry phase bioremediation can treat MGP site soils and deserves serious consideration for further study and/or full-scale implementation.

FIELD STUDY RESULTS

This section summarizes the results of two test runs conducted during the field study. Test 1 was operated for a total of 20 days beginning with a biological treatment followed by a 0.5% chemical treatment, then another biological treatment, and a final 2% chemical treatment. Test 2 was operated with a biological treatment, followed by a 2% chemical treatment, and a final biological polishing.

The Test 1 results are plotted in Figure 2, showing the overall decreasing trend of the PAH concentrations in the soil. Results of samples taken at day 0, 20, and 40 are shown in Table 6 and Figure 3. The 40-day sample was taken from the thickener 20 days after the slurry reactor treatment was completed. A 95% treatment efficiency was achieved for removal of total PAHs from the soil. More significant is the more than 90% disappearance of the carcinogenic PAHs from the soil. Tables 7 and 8 reflect the removal efficiency of PAHs with a different number of aromatic rings. These results indicate that the PAHs with a fewer (2 and 3) number of aromatic rings were degraded more effectively than those with more (4, 5, and 6) aromatic rings.

Table 6. PAH degradation data during Test 1 of the field study.

PAHs (mg/kg)	Day 1	Day 20	Day 40[a]	% Removal
Total	1164	157	53	95
Noncarcinogenic	974	86	33	97
Carcinogenic	191	71	20	90

[a]Samples taken from the thickener.

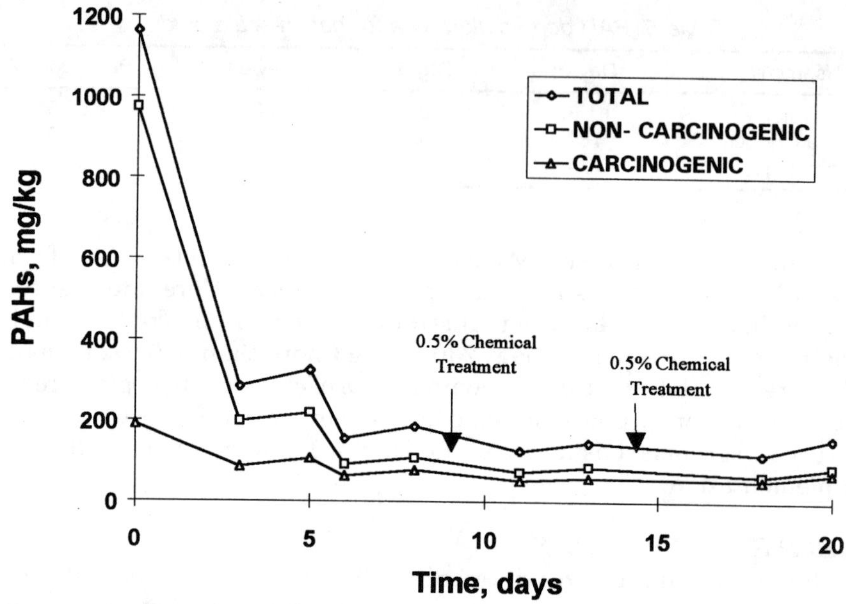

Figure 2. Time plot of residual PAH concentrations in the soil during Test 1 of the field study.

PAH removal results of Test 2 as summarized in Table 9, show the same trend as that of Test 1. Since Test 2 was conducted for a shorter duration, only 11 days versus 20 days for Test 1, its treatment efficiency was slightly lower that of Test 1.

TOXICITY TEST RESULTS

Soil samples for Tests 1 and 2 were tested to determine the effect of MGP-REM Process on reducing the toxicity levels in soil. The soils before treatment and after treatment were tested using the Microtox Toxicity Test as described in the Methods section. The toxicity results, reported on a dry soil basis, are presented in Table 10. The measure

Table 7. Degradation data of soil PAHs as grouped by ring numbers during Test 1 of the field study.

PAHs (mg/kg)	Day 1	Day 20	Day 40[a]	% Removal
Total	1164	157	53	95
2,3-Ring	718	49	21	97
4,5,6-Ring	447	108	32	93

[a]Samples taken from the thickener.

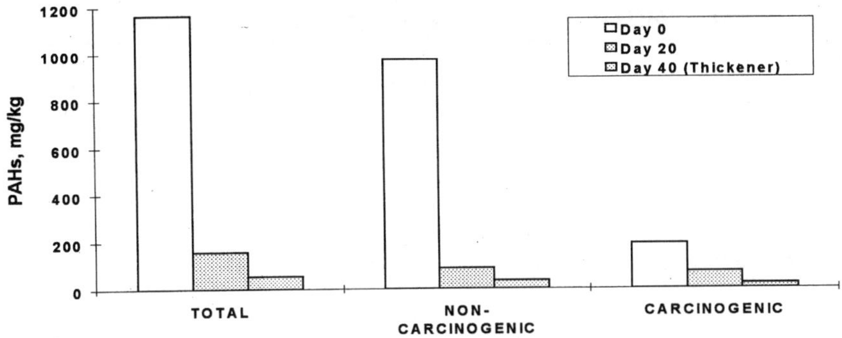

Figure 3. PAHs concentration in the soil for days 0, 20, and 40 during Test 1 of the field study.

of toxicity used here is EC50 (effective concentration 50), meaning the percentage of soil slurry that will inhibit 50% of the bacterial activity. Therefore, the lower the EC50 reading, the higher is the toxicity level in the soil.

As seen from Table 10, the MGP-REM process decreased the toxicity of the soil in Test 1 by more than 50 times. For Test 2, the original soil was not as toxic as Test 1, but still the MGP-REM process managed to reduce the toxicity by a factor of 5.

AIR MONITORING RESULTS

The compounds detected in the air emissions included BTEX and naphthalene, acenaphthene, fluorene, and phenanthrene among the PAHs. The highest emission rates of benzene and total hydrocarbons (BTEX + PAHs) in the exit airstream observed during the two test runs are shown in Table 11. These numbers are compared with the maximum emission rates permitted by the New Jersey Department of Environmental Protection (NJDEP). These results indicate that at any

Table 8. Comparison of individual ring PAHs concentrations in soil during Test 1 of the field study.

PAHs (mg/kg)	Day 1	Day 20	Day 40[a]	% Removal
2-Ring	202	21	7.6	96
3-Ring	516	29	13.2	97
4-Ring	337	41	18.2	95
5-Ring	85	46	11.3	87
6-Ring	24	21	2.8	88

[a]Samples taken from the thickener.

Table 9. PAHs degradation data during Test 2 of the field study.

PAHs, mg/kg	Day 1	Day 11	Day 27[a]	% Removal
Total	722	97	72	90
Noncarcinogenic	565	61	48	92
Carcinogenic	157	36	24	85
Group Ring PAHs				
2,3-Rings	359	36	33	91
4, 5, and 6-Rings	363	61	39	89
Individual Ring PAHs				
2-Ring	59	14	13	78
3-Ring	301	21	20	93
4-Ring	266	26	20	92
5-Ring	74	22	14	81
6-Ring	23	13	5	80

[a]Samples taken from the thickener.

time during the test runs, the emissions prior to and after the carbon treatment were significantly lower (by five orders) than the NJDEP limits of 0.05 and 0.5 lb/h for benzene and total hydrocarbons (BTEX + PAHs), respectively. These results suggest that the air emissions from both test runs do not pose a health hazard; therefore, this process requires minimal air pollution control and monitoring.

Conclusions

The following conclusions can be drawn from this study:

- The field-scale study demonstrated that the MGP-REM Process (the integrated chemical/biological treatment) in a slurry phase bioremediation system can effectively reduce over 90% of the total

Table 10. Results of the Microtox Toxicity Test conducted on soil samples from Test 1 and Test 2 of the field study.

Test	Soil Sample Description	EC50 @ 5 minutes
1	Before treatment	0.212
1	After treatment (day 20)	>10
2	Before treatment	0.803
2	After treatment (day 11)	3.35

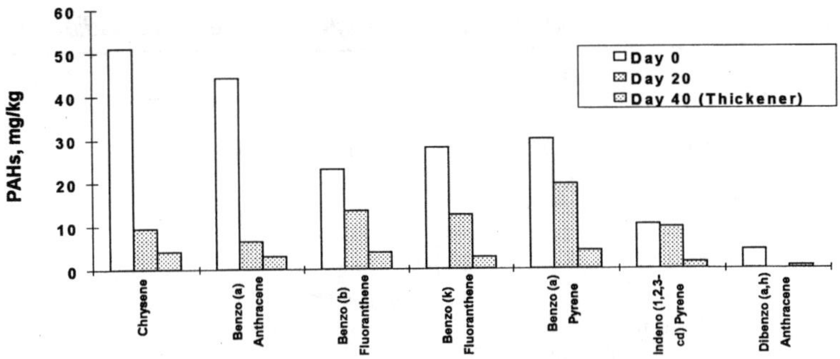

Figure 4. Carcinogenic PAH reduction data during Test 1 of the field study.

PAHs in MGP site soils. Results of the study showed that total carcinogenic PAHs in the soil can be reduced to below 25 mg/kg.

- As indicated by the Microtox Toxicity Test, this process reduced the toxicity levels present in the two soils tested by factors of 50 and 5, respectively.
- Air emissions from the two test runs were well in compliance with the NJDEP regulations; hence, minimal air pollution control and monitoring is required for this process.

ECONOMIC ANALYSES OF SLURRY PHASE BIOREMEDIATION

The financial benefits of the slurry phase bioremediation technology were recently reviewed by the Gas Research Institute (GRID, 1993) in comparison to incineration. Incineration of the MGP site soils generally costs between \$600 and \$1200 per cubic yard, while slurry phase bioremediation was estimated to be capable of remediating similar soil for \$85 to \$250 per cubic yard. When considering approximately 1500 MGP sites in the United States and assuming that an average of 10,000 cubic yards of contaminated soil may be present at each site, the potential savings that may be realized by remediating these soils using the

Table 11. Maximum air emission rates obtained during the field study.

Test Description	BTEX and PAHs, lb/h	Benzene, lb/h
Test 1	3.4×10^{-7}	Not detected
Test 2	3.3×10^{-6}	7.6×10^{-7}
NJDEP Limits	0.5	0.05

Table 12. Estimated treatment costs for slurry phase bioremediation.[a]

Case	Variation in Base Case	Cost/Cubic Yard
I	None	216
II	1-year operation with 210,000 gallon bioreactors	207
III	Continuous operations with CSTR[b]	210
IV	Biotreatment of <0.2-inch soil particles	212
V	Biotreatment at 40% solids concentration	195
VI a	MGP-REM Process (1% chemical treatment)	205
VI b	MGP-REM Process (2% chemical treatment)	220
VII	t-PAH removal rate = 1.0 day^{-1} c-PAH removal rate = 0.2 day^{-1}	173
VIII	Initial t-PAH removal rate = 1500 ppm Initial c-PAH removal rate = 400 ppm	175
IX	25% soil <40 mesh	166
X	Volume of contaminated soil = 50,000 cubic yards	83

[a]Data generated by Salameh et al., 1994.
[b]CSTR: continuous stirred tank reactor.

slurry phase bioremediation process rather than incineration is of the order of approximately $9 billion dollars.

A similar study was conducted by Remediation Technology, Inc. (RETEC) to evaluate the economics of the biological treatment of contaminated soils found at MGP sites in a bioslurry reactor (Salameh et al., 1994). In the cost analyses, only costs associated with treatment for the soils were evaluated. Not included in the analyses are the following costs: remedial investigation, feasibility studies, treatability studies, regulatory negotiations, site closure, and long-term monitoring. It has been assumed that the costs for these operations are not dependent on the remedial technology used. The following assumptions were made for the base case scenario:

- volume of contaminated soil = 10,000 cubic yards
- prescreening of >2-inch soil
- washing of <2-inch soil
- biotreatment of <40 mesh soil particles
- batch biotreatment at 25% solids concentration
- contaminant removal rate for total PAHs (t-PAH) = 0.5 day^{-1}
- contaminant removal rate for carcinogenic (c-PAH) = 0.1 day^{-1}
- initial t-PAH concentration = 3000 ppm
- initial c-PAH concentration = 800 ppm
- residual concentration after biotreatment = 20% for t-PAH and 40% for c-PAH

- assumed soil size fraction distribution: 25% is >2-inch; 25% is <2-inch, and >40 mesh; 50% is <40 mesh
- 2 years operation with 65,000 gallon bioreactors

The estimated costs for the base case and its variations are shown in Table 12.

REFERENCES

Brox, G.H. "A New Solid/Liquid Contact Bioslurry Reactor Making Bio-Remediation More Cost-Competitive," Paper presented at the *1989 Colorado Hazardous Waste Management Society Conference,* Denver, Colorado, November 6 and 7, 1989.

Brox, G.H. and Hanify, D.E. "Bioremediation of Hazardous Waste in a Slurry Reactor." Presented at Batelle's In-situ and On-site Bioreclamation Poster Session, March 19–21, 1991, San Diego, California.

GRID. "Cutting the Cost of Soil Cleanup With Bioremediation," Vol. 16, No. 3, Fall 1993.

Kelley, R.L. and Srivastava, V.J. "Field-Scale Evaluation of an Integrated Treatment for Remediation of PAHs in Manufactured Gas Plant Soils," Paper presented at *1992 Spring National A.I.Ch.E. Meeting,* New Orleans, LA, March, 1992.

Lauch, R.P., Herrmann, J.G., Mahaffey, W.R., Jones, A.B., Dosani, M., Hessling, J. "Removal of Creosote From Soil by Bioslurry Reactors." In *Environmental Progress,* Vol. 11, No. 4, pp. 265–271, November, 1992.

Lewis, R.F. Smith, M., Hessling, J., Dosani, M. "Emissions of Organics From Bioslurry Reactors Treating Soil Contaminated With Wood Preserving Waste." Paper presented at the *18th Annual Risk Reduction Engineering Laboratory Research Symposium,* 14–16 April, 1992, Cincinnati, Ohio.

Liu, B.Y., Pradhan, S.P., Srivastava, V.J., Pope, R., Hayes, T., Linz, D., Proulx, C., Jerger, D. and Woodhull, P. "MGP Soil Remediation in a Slurry-Phase System: a Pilot-Scale Test" in *Proceedings of Gas, Oil, and Environmental Biotechnology VI,* Des Plaines, Illinois: Institute of Gas Technology, 1994.

Ravikumar, J.X. and Gurol, M.D. "Fenton's Reagent as a Chemical Oxidant for Soil Contaminants," Paper presented at the *Chemical Oxidation Technology for the Nineties, Second International Symposium,* Vanderbilt University, Nashville, Tennessee, February 1992.

Ross, D. "Slurry-Phase Bioremediation: Case Studies and Cost Comparisons," *Bioremediation* 1:61–74, (1990–1991) Winter.

Salameh, M.F., Friday, D.D. and Nakles, D.V. Economic Evaluation of the Biological Treatment of MGP Soils in a Liquids/Solids Slurry Reactor. *Topical Report, Gas Research Institute, Chicago,* August 1994.

Srivastava, V.J. "Manufactured Gas Plant Sites: Characterization of Wastes and IGT's Innovative Remediation Alternatives," Paper presented at the *Symposium for Hazardous and Environmentally Sensitive Waste Management in the Gas Industry,* January 1993.

Srivastava, V.J., Kelley, R.L., Paterek, J.R., Hayes, T.D., Nelson, G.L. and Golchin, J. "A Field-Scale Demonstration of a Novel Bioremediation Process for MGP Sites," *Applied Biochemistry and Biotechnology,* Vol. 45/46, 1994.

Srivastava, V.J., Kelley, R.L. and Gauger, W.K. 1991. "Field-Scale Evaluation of an Integrated Treatment for Remediation of PAHs in Manufactured Gas Plant Soils," in *Proceedings of Gas, Oil, and Environmental Biotechnology IV,* Chicago, Illinois: Institute of Gas Technology.

VISITT Version 3.0 User Manual, USEPA Report Number: EPA 542-R-94-003. Diskettes, EPA/542/F-93/001, 1994.

Woodhull, P.M., Jerger, D.E. and Cady, D.J. 1993. "Economics of a Full-Scale Slurry-Phase Biological Treatment Process for Wood Preserving Wastes," Paper presented at the *I&EC Special Symposium,* American Chemical Society, Atlanta, Georgia.

Full-Scale Slurry Phase Biological Treatment of Wood-Preserving Wastes at a Superfund Site

DOUGLAS E. JERGER
OHM Remediation Services Corp.
16406 U.S. Route 224 East
P.O. Box 551
Findlay, OH 45839-0551

INTRODUCTION

Slurry phase biological treatment is a process in which contaminated soils are suspended in bulk water phase to increase the biodegradability of the contaminants. The successful development of a slurry phase process is based on data obtained through well-designed laboratory and pilot studies emphasizing process optimization.

The effect of process variables on contaminant biodegradability must be understood to adequately assess process performance. A lack of understanding of the interaction of these physical, chemical, and biological parameters on the rate and extent of process contaminant biodegradability can result in poor process performance. The impact of these factors on process operation and performance must be integrated into the engineering design phase to maximize treatment efficiency and achieve cleanup criteria.

Slurry phase biological treatment has been utilized to treat soils and sludges contaminated with polycyclic aromatic hydrocarbons (PAHs). These matrices have been associated with the industrial processes of creosote preservative wood treatment and former manufactured gas plants. Individual PAH compounds and isomers vary widely in their susceptibility to biodegradation. The more water-soluble, lower molecular weight PAH compounds are generally biodegraded at a faster rate and to a greater extent than the less soluble, higher molecular weight PAHs.

Full-scale slurry phase biological treatment was performed to treat approximately 10,500 tons of RCRA-K001 wood-preserving wastes and soils contaminated with PAHs at a former wood treating site. Waste characterization and laboratory slurry reactor studies were conducted on representative site samples to determine process configuration, operation, and performance. A quantitative knowledge of the physical, chemical, and biological parameters affecting the process engineering and optimization was necessary for process scaleup. The results from these studies were used to develop process design and operating conditions for bench-, field-pilot-, and full-scale slurry phase treatment systems.

Treatment efficiencies approaching 95% and 70% were achieved for total and carcinogenic PAHs contained in the wood-preserving waste material. Key parameters identified for successful process development were temperature, mixing, aeration (dissolved oxygen concentration), and type and concentration of PAH.

The use of slurry phase biological treatment for the remediation of PAH-contaminated soils/sludges was shown to be a cost-effective, destructive, on-site technology for treatment of materials contaminated with PAHs. The development and application of this technology was based on sound scientific and engineering principles. These principals were manipulated to achieve final cleanup concentrations within specified time periods. This understanding served to justify cleanup concentrations on the basis of system performance and land use considerations.

WOOD-PRESERVING SITES

The wood-preserving industry is involved in the manufacturing of insect- and fungus-resistant lumber products. This industry had its inception prior to the turn of the century, and over 450 facilities are in operation today. Due to the treatment processes employed and the history of these sites, the majority of wood-preserving plants suffer from some form of soil and groundwater contamination (Ryan and Smith, 1986).

The three primary types of wood-preserving chemicals include creosote, a coal tar distillate mixture of over 250 individual compounds; pentachlorophenol (PCP); and metallic salts. Creosote is typically 35% aliphatic hydrocarbons (oil) and 65% PAHs; pentachlorophenol, ranging from 4–8% by weight in a heavy oil carrier; and metallic salts, which are typically aqueous mixtures of copper, chromium, and arsenic (CCA).

Chronologically, creosote was the first of these wood preservatives to be used, predating the turn of the century. PCP began to be used in the 1940s and the 1950s. CCA came into general use in the 1960s. All three types of wood-preserving chemicals are still in use today, often at the same wood treatment plant.

Contamination of wood-preserving facilities occurs in three ways: (1) spillage of material from storage tanks, pipes, etc.; (2) wastewater streams sent to unlined holding or treatment ponds; and (3) drippage from treated material as it dries. The source of contamination is important when determining the regulatory issues concerning permitting, cleanup criteria, and disposal of the treated material.

The majority of contaminated groundwater, soil, and debris found at wood-preserving sites is Resource Conservation and Recovery Act (RCRA) *listed waste* (USEPA, 1992). The material can be categorized into one of the following five waste codes:

- K001—wastewater treatment sludges from wood-preserving processes
- U051—creosote
- F032—PCP wastewater, residuals, drippage
- F034—creosote wastewater, residuals, drippage
- F035—As/Cr wastewater, residuals, drippage

Current RCRA regulations require that soil and debris contaminated with hazardous waste must be subject to the same *best demonstrated available technology* (BDAT) treatment standards as the waste from which they are derived. Rules are under consideration by the EPA whereby soil and debris may be given separate treatment standards. The treatment standards applied to soils and debris will have a major impact on the market potential of on-site biological treatment technologies.

The K001 waste is the only waste that has established BDAT treatment standards. The treatment standard for K001 waste is as follows:

- naphthalene $\leq$ 1.5 ppm
- PCP $\leq$ 7.4 ppm
- phenanthrene $\leq$ 1.5 ppm
- pyrene $\leq$ 1.5 ppm
- toluene $\leq$ 28 ppm
- xylenes $\leq$ 28 ppm

This standard is based on the use of incineration as the treatment technology. Although these criteria can be realized using biological treatment techniques, a number of variables can adversely affect the process when treating to K001 standards. The most difficult criteria to meet are those for pyrene and phenanthrene. Both compounds are relatively immobile and therefore difficult to remove from soil, especially soil that has a high clay and humic content. These compounds also biodegrade at a slow rate.

EPA can also choose to use risk-based treatment criteria. A review of 28 records of decisions (RODs) indicates that cleanup criteria have ranged from 1.5 to 150 ppm for PCP and 1 to 700 ppm for creosote (USEPA, 1992).

EPA (USEPA, 1988) lists four methods of treating contaminated soil from wood-preserving sites:

- biological
- incineration
- soil washing
- stabilization

EPA has issued RODs on a number of wood treatment sites on the Superfund list. Biological treatment is specified in over 40% of the RODs. In the instances where soil washing was specified, biological treatment was typically recommended for the finer soil fractions. Bioremediation is by far EPA's treatment of choice at these sites.

SLURRY PHASE BIOLOGICAL TREATMENT

Soils or sludges that are highly contaminated (>5.0% organics) or soils containing toxic levels of a specific contaminant [e.g., polycyclic aromatic hydrocarbons (PAHs), creosote] are most suitable for slurry phase biotreatment (USEPA, 1988). Assuming that a manageable slurry can be obtained within a range of 15% to 40% (w/v) solids, then one can obtain an instantaneous dilution of the carbon load ranging from 0.75% to 2.0% from the preparation of the slurry suspension. In addition, the bioavailability of the waste organics is going to be a function of the solubility of the compounds and their adsorption characteristics on the soil. Thus, hydrophobic compounds such as high molecular weight PAHs will tend to have a low water solubility in the water phase of the slurry. These characteristics serve to further buffer the biological component of the slurry from shock loadings of organic waste.

A schematic of a slurry biodegradation process is shown in Figure 1. Waste preparation includes excavation and/or moving the waste material to the process where it is normally screened to remove debris and large objects. Particle size reduction, water addition, and pH and temperature adjustments are other important waste preparation steps that may be required to achieve the optimum inlet feed characteristics for maximum contaminant reduction.

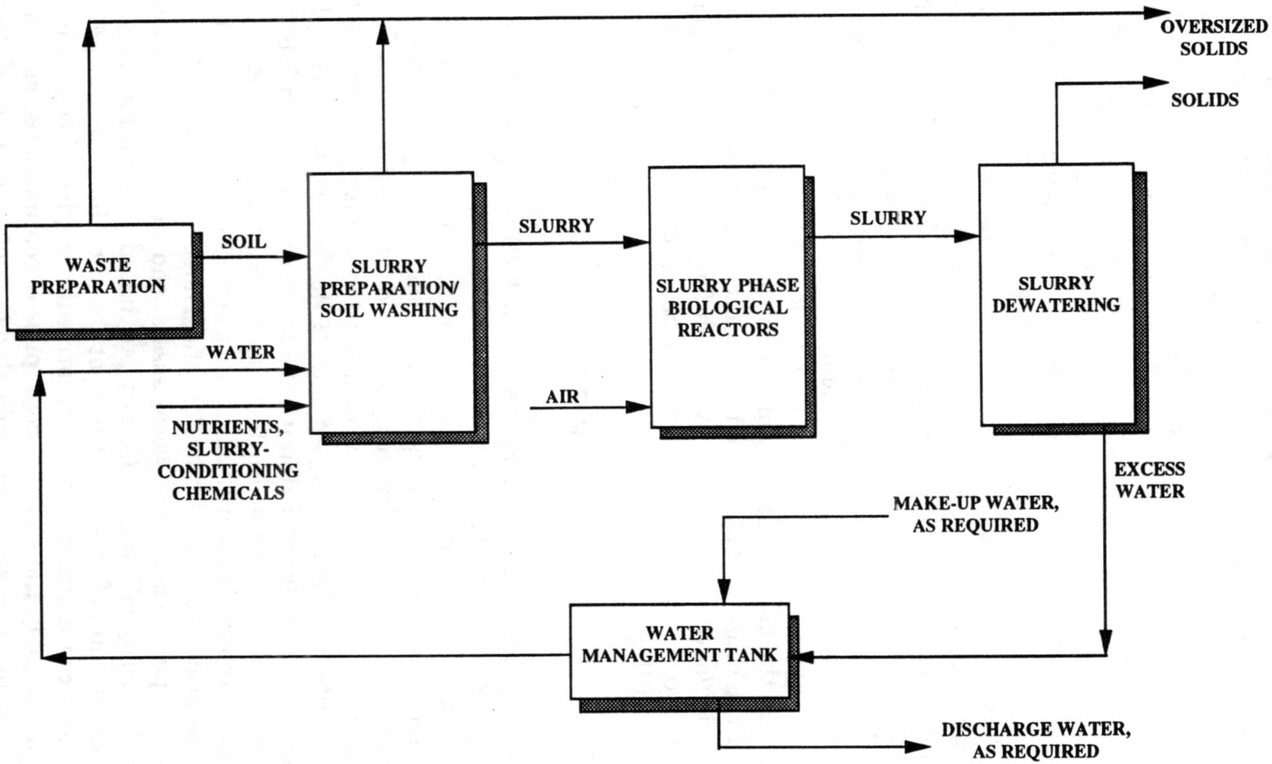

Figure 1. Full-scale slurry phase biological remediation process for K001 wastes.

After appropriate pretreatment, the wastes are suspended in a slurry form and mixed in a tank to maximize the mass transfer rates and contact between contaminants and microorganisms capable of degrading those contaminants. Aerobic treatment in batch mode has been the most common mode of operation. This process can be performed in contained reactors or in lined lagoons. Aeration is provided by floating or submerged aerators or by compressors and spargers. Mixing is provided by aeration alone or by aeration and mechanical mixing. Nutrients and neutralizing agents are supplied to optimize conditions for microbial activity. Other materials, such as surfactants, dispersants, and primary or cosubstrates, can be used to improve the materials' handling characteristics or increase substrate availability for degradation. Microorganisms may be added initially to seed the bioreactor or added continuously to maintain the correct concentration of biomass. The residence time in the bioreactor varies with the soil or sludge matrix, physical/chemical nature and concentrations of the contaminant, and the biodegradability of the contaminants. Once biodegradation of the contaminants is completed, the treated slurry is sent to a separation/dewatering system. A clarifier for gravity separation, or any standard dewatering equipment, can be used to separate the solid phase and the aqueous phase of the slurry.

SITE HISTORY

The Southeastern Wood Preserving (SEW) site is an abandoned wood-preserving facility operated from 1928 until early 1979. The site covers about 10–12 acres near downtown Canton, Mississippi, and is bordered by residential, industrial, and commercial properties.

The Canton Treating Company operated the site from 1928 to 1965. The Dickson Treating Company operated the facility from October 1965 until it filed for bankruptcy in late 1979. The assets were held by bankruptcy court from 1979 until 1982 when Southeastern Wood Preserving purchased the site. In 1984, Southeastern defaulted on its Small Business Administration Loan, having never operated the facility. At the foreclosure auction, White Pole and Timber Company of Kennedy, Alabama, purchased the physical assets, and the Industrial Development Authority of Madison County purchased the real property.

After 50 years of wood treatment at the site, there were large areas of obvious contamination in the treatment area, heavy contamination from drippage in the storage areas, piles of contaminated soil on-site, storage tanks containing creosote sludge, and three unlined lagoons covered over by the previous owner.

In 1986, EPA initiated the Phase I of a removal action at SEW to stabilize the unlined lagoons that had been filled in and covered. The lagoons contained creosote sludge that was leaching into the adjacent Bachelor Creek. Additional creosote contamination was reaching the creek through runoff from flooding of heavily stained and contaminated treatment areas. An EPA Emergency Response and Cleanup Services (ERCS) contractor excavated approximately 8000 cubic yards of sludge and contaminated soil from within and around the lagoons and treatment and storage areas. ERCS also dismantled the creosote storage tanks and treated water from flooded areas of the site. The excavated materials were stabilized using lime kiln dust and stockpiled on-site for future treatment or disposal during Phase II of the removal action.

Phase II of the removal action was planned to consist of either on-site treatment or off-site disposal of the material in the waste pile. Before this could be done, the quantity of soil to be treated had to be determined, disposal options and costs evaluated, and the site ranked for possible inclusion on the National Priority List (NPL). Because SEW received a preliminary NPL score of 21.9 in 1989, it became apparent that the site would not be proposed for the NPL and thus, no remedial followup could be anticipated. Additional funds to complete Phase II would be requested when funds were available and the amount needed could be accurately estimated.

The USEA Soil Conservation Service (SCS) contacted EPA in December 1988 concerning a soil erosion prevention plan that called for widening and lining Bachelor Creek. While surveying the creek, SCS personnel observed oily waste leaching into the creek from the SEW site. In April 1989, EPA Technical Assistance Team (TAT) personnel sampled the soils to be excavated during the planned widening. The soils contained concentrations of polynuclear aromatic hydrocarbons (PAHs) ranging from 199 to 856 mg/kg (ppm). Because of the high concentrations in the soil and creek sediment, the SCS requested that EPA perform the widening work in the portion of Bachelor Creek bordering the SEW site.

EPA and SCS signed an Interagency Agreement in August 1989, under which EPA agreed to perform the creek widening work. EPA's responsibilities included excavating the creosote-contaminated soil to form a 20-foot wide bottom with 2 : 1 side slopes, then lining the new creek channel with geotextile fabric and riprap (rock).

A small amount of visibly contaminated soil excavated during the widening was added to the existing waste stockpile. Analyses performed on a composite sample of the remaining excavated soil indicated that the concentration of total PAHs (<149 ppm) was below the action level for treatment of 1000 ppm recommended by the Agency for Toxic Substances and Disease Registry (ATSDR), so this excess soil was spread on-site.

OHM Remediation Services Corp. (OHM) conducted the full-scale remediation of creosote-contaminated soils and sludges at the Southeastern Wood Preserving Superfund site in Canton, Mississippi. The overall treatment process included materials excavation and handling, soil washing, slurry reactor treatment, and slurry dewatering. OHM performed laboratory treatability studies to provide site-specific information necessary for the selection and implementation of slurry phase treatment technology. The results from these studies were used to develop the process design and operating conditions for the bench-, field-pilot-, and full-scale treatment systems.

LABORATORY TREATABILITY TESTING

The material stockpile at the site was power-screened through a ½-inch grizzly screen. The screening process removed debris such as large stones, plastic sheeting, and railroad ties and produced material appropriate for soil washing, slurry preparation, and treatment. The material used for the laboratory testing was further sieved through a No. 200 mesh screen.

Laboratory-scale slurry phase bioremediation studies were performed in sealed 4-L glass bioreactors fitted with stainless steel baffles. A 1/30 hp fixed shaft mixer with a stainless steel blade impeller was operated at 1000 rpm to achieve a homogeneous slurry suspension. The pH of each slurry was adjusted to 7.2 and monitored daily. Concentrations of bioavailable nutrients, ammonia-nitrogen, and orthophosphorus were monitored three times weekly. The reactors were operated in a batch mode for 30 days at 24 $\pm$ 2°C. The slurry solids concentration was 20 wt %. The dissolved oxygen concentration was maintained at more than 2.0 ppm by delivering air directly beneath the impeller.

The bioreactors were sampled throughout the 30 days of operation. The samples were obtained by manually removing 25 ml of slurry through surface ports. Sample aliquots for PAH analysis were centrifuged, dried with sodium sulfate, and soxhlet extracted for 24 h following EPA Method 3540 (USEPA, 1989). PAH concentrations in the slurry solids were determined by GC/MS EPA Method 8270 (USEPA, 1989). The PAHs quantified were the 16 compounds established by EPA. Aqueous PAHs were measured by high-pressure liquid chromatography following filtration through a 0.45-micron filter. Effluent air was passed through a vapor phase trap containing a porous styrene-divinyl benzene copolymer (Supelco, Bellefonte, Pennsylvania). The traps were analyzed by NIOSH Method 5515 to account for PAH losses by volatilization. At the termination of the studies, any material adhering to the internal reactor surfaces was solvent-washed to determine PAH losses by adsorption. Abiotic controls accounted for PAH losses by nonbiological mechanisms.

Physical characterization of the power-screened material indicated 5% gravel (>200 mm, >10 mesh), 40% sand (>0.08 mm, >200 mesh), and 55% silts and clays (<0.08 m, <200 mesh) (Figure 2). The particle size distribution of the silts and clays treated in the full-scale biological slurry reactors indicated that 85% of the material was less than 30 microns (Figure 2). The smaller particle sizes typically contain higher concentrations of contaminants due to a higher surface/volume (mass) ratio and are more amenable to slurry suspension due to lower densities.

Characteristics of the slurry reactor feedstock (<95 microns) are presented in Table 1. The total organic carbon and humic concentrations were in the upper range representative of these types of soils. The total petroleum hydrocarbon concentration was high probably due to the petroleum hydrocarbon carrier for the creosote material. The PAH concentrations ranged from 8000 to 12,000 mg/kg with the seven EPA carcinogenic PAHs accounting for approximately 20%. Aerobic heterotrophic bacterial populations and PAH-degrading bacterial populations ranged from 10^7 to 10^8 and 10^5 to 10^6 cfu/g dry soil, respectively, indicat-

Table 1. Characterization of K001 waste feedstocks for slurry reactor treatment.

Physical	
▶ Particle size distribution (μm)	95% < 50
▶ Slurry density (20 wt %) g/ml	1.2
Chemical	
▶ Total organic carbon (mg/kg)	32,300
▶ Humus content (mg/kg)	44,700
▶ Total petroleum hydrocarbons (mg/kg)	11,000
▶ Total PAHs (mg/kg)	12,000
▶ Total CPAHs (mg/kg)	1700
Biological	
▶ Total aerobic heterotrophic bacteria (cfu/g)	10^7–10^8
▶ Total PAH degrading bacteria (cfu/g)	10^5–10^6

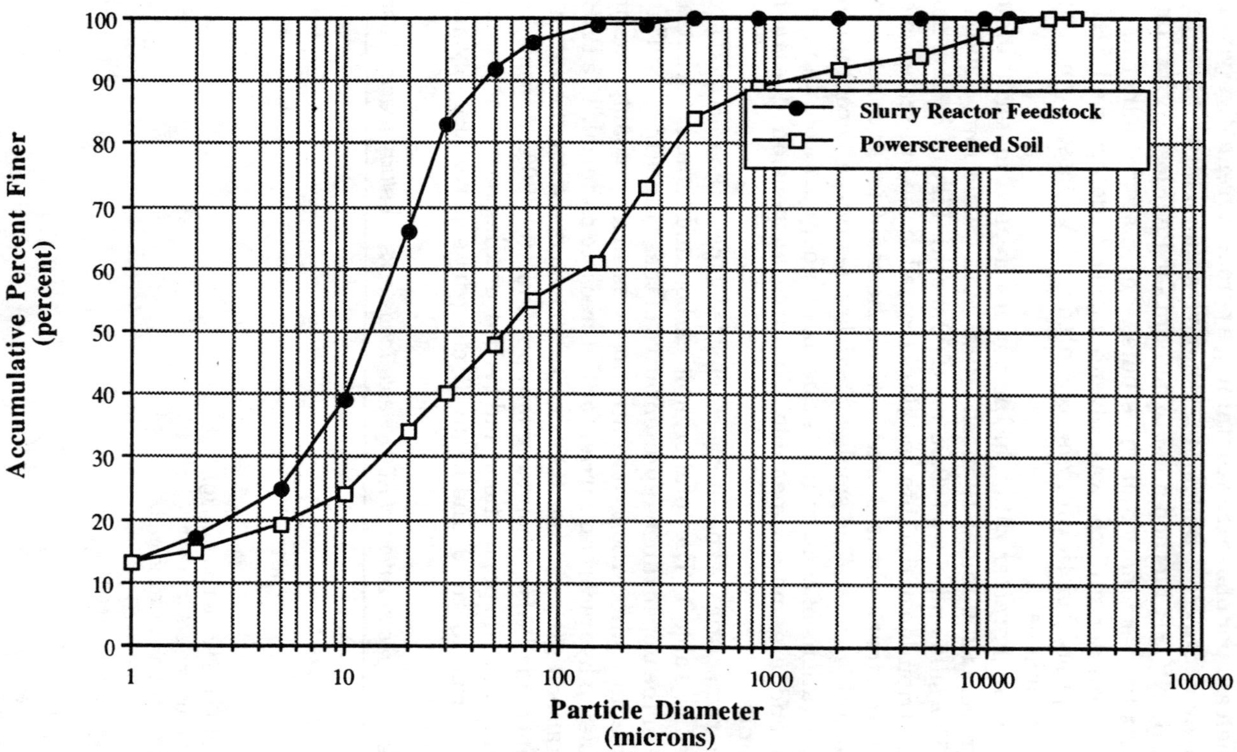

Figure 2. Particle size distribution of power-screened soil and slurry reactor feedstock.

ing an indigenous population for enhancement and degradation of the creosote constituents.

The laboratory slurry reactors and operating conditions for the laboratory slurry reactors are shown in Figure 3 and Table 2. Approximately 95% treatment efficiencies were achieved for total PAHs at 10 wt % and 20 wt % slurry solids in the laboratory slurry reactors following 30 days (Figure 4). The majority of the PAH biodegradation occurred during the initial 5 to 10 days incubation followed by a very slow rate of treatment. Degradation rates are comparable to those reported by Mueller et al. (1991) for laboratory studies and Lauch et al. (1992) for laboratory- and pilot-scale tests. Within 5 days an average PAH concentration of 10,300 mg/kg was reduced to 490 mg/kg. Total PAH concentration in the abiotic control reactor was 7800 mg/kg following 23 days of incubation. In addition, the final concentrations of phenanthrene and pyrene measured in the abiotic control reactors were 80% and 100% of the initial concentration, respectively, indicating that biodegradation was the mechanism responsible for treatment of PAHs.

The mass of PAHs was less than 0.010 mg in the vapor phase trap material following 10 days of incubation, indicating that less than 0.0004% of the PAHs were volatilized during treatment. Analysis of reactor surfaces to account for PAH adsorption indicated that less than 0.002% of the PAHs were adsorbed onto the vessel walls.

Three-ring PAHs such as phenanthrene, anthracene, and fluorene were extensively biodegraded in the slurry reactors as a reduction of 98 to 99% from initial concentrations was achieved (Figure 5). Phenanthrene was most rapidly degraded, followed by anthracene and fluorene. No lag phase was observed for any of these compounds. All of these PAHs have been reported to be utilized by microorganisms as sole carbon and energy sources (Heitkamp and Cerniglia, 1988). The

Table 2. Laboratory reactor operating conditions.

Volume (L)	4
Impeller speed (rpm)	1000
Air flow rate (scfm)	2.6×10^{-3}
Solids loading (%)	20
Initial defoamer (mg/L)	200
Initial dispersant (mg/L)	1000
Temperature (°C)	24 ± 2
pH (SU)	7.2 ± 0.4
Dissolved oxygen (mg/L)	>2.0
$NH\text{-}_4\text{-}N$ (mg/L)	50 ± 5
PO_4P (mg/L)	10

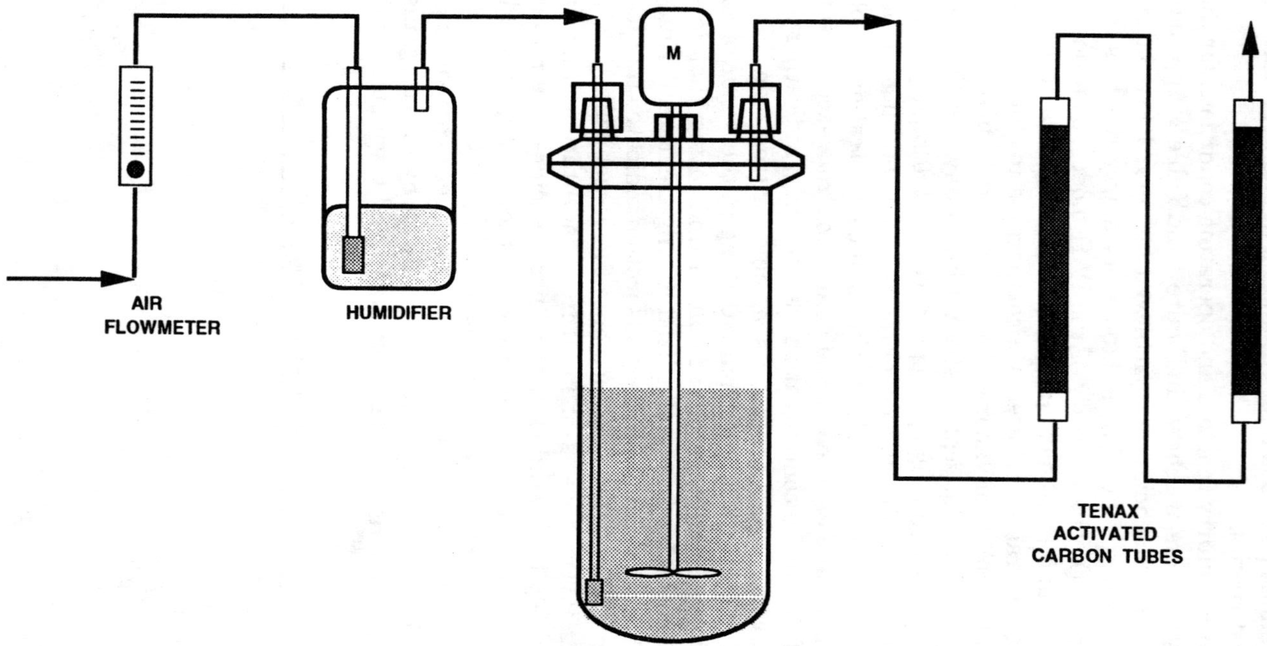

Figure 3. Slurry bioreactor.

4-ring PAHs such as fluoranthene, pyrene, chrysene, and benzo(*a*)anthracene were also biodegradable (Figure 6). The overall treatment efficiency for this group of compounds ranged from 85 to 90%. The concentration of fluoranthene was reduced by 97%, while 90% reductions were observed for benzo(*a*)anthracene and pyrene. The treatment efficiency for chrysene was 84%. A longer lag period was observed prior to degradation of the 4-ring PAHs. Microorganisms that utilize benzo(*a*)anthracene, fluorene, and pyrene as sole carbon and energy sources have been reported (Heitkamp and Cerniglia, 1988). The first-order degradation rate for the total PAHs was 0.37 $days^{-1}$. The first-order degradation rates for phenanthrene were 0.38 and 0.73 at 10% and 20% slurry solids concentrations, respectively (Figure 7). The increased degradation rate at the 20% slurry concentration was probably due to the increased organic loading rate (substrate concentration). The organic loading rates for phenanthrene were 0.038 and 0.090 g/L/day at the 10 and 20% slurry solids concentrations, respectively. The decay rate for total PAHs was in the range of 0.19 to 0.54 $days^{-1}$ reported for sediments containing higher concentrations of PAHs in slurry reactors (Middleton et al., 1991). Treatment efficiencies and rates for the 4-ring compounds increased at the higher concentrations,

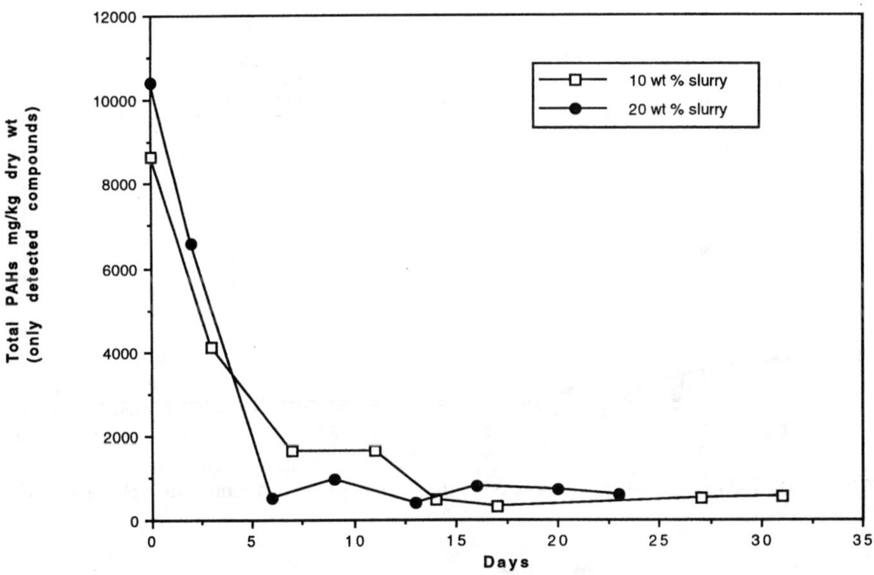

Figure 4. Laboratory slurry phase biological treatment of polycyclic aromatic hydrocarbons in soil.

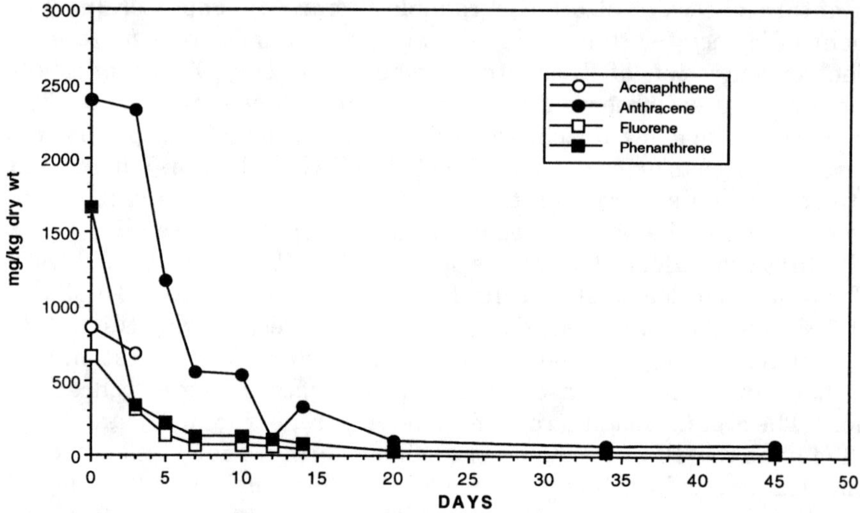

Figure 5. Laboratory slurry phase biological treatment of 3-ring polycyclic aromatic hydrocarbons.

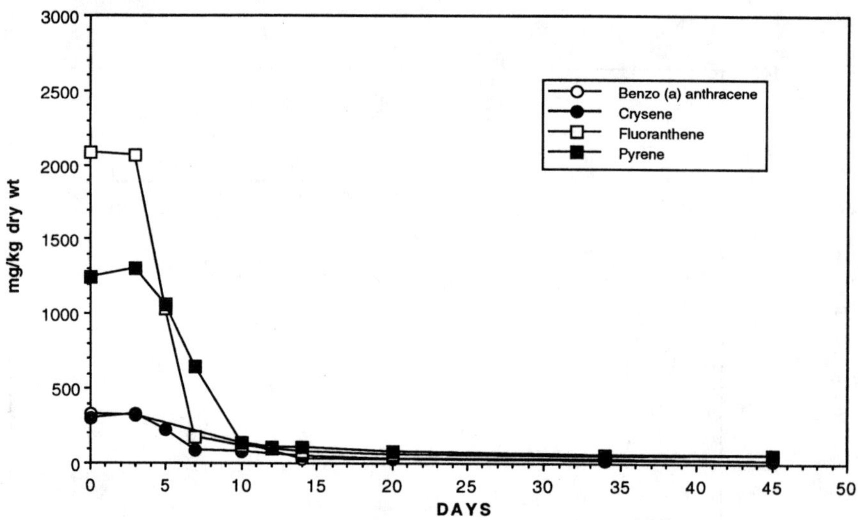

Figure 6. Laboratory slurry phase biological treatment of 4-ring polycyclic aromatic hydrocarbons.

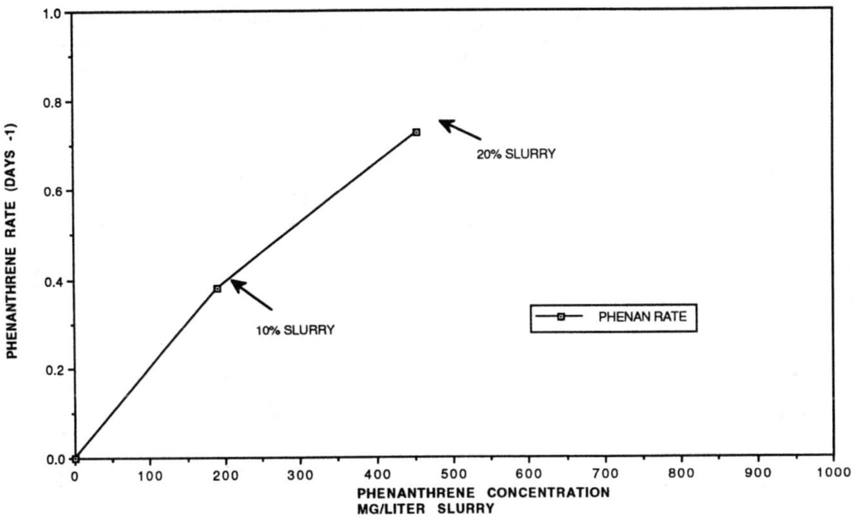

Figure 7. The effect of organic loading rates on the degradation of phenanthrene in slurry phase reactors.

which the authors attributed to catabolic activities on the 2- and 3-ring PAHs and co-metabolism. Up to 60% degradation of 5- and 6-ring PAHs present in the soil, including benzo(*b*)fluoranthene, benzo(*k*)fluoranthene, and benzo(*a*)pyrene, was also reported.

BENCH-SCALE PILOT TESTING

The treatment vessel was a 25-gallon cylindrical stainless steel tank with cover. A Lightnin' shaft-mounted mixer provided slurry suspension. Mechanical foam breakers were installed on the mixer shaft.

Air was added to the reactor through metering valves to control flowrates. The dissolved oxygen concentration of the reactors was monitored and the air flowrate adjusted to maintain a minimum of 2.0 to 4.0 ppm dissolved oxygen in the slurry. In order to minimize foam formation, a dispersant and a defoamer were added to the slurry reactors and allowed to mix for 2 hours before the addition of air to the slurry. The operating conditions are shown in Table 3.

The slurry for testing was obtained from the slurry preparation process at the site and shipped to the laboratory in 55-gallon drums. A single drum was thoroughly mixed for 12 to 14 hours to obtain a homogeneous sample. Nitrogen and phosphorus concentrations were

Table 3. Bench pilot slurry reactor operating conditions.

Volume	25 gallon
Mixing speed	765 rpm
Impeller	3.4 inch A-310
Slurry	25 gal (approximately 25% solids)
Defoamer	1000 ppm dispersant, 200 ppm defoamer
Ammonium nitrate	150 g (dissolved in water before addition)
Triple phosphate	75 g (dissolved in water before addition)
pH	7.1 to 7.3 (adjust as required)
Air flowrate	3,000 ml/min (adjust as required to maintain 4.0 ppm D.O.)

adjusted by the addition of ammonium nitrate and triple phosphate, respectively.

The pH, air flowrate, temperature, and dissolved oxygen were monitored daily. The phosphate-phosphorus and ammonia-nitrogen were also monitored daily. Pilot reactor sampling was conducted on day 0, 1, 4, 7, 10, 14, 17, 20, 24, 27, and 30.

Samples were collected from the center sampling valve of each pilot reactor while the mixer or rake/airlift was operating. The sampling port was flushed prior to sample collection. Approximately 10 g of slurry from the reactors were collected in a disposable centrifuge tube at each sampling event for PAH analysis. The slurry was then transferred directly into a soxhlet extraction thimble containing 0.25 inches anhydrous sodium sulfate resting in an 8-ounce sample jar. The jar, thimble, and sodium sulfate were placed on a pan balance and tared before slurry transfer. Additional sodium sulfate was added to the thimble and mixed with a stainless steel spatula. A glass wool plug was added to the thimble, and an aliquot of acetone was poured into the 8-ounce jar.

An additional 10 g were collected during sampling for percent solids (weight/weight) analysis. The percent solids were determined by drying samples at 105°C overnight. The average percent solids were used to determine PAH concentrations on a dry weight basis. A portion of this sample was used for a plate count analysis of the slurry to determine microbial population density.

Overall degradation of 3-, 4-, and 5-ring PAHs in the bench-scale reactors is shown in Figures 8, 9, and 10. The 3-ring PAHs were most rapidly degraded, and the rates of degradation were similar to the rates observed in the laboratory-scale reactors. Treatment efficiencies were also similar, ranging from 96 to 99% for phenan-

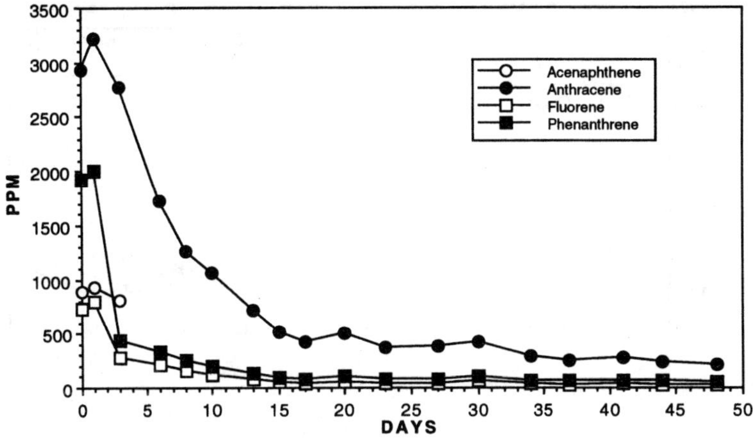

Figure 8. Biological treatment of 3-ring PAHs in the bench pilot-scale slurry phase reactor.

threne for example. The 4-ring PAHs were degraded more slowly in comparison to the 3-ring PAHs. Pyrene degradation kinetics were slower in the bench pilot reactors in comparison to the laboratory reactors, possibly due to the different mixing characteristics such as the longer tank turnover time. The 5- and 6-ring PAHs were not degraded to any appreciable extent.

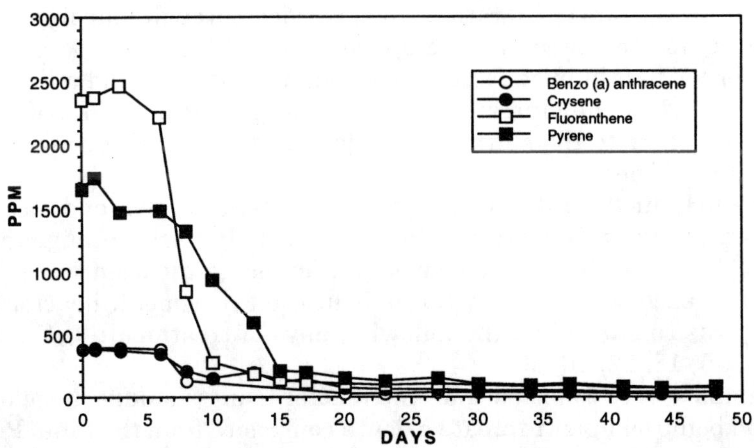

Figure 9. Biological treatment of 4-ring PAHs in the bench pilot-scale slurry phase reactor.

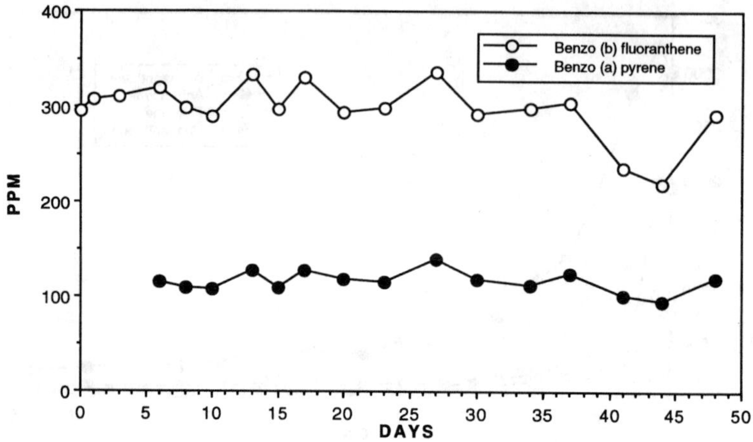

Figure 10. Biological treatment of 5- and 6-ring PAHs in the bench pilot-scale slurry phase reactor.

FIELD-SCALE PILOT TESTS

The field-scale treatment vessel was an open-top, 350-gallon cylindrical stainless steel tank, 48 inches in diameter with a 45-inch effective slurry depth (Table 4). Four stainless steel baffles were attached to the tank walls. A Lightnin' mixer provided power to suspend the slurry.

Air was added to the reactor through a metering valve. The dissolved oxygen concentration in the reactor was monitored and the air flowrate adjusted, as required, to maintain a minimum of 4.0 ppm dissolved oxygen in the slurry. Aeration was provided through four fine bubble diffusers located beneath the impeller.

In order to minimize foam formation, a dispersant and a defoamer were added to the slurry and allowed to mix for 2 hours before the addition of air to the slurry. The defoamer was added during reactor operation, if necessary.

The pH, air flowrate, temperature, dissolved oxygen, and dissolved oxygen uptake rate were monitored on a daily basis throughout the tests. The phosphate-phosphorus and ammonia-nitrogen were monitored on each sampling day throughout the experiment. Reactor sampling was conducted on the following days of treatment: 0, 1, 2, 3, 6, 9, 12, 15, 18, 22, 26, 30.

Samples were collected from the reactor using a sludge judge or the Bacon bomb sampler to create a depth composite from the tank. Proper decontamination and sampling handling procedures were in effect at all times. Two quart jars approximately two-thirds full of slurry from each reactor were collected at each sampling event for PAH analysis, nutrient, total solids, and plate count analyses.

The operating parameters for the field pilot test are presented in Tables 5 and 6. The dissolved oxygen was maintained at concentrations greater than 2 ppm in the reactors. A 12% solids concentration was maintained in the 900 series test and was increased to 28% in the 1000 series test. Temperature, pH, and nutrients were maintained in a range conducive for microbial activity.

The slurry temperatures in the reactors during the operational period ranged from 22°C to 25°C in the 900 series and from 25°C to 33°C in the 1000 series test (Figure 11). The dissolved oxygen concentration was maintained above 2 ppm to insure a nonlimiting oxygen supply for microbial degradation. The dissolved oxygen uptake rate (DOUR) was inversely proportional to the dissolved oxygen concentration. The DOUR was highest at the onset of the treatment cycle and decreased with time. Two maximum DOURs were measured that corresponded to low dissolved oxygen concentrations in the slurry, indicating increased microbial activity.

Table 4. Field pilot reactor—operating conditions.

	Reactor	
Parameter	902	1002
Mixing		
Type	Agitator	Agitator
Speed, rpm	550	~1000
Slurry volume, gallon	350	350
Aeration		
Air flowrate, cfm	~4	~4
Dissolved oxygen, mg/L	>2.0	>2.0
Solids		
Total solids, percent	~12	~28
Particle size, microns	<200	<3000
Temperature °C	20–25	25–30
pH (SU)	6.8–7.4	6.8–7.4
Nutrients		
Ammonia-nitrogen, mg/L	>100	>100
Phosphate-phosphorus, mg/L	>5	>5

Table 5. Operating parameters for field pilot reactor (902) during slurry phase biodegradation of polycyclic aromatic hydrocarbons.

Day	Total Solids (%)	NH_3-N (mg/L)	PO_4-P (mg/L)	pH (SU)	Temperature[a] (C)	Plate Counts (cfu/mL)	Diss Oxy[a] (mg/L)	DOUR[a] (mg/L min)
0	12.0%	NA	NA	7.0	19.5	1.67×10^9	2.2	0.38
1	11.3%	302	36	6.5	21.5	1.58×10^9	1.7	0.63
2	11.5%	306	81	6.5–7.0	20	4.33×10^9	8.1	0.32
3	11.1%	312	144	6.5–7.0	20	1.69×10^9	7.1	0.34
4	12.8%	NA	NA	6.1–7.2	22	8.08×10^9	5.6	0.60
4	13.0%	NA	NA	6.1–7.2	22	5.29×10^9	5.6	0.60
4	12.9%	NA	NA	6.1–7.2	22	6.43×10^9	5.6	0.60
5				6.3–7.5	23		3.3	0.80
6	12.6%	NA	NA	7.1	23	2.45×10^9	5.5	0.54
7				7.2	24		6.1	NA
8				7.3	24		7.4	0.22
9	12.6%	144	65	7.5	25	3.98×10^9	6.9	0.26
10				7.6	25.5		7.2	0.26
11				7.6	25		7.3	0.12
12	12.9%	170	115	7.2	25	1.33×10^9	6.9	0.14
13				7.2	25		6.9	0.09
14	11.9%	NA	NA	NA	NA	6.44×10^8	NA	NA

[a]Analysis performed on-site.
NA = Not analyzed.

Table 6. Operating parameters for field pilot reactor (1002) during slurry phase biodegradation of polycyclic aromatic hydrocarbons.

Day	Total Solids (%)	NH_3-N (mg/L)	PO_4-P (mg/L)	pH (SU)	Temperature[a] (C)	Plate Counts (cfu/mL)	Diss Oxy[a] (mg/L)	DOUR[a] (mg/L min)
0	27.9%	147	3.2	6.87–7.13	24	1.25×10^9	3.0	0.92
1	28.0%	155	3	6.8	25.5	1.54×10^9	3.1	2.29
2	28.6%	151	4.5	7.1	26	1.22×10^9	4.4	0.56
3	29.2%	NA	NA	6.67–7.12	31	9.62×10^8	1.9	2.20
4				6.6	31		3.6	3.40
5				6.6	34		1.2	7.20
6	29.2%	NA	NA	6.8–7.15	33	4.80×10^9	2.5	1.20
7				7.2	29		5.5	0.30
8				5.9–6.9	26		5.7	0.35
9	29.9%	750	110	6.5-n7.2	25	1.44×10^{10}	7.4	0.32
10				7.0	33		8.0	0.27
11				7.1	33		7.2	0.48
12	31.1%	NA	NA	6.95–7.25	33	9.20×10^9	6.3	0.39
13				7.2	28		6.5	0.25
14				7.2	29		6.5	0.16
15				7.2	29		7.0	0.20
16	31.90%	625	58.8	7.0	29	4.63 × 10;9	6.9	0.27

[a]Analysis performed on-site.
NA = Not analyzec.

The biological treatment of the PAHs in the field pilot reactor is presented in Table 7. Treatment rates and treatment efficiencies for the total and carcinogenic PAHs were comparable to the laboratory- and bench-scale reactors.

FULL-SCALE TREATMENT SYSTEM

Process Description

SLURRY PREPARATION

The stockpiled material at the site was power-screened through a series of screens to prepare the material for treatment. The power-screen separated the material into a debris (greater than 6-inch) pile, two power-screen reject piles (greater than 2-inch and greater than 0.5-inch), and the material requiring further treatment (less than 0.5-inch).

The slurry preparation system consisted of a three-chambered slurry mix tank with a vibrating shaker screen and a hydrocyclone. The screened material was loaded into the first chamber with a tracked excavator, and water was added to achieve the operating slurry solids concentration (18 to 20 wt %). Two 10-hp mixers in the first chamber provided mixing and homogenization of the slurry. A 50-hp slurry pump transferred the slurry from the first chamber to the 12-mesh shaker

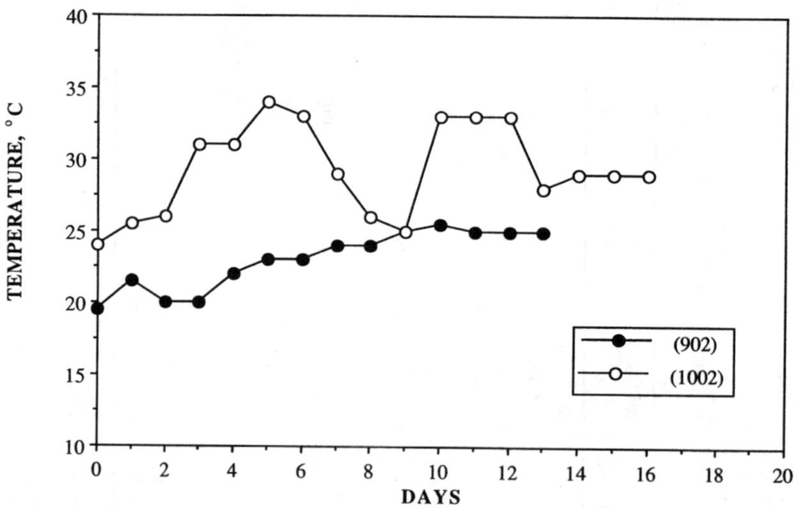

Figure 11. Temperature profiles for field pilot reactors during slurry phase biodegradation of polycyclic aromatic hydrocarbons.

Table 7. Slurry phase biological treatment of polycyclic aromatic hydrocarbons in wood-preserving wastes (Reactor 1002).

Compound	Units	Day 0	Day 1	Day 2	Day 3	Day 6	Day 9	Day 12	Day 16
Acenaphthene	mg/kg	532	535	512	544	528	<27.2	<26.9	<32.5
Acenaphthylene	mg/kg	<137	<138	<139	<140	<141	<142	<143	<144
Anthracene	mg/kg	1440	1510	1670	1480	1540	319	304	217
Fluoranthene	mg/kg	1300	1340	1430	1620	1510	126	110	74.8
Fluorene	mg/kg	543	492	273	195	225	59	51.2	<32.5
Naphthalene	mg/kg	<137	<138	<139	<140	<141	<142	<143	<144
Phenanthrene	mg/kg	1060	664	286	243	276	92.8	85.6	55.5
Pyrene	mg/kg	920	942	989	1220	1170	63.8	66.7	39.6
Benzo(*a*)anthracene	mg/kg	217	237	239	313	302	27.3	<26.9	<32.5
Benzo(*a*)pyrene	mg/kg	<137	<137	<152	<174	<153	70.1	70.5	55
Benzo(*b*)fluoranthene	mg/kg	236	241	259	278	264	277	237	184
Benzo(*g,h,i*)perylene	mg/kg	<137	<138	<139	<140	<153	<27.2	<26.9	<32.5
Benzo(*k*)fluoranthene	mg/kg	<137	<137	<152	<174	<153	<27.2	<26.9	<32.5
Chrysene	mg/kg	220	229	247	303	295	67.8	98	43.7
Dibenzo(*a,h*)anthracene	mg/kg	<137	<138	<139	<140	<141	<142	<143	<144
Indeno (1,2,3-*c,d*)pyrene	mg/kg	<137	<137	<152	<174	<153	34.1	<26.9	<32.5

Samples were soxhlet extracted and analyzed by EPA method 8270.
Values are reported on a dry-weight basis.

screen. The greater than 12-mesh reject was stockpiled on-site and the remaining slurry flowed into the second mix chamber where two 10-hp mixers provided mixing and further homogenization of the slurry. A 50-hp slurry pump transferred the slurry to a hydrocyclone that removed and stockpiled material greater than 200 mesh (80 μm). A single 10-hp mixer provided mixing and blending of the slurry, nutrients, and slurry conditioning chemicals in the third chamber. A 75-hp slurry pump transferred the slurry to the four slurry phase bioreactors for further treatment. Approximately 25 to 30% of the material fed to the slurry preparation system was removed from the slurry and stockpiled on-site.

SLURRY TREATMENT

The prepared soil slurry was pumped to one of four 210,000-gallon, circular, open-top, bolted-steel biological slurry phase reactors with an operating capacity of 180,000 gallons. Aeration of the slurry was provided by a circular layout of 66 fine-bubble diffusers mounted on a 4-inch PVC header on the reactor floor. A 65-hp blower supplied approximately 650 cubic feet per minute of air to the diffuser system. A floating, direct-drive, 40-hp mixer maintained the slurry in suspension.

The reactors were monitored daily for pH, temperature, dissolved oxygen, and DOUR. Discrete slurry samples were collected and composited for PAH analysis to monitor biological treatment. The reactors were operated in a batch mode for 8 to 11 days to achieve the treatment criteria. At the completion of biological treatment, the slurry was transferred to the slurry dewatering unit with a 75-hp slurry pump.

SLURRY DEWATERING

The slurry dewatering unit was a 425-ft long, 160-ft wide, and 6-ft deep HDPE-line cell. A leachate collection system on the floor of the cell and a collection sump facilitated removal of the excess water. The excess water was pumped to the 350,000-gallon water management tank for reuse as water in the slurry preparation system. The soil in the dewatering unit was worked with an excavator and tilled with a disc to further dewater and dry the treated material.

Process Operation and Performance

The full-scale treatment system was extensively monitored during 1992 field operations. Thirteen batches were treated in two reactors during this time period, slurry reactor 1, batches 4 through 10, and slurry reactor 2, batches 5 through 10. The treatment time for these

Table 8. Full-scale slurry phase reactor treatment batches.

Batch	Start Date	Finish Date	Batch Cycle Treatment Time/Days
Slurry Reactor 1			
4	5/13/92	6/15/92	33
5	6/22/92	7/21/92	30
6	7/30/92	8/10/93	12
7	8/20/92	9/9/92	20
8	9/15/92	10/1/92	18
9	10/9/92	10/25/92	16
10	10/30/92	11/23/92	25
Slurry Reactor 2			
5	5/23/92	6/12/92	21
6	7/10/92	7/27/92	17
7	8/4/92	8/17/92	13
8	8/26/92	9/28/92	32
9	10/4/92	10/22/92	19
10	10/27/92	11/23/92	26

batches typically ranged from 15 to 30 days (Table 8). Although the majority of the treatment occurred within the initial 5 to 10 days, the treatment period was extended to confirm the lack of any further appreciable contaminant reduction.

The full-scale reactors were 250,000 gallons with an operating volume of 180,000 gallons (Table 9). The air flowrate was maintained in a range from 300 to 450 scfm to supply sufficient oxygen for microbiological activity. The remaining operating conditions were maintained in a range similar to the laboratory and pilot reactor tests (Table 9). The reactor slurry temperatures ranged from 25°C to 40°C, depending on

Table 9. Field reactor operating conditions.

Volume (L)	750,000
Impeller speed (rpm)	900
Air flow rate (scfm)	350 ± 100
Solids loading %	20
Initial defoamer (mg/L)	200
Initial dispersant (mg/L)	1000
Temperature (°C)	30 ± 10
pH (SU)	7.2 ± 1.0
Dissolved oxygen (mg/L)	>2.0
NH_4-N (mg/L)	60 ± 20
PO_4P (mg/L)	20 ± 10

the operating period (Figure 12). The decrease in temperature observed during the final stages of batch operation was probably due to a reduction in the air flowrate. The temperature of the air delivered to the reactors from the blowers was generally 100°C due to frictional heating.

The dissolved oxygen concentrations in the slurry ranged from less than 2 mg/L during the initial days of treatment to concentrations approaching dissolved oxygen saturation during the final stages of treatment. The pH of the slurry was maintained in the neutral range between 6.5 and 7.5. The dissolved oxygen uptake rate was a good indicator of microbial activity. A higher DOUR was measured during the initial days of the batch treatment cycle (Figure 13). This indicated a higher rate of microbial activity, which correlated to a higher rate of PAH degradation (Figures 14 and 15). The rate and extent of PAH degradation in the individual batches are shown in Figures 16 and 17.

The treatment performance for six batches is summarized in Table 10. The 3-ring PAHs accounted for approximately 50% of the total PAHs and were the most extensively degraded. Initial concentrations of 909, 630, and 1031 mg/kg were reduced to 6, 14, and 34 mg/kg for acenaphthene, fluorene, and phenanthrene, respectively, representing treatment efficiencies exceeding 97%. Anthracene, the 3-ring PAH present at the highest concentration of 1950 mg/kg, was treated to a final concentration of 121 mg/kg, representing a 94% treatment efficiency. Acenaphthylene was present at 93 mg/kg and was readily biodegraded. The data from selected batches are presented in Tables 11 through 16.

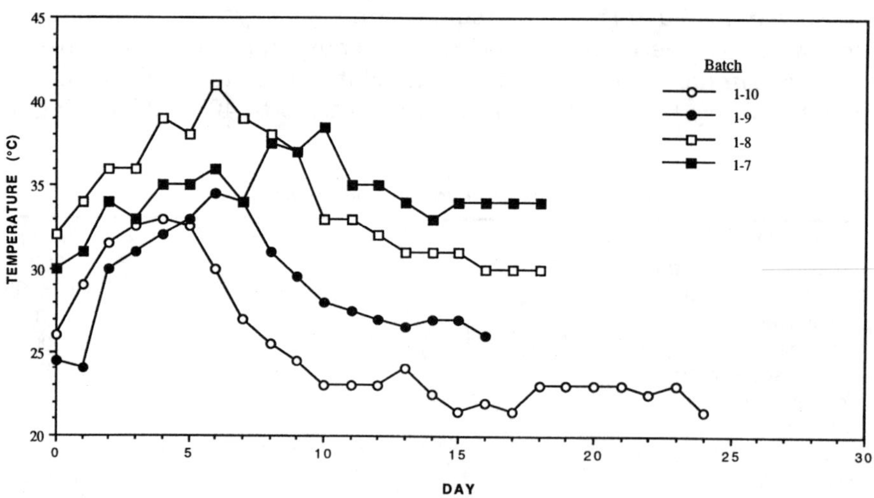

Figure 12. Slurry phase reactor 1—operating temperatures.

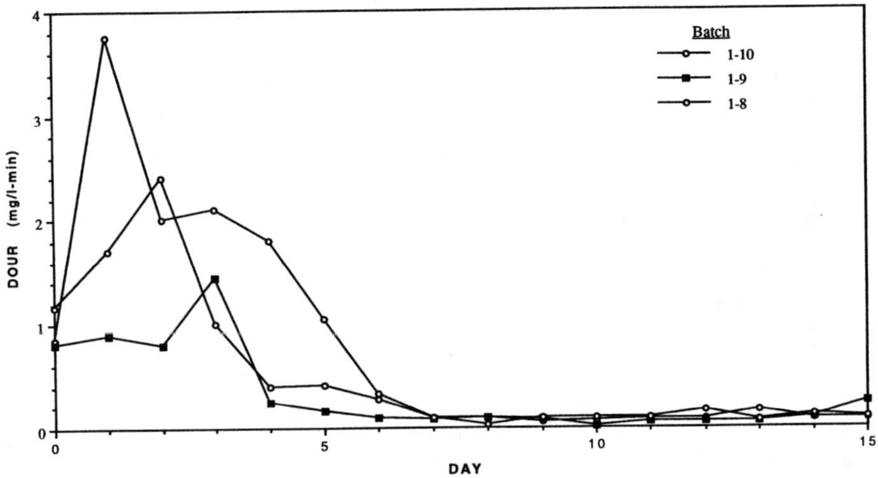

Figure 13. Slurry phase reactor 1—dissolved oxygen uptake rate.

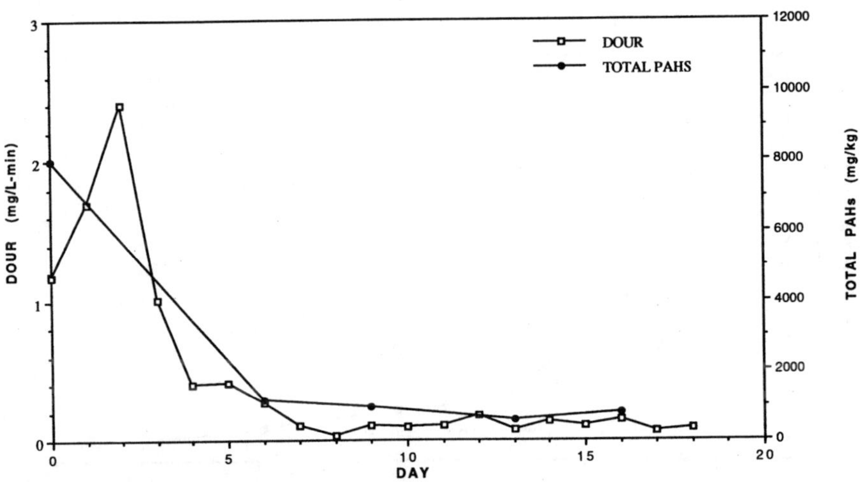

Figure 14. Comparison of dissolved oxygen uptake rate (DOUR) and total polycyclic aromatic hydrocarbon degradation during slurry phase reactor operation (Reactor 1, Batch 8).

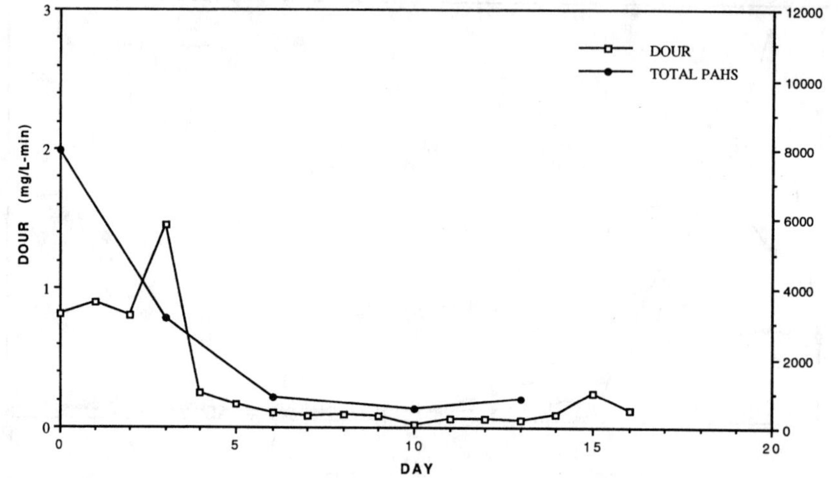

Figure 15. Comparison of dissolved oxygen uptake rate (DOUR) and total polycyclic aromatic hydrocarbon degradation during slurry phase reactor operation (Reactor 1, Batch 9).

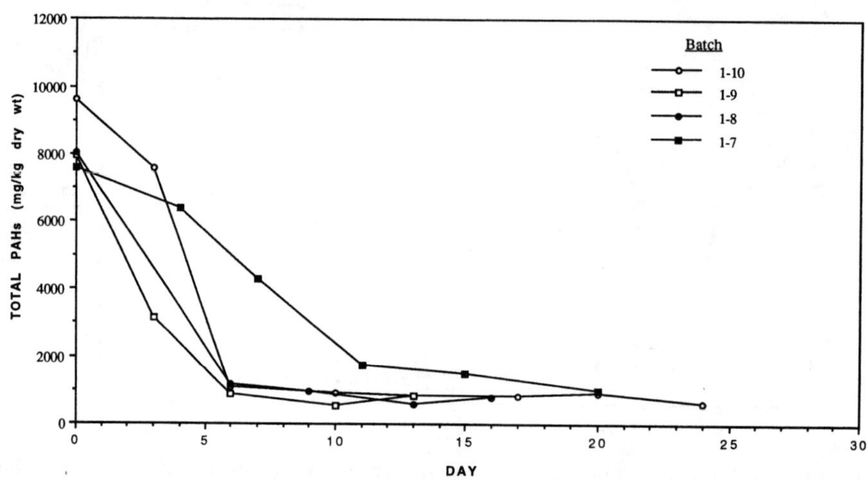

Figure 16. Slurry phase reactor 1—treatment of total polycyclic aromatic hydrocarbons.

Treatment efficiencies were 97 and 98% for fluoranthene and pyrene, the 4-ring PAHs present at the highest concentrations, 1708 and 1148 mg/kg, respectively. The two carcinogenic 4-ring PAHs, benzo(*a*)anthracene and chrysene present at low concentrations (approximately 290 mg/kg) were readily biodegraded at greater than 90% treatment efficiencies. A longer time was generally required to achieve these treatment efficiencies in comparison to the 3-ring PAHs.

The biodegradability of the 5- and 6-ring PAHs was substantially less than the 3- and 4-ring compounds. These compounds are not utilized by the microorganisms as sole carbon and energy sources. Five- and 6-ring PAHs have a lower aqueous solubility and are therefore less available to the microorganisms. Although the treatment efficiency ranged from 40 to 78% for this group of compounds, the benzo(*b*) and (*k*) fluoranthene isomers were the only compounds that exhibited a statistically significant difference in concentrations before and after treatment.

Process Economics

PROCESS COMPONENT COSTS

The treatment costs presented in the following sections are based on the labor, equipment, materials and supplies, and installation costs utilized at the Southeastern Wood Preserving Superfund site. The unit costs were calculated based on the total cost of labor, equipment, and supplies required to process one 180,000-gallon batch of material

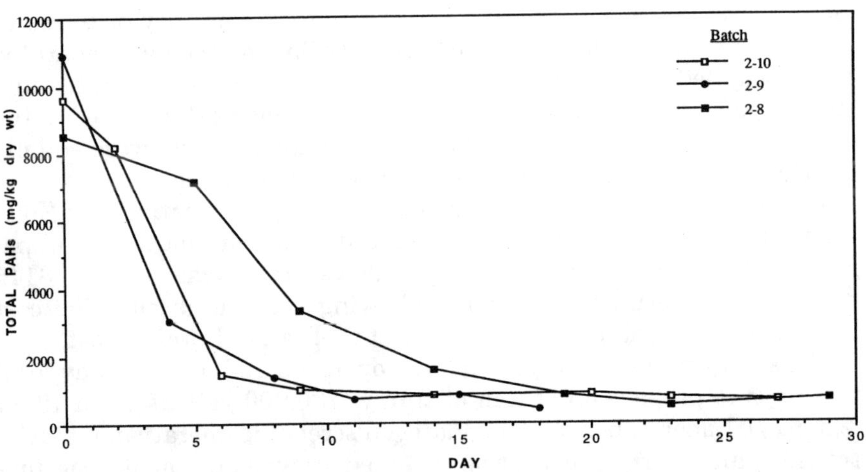

Figure 17. Slurry phase reactor 2—treatment of total polycyclic aromatic hydrocarbons.

Table 10. Full-scale slurry phase biological treatment of polycyclic aromatic hydrocarbons in wood-preserving wastes (n = 6 batches).

	Concentration (mg/kg)		
Compound	Initial Mean ± Std Dev.	Final Mean ± Std Dev.	Treatment Efficiency
3-Ring PAHs			
Acenaphthene	909 ± 230	6 ± 3	99
Acenaphthylene	93 ± 81	15 ± 5	82
Anthracene	1,950 ± 530	121 ± 59	94
Fluorene	630 ± 283	14 ± 6	97
Phenanthrene	1,031 ± 661	34 ± 23	96
4-Ring PAHs			
Benzo(*a*)anthracene[a]	280 ± 51	12 ± 5	95
Chrysene[a]	296 ± 59	36 ± 11	90
Fluoranthene	1,708 ± 395	32 ± 7	98
Pyrene	1,148 ± 252	33 ± 12	97
5- and 6-Ring PAHs			
Benzo(*b*)fluoranthene[a]	321 ± 34	208 ± 54	52
Benzo(*k*)fluoranthene[a]	[combined with benzo(*b*)fluoranthene]		
Benzo(*g,h,i*)perylene	92 ± 82	18 ± 12	43
Benzo(*a*)pyrene[a]	130 ± 52	79 ± 15	34
Dibenzo(*a,h*)anthracene[a]	92 ± 82	9 ± 6	78
Indeno(2,3,*c,d*)pyrene[a]	94 ± 79	31 ± 5	46

[a]Carcinogenic PAHs.

through the treatment system divided by the amount of contaminated material treated per batch. Typically, the total quantity of contaminated material per batch was 230 tons, with 35 tons of debris separated and stockpiled by the shaker screen, 35 tons of debris separated and stockpiled by the hydrocyclone, and 160 tons biologically treated in the slurry reactors. One-time installation and supplies costs were equally distributed based on the total number of batches processed.

The cost for the full-scale slurry phase biological treatment of K001 wastes to achieve risk-based treatment criteria, including slurry preparation, slurry treatment, and slurry dewatering, was $120 to $145 per ton of material treated. The following sections discuss the cost breakdown for each treatment section and the total project cost.

The slurry preparation process was operated 12 hours per day for 3 to 4 days to prepare one batch of slurry (180,000 gallons of an 18 wt % slurry). Labor to perform the material screening operation, material transfer, and slurry preparation included heavy equipment operators and field operations personnel. In addition to the slurry tank and its

Table 11. Full-scale slurry phase biological treatment of polycyclic aromatic hydrocarbons in wood-preserving wastes (Reactor 1, Batch 5).

Compound	Concentration (mg/kg)									Treatment Efficiency
	Day 0	Day 3	Day 7	Day 10	Day 14	Day 17	Day 21	Day 24	Day 29	
3-Ring PAHs										
Acenaphthene	642	<82	<20	3	<13	<33	<22	<8	<8	100
Acenaphthylene	34	<82	23	11	19	<33	<22	14	15	68
Anthracene	1050	1010	59	104	58	98	140	67	61	90
Fluorene	181	150	<20	14	<13	<33	<22	<8	<8	92
Phenanthrene	220	256	<20	27	16	24	31	14	19	88
4-Ring PAHs										
Benzo(*a*)anthracene[a]	224	<82	<20	10	<13	<33	<22	<8	<8	96
Chrysene[a]	228	204	53	24	28	49	44	25	24	89
Fluoranthene	1060	327	31	25	27	39	36	23	36	98
Pyrene	878	196	27	25	26	32	33	22	28	97
5- and 6-Ring PAHs										
Benzo(*b*)fluoranthene[a,b]	367	731	353	155	214	26	209	155	184	58
Benzo(*k*)fluoranthene[a]										
Benzo(*g,h,i*)perylene	21	<82	44	23	35	45	38	24	28	—
Benzo(*a*)pyrene[a]	92	211	120	52	71	82	74	53	57	43
Dibenzo(*a,h*)anthracene[a]	15	<82	<20	10	<13	<33	<22	<8	14	—
Indeno(2,3,*c,d*)pyrene[a]	30	<82	57	28	44	54	47	31	35	—

[a]Carcinogenic PAHs.
[b]Sum of (*b*) and (*k*) isomers.

Table 12. Full-scale slurry phase biological treatment of polycyclic aromatic hydrocarbons in wood-preserving wastes (Reactor 1, Batch 8).

Compound	Concentration (mg/kg)					Treatment Efficiency
	Day 0	Day 6	Day 9	Day 13	Day 16	
3-Ring PAHs						
Acenaphthene	968	<13	<30	7	<14	99
Acenaphthylene	<163	<13	<30	13	20	92
Anthracene	1560	192	94	55	85	96
Fluorene	669	37	<30	7	<14	99
Phenanthrene	1250	57	<30	14	22	99
4-Ring PAHs						
Benzo(*a*)anthracene[a]	287	<13	<30	7	<14	98
Chrysene[a]	302	78	84	55	84	82
Fluoranthene	1570	213	51	32	41	98
Pyrene	1080	42	<30	18	27	98
5- and 6-Ring PAHs						
Benzo(*b*)fluoranthene[a,b] Benzo(*k*)fluoranthene[a]	337	321	573	240	300	29
Benzo(*g,h,i*)perylene	<163	23	<30	26	31	—
Benzo(*a*)pyrene[a]	<163	106	99	80	108	—
Dibenzo(*a,h*)anthracene[a]	<163	<13	<30	7	19	—
Indeno(2,3,*c,d*)pyrene[a]	<163	32	36	33	40	—

[a]Carcinogenic PAHs.
[b]Sum of (*b*) and (*k*) isomers.

Table 13. Full-scale slurry phase biological treatment of polycyclic aromatic hydrocarbons in wood-preserving wastes (Reactor 1, Batch 9).

Compound	Concentration (mg/kg)					Treatment Efficiency
	Day 0	Day 3	Day 6	Day 10	Day 13	
3-Ring PAHs						
Acenaphthene	692	<110	<11	<8	<8	99
Acenaphthylene	28	<110	20	<8	17	71
Anthracene	2140	385	109	102	152	95
Fluorene	499	<110	11	8	16	98
Phenanthrene	395	84	25	23	35	94
4-Ring PAHs						
Benzo(*a*)anthracene[a]	283	<110	15	<8	12	97
Chrysene[a]	301	175	63	30	32	90
Fluoranthene	1950	626	38	26	23	99
Pyrene	1220	1210	65	30	35	98
5- and 6-Ring PAHs						
Benzo(*b*)fluoranthene[a,b]	278	247	372	131	185	53
Benzo(*k*)fluoranthene[a]						
Benzo(*g,h,i*)perylene	33	<110	<11	25	51	—
Benzo(*a*)pyrene[a]	105	180	126	74	104	—
Dibenzo(*a,h*)anthracene[a]	<40	<110	<11	<8	22	—
Indeno(2,3,*c,d*)pyrene[a]	<40	<110	34	31	66	—

[a]Carcinogenic PAHs.
[b]Sum of (*b*) and (*k*) isomers.

Table 14. Full-scale slurry phase biological treatment of polycyclic aromatic hydrocarbons in wood-preserving wastes (Reactor 1, Batch 10).

Compound	Concentration (mg/kg)								Treatment
	Day 0	Day 3	Day 6	Day 10	Day 13	Day 17	Day 20	Day 24	Efficiency
3-Ring PAHs									
Acenaphthene	892	599	10	9	<28	<11	<27	<11	99
Acenaphthylene	<59	44	12	19	<28	16	<27	19	68
Anthracene	2280	2490	320	230	309	174	281	102	90
Fluorene	661	326	34	25	44	23	39	18	96
Phenanthrene	2030	504	56	79	70	42	65	33	96
4-Ring PAHs									
Benzo(*a*)anthracene[a]	237	253	22	20	<28	18	<27	15	92
Chrysene[a]	247	298	31	41	36	39	32	26	83
Fluoranthene	1850	1550	38	31	44	44	47	34	98
Pyrene	1010	1020	77	48	42	46	43	32	95
5- and 6-Ring PAHs									
Benzo(*b*)fluoranthene[a,b]	296	301	301	254	227	247	240	170	42
Benzo(*k*)fluoranthene[a]									
Benzo(*g,h,i*)perylene	<59	<29	21	3	<28	31	28	30	—
Benzo(*a*)pyrene[a]	100	114	93	95	72	93	90	72	28
Dibenzo(*a,h*)anthracene[a]	<59	<29	12	3	<28	19	<27	19	—
Indeno(2,3*c,d*)pyrene[a]	<59	35	30	31	<28	41	38	39	—

[a]Carcinogenic PAHs.
[b]Sum of (*b*) and (*k*) isomers.

Table 15. Full-scale slurry phase biological treatment of polycyclic aromatic hydrocarbons in wood-preserving wastes (Reactor 2, Batch 9).

Compound	Concentration (mg/kg)						Treatment Efficiency
	Day 0	Day 4	Day 8	Day 11	Day 15	Day 18	
3-Ring PAHs							
Acenaphthene	1280	277	<40	3	<16		100
Acenaphthylene	<223	<39	<40	14	<16	12	—
Anthracene	2340	446	327	135	251	48	94
Fluorene	1040	74	49	16	60		98
Phenanthrene	1300	110	77	30	84	12	98
4-Ring PAHs							
Benzo(*a*)anthracene[a]	370	93	<40	10	<16	7	97
Chrysene[a]	397	160	89	33	28	14	92
Fluoranthene	2210	832	116	37	48	17	98
Pyrene	1610	786	93	46	40	20	97
5- and 6-Ring PAHs							
Benzo(*b*)fluoranthene[a,b]	345	226	371	259	196	148	25
Benzo(*k*)fluoranthene[a]	—	—	—	—	—	—	—
Benzo(*g,h,i*)perylene	<223	<39	42	3	24	20	52
Benzo(*a*)pyrene[a]	<223	95	137	91	72	59	59
Dibenzo(*a,h*)anthracene[a]	<223	<39	<40	3	<16	8	—
Indeno(2,3*c,d*)pyrene[a]	<223	<39	52	24	31	28	46

[a]Carcinogenic PAHs.
[b]Sum of (*b*) and (*k*) isomers.

Table 16. Full-scale slurry phase biological treatment of polycyclic aromatic hydrocarbons in wood-preserving wastes (Reactor 2, Batch 10).

Compound	Concentration (mg/kg)								Treatment Efficiency
	Day 0	Day 2	Day 6	Day 9	Day 14	Day 20	Day 23	Day 27	
3-Ring PAHs									
Acenaphthene	981	618	<33	7	<10	<30	<30	6	99
Acenaphthylene	<51	<51	<33	12	16	<30	<30	23	55
Anthracene	2330	2720	266	334	200	253	157	100	96
Fluorene	732	315	37	32	23	<30	<30	15	98
Phenanthrene	988	484	61	50	48	51	38	31	97
4-Ring PAHs									
Benzo(*a*)anthracene[a]	277	285	<33	16	20	<30	<30	16	94
Chrysene[a]	302	312	101	39	44	39	38	31	90
Fluoranthene	1610	1970	240	35	64	52	49	43	97
Pyrene	1090	1080	215	63	61	40	35	33	97
5- and 6-Ring PAHs									
Benzo(*b*)fluoranthene[a,b]	304	330	359	278	244	287	243	213	30
Benzo(*k*)fluoranthene[a]	—	—	—	—	—	—	—	—	—
Benzo(*g,h,i*)perylene	<51	<51	<33	16	<10	<30	36	29	—
Benzo(*a*)pyrene[a]	98	101	131	73	88	88	86	82	16
Dibenzo(*a,h*)anthracene[a]	<51	<51	<33	6	<10	<30	<30	20	—
Indeno(2,3*c,d*)pyrene[a]	<51	<51	49	24	28	37	44	49	—

[a]Carcinogenic PAHs.
[b]Sum of (*b*) and (*k*) isomers.

associated equipment (i.e., pumps, vibrating screen, hydrocyclone, mixers, etc.), a tracked excavator and rubber-tire loader were required. Supplies and materials included equipment fuel, nutrients, and slurry conditioning chemicals. The total cost to prepare the slurry, on a per-ton basis, was $50 to $60, with the cost by category summarized in Table 17.

The slurry phase biological treatment section was operated 24 hours per day for 8 to 12 days for each slurry batch. Labor to operate the reactors consisted of a field technician to sample the four operating reactors and to perform the daily pH, dissolved oxygen, temperature, and other biological process monitoring and sampling. Equipment costs were the slurry reactor, mixer, and air blower for each reactor. Supplies included diffusers, sampling supplies, and replacement impellers for the mixers. The total cost to biologically treat the slurry, on a per-ton basis, was $40 to $55, with the cost by category summarized in Table 17.

The slurry dewatering system operation consisted of aerating and drying the treated slurry and recycling the excess water to the water management tank. Supplies and materials consisted of the HDPE-lined drying cell, the liner for the water management tank, and equipment fuel. The total cost to dewater the treated slurry, on a per-ton basis, was $20 to $40, with the cost breakdown by category summarized in Table 17.

TOTAL PROJECT COSTS

In addition to the costs directly associated with the treatment of the contaminated material, additional project costs included design engineering, treatability and the pilot-scale testing, site closure, and project administration and support. Process, electrical, and civil engineering design, equipment specification and fabrication, and system

Table 17. Cost breakdown for the full-scale slurry phase biological remediation of RCRA K001 Wastes.

Cost Category	Slurry Preparation	Slurry Biotreatment	Slurry Dewatering
Labor and equipment	$30 to $35	$10 to $15	$5 to $10
Supplies, materials, utilities	$20 to $25	$25 to $30	$15 to $20
Analytical	<$5	$5 to $10	<$5
Subtotal	$50 to $60	$40 to $55	$20 to $30
Total cost for the full-scale slurry phase biological remediation of K001 wastes		$120 to $145	

Reproduced from Jerger and Woodhull (1995), copyright © 1995, John Wiley & Sons, Inc.

procurement were performed as part of the design engineering task. Treatability and pilot-scale testing were performed as part of the design engineering task. Treatability and pilot-scale testing were performed to characterize the contaminated material, determine treatment efficiencies, obtain slurry reactor operating conditions, and optimize full-scale operating parameters. Site closure tasks included placement of the debris, shaker screen rejects, and hydrocyclone rejects in the dewatering cell, construction of a clay cap on the cell, and final site grading and seeding. Project management and engineering during system operation, site administration, field office facilities and supplies, and support costs were totaled as part of project administration and support.

The total project cost, including engineering, treatability and pilot-scale testing, slurry treatment, site closure, and administration for the full-scale slurry phase biological treatment of 14,140 tons of K001 waste material, was approximately $2,800,000 or $190 to $200 per ton of material treated (Table 18).

Factors Influencing Costs

Three variables had a significant impact on the total project cost of the treatment system. Treatment costs were affected by the slurry phase reactor solids concentration, biological reactor batch treatment time, and the percentage of material removed in the slurry preparation system. The following sections discuss the impact of each of the variables on total project cost. The project costs were calculated based on the assumption that each variable was independent of the other two variables, and the treatment system required minimal process modifications.

Table 18. Total project cost for the full-scale slurry phase biological treatment of RCRA K001 wastes.

Description	Cost
Treatability and pilot-scale testing	$200,000
Design engineering	$100,000
Slurry preparation	$800,000
Slurry treatment	$700,000
Slurry dewatering	$400,000
Site closure	$500,000
Project administration and support	$200,000
Total project cost	$2,800,000
Total project cost per ton of material treated	$190 to $200

REACTOR SOLIDS CONCENTRATION

By increasing the slurry solids concentration in the slurry phase reactor, the amount of soil treated increases, decreasing the cost per ton of material treated (Figure 18). Increasing the slurry solids concentration in the reactors from 15 wt % to 30 wt % decreases the treatment cost from approximately \$215 per ton to \$150 per ton. Increasing the slurry solids concentration above 25 to 30 wt % may require a modification of the slurry reactor design. The impact of the increased design, equipment, and operating costs for the modified reactors must be determined prior to system design. In general, for a given reactor configuration, the greater the slurry solids concentration, the lower is the treatment cost on a per-ton basis.

REACTOR BATCH TREATMENT TIME

As batch treatment time (residence time) for the reactors increases, the cost per ton increases (Figure 19). Increasing the batch treatment time from 11 days to 33 days increases the treatment cost from \$190 to \$200 per ton to approximately \$240 per ton. The batch treatment time was dependent on initial PAH concentrations of the contaminated material and reactor slurry temperature. The minimum slurry temperature required in the reactors was approximately 20 to 25°C, and the biological treatment rate doubled for each 10°C temperature increase. Prior to sys-

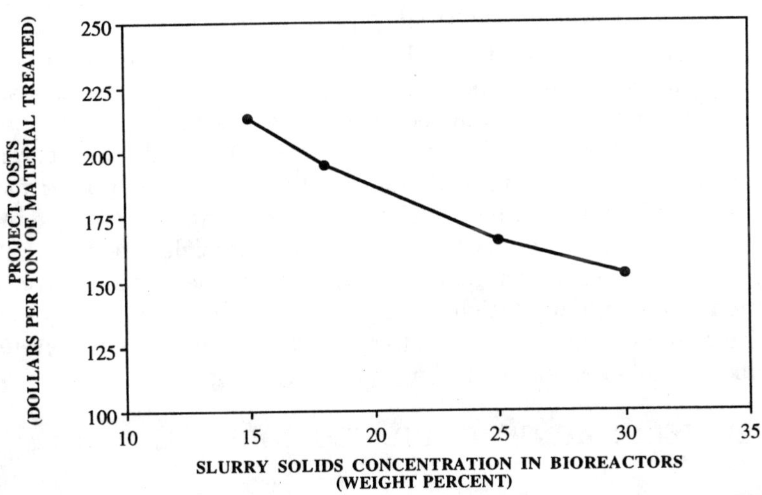

Figure 18. Effect of solids concentration in slurry phase biological reactors on project costs. [Reprinted with permission from Jerger and Woodhull (1995), copyright 1995, John Wiley & Sons, Inc.]

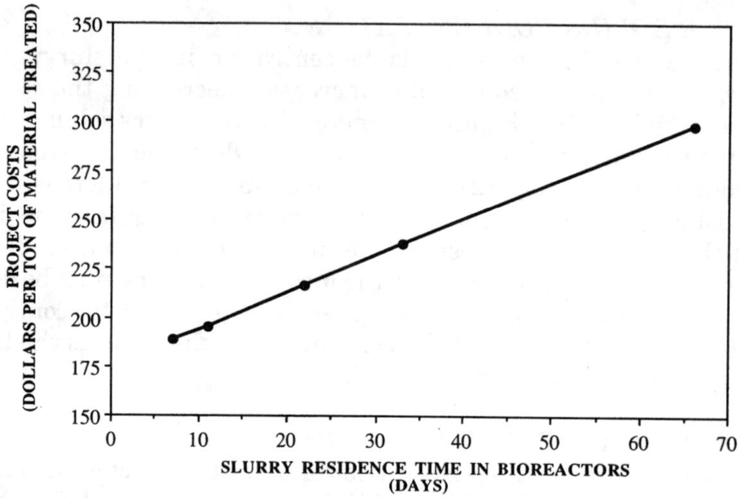

Figure 19. Effect of residence time in slurry phase biological reactors on project costs. [Reprinted with permission from Jerger and Woodhull (1995), copyright 1995, John Wiley & Sons, Inc.]

tem design, the cost impact for heating the reactors must be compared versus the schedule impact of temporary shutdown of the treatment system during periods of low slurry temperatures in the reactors.

SOIL PARTICLE SIZE DISTRIBUTION

For a fixed soil volume treated in the slurry phase reactors, as the percent of soil removed in the slurry preparation system increases, the cost per ton of material treated decreases (Figure 20). Increasing the soil removed in the slurry preparation system from 10 to 50 wt % decreases the treatment cost from approximately $225 per ton to approximately $165 per ton of material treated. The percent of material removed in the slurry preparation system is dependent on the particle size distribution of the material, contaminant concentration versus particle size, and treatment criteria. The particle size fraction treated during slurry preparation and in the slurry phase reactors was determined during extensive treatability and laboratory- and pilot-scale testing.

DEVELOPMENT OF SOIL CLEANUP STANDARDS

The primary goal of hazardous waste site remediation is to achieve a set of conditions that are environmentally safe for organismal receptors and will prevent any further action on-site. Establishing these conditions raises the issue of "how clean is clean" or "what concentration of a contaminant in soil is environmentally acceptable" for contaminated sites. This debate is occurring on a national scale as part of the Superfund reauthorization.

Historically, cleanup goals have not explicitly addressed whether the remedial action results in a safe site. Treatment criteria for soil contaminants have been based on limits of detection (LODs), background concentrations or technical capabilities (best demonstrated available technology, or BDAT). Selecting an appropriate value to represent background concentrations can be problematic in that there is a substantial amount of variability in the natural chemical composition of soils, and many chemicals may have both natural and anthropogenic sources (LaGoy and Quirk, 1994).

The use of Resource Conservation and Recovery Act (RCRA) BDAT, or LOD, as a cleanup criteria may also be inappropriate if the remediation goal is to protect human health and the environment. BDAT-based criteria may be over- or underprotective of human health and the environment, preventing the use of a more cost-effective treatment technology such as bioremediation. BDAT-based treatment criteria may also focus on a subset of chemicals of limited toxicity, making it difficult to determine if the remedial action removed the health threat posed by other chemicals in the mixture.

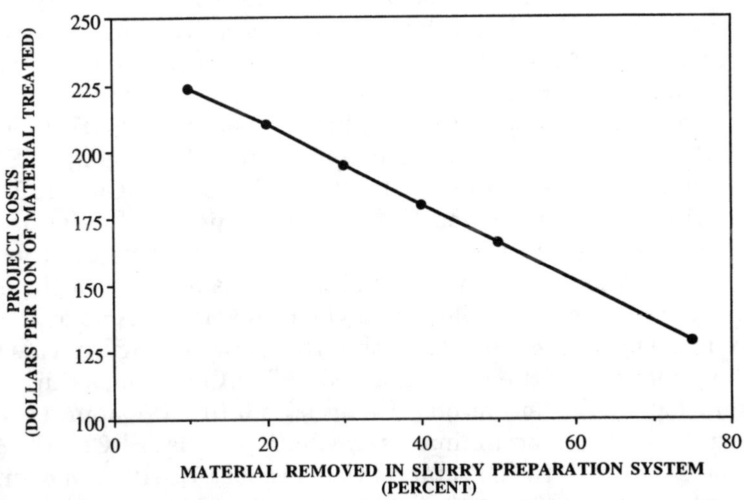

Figure 20. Effect of percent of material removed in slurry preparation system on project costs. [Reprinted with permission from Jerger and Woodhull (1995), copyright 1995, John Wiley & Sons, Inc.]

The performance of the bench-scale and field-pilot-scale tests and full-scale operation of the slurry phase biological treatment system showed that a substantial level of PAH reduction could be achieved by bioremediation. The reductions in total PAHs and carcinogenic PAHs ranged from 85 to 95% and 50 to 70%, respectively. These treatment efficiencies represented final concentrations of less than 950 mg/kg dry weight soil solids total PAHs and less than 180 mg/kg dry weight soil solids of benzo(*a*)pyrene equivalent carcinogenic PAHs.

As a result of the laboratory and field efforts—which were verified by an extensive literature review—it was determined that present state-of-the-art knowledge in the field of bioremediation could still not allow it to consistently achieve the BDAT levels met by incineration for the higher-ring PAHs of pyrene and phenanthrene. Thus, it became necessary to modify the contract acceptance criteria to a level based upon maximum achievable contaminant removals, as well as protection of human health and the environment.

Bench- and pilot-scale slurry reactor treatment of wood-preserving wastes has been unsuccessful in achieving incineration-based RCRA BDAT treatment standards (USEPA, 1993). Thus, because EPA must generally employ fixed-price contracts or subcontracts for fund-led cleanups, it must be more flexible in cleanup criteria if it wishes to employ a variety of innovative technologies. It must also assure that a reasonable level of evidence is available to demonstrate that the technology chosen is likely to be able to achieve the criteria specified.

The final action levels for treatment of creosote wastes from the Southeastern Wood Preserving site were established based on a consideration of technical capability, protection of human health, and risk. Target remediation goals included values for both total PAHs (950 mg/kg was the target level) and B[*a*]P-equivalents (180 mg/kg of B[*a*]P-equivalent PAHs was the target level) (Nisbit and LaGoy, 1992). In general, final levels were well below these values. Unlike either the RCRA LDR values or LDR Guide 6A, the target levels used for the Southeastern Wood Preserving Site allow a consideration of the effect of treatment on site risks. B[*a*]P-equivalent levels of PAHs at the site were reduced by bioremediation to slightly below 100 mg/kg. This level is in the upper end of the range allowed by EPA at various Superfund sites and for similar situations. Using standard EPA approaches to calculating allowable preliminary remediation goals (PRGs), the level of 100 mg/kg is above the EPA target risk level of 10^{-4}–10^{-6}. In addition, site-specific factors suggest that actual risks would be lower than predicted using the PRG approach, particularly if direct contact exposure were to be controlled.

Total PAHs were degraded substantially by the slurry phase biological treatment, but the carcinogenic PAHs were less affected than the noncarcinogenic compounds. The noncarcinogenic PAHs tend to be lower molecular weight compounds than carcinogenic PAHs. In addition to the PAHs, other compounds such as petroleum hydrocarbons will also be reduced, as shown by the substantial reduction in limits of detection. Biodegration is likely to have a substantial influence on the reduction of site-specific risks.

The more soluble compounds, such as the lower molecular weight aliphatic and aromatic petroleum hydrocarbons, can act as carriers for the carcinogenic PAHs that have very low water solubility. Consequently, destruction of these soluble compounds will greatly reduce the mobility of the carcinogenic PAHs. Low molecular weight PAHs that could act as carriers for dermal or oral absorption will be reduced, possibly reducing the potential bioavailability of the carcinogenic PAHs. Some reduction in carcinogenic PAH levels is being achieved; the residual material is apparently not readily available for biological action and may be less bioavailable as well.

Considerable work remains to be done to decrease the uncertainties in risk assessment, and some uncertainties are inherent to the process and will persist. However, risk assessment is a valuable tool that can be used to encourage the use of innovative, but often non-totally destructive, remediation technologies, by providing regulators with a method to judge the adequacy of a cleanup action. Totally destructive actions such as excavation and off-site incineration are effective treatment technologies, but their costs are substantial. Alternative treatment methods, such as bioremediation, can offer a cost-effective alternative to these more destructive technologies.

CONCLUSIONS

The experience with the Southeastern Wood Preserving project demonstrates that slurry phase bioremediation is a viable treatment alternative for creosote-contaminated soil. Although the mandated treatment standard for K001 waste was not achieved by the process, significant destruction of the contaminants did occur. The potential for wide-scale use of this technology is high, particularly for CERCLA actions where the LDR treatability variance process and/or establishment of site-specific, risk-based treatment standards are an option. In comparison to incineration, the only readily available technology that can consistently meet the K001 treatment standards, slurry phase bioremediation has a lower cost and is more readily accepted by the public. These factors, cost, and community acceptance are extremely important in bringing a project to remediation, particularly for public sector projects.

REFERENCES

Heitkamp, M.A. and C.E. Cerniglia. 1988. "Mineralization of Polycyclic Aromatic Hydrocarbons by a Bacterium Isolated from Sediment Below an Oil Field.," *Applied and Environ. Micro.,* 54: 1612–1614.

Jerger, D.E. and P.M. Woodhull. 1995. "Economics of a Commercial Slurry-Phase Biological Treatment Process.," *Biological Unit Processes for Hazardous Waste Treatment,* proceedings of the 1995 conference, San Diego, CA. New York: John Wiley & Sons, Inc.

LaGoy, P.K. and Quirk, T.C. 1994. "Establishing Generic Remediation Goals for the Polycyclic Aromatic Hydrocarbons: Critical Issues,." *Environ. Health Perspective*, 102(4): 348–352.

Lauch, R.P., J.G. Herrmann, W.R. Mahaffey, A.B. Jones, M. Donsani and J. Hessling. 1992. "Removal of Creosote from Soil by Bioslurry Reactors.," *Environmental Progress* 11(4): 265–271.

Middleton, A.C., D.V. Nakles and D.G. Linz. 1991. "The Influence of Soil Composition on Bioremediation of PAH-Contaminated Soils." *Remediation*, 1: 391–407.

Mueller, J.G., S.E. Lantz, B.O. Blattmann and P.J. Chapman. 1991. "Bench-Scale Evaluation of Alternative Biological Treatment Processes for the Remediation of Pentachlorophenol- and Creosote-Contaminated Materials: Slurry-Phase Bioremediation.," *Environmental Science Technology* 25: 1055–1061.

Nisbet, I.C.T. and P.K. LaGoy. 1992. "Toxic Equivalency Factors (TEFs) for Polycyclic Aromatic Hydrocarbons (PAHs).," *Regulatory Toxicology and Pharmacology,* 21: 32–46.

Ryan, J.R. and J.R. Smith. 1986. "Treatment of Wood Preserving Wastes." *Proceedings of the National Conference on Hazardous Wastes and Hazardous Materials.* Hazardous Materials Control Research Institute. Washington, D.C., pp. 80–86.

USEPA. 1988. *Technology Screening Guide for Treatment of CERCLA Soils and Sludges.* EPA/540/2-88/004.

USEPA. 1989. *Test Methods for Evaluating Solid Waste, Physical/Chemical Methods,* 3rd Edition, SW-846, Update 1. Washington, D.C.: Office of Solid Waste and Emergency Response.

USEPA. 1992. *Contaminants and Remedial Options at Wood Preserving Sites.* EPA/600/R-92/182. Cincinnati, Ohio.

USEPA. 1993. *Applications Analysis Report: Pilot-Scale Demonstration of a Slurry-Phase Biological Reactor for Creosote-Contaminated Soil,* EPA/540/A5-91/009, Cincinnati, Ohio.

Biofilters in Practice

GERO LESON
Environmental Consulting
2324 30th Street
Santa Monica, CA 90405

INTRODUCTION

For decades, microorganisms have been used for the removal of dissolved organic pollutants from industrial and municipal wastewater. Since the early 1980s microbiological technologies have also been developed for the cleanup of soils and groundwater contaminated by petroleum hydrocarbons and synthetic chemicals. During the same period the treatment of air contaminants in off-gases from municipal and industrial operations by biological air pollution control (APC) technologies, most notably biofiltration, has been well established in several European countries (*VDI Berichte 735,* 1989; *Biotechniques,* 1992; *VDI Berichte 1104,* 1994). Initially, biofilters were used for the control of odors from wastewater, composting, food processing, and livestock breeding operations. More recently, they are also increasingly used as an energy-efficient, cost-effective, and environmentally friendly means of destroying low concentrations of volatile organic compounds (VOC) and air toxics present in off-gases from various industrial and commercial sources.

In the United States the interest in biofiltration has grown rapidly since the late 1980s, mainly due to the technology's perceived simplicity and cost advantages over competing technologies. This growth is reflected in the increasing number of suppliers of biofilter systems and their pilot testing activities in a variety of industries. The numbers of simple, open biofilter systems for the control of odors from wastewater treatment and composting operations and of more sophisticated full-scale biofilters for the control of VOC and air toxics also has increased gradually.

Copyright Gero Leson.

Overall, the experience gained from installations in Europe indicates that biofilter systems have provided reliable and cost-effective pollutant removal in a variety of industries. In a number of systems, failure to maintain proper operating conditions in the biologically active filter material or its rapid compaction have also caused poor performance and need for extensive repairs. In response, several designs intended to improve reliability have been developed by suppliers of biofilter systems.

This chapter will summarize the basics of biofiltration, provide a brief overview of the history of biological air pollution control, summarize experiences with biofilters in their main areas of application, and review economic aspects and trends in system design. While biofiltration is now used in many countries worldwide, the discussion focuses on developments in those countries where biofilters were mainly developed, the Netherlands and Germany, and on the United States, the largest national market for APC equipment. Biofiltration now is the most commonly applied biological APC technology and will be the focus of this chapter. Bioscrubbers and biotrickling filters (BTF) represent biological APC alternatives that are also increasingly used in specific applications and will be introduced briefly.

OVERVIEW OF BIOFILTER TECHNOLOGY

The following section provides an overview of the technological principles of biofiltration and of their ramifications for field applications. More quantitative descriptions of the underlying physical and biochemical processes are provided elsewhere in this volume and in the literature (Ottengraf 1986; *VDI Berichte,* 735, 1989; Leson and Winer, 1991; *Biotechniques,* 1992; *VDI Berichte 1104,* 1994).

Principles of Biofiltration

A biofilter is a biochemical reactor for the degradation of biodegradable air pollutants. It contains a porous, solid-state filter material or packing, whose surface is covered with a wet layer, the biofilm, containing immobilized microorganisms. The basic operating principles of a biofilter involve the continuous venting of a contaminated gas stream through the reactor where target air pollutants diffuse into the biofilm. The contaminants are then aerobically biodegraded by resident microorganisms, predominantly bacteria. Figure 1 shows a conceptual model of the filter material. Products of the biooxidation of gaseous air pollutants in filter material include

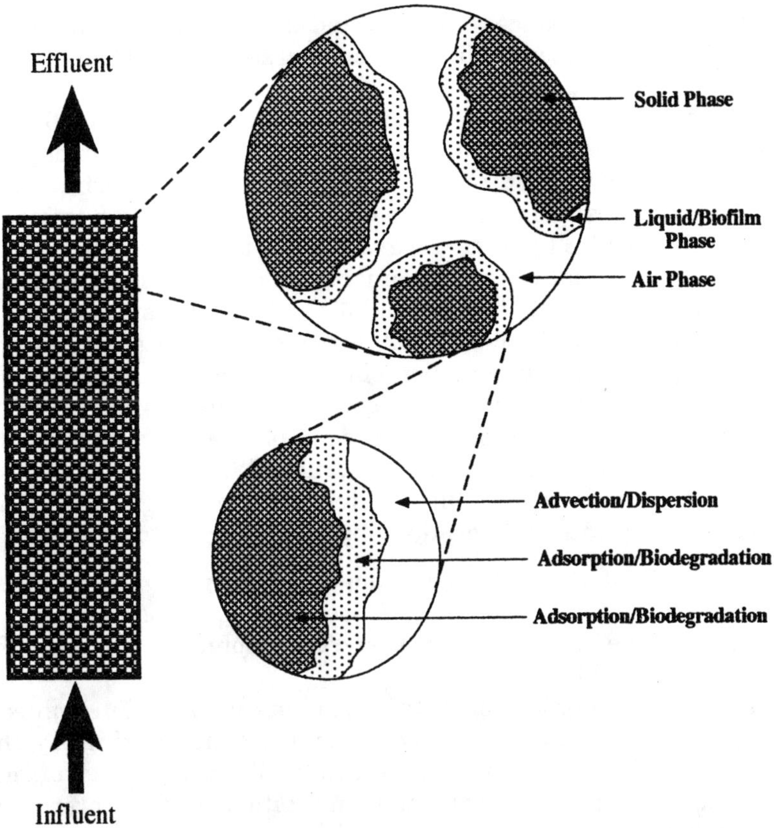

Figure 1. Conceptual model for biofilter material. (Adapted from Hodge, 1993.)

- water and carbon dioxide (CO_2), which is released into the treated off-gas stream and discharged to atmosphere
- microbial biomass
- inorganic acids from the oxidation of VOC species containing chlorine (Cl), sulfur (S), and nitrogen (N)
- intermediates from the incomplete biooxidation of VOC, such as acetic acid
- heat

Compost derived from yard waste, municipal solid waste, wood chips, bark, sewage sludge, and other organic material such as peat and heather has generally been the basis of the filter material. More recently, the use of inert additives such as polystyrene pellets, granular activated carbon (GAC), and ceramic material has become popular because of their resistance to biodegradation and compaction. Mineral soils have also been used as filter material in several "soil bed" installations in the United States.

The heterotrophic microorganisms immobilized in the biofilm use the VOC's organic carbon content as their sole source of energy and carbon for cell growth. Typically, only 50% of the organic carbon offered to the microorganisms is directly metabolized. The balance is used for the growth of new cell mass. However, under steady-state conditions, concurrent die-off and lysis of organisms keep cell mass constant and result in the complete net oxidation of the VOC transferred into the biofilm (Ottengraf and Diks, 1992). Oxygen and inorganic nutrients are also required for proper microbial growth and metabolism. The former must be present in the off-gas and diffuses into the biofilm together with the pollutants. Nutrients are present, organically bound in the packing, or added to it in a mineralized form.

The effectiveness of a given filter material in removing and degrading specific air pollutants is determined by two major factors:

1. The rate of pollutant transfer from the off-gas into the biofilm: It depends mainly on vapor phase concentration, water solubility and volatility of the target pollutants, and the amount of effective biofilm surface in the filter material.
2. The rate of biodegradation of target pollutants: It is determined by the presence of suitable microorganisms in the packing, by their environmental conditions (temperature, pH, availability of moisture, nutrients and oxygen), and by inhibition or other interference effects from pollutants.

At lower pollutant concentrations, typically <1000 parts per million measured as carbon (ppmC),[1] particularly of poorly water-soluble compounds, pollutant elimination is limited by their rate of diffusion into and within the biofilm (diffusion limited regime). At higher concentrations of more water-soluble, oxygenated compounds, the biofilm is fully penetrated by the target pollutants. Their elimination becomes limited

[1] The use of ppmC, also referred to as "ppm measured as methane," allows for quantification and comparison of volume-based concentrations of gaseous organic carbon independent of the size of the VOC molecules involved. For a given VOC species, ppmC and ppm_v are related by ppmC = ppm_v · number of carbon atoms per molecule. ppmC values can be converted into mass-based TOC concentrations in mg/m^3, commonly used in the European literature, by multiplying with a factor of 0.5 (1 ppmC = 0.5 mg/m^3 of TOC at 20°C). Due to the contribution from hydrogen, oxygen, and other species, mass-based concentrations of total VOC are always higher than those for TOC. Example: For propanol (C_3H_7OH) 500 ppm_v = 1,500 ppmC = 750 mg/m^3 of TOC = 1250 mg/m^3 of VOC at 20°C.

by the degradation capacity of the microorganisms, often as a result of insufficient transfer of oxygen into the biofilm (Ottengraf, 1986).

A common way of characterizing the effectiveness of a given packing in removing specific pollutants is the (empty bed) residence time required to remove target pollutants with a desired control efficiency. It relates the off-gas flowrate for a source to the required volume of biofilter material[2] and depends on the nature and concentration of the target pollutants, the off-gas temperature, the overall effectiveness of the packing with respect to pollutant transfer and biodegradation, and the desired control efficiency. Since existing engineering models do not permit accurate prediction of the required residence time for multicomponent off-gases, it is usually determined in a pilot test, unless biofilter performance data for a similar off-gas stream exist.

As a general rule, residence times of less than 1 minute are desirable for full-scale biofilters. They will result in a sufficiently small filter volume per volume of treated off-gas and indicate potential capital cost advantages of biofilters (see below). In more recently installed full-scale systems, residence times range typically between 15 seconds and 2 minutes. Short residence times will suffice if only low concentrations, typically less than 1000 ppmC, of water-soluble VOC (alcohols, ketones, aldehydes) must be removed. This will often be the case for odor control applications. Longer residence times are required if poorly water-soluble and/or highly volatile compounds (light petroleum hydrocarbons and chlorinated VOC), which are not efficiently transferred into the biofilm, are present or if high concentrations (>1000 ppmC) of water-soluble compounds must be removed with high removal efficiencies.

While the majority of the compounds typically treated in a biofilter can be classified as VOC, several inorganic compounds, most notably hydrogen sulfide (H_2S) and ammonia (NH_3), can also be removed. Table 1 provides a partial list of common biodegradable, gaseous air pollutants and a relative rating of their elimination in a biofilter. The table suggests that oxygenated VOC are, in general, efficiently removed while particularly light aliphatics show only minimal removal. The actual biofilter reactor contains one or more beds of biologically active filter material. It may be partially open or fully enclosed. Concrete walls, steel frames with sheet metal panels, or fiberglass are most commonly used as materials of construction. Beds are typically 0.9–1.5 m (3–5 ft) in height. Filter beds may be supported by off-gas distribution systems of various designs, such as slotted concrete slabs, grates covered with

[2] Residence time [seconds] = 3600 · volume of filter material [m^3]/off-gas flow rate [m^3/h].

Table 1. Relative removal of common air pollutants in biofilters.

Organic Compounds		HAP[a]
Aliphatic Hycrocarbons		
Methane	(+)	
Pentane	(+)	
Hexane	+	X
Octane	+	
Butadiene	++	X
Cyclohexane	(+)	
Terpenes	+	
Aromatic Hydrocarbons		
Benzene	+	X
Toluene	+	X
Xylene	+	X
Styrene	+	X
Ethyl benzene	+	X
Oxygenated Compounds		
Alcohols	++	
Methanol	++	X
Ethanol	++	
Propanol	++	
Butanol	++	
Ethers		
Tetrahydrofuran	++	
Diethyl ether	(+)	
Dioxane	(+)	X
Aldehydes	++	
Formaldehyde	++	X
Acetaldehyde	++	X
Butyraldehyde	++	X
Ketones	+	
Acetone	++	
Methyl ethyl ketone	++	X
Carboxylic acids	++	
Butyric acid	++	
Carboxylic acid esters	+	
Ethyl acetate	+	
Phenols	++	X
Cresols	++	X
Sulfur Containing Compounds		
Carbon disulfide	+	X
Sulfides	+	
Dimethyl sulfide	+	
Thiocyanates	+	
Sulfur containing heterocyclic compounds	+	
Thiophene	+	
Mercaptans	+	
Methyl mercaptan	+	

Table 1. (continued).

Organic Compounds		HAP[a]
Nitrogen Containing Compounds		
Amides	+	
Amines	++	
Trimethylamine	++	
Nitrogen containing heterocyclic compounds	+	
Pyridine	+	
Nitro compounds	(+)	
Nitriles	+	
Acetonitrile	+	
Isonitriles	+	
Chlorinated Hydrocarbons		
Methylene chloride	+	X
1,1,1-Trichloroethane	(+)	X
Trichloroethylene	(+)	X
Perchloroethylene	(+)	X
Chlorophenols	+	
Inorganic Compounds		
Hydrogen sulfide	+	
Ammonia	+	
Carbon monoxide	+	
Nitrogen oxides	+	

[a]HAP: Listed as Hazardous Air Pollutant under 1990 CAA Amendments, Title III.
Legend:
++ High removal
\+ Medium removal
(+) Minimal removal
Sources: *VDI Richtlinie 3477: Biological Waste Gas/Waste Air Purification, Biofilter.* 1991; various sources of bench-, pilot- and full-scale test data.
Note: Rating refers to the removal of single components. Behavior in mixtures may vary. The relative rating of compounds may also vary with their concentration and with off-gas temperature.

mesh wire, and slotted PVC pipes. To avoid dry-out of the packing by unsaturated off-gas, the contaminated off-gas is usually prehumidified in a spray-nozzle or packed-column humidifier. If readily available, steam may also be injected into the off-gas. In particular, enclosed systems usually provide for additional direct spray humidification of the packing to compensate for moisture losses caused by incomplete prehumidification and temperature increases at higher VOC concentrations (see below).

Most biofilters to date have been built as single-level, open systems. In recent years fully enclosed systems with stacked beds, which produce smaller footprints and provide higher levels of process control with more reliable monitoring, are increasingly used, especially in VOC and air toxics control applications. Examples of typical biofilter installations, a fully enclosed three-level unit and an open single-level biofilter, are shown in Figures 2 and 3. Regardless of their specific, often proprietary design features, almost all biofilters currently installed involve the major elements displayed in Figure 4.

Energy-related operating cost for biofilters depend primarily on the pressure differential required to vent off-gases through filter bed, duct work, and prehumidifier. For large off-gas volumes the related cost of electricity becomes significant.[3] The typically largest contributor, the pressure drop across the filter material itself, is a function of the original composition, degree of compaction, and moisture content of the packing and the off-gas face velocity. Figure 5 shows this relation for two different filter materials. The graph suggests that, particularly at the shorter residence times and correspondingly higher face velocities appropriate for many odor control applications, the difference in energy consumption between different filter materials becomes significant, and the use of a less compact, more durable filter material will generate energy savings.[4]

Since biofilters have been applied to a wide range of contaminated gas streams, their main characteristics, such as off-gas flow rate, residence time, filter size, and pressure drop vary widely. Table 2 provides ranges for several characteristics of existing full-scale installations.

In addition to biofilters, biotrickling filters (BTF) have recently emerged as a biological treatment alternative for off-gases containing acidifying substances, such as reduced sulfur and nitrogen compounds. In a BTF, inert packing material, usually made from plastics or ceramics, supports the growth of the active biofilm, and liquid is continuously recirculated over the packing. This allows for continuous removal of acidic degradation products and salts and for the pH control without deterioration of the filter material. Compost-based filter materials that are periodically flushed out for acid and salt removal are prone to compaction. Bioscrubbers, the combination of a packed-bed scrubber

[3] Electric power demand is given by P [kW] = $2.72 \cdot 10^{-7}$ Δp [Pa] Q[m^3/h] / ϵ = $1.17 \cdot 10^{-4}$ Δp [″wg] Q[cfm] / ϵ where Δp is the pressure drop across the entire biofilter system, Q the volumetric off-gas flow, and ϵ the combined motor/fan efficiency.

[4] For example, for a flow of 50,000 m^3/h (29,400 cfm), a residence time of 15 sec, a bed height of 1 m, and $\epsilon = 0.6$, the resulting face velocity is 208 m/h. This corresponds to pressure drop for the two filter materials of 230 and 1040 Pa, respectively. The respective power requirements for the blower are 5 and 24 kW.

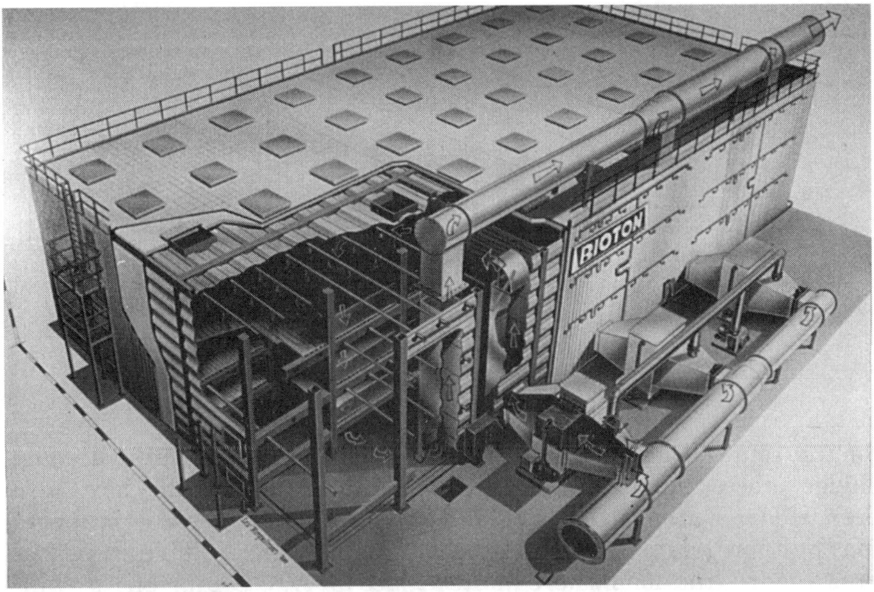

Figure 2. Schematic of an existing, enclosed biofilter for the control of VOC from an industrial wastewater treatment plant. Filter volume: 2750 yd^3 (2100 m^3), filter area: 7500 ft^2 (700 m^2), off-gas flow rate: 45,000 scfm (75,000 m^3/h) (*Source:* ClairTech b.v.).

Figure 3. Open, single-level biofilter for the control of VOC from vegetable oil production. Filter area: 3200 ft^2 (300 m^2), off-gas flow rate: 23,000 scfm (39,000 m^3/h) (*Source:* Braintech GmbH).

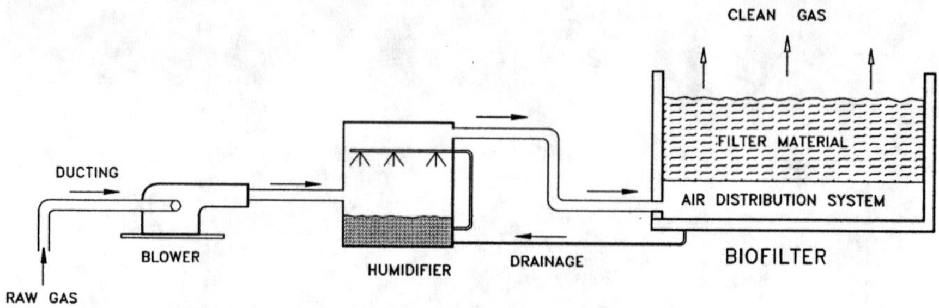

Figure 4. Conceptual design of an open biofilter.

for the removal of VOC from an off-gas stream and of an activated sludge process for the regeneration of the scrubber fluid, have also been used in Europe in a variety of full-scale applications. Bioscrubbers may be appropriate for the treatment of highly water-soluble air pollutants and in situations where limited space for APC equipment is available. More detailed reviews of these technologies can be found in the literature (*VDI Berichte 735,* 1989; *Biotechniques,* 1992; *VDI Berichte 1104,* 1994).

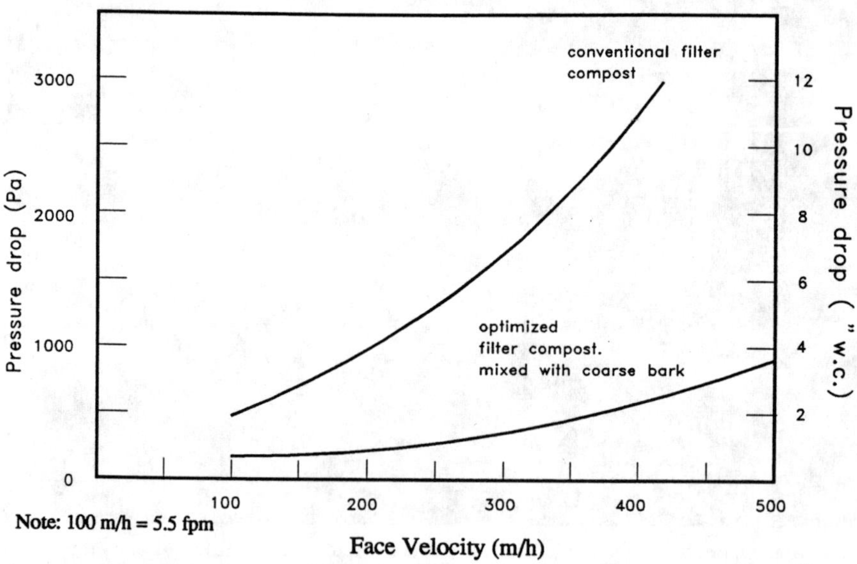

Figure 5. Pressure drop for two filter materials as a function of off-gas velocity. (Adapted from Don and Feenstra, 1984.)

Table 2. Typical characteristics of biofilter systems.

Off-gas flow rate:	100–250,000 m^3/h	50–150,000 cfm
Filter volume:	10–3,000 m^3	15–4,000 yd^3
Pressure drop across filter bed	125–3000 Pa	0.5–12″ wg
Number of levels:	1–5	
Residence time:	10 sec–3 min[a]	

[a]Residence times of up to 30 minutes may be required in soil bed installations.

Operating Requirements for Biofilter Materials

As with other biotechnologies, the filter material must provide a favorable environment for resident microorganisms for effective operation of the biofilter over a period of several years before replacement of the filter material becomes necessary. The major factors of concern include

- pollutant transfer and structural integrity: For efficient pollutant transfer, biofilter materials must provide large, wetted internal surface areas and a sufficiently open pore structure that results in a low pressure drop. Compaction of the packing over time, caused by its gradual degradation and settling of fine particles, must be kept to a minimum by proper selection of raw materials. Compaction will reduce treatment efficiency, increase pressure drop and energy consumption, and require the early replacement of the packing.
- moisture content: The presence of sufficient moisture in the filter material is essential for the establishment of a large internal surface for efficient pollutant transfer from off-gas to packing, the formation of a durable biofilm, and the metabolic activity of the microorganisms. A moisture content of 50–60% by weight is desirable for most packing materials. Contaminated off-gases that are incompletely prehumidified or contain high concentrations of VOC (the exothermic biodegradation of which results in a temperature increase across the filter bed) will remove moisture from the filter material. In those cases the additional spraying of water onto the filter material will be required to prevent drying.
- temperature: Biofilters most commonly rely on the activity of mesophilic microorganisms. Accordingly, bed temperatures should be maintained at 20–40°C (70–105°F) for optimum metabolism. Higher temperatures may cause a shift to thermophilic populations capable of degrading the fibrous materials in the packing. This results in the accelerated mineralization and

compaction of organic filter materials. At temperatures less than 10°C (50°F) performance decreases rapidly, mainly as a result of reduced rates of metabolism.

- pH: Most microorganisms capable of degrading VOC show optimum growth and pollutant removal at pH values in the biofilm between 6.5 and 7.5. At pH above 8.5 and below 5, microbial growth rates decrease. The formation of acidic by-products (mainly from the degradation of sulfur-, nitrogen-, and chlorine-containing organic compounds, but also from high concentrations of nonsubstituted VOC) will cause the consumption of alkalinity in the filter material and subsequent reduction of the pH. Unless properly buffered by alkaline material, the low pH will usually reduce the biofilter's capacity for VOC removal. In some applications, the operation of a biofilter at lower pH may be possible. For example, the removal of H_2S by thiobacilli continues at pH values of less than 3.
- nutrients: Microorganisms require the presence of inorganic nutrients such as nitrogen, phosphorus, and potassium in the packing for cell growth. Fibrous raw materials (wood chips, bark, peat) are generally deficient in these nutrients and may require the addition of nutrient-rich compost or nutrients in water-soluble form. However, the latter must be practiced with caution. Otherwise, at high VOC loads per volume of filter material, the free availability of nutrients will result in increased performance but also will cause excessive biofilm growth and clogging of the media.

Advantages and Disadvantages of Biofiltration

In most full-scale installations to date, biofilters treat off-gas containing odorous substances or VOC in low concentrations, typically less than 1000 ppmC. In these applications, biofiltration competes with other APC technologies, such as incineration, wet scrubbing, and adsorption.

Based on two biofilter characteristics, i.e. the destruction of pollutants without the need for additional energy and a comparatively simple design, biofilters provide the following advantages

- lower capital cost: Total installed capital costs for biofilters are frequently lower than for incineration or adsorption systems, particularly when residence times of less than 60 seconds are sufficient.

- lower operating cost: Electricity for blower, pumps and controls, and the periodic replacement of packing are the main operating costs for a properly designed biofilter. At VOC inlet concentrations of less than 1000 ppmC, the cost of supplementary fuel to operate even highly energy-efficient incineration systems such as regenerative thermal and catalytic oxidizers is substantial. The corresponding potential cost savings for biofilters is significant.
- environmental benefits: Biofilters usually nearly completely destroy target pollutants without supplemental energy, formation of combustion emissions such as carbon monoxide (CO) and nitrogen oxides (NO_x), and use of potentially hazardous chemicals such as sodium hydroxide or sodium hypochlorite.
- low maintenance requirements: Except for the periodic replacement of the packing, maintenance for well-designed biofilters with high levels of passive process control is typically lower than that for incineration, adsorption, or scrubbing systems. The operation involves routine inspections of the blower, humidifier, and filter material and the review of recorded operating parameters.

Biofilters may also exhibit the following disadvantages when compared to established APC technologies

- larger space demand: Compared to incineration or adsorption systems, biofilters generally require a significantly larger footprint.
- low level of process control: The low level of available process control and monitoring in a biofilter also complicates identification and elimination of system malfunctions caused by a flawed design or by off-gases with challenging characteristics, such as excessive or strongly fluctuating concentrations of VOC or acidifying compounds, elevated temperatures, and high concentrations of particulates.
- sensitivity to environmental conditions: In particular, open biofilters are sensitive to changes in environmental conditions, such as temperature, relative humidity, and precipitation. Such changes may produce drying of the filter material or formation of condensate and excessive drainage producing poor moisture control in the filter material.
- replacement of filter material: With large full-scale biofilter installations, the periodically required replacement of filter material is a considerable task requiring on-site space and labor.

- discharge of wastewater: The need for the periodic discharge of drainage from the humidifier and filter beds in large systems usually requires a wastewater discharge connection unless on-site treatment and discharge into surface waters is feasible.
- no recovery of raw materials: As with other destructive APC technologies, biofilters do not recover potentially valuable raw materials.

The potential advantages and disadvantages of biofiltration suggest that its suitability and competitiveness will strongly depend on off-gas and site characteristics and must be evaluated on a case-by-case basis.

HISTORY OF BIOFILTRATION

Suggestions to treat odorous off-gases by biological methods can be found in the literature as early as 1923 when Bach (1923) introduced the concept of controlling airborne H_2S emissions from sewage treatment plants. Reports on early applications of this concept dating back to the 1950s were published in the United States and West Germany. Pomeroy received a U.S. patent for a soil bed concept in 1957. He also describes a successful soil bed installation in California (Pomeroy, 1963). Around 1959, a soil bed for the control of odors from an incoming sewer main was also installed at a municipal sewage treatment plant in Nuremberg, West Germany (Hartmann, 1976).

In the United States, the first systematic research on the biofiltration of H_2S was conducted by Carlson and Leiser (1966) in the early 1960s. Their work included the successful installation of several soil beds at a wastewater treatment plant near Seattle and demonstrated that biodegradation rather than sorption accounted for the odor removal. During the 1970s and 1980s, several researchers in the United States further studied the soil bed concept and demonstrated its usefulness in several full-scale applications (Bohn, 1975; Bohn and Bohn, 1986), including the control of odors from rendering plants (Prokop and Bohn, 1985) and the destruction of propane and butane released from an aerosol can filling operation (Kampbell et al., 1987). However, use of the technology in the United States remained limited, mainly due to lenient regulatory requirements, separation of industrial and residential areas in new developments, and comparatively low-cost energy.

At the same time, the need for cost-effective odor control increased in Germany and the Netherlands, driven by a high density of urban development and growing environmental awareness among the population. As a result, simple biofilters were increasingly used in odor control.

The biofilters built in the 1960s consisted of perforated pipes for off-gas distribution covered with mineral soils as the filter media. While such systems provided sufficient control of odors in many applications, they did not include process control and were subject to frequent operational problems such as the clogging of the air distribution system, uneven off-gas distribution, and drying of the beds.

In the 1970s, more advanced open biofilters were developed. These systems, now widely used for odor control, consisted of an open bed, typically enclosed by concrete walls, and an advanced air distribution system supporting the filter material. The filter material usually consisted of a mixture of compost or peat and structural support (branches, wood chips, bark, or mineral granulates) to prevent the filter material from compacting, to reduce pressure losses across the bed, and to minimize channeling of untreated off-gas.

With the emergence of these systems, biofiltration became increasingly popular in Germany and the Netherlands, still mostly for odor control, but increasingly for removal of higher concentrations of VOC and air toxics. Some of these installations experienced operational problems, including drying and overwatering of the filter material and rapid bed compaction, resulting in high pressure drops and energy costs and shorter bed life. These experiences and regulations requiring efficient and reliable operation of air pollution control systems, particularly for VOC and air toxics control applications, stimulated the development of a new generation of more advanced biofilters. These systems are usually fully enclosed, employ a filter material designed for low pressure drop and extended useful life, and include more process control, particularly for the filter material's moisture content. After almost a decade of development and full-scale demonstration in a variety of applications, these systems are now generally accepted by regulatory agencies in several European countries as reliable, cost-effective, and environmentally benign control technology for a range of off-gas streams containing low concentrations of biodegradable, gaseous air pollutants.

It can be inferred from the lack of literature on biofiltration published in the United States during the 1970s and 1980s that little attention was paid to these European developments. Since the late 1980s, increasing regulatory pressure to control odors and dilute emissions of VOC and air toxics, the establishment of European biofilter firms in the United States, and an increasing number of publications in English have raised awareness of the technology. The technology in the United States is now marketed by numerous vendors. Simple, open systems are now increasingly used for odor control at wastewater and composting facilities (Williams, 1994). By late 1995, approximately 20 large-scale biofilters in industrial applications, treating more than 17,000 m^3/h (10,000 cfm) of off-gas have been installed. More than 50 small systems have been installed since 1993 for the removal of VOC from soil vapor extraction (SVE) off-gases. Although with significant delay, operational experiences similar to those in Europe are now occurring in the United States and other industrialized countries and are producing similar trends in system design and areas of application.

CRITERIA FOR THE APPLICABILITY OF BIOFILTRATION

This review of fundamentals, advantages, and disadvantages of biofiltration indicates that biofiltration will generally not apply to entire industries but may be a cost-effective control alternative for specific off-gas streams and site-specific and regulatory settings. This section provides several screening criteria to aid in the initial assessment of the suitability of biofiltration for a given off-gas source.

Off-Gas Specific Criteria

OFF-GAS COMPOSITION

Table 1 includes information on the relative removal of common biodegradable air pollutants in biofilters. Water-soluble, or polar, VOCs are generally removed most efficiently and are preferable for biofilter treatment. Aromatic hydrocarbons, such as BTEX, are removed at somewhat lower rates, while removal of short-chained, aliphatic compounds (methane-pentane) is particularly poor. Hydrogen sulfide and ammonia are removed efficiently. In mixtures of VOC, the presence of significant proportions of poorly removed compounds will limit the percent removal efficiency for total VOC that can be achieved at a reasonable residence time.

POLLUTANT CONCENTRATIONS

Total VOC concentrations <1000 ppmC are preferable and improve competitive advantages of biofiltration over incineration technologies. Particularly for water-soluble compounds, average concentrations >–4000 ppmC should be avoided in order to minimize the potential for dry-out at the biofilter inlet and the formation of acidic by-products. Excessive concentrations of compounds toxic to microorganisms, such as formaldehyde and SO_2, must also be avoided. Concentrations of H_2S and NH_3 of <10 ppm are preferable to control the rate of acidification of the filter material.

OFF-GAS TEMPERATURE

Temperatures of 20–40°C (70–105°F) in the off-gas entering the filter material are preferable. Higher temperatures in the off-gas vented from the source are acceptable if the off-gas is dry and permits adiabatic cooling, i.e., if its wet bulb temperature is less than 40°C.

PARTICULATES

High loads of particulates related to metals, natural or synthetic resins, oil, and grease will clog the air distribution system and packing and must be removed in a pretreatment step. Acceptable particulate concentrations in the off-gas entering the filter bed depend on the nature of the particulates and the off-gas face velocity. As a guideline, total particulate concentrations in the off-gas entering the bed should be less than 5 mg/m^3 (0.002 gr/cf).

SOLVENT RECOVERY

Solvent recovery may provide economic advantages where large quantities of pure solvents with a high retail value are emitted from a process. Where mixtures of inexpensive solvents are emitted at low concentrations, on-site recovery is usually not cost-effective, and the use of a destructive technology, such as biofiltration, becomes more attractive.

Site-Specific Criteria

AVAILABILITY OF SPACE

Particularly for off-gas flows >10,000 m^3/h, biofilters may occupy a considerable footprint. Depending on residence time and system design, a biofilter reactor will typically require 4–22 $m^2/1000\ m^3/h$ (75–400 $ft^2/1000$ cfm). The blower and humidifier will require additional space. The availability of rooftop space will aid in space-constrained settings but also will increase construction costs.

COST OF ENERGY

Savings in operating cost and, consequently, the cost-competitiveness of biofiltration varies with the local costs of electricity and natural gas. Energy cost considerably less than $0.07/kWh for electricity and $0.014/kWh ($4/MMBTU) for natural gas will diminish biofiltration's operating cost advantages.

Regulatory Control Requirements

Where odor control is required for sources operating at least semi-continuously, biofilters should always be considered. In cases where VOC control is the main regulatory goal, VOC removal efficiencies in excess of 90% can be achieved at reasonable residence times for oxygenated VOC and for some aromatic hydrocarbons. For other petroleum hydrocarbons, particularly light aliphatics, target removal efficiencies of >90% will require residence times of several minutes and result in high capital cost. In general, mass-based emission limits for VOC or risk-based limits for air toxics are preferable for biofiltration and tend to improve its economics since these limits usually allow systems to be designed for lower percent control efficiencies and correspondingly shorter residence times.

FULL-SCALE APPLICATIONS OF BIOFILTRATION

Biofilters have now been installed in considerable numbers worldwide. The estimated total number of biofilters treating force-vented off-gases is between 500 and 1000. Exact estimates are difficult because many of these systems are open, simple, "home designs" for odor control. The number of globally installed advanced filters, i.e., fully enclosed systems with some level of moisture control and designed filter material, is approximately 200.

Biofilters have treated off-gases from a wide range of source categories. Table 3 provides an overview of industries and specific off-gas sources where biofilters have been applied. Most of these applications have been demonstrated in Europe on the full-scale level. However, biofilters will rarely constitute the technology of choice for entire source categories. Instead, their applicability will, as discussed above, depend on off-gas characteristics, site conditions, and regulatory requirements. The table suggests that biofilters no longer mainly treat off-gases from waste management operations, but from a wide variety of industries that will be reviewed in the following.

Odor Control Applications

Almost all of the early applications of biofilters were used to remove low concentrations of odorous, biodegradable compounds. The target pollutants in these applications included volatile reduced sulfur and nitrogen compounds and several groups of nonsubstituted, usually oxygenated VOC, such as alcohols, ketones, ethers, esters, aldehydes, and

Table 3. *Examples of biofilter applications.*

Industrial and Commercial Sources	Food Production and Processing	Waste Management
Coating and Printing	Baking processes	Wastewater treatment plants (residential and industrial)
Automotive coating	Bread	Waste transfer stations
Can and drum coating	Eggroll	Composting operations
Photographic film coating	Broth production	Sludge and manure drying
Piano manufacturing	Fish flour production	Hazardous waste storage
Plastics coating	Gelatin production	Landfill gas extraction
Wire coating	Meat and fish frying	Plastics recycling
Wood coating	Slaughter houses	Soil vapor extraction
Off-set printing	Smoke houses	Waste oil recycling
Screen printing	Sugar production	Meat and fish rendering
Manufacturing	Vegetable oil extraction	Bone processing
Adhesive production	Yeast production	Blood drying
Brake pad and clutch production	Coca roasting	
Coatings production	Coffee processing	
Fiberglass applications	Corn processing	
Flavors and fragrances production	Tobacco processing	
Furniture manufacturing	Animal food production	
Ink production	Hay drying	
Iron foundries	Soybean processing	
Investment foundries	Pet food production	
Pharmaceutical industry	Feed lots	
Pill coating		
Plastics production		
Polymer production		
Acrylic glass production		
Foam blowing		
Polyester resin production		
Polyethylene production		
Styrene-butadiene rubber (latex) production		
Rayon production		
Sanitation products		
Solvent production		
Wood products manufacture (particle board, plywood, oriented strand board)		
Storage and Transfer of Organics		
Chemical storage		
Crude oil storage		
Tanker cleaning		

organic acids. Biofilters can be found as odor control devices in the following source categories:

- wastewater treatment: Biofilters frequently treat emissions from headworks, primary sedimentation, sewer lines, and sludge management processes. The systems, usually open, are used particularly for the control of streams where much of the odor is caused by nonsubstituted VOC, rather than H_2S or other reduced sulfur compounds.
- composting: The storage and aerobic conversion of organic wastes to compost generate a multitude of highly odorous organic intermediates. In particular, composting facilities in urban areas now routinely collect their process and ventilation off-gases and treat them in biofilters. Off-gas flows frequently exceed 100,000 m^3/h. While open systems still prevail, enclosed systems are now frequently used in space-limited locations.
- food processing: Many food processing operations release odors, often offensive, from cooking, frying, roasting, baking, and steaming equipment. While simple, open systems were initially predominant, enclosed, more controlled systems are now increasingly used. Hot and particulate-laden streams also frequently require pretreatment.
- flavors and fragrances: Off-gases from the production of flavors and fragrances usually contain low concentrations of the (odorous) products and process solvents. More than 10 fully enclosed biofilters have now been installed for odor control at facilities in Europe and the United States. Residence times of less than 30 seconds are usually sufficient and make these applications economically attractive.

Although petroleum refineries and kraft pulp mills are known as frequent sources of odor complaints, biofilters have not been applied in these industries to any significant degree. This is mainly due to the typically high concentrations of reduced sulfur and/or petroleum hydrocarbons in the target streams and the common batch-type release of emissions.

VOC and Air Toxics Control Applications

In addition to these "traditional" odor sources, other industrial and commercial sources of dilute air pollutants increasingly have installed biofilters. Depending on the local regulatory focus, these applications have been brought about by odor control or, since the late 1980s, also by the need to control specific air toxics and high loads of dilute VOC. The latter continues to generate biofilter installations, particularly in Germany and the Netherlands, usually involving more controlled, fully enclosed systems with a designed filter material, which have eased concerns among regulators over the operational reliability and the performance monitoring of biofilters. In the United States, much of the growing interest in biofiltration, experienced since 1989, has been fueled by the search for cost-effective VOC and air toxics control alternatives. Table 3 suggests that most of these industrial applications fall into one of the following source categories:

- coating and printing: Biofilters have been used occasionally to eliminate solvents emitted from coating and printing equipment, such as presses and ovens, and from ventilation air. While the shift to a low-VOC formulation is often economically and environmentally preferable, product quality requirements and the lack of substitute materials sometimes dictate continued use of a high-VOC material and the need for end-of-pipe treatment. Examples include the use of biofilters in specialty applications such as piano manufacturing, photographic film coating, and paint manufacturing. In the United States, a large-screen printing operation is currently investigating the use of biofiltration for control of solvent emissions.
- chemical industry, polymers, and pharmaceutical production: This category encompasses a large variety of processes, such as separation, drying, curing, processing, and coating of synthetic fibers, bulk polymers and other petrochemicals, and pharmaceuticals. In particular, emissions of odorous and toxic compounds, such as styrene, methyl methacrylate, and ethyl acetate, from the production of synthetic resins such as ABS and Plexiglas have been treated in several biofilters in Europe and the United States.
- petroleum products: The number of full-scale biofilters for the treatment of hydrocarbons from petroleum operations has been limited to date. This is mostly due to the poor removal of petroleum hydrocarbons, particularly aliphatics, in biofilters and the typically strong fluctuations in off-gas characteristics. Several smaller systems provide odor control for crude oil storage and handling equipment. In the United States, the commonly required removal of gasoline vapors from soil vapor extraction (SVE) off-gas has rapidly created the most popular biofilter application to date, if measured by the considerable

number (more than 50) of smaller systems, usually treating less than 250 cfm of SVE off-gas.

- other manufacturing processes: This group involves a variety of production processes with particular off-gas characteristics amenable to biofiltration. They include iron foundries emitting odors and air toxics (formaldehyde, benzene, phenols) from binders used in sand molds, mold production in precision or investment casting operations generating ethanol emissions, and the manufacturing of brake pads and clutches, which releases odorous emissions from resin usage. In the United States at least one large biofilter treats dilute VOC emissions from the press vents in a particle board line that emits alcohols, aldehydes, and terpenes. If successful, this application may become one of the domains of biofiltration in the United States.

PERFORMANCE AND OPERATIONAL PROBLEMS

Given the wide range of off-gas sources and biofilter designs, it comes as no surprise that operational experiences with biofiltration during its development and maturation periods also varied considerably. Many operators of biofilter systems confirmed that the installed system generally achieved the desired level of treatment. For odor control applications this often meant more than 95% odor removal. Where the removal of total VOC was required, removal efficiencies of 80–95% were routinely achieved, particularly for the more water-soluble, oxygenated VOC. The useful life of filter materials varied from only several months for compact filter materials or off-gases containing high loads of acidifying compounds (H_2S) to more than 5 years for systems employing designed filter material with inert ingredients. Well-designed systems also required little maintenance other than routine inspections and the replacement of the filter material.

Failure to control one or several of the parameters affecting the operation of a biofilter has also caused performance problems or complete failure in a significant number of full-scale biofilters. The following list of common problems is based on available project experience:

- moisture balance: Gradual drying of the packing by the off-gas with subsequent loss in active surface and microbial activity is the most frequent cause of biofilter performance problems. It results from insufficient prehumidification of the off-gas and/or inadequate spray-irrigation of the packing. The likelihood of drying increases for off-gases with VOC concentrations exceeding

100 ppmC. Overwatering of poorly drained filter material, on the other hand, produces wet, anaerobic zones, increases in pressure drop, wash-out of smaller particles, and potentially irreversible structural changes to the filter material leading to the formation of channels (see below). The same effect may be caused by an off-gas with wet bulb temperatures above ambient levels. Such moist gases routinely cause condensation of moisture in the filter bed, particularly in systems that are poorly insulated.

- compaction of filter material: Installation of a filter material that is readily biodegraded and has low initial porosity and high content of fine particles less than 3 mm in size routinely causes a swift increase in pressure drop, increasing energy demand and the formation of channels in the filter material. Channeling leads to the discharge of only partially treated off-gas and reduces the system's overall removal efficiency for target pollutants.
- acidification: Acidic degradation products are formed in a biofilter during oxidation of VOC containing nitrogen, sulfur, or chlorine atoms. Formation and release of acidic intermediates, such as acetic acid, has also been observed for streams with high concentrations of water-soluble compounds, such as ethanol (Ottengraf, 1986; Leson, 1993; Leson, 1995; Devinny and Hodge, 1995). In many instances, the release of acidic degradation products into the packing has produced gradual acidification, disappearance of microbial populations sensitive to low pH, and corrosion of construction materials such as concrete and galvanized steel. The release of such intermediates has also caused local odor problems, particularly for open biofilters. The open filters discharge at low gas velocities, typically 40–200 m/h (2–10 fpm), and on grade level, which results in poor dispersion of the treated off-gas.
- particulates: The presence of high concentrations of particulates in an off-gas has repeatedly caused gradual clogging of the off-gas pretreatment air distribution system and, in the layer of filter material, an increase in pressure drop and the formation of channels. This problem is particularly difficult to remedy in upflow systems since the biofilter inlet is not easily accessible.
- undersizing: Lower than desired removal efficiencies for target pollutants have frequently been caused by undersizing of the biofilter, which produces insufficient off-gas residence time in the filter material. Undersizing is often caused by insufficient knowledge of the composition and other characteristics of an off-gas stream or by pilot-testing of a nonrepresentative slip-stream.

While the number of publications in English describing biofilter research on the bench-scale has increased rapidly since 1990, the above summarized operational experiences with full-scale biofilters are not well documented, particularly in the English literature.

BIOFILTER DESIGN OPTIONS

Various design concepts and features for biofilters are now offered by suppliers in Europe and the United States or can be considered for in-house design and construction. Several of these options have been developed in order to minimize the risk of operational problems discussed above. Available biofilter options are best characterized by their design and type of construction, their choice of filter material, and their level of process control. Summaries of the major options and their potential advantages and disadvantages follow.

Design and Construction

OPEN VERSUS ENCLOSED SYSTEMS

This option refers to the main conceptual difference between biofilter systems. Open single-level biofilters are comparatively simple in design, are less expensive to build on a unit volume basis, and provide easy access to the packing for monitoring and replacement. They are also sensitive to changes in climatic conditions, such as temperature and precipitation, and allow only limited control of major operating parameters. The performance of open systems is difficult to monitor, and the low exit velocities of the treated off-gas may result in fogging. Enclosed systems provide the possibility of more efficient process control by reducing or eliminating the impact of ambient temperature, precipitation, wind, deposits of particulates, and plant growth in the filter material. They allow for single point discharge and, consequently, more reliable monitoring of the treated off-gas and more efficient dispersion.

SINGLE- VERSUS MULTI-LEVEL SYSTEMS

Multi-level systems typically containing two, in some cases up to five, filter beds in series or parallel have been used in more than 100 biofilter installations. The cost for design and construction of single-level biofilters per volume of filter material are generally lower than for multi-level systems. However, the latter can provide significant space savings. They are usually built as enclosed systems, which also permits more efficient process control.

MATERIAL OF CONSTRUCTION

To date, the majority of biofilters have been built as concrete structures or as steel frame units with wall panels. The use of concrete, common in single-level systems, results in lower capital cost. Standardized steel frame units can be erected faster and provide more flexibility with respect to necessary system modifications and potential relocation. Construction of an open single-bed filter in an excavation with earthen containment usually is the least expensive option on a per volume basis. Limited access to filter material and air distribution system, the generally lower per-volume efficiency, and perceived and real impacts on groundwater quality are its disadvantages.

Filter Material

The original composition of filter material is the single most important factor determining its longevity, the effectiveness of the biofilter, and its operating cost. Available choices include the following. Organic materials and wastes from a variety of sources (wood, bark, leaves and other green waste, municipal solid waste, sewage sludge, heather, peat) usually account for the bulk of the filter material. They are generally composted prior to installation, which provides for more active surface, availability of nutrients to microorganisms and a reduced rate of mineralization and compaction following installation. Inert materials, such as ceramic or styrofoam pellets and granular activated carbon (GAC), are frequently added to improve long-term structural stability, reduce flow resistance and potential for channeling, and provide buffer capacity for peak pollutant loads. Because they provide a slowly degradable "skeleton" to the packing, hardwood chips are now widely used in European biofilter applications. Since finer particles tend to migrate in the filter material and fill up available void space, particularly if frequently irrigated, their amount is often controlled by screening the packing prior to its installation.

The inoculation of filter material with a microbial culture fed on mixtures of the target pollutants is practiced by several vendors. It tends to reduce acclimation periods and may improve performance in the case of off-gases containing complex mixtures, particularly of xenobiotic VOC. Inorganic nutrients are frequently added to filter materials that provide insufficient levels to resident microorganisms. Their addition in a water-soluble form, such as in commercial fertilizers, must be practiced carefully since it may result in excessive biomass growth. Lime is often added to the filter material to improve its alkalinity and to buffer release of acidic by-products of biodegradation. The addition of calcium carbonate lime is preferable since it avoids pH excursions.

Designed filter materials are considerably more expensive, typically \$100–300/m^3, than composts based on unprocessed waste materials, which may cost as low as \$30/m^3. The high cost applies in particular to packings containing significant amounts of inert additives and involving extensive processing. Their longer useful life, which may exceed 5 years as compared to 1–3 years for unprocessed materials; their more efficient removal of pollutants, resulting in a smaller required filter volume; and their lower specific energy consumption should be considered in evaluating the true cost of filter material replacement.

Upon replacement, the spent filter material needs to be disposed of in a cost-effective, environmentally safe, and convenient manner. Based on available experience, the presence of VOC residues or heavy metals introduced with the fresh material is not likely to result in its classification as a hazardous waste. This allows disposal of spent filter material in a sanitary landfill, potentially as landfill cover. Filter material acceptability for cost-effective and beneficial reuse in landscaping applications, as is often practiced, will primarily depend on the presence of inert additives, such as ceramic or plastic materials. Vendors utilizing such additives may offer their recovery and reuse under a service contract.

Process Control

In addition to the passive process control from filter design and selection of filter material, some biofilter designs involve active control of two major operating parameters, the packing's moisture content and the off-gas temperature.

- moisture control: The moisture content in the packing, depending on off-gas characteristics, is subject to significant fluctuations. As already mentioned, drying causes poor system performance and makes moisture control the single most important control issue. Prehumidification of off-gases with relative humidities below 90% is generally practiced. Prehumidifiers are commonly installed between blower and filter bed because the temperature increase across the blower would otherwise reduce the relative humidity of the off-gas typically to less than 90% before entering the biofilter. Incomplete prehumidification, VOC concentrations in excess of 100 ppmC, and ambient temperatures in excess of the off-gas inlet temperature will result in net moisture removal from the filter bed and require additional over-bed spray humidification. Manually or timer controlled irrigation are potential solutions for off-gases with fairly consistent characteristics and are now used by vendors of both open and enclosed systems. A possible alternative, frequently applied to improve reliability of performance and reduce maintenance, is the use of a fully automatic moisture control system, such as the one developed by a Dutch biofilter vendor. It continuously determines the packing's bulk moisture content by measuring its weight and provides over-bed irrigation as necessary. Proper operation requires the use of a fully enclosed system.

- temperature control: Depending on raw-gas temperature and humidity, off-gas precooling or heating may be required to maintain bed temperatures within the desirable range of 20–40°C. Adiabatic cooling of dry off-gas, waste heat recovery from hot gases, and use of available waste heat for off-gas preheating are potential energy-efficient options that have been applied in several applications. The need to preheat with fossil fuels or to refrigerate a hot and moisture-saturated off-gas results in additional energy cost but has been applied for several smaller off-gas streams. Use of ambient dilution air for cooling purposes is generally not desirable since it increases off-gas volume and required system size.
- direction and distribution of off-gas flow: Most biofilter systems are operated in an upflow mode where the off-gas enters the filter bed from the bottom. Several vendors now offer downflow systems that require the use of an enclosed biofilter structure. Their main objective is to provide effective irrigation to the initial contact zone between off-gas and packing where the drying process caused by incomplete prehumidification and/or exothermic biodegradation will be most rapid. Various off-gas distribution and packing support alternatives are available. They include simple designs based on slotted PVC pipes, as well as patented systems involving slotted concrete slabs or perforated concrete blocks that allow access for loading equipment. Performance of these systems has varied considerably. The potential for clogging of air distribution slots, interference with the collection of drainage, and corrosion of concrete elements should be evaluated for each filter material and off-gas.

In addition to these major options, biofilter designs may vary with respect to the collection, routing, and disposal of drainage from the beds; the selection of bed height; and the degree to which off-gas and system parameters (temperature, relative humidity, pressure drop, VOC concentration, moisture content) are monitored.

ECONOMIC CONSIDERATIONS

The potential savings in capital and operating costs are considered by industry to be biofiltration's main advantage over competing technologies. Their extent will vary mainly as a function of the composition of the target off-gas, the biofilter design, and the local cost of electricity and natural gas. The following provides a brief review of the major economic considerations applicable to biofiltration.

Capital Cost

The previous discussion of available biofilter design suggests that the total capital investment (TCI) for biofilter systems per volume of filter material may vary considerably, depending on bed volume, design of the filter vessel, and the system's level of sophistication. The biofilter market in the United States, particularly for an enclosed system, is still in its infancy and characterized by strong competition. As a result, TCI levels continue to fluctuate. Available European information on system cost does not translate well, due to differences in the structure of equipment and labor costs and fluctuations in the dollar exchange rate. In order to develop initial information on TCI for vendor-installed biofilters in the United States, available quotes for full-scale systems with more than 50 m^3 of packing were compiled. Since most of these estimates are budget quotes and do not reflect the cost of actually installed systems, they must be considered preliminary. We have also excluded systems with filter volumes of less than 50 m^3, since cost and performance per volume vary considerably as a function of the degree of technical sophistication.

Figure 6 shows ranges for TCI as a function of the volume of packing. TCI includes the costs for turnkey installation of biofilter, auxiliary equipment, and controls. Not included are the costs for ductwork between emitting source and biofilters. Figure 6 reflects the economy of scale for larger biofilter systems. As a general rule, fully enclosed, more controlled systems of steel or fiberglass construction will be priced closer to the upper limit and open, single-level concrete systems with a low level of control to the lower one. However, competitive pricing and differences in experience between vendors may offset this rule.

To calculate the required volume of packing for a given flowrate and, using Figure 6, the corresponding TCI, the required empty bed off-gas

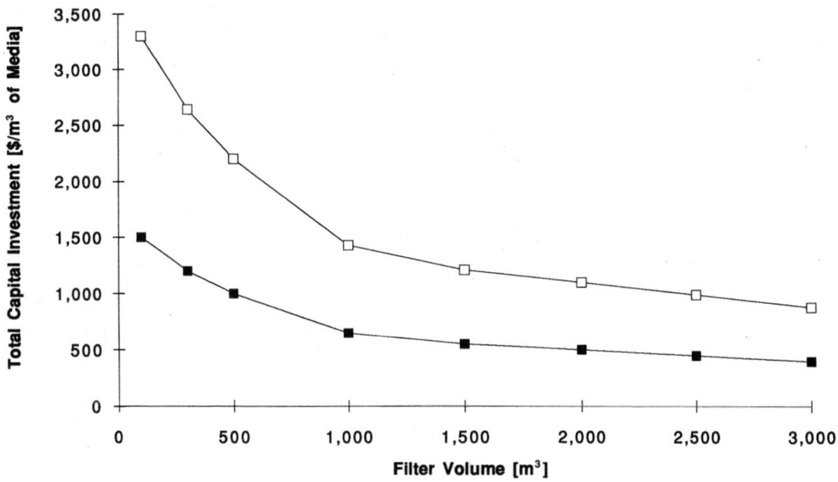

Figure 6. Estimated upper and lower limits of total capital investment (TCI) for biofilters.

residence time must first be determined, usually in a pilot test. The corresponding filter volume for a full-scale biofilter can be calculated as filter volume [m^3] = off-gas flowrate [m^3/h] · residence time [sec] / 3600. The TCI per unit of off-gas flow, obtained from Figure 6 and measured in $/($m^3$/h), permits a comparison with flowrate-based technologies such as incinerators.[5] Figure 7 shows this relationship for three residence times. Curves are based on the midpoint between minimum and maximum cost per volume from Figure 6 and represent average priced systems. Cost per volume of off-gas treated and TCI obviously increase, for a given flow, with an increase in required residence time. The figure also shows that the capital cost advantage of biofilters over regenerative thermal oxidizers (RTO), generally the most competitive established destructive technology for low VOC concentrations, is considerable for residence times <30 seconds but may be inverted if residence times become considerably longer than 60 seconds.[6,7]

[5] To obtain the specific TCI expressed in $/cfm, multiply the $/($m^3$/h) value by 1.7.

[6] The selected residence times are characteristic of odor control applications with low pollutant loads (15 seconds), moderate concentrations (several hundred ppmC) of water-soluble compounds (30 seconds), and higher concentrations of such compounds or low concentrations of aromatics that require removal of more than 90% (60 seconds). The removal of poorly water-soluble compounds, such as aliphatic petroleum hydrocarbons with removal efficiencies >90%, may require residence times of several minutes.

[7] TCI figures for RTO are based on recent cost ranges provided by vendors and users of the technology.

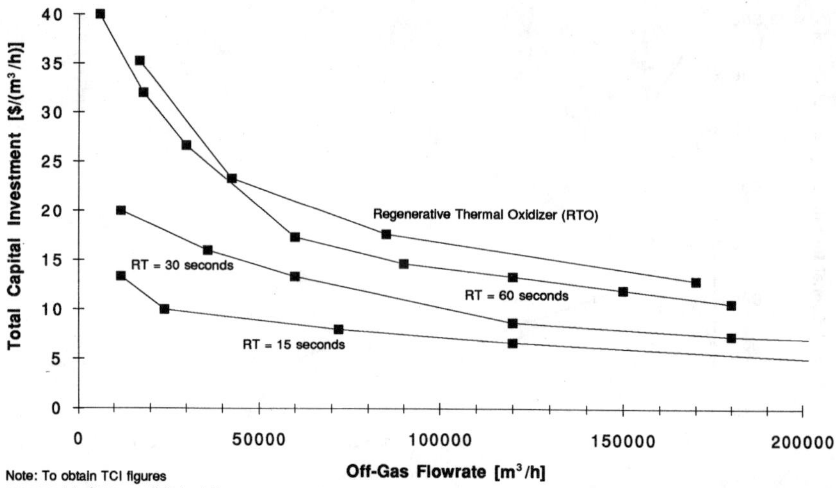

Note: To obtain TCI figures expressed in $/cfm, multiply by 1.7.

Figure 7. Comparison of total capital investment (TCI) for biofilters (three resistance times) and RTO.

Operating Cost

The high energy efficiency of biofiltration compared to incineration technologies and resulting savings in energy cost is generally considered the major source of its economic advantages. The extent of these savings depends mainly on the following:

- pressure drop across biofilter and incineration system and the resulting energy requirements for off-gas transport
- VOC concentration, off-gas temperature, and moisture content in the off-gas; energy efficiency of the incineration system; and the resulting need for supplementary fuel
- cost of electricity and natural gas

Figure 8 compares energy-related hourly operating costs for a biofilter and an RTO under two energy cost scenarios.[8] It indicates that biofilters may achieve considerable savings in energy cost at VOC inlet concentrations <1.5 g/m^3 (~2000–3000 ppmC). Even at higher concentrations

[8] Assumptions include an off-gas flowrate of 34,000 m^3/h (20,000 cfm), a (conservatively high) Δp across the biofilter system of 2500 Pa (10″ wg), a 94% thermal efficiency of the RTO, and a Δp of 5750 Pa (23″ wg) across the RTO. High energy cost scenario: $0.024/kWh ($7/MMBTU) for natural gas, $0.10/kWh for electricity; low energy cost scenario: $0.014/kWh ($4/MMBTU), $0.07/kWh.

where no supplementary fuel is required, the RTO's shorter residence time and correspondingly higher pressure drop result in considerable savings in electric costs by a biofilter. Figure 8 also reveals the increase in potential savings resulting from higher energy rates. In addition to energy costs, expenses for operation and maintenance and for the period replacement of the filter material must be included in the operating costs. Given the wide cost range for filter materials ($30–300/m^3) and their expected useful lives (1–5 years), annualized replacement cost may vary considerably. Based on our review, these costs can be estimated to add 50–100% to the energy costs of a biofilter, while O&M expenses for a properly designed and operated biofilter are typically less than for incinerators. Consequently, a biofilter treating 34,000 m^3/h (20,000 cfm) of off-gas may achieve typical hourly operating cost savings of $8–12 over RTO at VOC inlet concentrations of less than 1.5 g/m^3.

This preliminary economic comparison supports two of the applicability criteria for biofilters presented previously. It suggests that biofilters will typically become cost-competitive with energy-efficient incineration systems if VOC concentrations are <1.5 g/m^3 and if residence times of less than 60 seconds are sufficient to achieve the regulatory control objective. The comparison is intended to provide an initial understanding of the parameters affecting the cost of biofilter systems. For a more

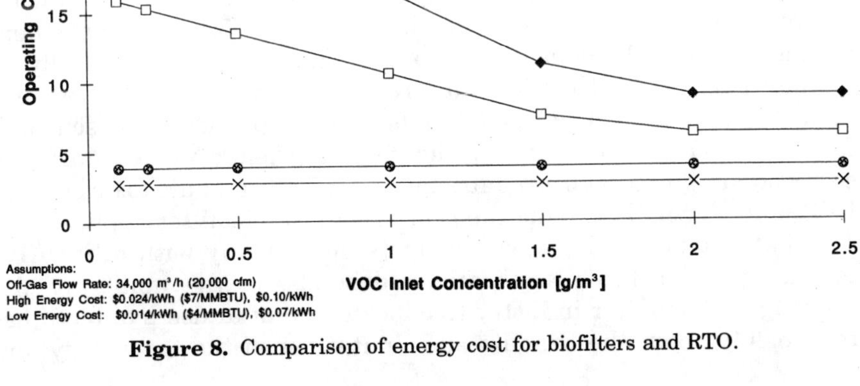

Figure 8. Comparison of energy cost for biofilters and RTO.

detailed economic comparison with established APC technologies, the reader is referred to the literature (USEPA, 1990; Mukhopadhyay and Moretti, 1993).

FORECAST

Past European developments and more recent trends in the United States suggest that a wide range of biofilter designs, from simple open biobeds to fully enclosed and controlled systems, will continue to be installed by industry. Open systems will be applied predominantly in odor control applications without stringent performance monitoring requirements and where sufficient space is available. These systems are particularly attractive to composting operations where fresh compost and machinery for the replacement of filter material are readily available. On the other end of the spectrum, fully enclosed and controlled systems will, despite their somewhat higher initial capital cost, be preferred by industry for VOC and air toxics control applications with enforceable regulatory performance and monitoring requirements, off-gas streams with fluctuating VOC concentrations that frequently exceed 1000 ppmC, and facilities with limited space and resources for maintenance and filter material replacement. Particularly among larger corporations, there appears to be an increasing willingness to consider and purchase more advanced biofilters and filter materials if they are perceived as more reliable, are likely to require less maintenance, and have lower life-cycle cost.

The previous section suggests that biofilters, if properly designed and built, will provide significant economic advantages and additional environmental benefits for specific off-gas streams. The recent development of biotrickling filters for air pollution control and advances in bioscrubber technology will likely extend the range of off-gas sources that are suitable for biological APC technologies. Yet to what extent biofiltration will be applied in the United States over the next decade will crucially depend on two main factors. Since essentially all installations of APC equipment are driven by regulatory requirements, the yet to be developed details of the implementation schedule for the 1990 amendments to the federal Clean Air Act and the corresponding regulatory requirements for VOC and air toxics (Titles I, III, and V) will determine the pool of potential biofilter applications. To what extent biofilters will compete successfully with other APC technologies for these applications depends on the ability of the growing U.S. biofilter industry to educate potential users of biofilters responsibly about the benefits and limitations of the technology and to back up their claims with successful and reliable demonstration projects.

ACKNOWLEDGEMENT

The findings summarized in this chapter are based in part on research funded by the South Coast Air Quality Management District's (SCAQMD) Technology Advancement Office and by Southern California Edison's Research, Development and Demonstration Department.

REFERENCES

H. Bach. "Schwefel im Abwasser," *Gesundheits-Ingenieur,* 46: 370 (1923).

Biotechniques for Air Pollution Abatement and Odour Control Policies, A.J. Dragt, J. van Ham, Eds.; Elsevier, Amsterdam, 1992.

H.L. Bohn. "Soil and compost filters of malodorant gases," *JAPCA,* 25: 953 (1975).

H.L. Bohn, R.K. Bohn. "Soil bed scrubbing of fugitive gas releases," *J. Environ. Sci. Health,* A21: 561 (1986).

D.A. Carlson, C.P. Leiser. "Soil beds for the control of sewage odors," *J. Wat. Poll. Contr. Fed.,* 38: 829 (1966).

J. Devinny, D. Hodge. "Formation of acidic and toxic intermediates in overloaded ethanol biofilters," *JAWMA,* 45: 125 (1995).

J.A. Don, L. Feenstra. "Odour abatement through biofiltration," lecture given in Louvain-La Neuve (Belgium) in April 1984, *Symposium on Characterization and Control of Odoriferous Pollutants in Process Industries,* 1984.

H. Hartmann. "Geruchsbekämpfung in Kläranlagen am Beispiel der Klärwerke der Stadt Nürnberg," *Korrespondenz Abwasser,* 23: 275 (1976).

D. Hodge. Biofiltration: Application for Treatment of Ethanol Vapors, Doctoral Thesis, University of Southern California, 1993.

D.H. Kampbell, J.T. Wilson, H.W. Read, T.T. Stocksdale. "Removal of volatile aliphatic hydrocarbons in a soil bioreactor," *JAPCA,* 37: 1236 (1987).

G. Leson, A. Winer. "Biofiltration: an innovative air pollution control technology for VOC emissions," *JAWMA,* 41: 1045 (1991).

G. Leson. Biofiltration—Investigation of a Novel Air Pollution Control Technology for the South Coast Air Basin, D. Env. Thesis, University of California, Los Angeles, 1993.

G. Leson, R. Chavira, A. Winer, D. Hodge. "Experiences with a Full-Scale Biofilter for Control of Ethanol Emissions," *Proceedings of the 88th Annual Meeting of the Air and Waste Management Association,* San Antonio, TX, 1995.

N. Mukhopadhyay, E.C. Moretti. *Current and Potential Future Industrial Practices for Reducing and Controlling Volatile Organic Compounds,* Center for Waste Reduction Technologies, American Institute of Chemical Engineers, New York, 1993.

S.P.P. Ottengraf. "Exhaust Gas Purification," in *Biotechnology, Vol. 8.;* H.J. Rehm, G. Reed, Eds., VCH Verlagsgesellsch., Weinheim, Germany, 1986.

S.P.P. Ottengraf, R.M.M. Diks. "Process technology of biotechniques," in *Biotechniques for Air Pollution Abatement and Odour Control Policies,* A.J. Draft and J. van Ham, Eds., Elsevier, 1992.

R.D. Pomeroy. "Controlling sewage plant odors," *Consulting Engineer,* 20: 101 (1963).

W.H. Prokop, H.L. Bohn. "Soil bed system for control of rendering plant odors," *JAPCA,* 35: 1332 (1985).

USEPA, *OAQPS Control Cost Manual,* Fourth Edition, EPA-450/3-90-006, 1990.

VDI Berichte 1104: Biologische Abgasreinigung, VDI Verlag, Düsseldorf, Germany, 1994.

VDI Berichte 735: Biologische Abgasreinigung (Reports of the German Society of Engineers: Biological Air Pollution Control), VDI Verlag, Düsseldorf, Germany, 1989.

T.O. Williams. "Biofiltration for control of odorous emissions & VOCs from wastewater & sludge processing facilities." *Proceedings of the WEF Specialty Conference on "Odor and VOC Emission Control for Municipal and Industrial Treatment Facilities,"* Jacksonville, FL, 1994.

Metal Biosorbents from Waste Fungal Biomass: A New Bioremedial Material for Control of Heavy Metal Pollution

K. M. Paknikar, P. R. Puranik and A. D. Agate
MACS Agharkar Research Institute
Pune 411 004, India

S. R. Naik
Hindustan Antibiotics Ltd.
Pimpri, Pune 411 018, India

INTRODUCTION

The control of environmental pollution has assumed greater importance due to the increasing public awareness of the deleterious effects pollution has on human life. With rapid industrialization, the problem of water pollution has escalated severalfold. The contamination of water bodies with difficult-to-degrade organic compounds and other toxic substances like heavy metals has become a challenging problem requiring immediate attention of scientists and technologists. Since metals are a nonrenewable, finite natural resource, the challenge is not limited only to their removal from the water streams, but also extends to finding efficient and economical ways of recovery and recycling. One of the major sources of heavy metal pollution are electroplating, pigment producing, metal finishing, printed board manufacturing, and other electrical equipment manufacturing industries. When wastewaters emanating from such industries are released into rivers or other natural water bodies, they prove to be harmful to the biota due to the presence of metals like lead, cadmium, chromium, copper, zinc, silver, etc. While passing along the food chain, these metals get bioaccumulated in small aquatic organisms and finally reach humans by the process of biomagnification. Sometimes the levels of metals found in algal/plant biomass serve as an index to detect the water contamination by metallic species. These studies, however, are predominantly of an ecological interest (Vymazal, 1990; Wong and Wong, 1990). A majority of heavy metals are toxic at high concentrations and, as such, pose serious health hazards to humans and livestock. Copper, zinc, and nickel are phytotoxic if their concentrations increase in soil, whereas cadmium is zootoxic at a comparatively lower concentration. Table 1 gives the toxicity data for various metals and their permissible limits as defined by the Indian Standards Institution.

Removal of heavy metals from industrial wastewaters becomes a necessity due to their toxic effects. The currently available technologies for removal of heavy metals from wastewaters include physicochemical processes such as chemical precipitation, solvent extraction, use of ion exchange resins, reverse osmosis, application of chelating agents, etc. These processes become economical on a practical scale only if the metal concentrations in the effluents are sufficiently high (in general, above $100\,mg\,L^{-1}$). There are no economically viable processes for the treatment of dilute (i.e., high volume–low tenor) wastes (Ting *et al.,* 1989). Thus, there is a big technological gap that needs to be filled immediately.

It is widely known that microorganisms have the ability to bind metals, in some cases selectively, from aqueous solutions. This phenomenon is termed *metal biosorption,* and the microorganisms responsible for the process are known as *biosorbents.* The most recent development in environmental biotechnology is therefore the use of microorganisms or microbial biomass, either living or dead, or their cell components for passive heavy metals uptake from aqueous solutions (Volesky, 1990; Holan and Volesky, 1994).

Fermentation industries, specifically antibiotics, generate huge amounts of mycelial wastes of microbial (mainly fungal) origin in India, which may be on the order of 5000 tons a day. At present, the waste is used in poultry feed or as an organic manure. Its use in processes like biomethanation or in metal recovery are nonexistent due to the lack of viable technologies. With this background in mind and spurred by the efforts pursued by some Indian researchers (Paknikar *et al.,* 1993), we overview the prospects of metal biosorbents specially developed from mycelial wastes as an important biological material and its application in bioremediation processes in the area of environmental biotechnology.

ADVANTAGES OF USING WASTE FUNGAL BIOMASS

Waste mycelial biomass can be employed usefully for removal of metals from wastewaters and also to claim them because this process does not depend on the life function of the microorganisms (Brierly and Brierly, 1983; Schumate and Strandberg, 1985; Tsezos and Volesky, 1981). The use of dead biomass for metal binding circumvents toxicity

Table 1. Toxicity data for various metallic pollutants.

Metal	Limits per Indian Standards Institution	Emanating Industries	Toxicity Effects on Humans
Arsenic	0.2 mg/L	Phosphate fertilizer paint, textiles	Cancer
Cadmium	1.0 mg/L	Phosphate fertilizer paint, batteries electroplating	Hypertension, renal damage cancer
Chromium	0.1 mg/L	Steel and alloys dyes and paints metal plating, tanneries rubber	Ulcers, skin diseases dementia, cancer
Copper	3.0 mg/L	Pesticides, electrical electroplating, rayon	Jaundice, gastrointestinal disturbances
Lead	0.1 mg/L	Paint, batteries, gasoline	Anemia
Mercury	0.01 mg/L	Chlor-alkali, electrical paint, fungicides, cosmetics pharmaceutical, thermometer	Damage to nervous system, liver, eyes
Nickel	3.0 mg/L	Electroplating, steel	Cancer
Selenium	0.05 mg/L	Paint, electronics, rubber	Gastrointestinal disturbances
Silver	NA	Photography, jewelry	Skin damage
Zinc	5.0 mg/L	Galvanizing, electrical pesticides, building material	Gastrointestinal irritation

problems due to high metal concentrations or extreme pH's and eliminates the economic component of nutrient supply, and subsequent metal and biomass recovery is often easier.

WHAT IS AN IDEAL METAL BIOSORBENT?

1. The ideal biosorbent should be produced at a low cost and should be reusable.
2. Particle size, shape, and mechanical properties of the biosorbent should be suitable for use in a continuous flow system in a completely mixed, packed, or fluidized-bed configuration.
3. Uptake and release of the metal from the biosorbent should be efficient and rapid.
4. It is desirable that the biosorbent should preferably be metal selective in order to separate single metal from a mixed solution.
5. Biosorbent should be in a physical state that can be reusable.

METHODOLOGY USED IN THE METAL BIOSORPTION STUDIES USING THE FUNGAL MYCELIAL WASTE

Preparation of Metal Biosorbent from Mycelial Waste

Mycelial waste of penicillin *(Penicillium chrysogenum)* fermentation having the physical and chemical characteristics shown in Table 2 was

Table 2. Physical and chemical characteristics of Penicillium chrysogenum *mycelial waste.*

Characteristics	*Penicillium chrysogenum* Biomass
Moisture percent in wet mycelia	75–80
Ash percent in dry mycelia	11.38
Protein percent	
Dried	29.38
Dried after boiling	21.22
Percent weight loss in boiling	20–22

used. Studies on a fungal adsorption system by Galun *et al.* (1983) had shown that uptake of uranium by *Penicillium digitatum* increased significantly by boiling and other chemical pretreatments. Similarly, the potential of some freshwater algal biomass for metal binding could be enhanced by chemical pretreatments (Darnall, 1991; Kubiak *et al.*, 1989). Some published reports suggested that biomass reinforcement by cross-linking may significantly affect the physical properties of natural biomass (viz., strength, hardness, swelling characteristics, etc.) and biosorption performance (Holan *et al.*, 1993). Taking into consideration these technological aspects, the mycelial waste was developed into a biosorbent preparation by various treatments and processing (see Figure 1). The mycelial waste obtained was dried in shallow trays in an oven at 60°C, crushed, and sieved to obtain granules <1.25 mm (–14 mesh). The granulated biomass (50 g) was pretreated with 100 ml of chemicals (viz., 1 N sodium hydroxide, 1 N hydrochloric acid, 1 N sulfuric acid, 1 M ammonium sulfate, 80% ethanol, 1% triton X-100 and dimethyl sulfoxide) for 1 hour at room temperature and kept in an orbital shaker incubator. For boiling water treatment 50 g of mycelia were in contact with 100 ml of boiling water for 15 minutes. The pretreated biomass was washed thoroughly with glass distilled water of pH 2.0, 4.0, and 6.0 in order to remove the residual chemicals and to condition the biomass at pH 2.0, 4.0, and 6.0, respectively.

Preparation of Standard Metal Solutions

The metal removal studies were carried out using standard solutions. The standard metal stock solutions (1000 mg L^{-1}) were prepared by dissolving 1 g metal powder in 1 + 1 nitric acid for copper and 1 + 1 hydrochloric acid for zinc and cadmium. For chromium and lead, stock solutions were prepared using potassium chromate in glass distilled water and lead nitrate in 1% nitric acid, respectively. The working metal solutions (100 mg L^{-1}) were prepared in glass distilled water; the pH of metal solution was adjusted with 0.1 N sodium hydroxide and 0.1 N hydrochloric acid.

Method for Contacting the Metal Solutions

The granulated biosorbent prepared from mycelial waste (–14 mesh) was packed in glass/PVC columns having a 1.5-cm diameter and 15-cm height. The standard solutions, having an optimum concentration of the desired metal, were passed through columns with the help of peristaltic pumps. The flowrate of the solution was adjusted in such a

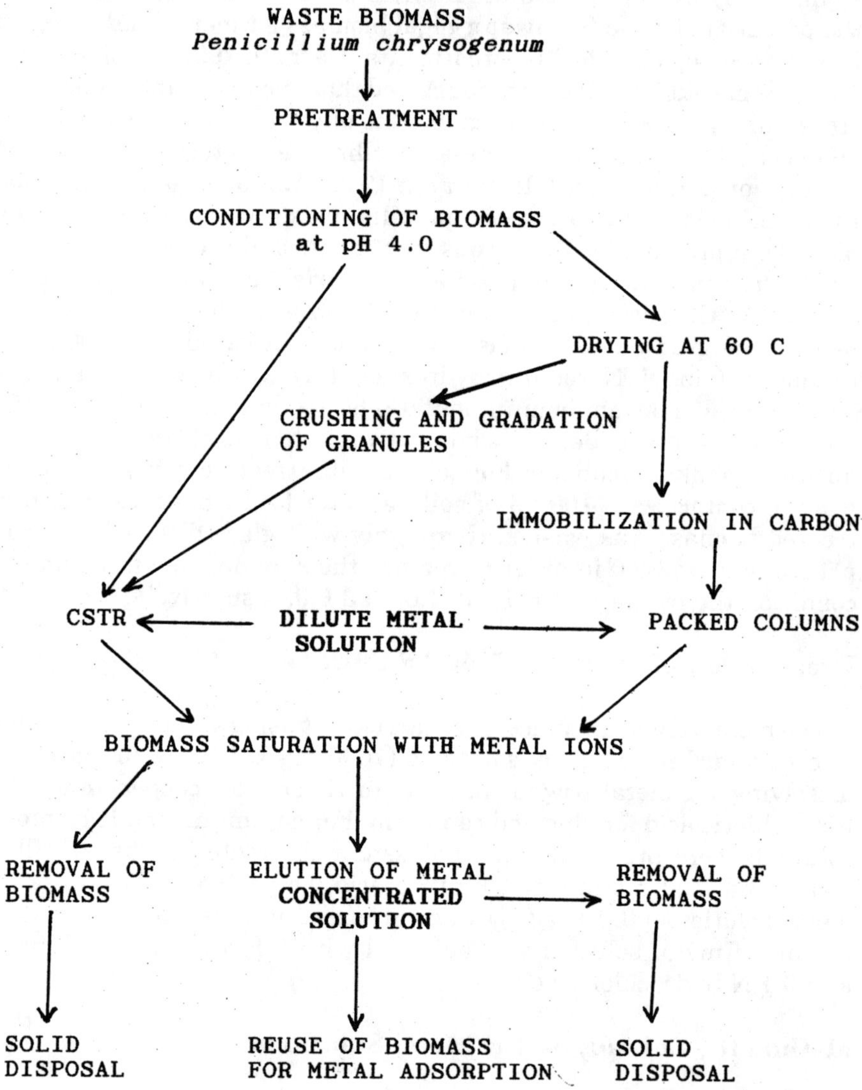

Figure 1. Flow sheet of biosorption process.

way to get a mean residence time of 10 minutes. The mean residence time was calculated as a reciprocal of volumetric flowrate (bed volumes per hour).

Effect of pH on Biosorption of Metals

The effect of pH on uptake of copper, cadmium, zinc, lead, and chromium was checked using granulated pretreated *P. chrysogenum* waste conditioned at different pH's (2.0, 4.0, and 6.0). Metal solutions used in the studies also had the requisite pH values, viz., 2.0, 4.0, and 6.0.

Metal Analysis

The concentration of metal ions in solutions before and after the contact with biosorbent was determined using an atomic absorption spectrophotometer (Perkin Elmer, USA, model 2380). The equipment was calibrated in the linear range of 5–30 mg L^{-1}. The minimum detectable concentration of the metals under standard conditions was 0.1 mg L^{-1}.

Adsorption Isotherms

The metal uptake value was calculated per following equation:

$$Q = V(C_i - C_f)/1000\ M$$

where

Q = milligram metal per gram of biomass
V = volume of metal solution (ml)
C_i = initial concentration of metal in solution
C_f = final concentration of metal in solution
M = mass of biomass (g)

The (Q) value thus obtained was used to plot an adsorption isotherm per Freundlich and Langmuir equations:

Freundlich equation

$$\log Q = \log K + 1/n \log C_{eq}$$

where

K and $1/n$ are constants
C_{eq} = equilibrium concentration of metal

Langmuir equation

$$C_{eq}/Q = 1/(b \cdot Q_{max}) + C_{eq}/Q_{max}$$

where

Q_{max} = maximum metal uptake

b = Langmuir constant

Determination of Metal Loading Capacity

In order to determine the metal loading capacity of the biomass, glass columns (1.5 cm diameter and 15 cm height) were packed with 5 g biosorbent material. The standard solutions containing 100 mg L^{-1} of the desired metals were passed through columns continuously with a flowrate of 1 ml min^{-1}, until there was no further uptake of the metal from the solution. The loading capacity was calculated as the total amount of metal adsorbed per gram dry weight of the biosorbent. All the experiments were carried out in triplicate.

Desorption of Metal Ions

Granulated mycelial biomass of *P. chrysogenum* (5 g) loaded with lead and zinc metal was used in the desorption of these metals. The elution of the metal ions from columns was carried out using 0.1 M EDTA and 0.1 M hydrochloric acid. The aliquotes (20 ml) were removed at an interval of 20 minutes and analyzed for the metal content by atomic absorption spectrophotometry.

Effect of Competing Cations on Metal Uptake

The effect of competing cations (viz., copper, zinc, cadmium, and iron) on uptake of lead metal ions was checked at various concentrations of competing cations (5, 10, 20, 40, and 80 mg L^{-1}). The concentration of lead metal in solution was 100 mg L^{-1}. The uptake of lead, as well as that of the competing cation by the biosorbent, was monitored by analyzing the residual metal contents after contacting for 10 minutes.

RESULTS, DISCUSSION, AND COMMENTS

During the initial screening work, the biosorption efficiency was judged with reference to metals like copper, cadmium, chromium, lead, and zinc. Subsequently, attempts were made to enhance the biosorption efficiency by various chemical pretreatments (Table 3). The results showed that pretreated mycelial biomass had elevated copper, cadmium, zinc, lead, and chromium uptake capability in comparison to the untreated mycelia. Metal uptake was enhanced by 37% in the case of both copper and zinc to obtain 75% and 97% removal from solution of pH 4.0 when mycelial pretreatments with dimethyl sulfoxide and boiling water, respectively, were employed. Pretreatment of the mycelia with boiling water enhanced lead and cadmium uptake only marginally (3–4%), as compared to the untreated mycelia. Maximum observed chromate uptake was only 37% by ethanol-treated mycelia. The treatments leading to elevated uptake are associated with solvent or denaturant properties. It has been reported that the pretreatments may expose metal binding sites either by configurational changes or by removal in solution of masking groups (Galun *et al.*, 1983). Ammonium

Table 3. Effect of pretreatments of the Penicillium chrysogenum *mycelial biomass on uptake of copper, cadmium, zinc, lead, and chromium.*

Metal	Pretreatments	Mean Percent Removal of Metal from 100 mg L^{-1} (Metal Solution of pH 4.0)
Cu	Untreated	38.0 (2.01)
	Ammonium sulfate	66.0 (2.12)
	Sodium hydroxide	56.0 (3.24)
	Dimethyl sulfoxide	74.9 (0.98)
Pb	Untreated	94.7 (4.43)
	Boiling	98.1 (1.41)
	Triton X-100	97.9 (1.77)
	Ethanol	97.1 (1.30)
Cr	Untreated	26.0 (2.95)
	Triton X-100	35.4 (2.90)
	Ammonium sulfate	33.5 (2.55)
	Ethanol	36.7 (2.40)
Zn	Untreated	60.0 (3.37)
	Boiling	97.1 (1.76)
	Ethanol	89.9 (2.05)
Cd	Untreated	93.6 (2.13)
	Boiling	97.7 (1.26)
	Triton X-100	96.9 (1.16)

Data from Paknikar et al. (1993).
Data shown only for the pretreatments giving maximum increase over the untreated biomass.
Figures in parentheses are standard deviation values.

sulfate treatment results in the formation of electrostatic bonds with protonated amino groups of chitosan in a fungal cell wall, forming NH_3SONH_4 groups, and the introduction of sulfate in chitosan includes a novel crystallinity in the biopolymer, while at the same time increasing the biopolymer's metal uptake capacity (Tsezos, 1984). Increased metal uptake of dimethyl sulfoxide-treated mycelial biomass could be attributed to the formation of stable complexes with metals.

Biosorption of the metals was found to be pH-dependent, which varied with the nature of pretreatment (Table 4). Copper was adsorbed maximally by dimethyl sulfoxide-treated mycelia at pH 4.0; chromium by triton X-100-treated mycelia at pH 2.0; and cadmium, lead, and zinc by boiling water-treated mycelia at pH 6.0, 2.0, and 6.0, respectively. The pH of the solution is known to affect the activity of the binding groups on the cell surface. It has been reported that pH effects are

Table 4. Effect of pH on uptake of metals by untreated and treated Penicillium chrysogenum *mycelium.*

Metal	Pretreatments	Mean Percent Removal of Metals by Mycelia from 100 mg L^{-1} (Metal Solutions of pH)		
		2.0	4.0	6.0
Cu	Untreated	40.2 (3.39)	35.1 (1.36)	28.3 (1.74)
	Amonium sulfate	37.0 (3.90)	53.2 (2.13)	40.1 (2.48)
	Sodium hydroxide	31.6 (1.32)	44.8 (5.15)	34.6 (2.35)
	Dimethyl sulfoxide	76.7 (2.73)	74.9 (1.62)	65.3 (1.82)
Cd	Untreated	60.1 (4.15)	72.8 (5.13)	73.0 (2.41)
	Boiling	85.7 (3.11)	89.7 (2.90)	90.4 (2.04)
	Triton X-100	68.3 (2.46)	79.5 (4.90)	79.9 (6.53)
Zn	Untreated	48.2 (5.90)	54.9 (5.17)	51.2 (5.20)
	Boiling	71.8 (3.50)	77.5 (1.86)	76.8 (4.98)
	Ethanol	66.7 (3.40)	66.5 (2.20)	68.2 (5.63)
Pb	Untreated	97.6 (1.50)	90.6 (2.83)	91.9 (2.55)
	Boiling	98.7 (0.47)	96.8 (1.20)	98.5 (0.42)
	Triton X-100	98.0 (0.77)	97.5 (0.59)	98.0 (0.37)
	Ethanol	98.2 (1.26)	96.8 (1.44)	95.6 (2.05)
Cr	Untreated	27.1 (3.09)	20.5 (3.23)	16.6 (2.28)
	Triton X-100	51.3 (3.90)	33.2 (5.01)	27.1 (3.48)
	Ammonium sulfate	40.4 (8.22)	30.4 (4.01)	37.8 (4.90)
	Ethanol	30.6 (2.28)	30.2 (1.61)	36.8 (3.21)

Data from Paknikar et al. (1993).
Figures in parentheses are standard deviation values.

greatest for copper, lead, zinc, cadmium, nickel, chromium, mercury, and silver (Ramelow *et al.*, 1991). If binding of metals occurs by an ion exchange process, it is expected that H^+ ions will compete with metal ions. At low pH values, positive sites predominate on cell walls; at higher pH values, negative sites predominate. With an increase in external pH, increased binding has been observed for a variety of different microorganisms (Gadd, 1986). Polysaccharides of the cell wall of biomass could provide binding groups, including amino, carboxyl, phosphate, and sulfate radicals. The amino and carboxyl groups and the nitrogen and oxygen of the peptide bond could be available for characteristic coordination bonding with metallic ions like lead, copper, and chromium. Such bond formation could be accompanied by displacement protons, dependent in part upon the extent of protonation as determined by pH. In addition to these functional binding groups, polysaccharides often have ion exchange properties (Aksu *et al.*, 1990; Muraleedharan and Venkobacher, 1990; Kuyueak and Volesky, 1988).

In general, it was observed that the *P. chrysogenum* mycelial waste adsorbed lead and zinc most efficiently. Therefore, biosorption kinetics for lead and zinc was studied in detail. Since the data could be fit in both Freundlich and Langmuir models (Figures 2–5 and Tables 5 and 6) for adsorption processes, it could be said that the experimental

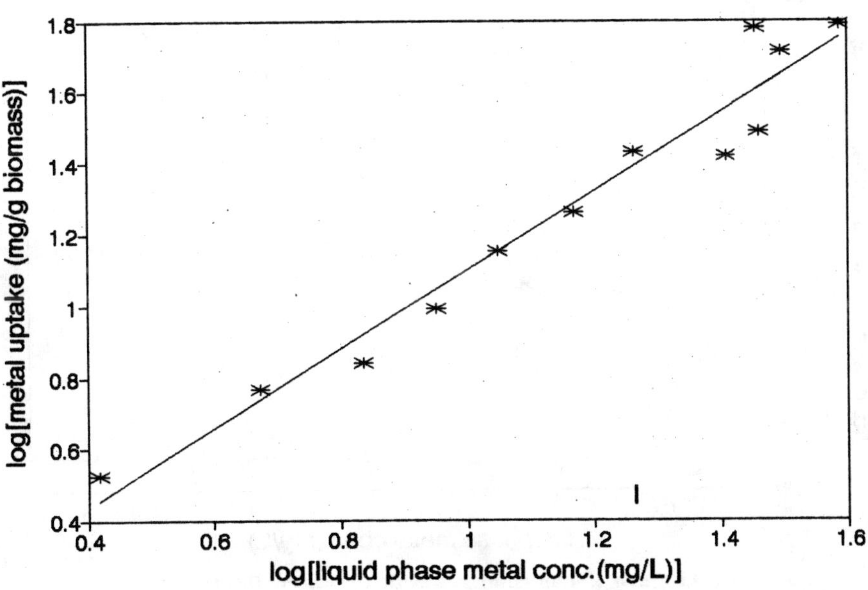

Figure 2. Freundlich adsorption isotherm (lead).

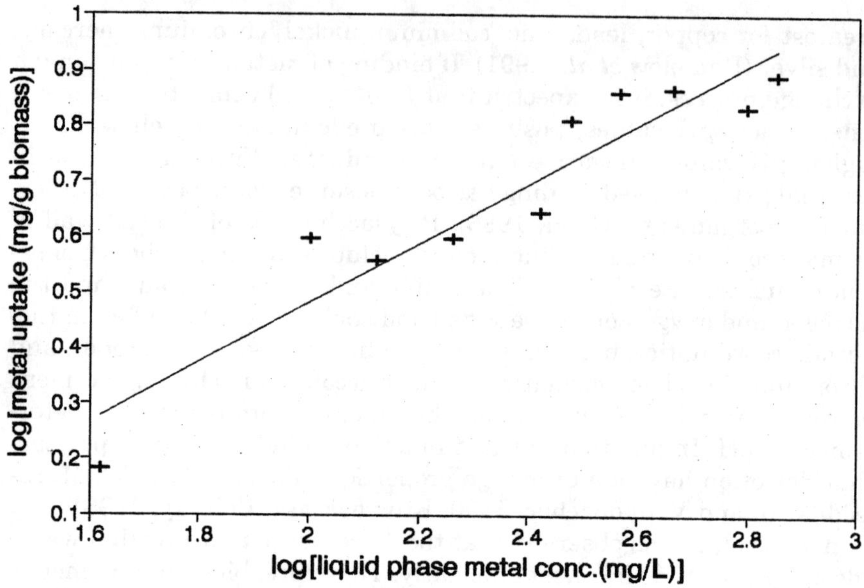

Figure 3. Freundlich adsorption isotherm (zinc).

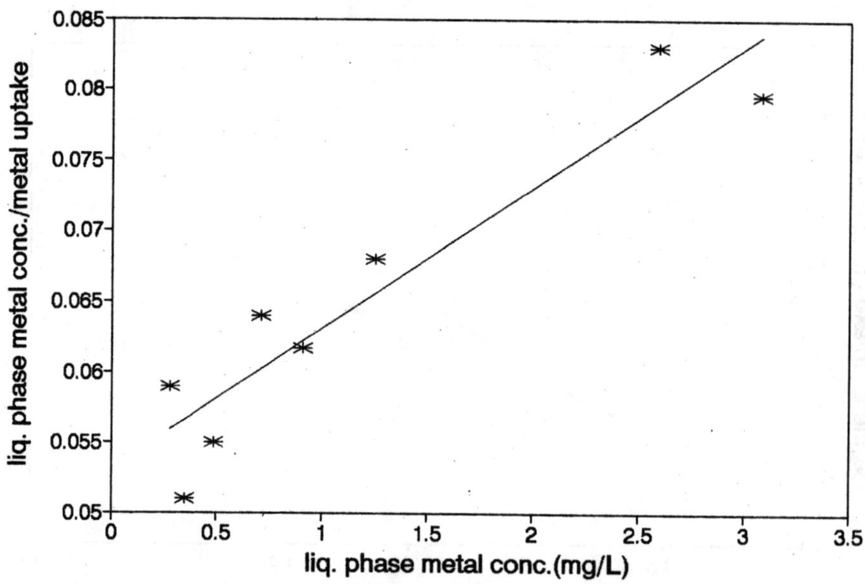

Figure 4. Langmuir adsorption isotherm (lead).

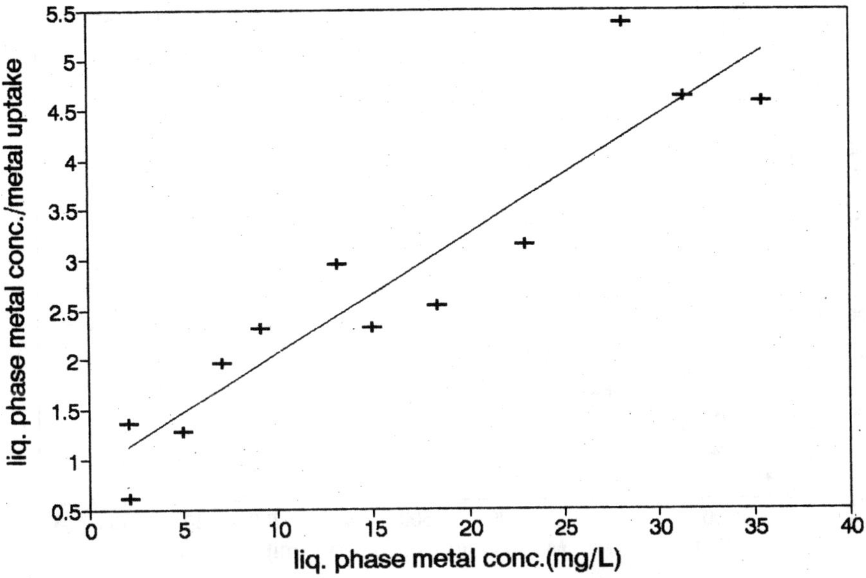

Figure 5. Langmuir adsorption isotherm (zinc).

results obeyed the basic principles underlying these models, viz., heterogeneous surface adsorption and monolayer adsorption at constant adsorption energy. The loading capacities for lead and zinc were 40 mg g^{-1} and 10 mg g^{-1}, respectively (Figure 6).

As regards desorption of the metals from loaded biomass, it was observed (Figure 7) that 60 ml 0.1 M EDTA could elute 90% of the adsorbed metal, while the same volume of 0.1 M hydrochloric acid could elute 80% of the adsorbed metal. In our experiments EDTA (0.1M) was found to be highly suitable for elution of loaded metals because it did not affect the biosorbent structure.

The presence of other metal ions did not influence the overall efficiency of lead biosorption (see Table 7) in the batch process. It was further noticed that, from among the competing cations tested, adsorp-

Table 5. Freundlich parameters for sorption by P. chrysogenum *waste biomass.*

Metal	$\log_{10} K \pm$ 95% CI	$1/n \pm$ 95% CI	K	r^2
Lead	−0.00813 ± 0.2042	1.109 ± 0.1675	0.9815	0.9561
Zinc	−0.592 ± 0.1765	0.5363 ± 0.154	0.2559	0.8903

$\log_{10} K$ is the intercept and $1/n$ is the slope of the regression of the metal sorbed (Q, mg g^{-1}) against the equilibrium concentration (Ceq, mg L^{-1}).

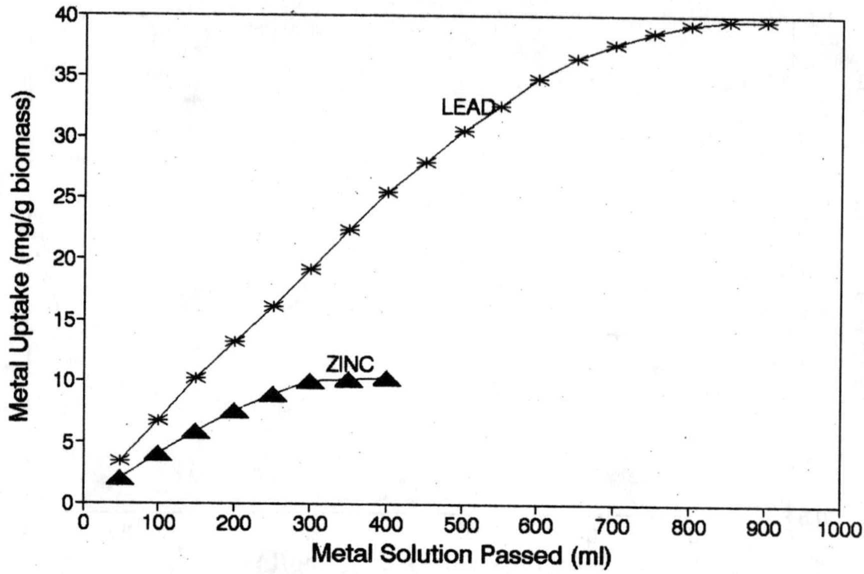

Figure 6. Metal loading capacity.

tion of iron on the biomass was the highest. This result is of importance in the application of the process in continuous mode where the presence of iron might significantly hamper the loading capacity of the biosorbent with respect to the desired metal (in this case lead).

It was observed that over the time (after five to six repeated usages) the column became compact, which adversely affected the flowrate and process operation. The compaction of the column was due to disintegration of the granular structure of the biomass. The presence of iron or any of the competing ions had no effect on compaction.

UPSCALING THE PROCESS TECHNOLOGY

Upscaling of column experiments was carried out at 50- and 700-g levels of lead and zinc solutions (100 mg L^{-1} concentration), maintaining a mean residence time of 10 ± 2 minutes for the adsorption process (Table 9). In order to obviate column-packing problems, a coarse biopolymer with increased hardness and less swelling was employed. For this purpose the biosorbent preparation (30% w/w) was mixed with carbon granules and a chemical cementing material. The carbon-biomass discs (4 cm diameter × 1.5 cm thickness) thus obtained were used in columns for metal sorption studies. It was observed that, with the

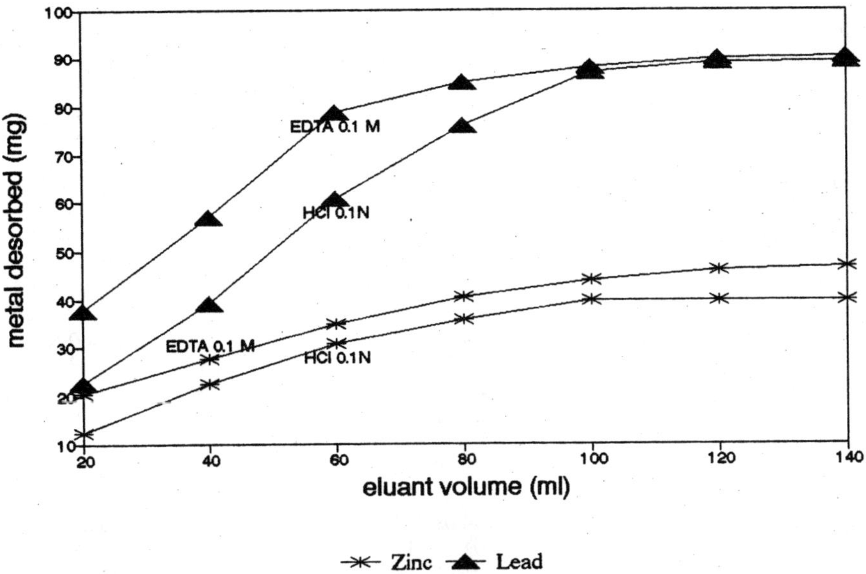

Figure 7. Desorption of metals.

use of carbon-biomass discs, the loading capacity for lead and zinc improved significantly (see Table 8). EDTA was found to be the best eluant for the metals. Typically, 500 ml 0.1 M EDTA was required to elute metal values with 90% efficiency from 100 g biomass loaded with lead or zinc.

The pilot experiments are being initiated using a stirred tank reactor of 2000-L capacity with a top-driven agitator and Rushton turbine impeller (see Figure 8); a 500-kg biosorbent preparation is used in the reactor. The agitation speed maintained during the operation is 40–60 rpm to achieve effective mixing and also to minimize structural in the biosorbent preparation. The flowrate is adjusted at 2.5 L min^{-1} to 10 L min^{-1} (0.3–1.2 bed vol h^{-1}) (see Table 9).

Table 6. Langmuir parameters for sorption by P. chrysogenum *waste biomass.*

Metal	$1/b.Q_{max}$ ± 95% CI	$1/Q_{max}$ ± 95% CI	b	Q_{max}	r^2
Lead	0.053 ± 0.010	0.00999 ± 0.0037	0.188	100.1	0.882
Zinc	0.875 ± 1.236	0.11882 ± 0.0324	0.136	8.416	0.874

$1/b \cdot Q_{max}$ is the intercept, and $1/Q_{max}$ is the slope of the regression of the ratio of equilibrium concentration (Ceq, mg L^{-1}) to the metal sorbed (Q, mg L^{-1}) against equilibrium concentration (Ceq, mg L^{-1}).

Table 7. Effect of competing cations on uptake of lead by Penicillium chrysogenum *biomass.*

Cation conc. (mg L^{-1})		Mean Percent Pb Adsorption	Mean Percent Cation Adsorption
Control		100.00	—
Cu	05	90.90 (1.23)	32.67 (2.04)
	10	93.78 (1.98)	38.21 (2.82)
	20	93.10 (1.76)	27.50 (0.46)
	40	93.66 (3.54)	06.37 (0.37)
	80	92.60 (2.23)	10.32 (2.54)
Zn	05	92.32 (2.88)	62.55 (1.76)
	10	92.66 (3.08)	32.58 (2.22)
	20	93.30 (5.01)	28.05 (2.04)
	40	92.84 (3.49)	14.25 (1.44)
	80	94.09 (2.28)	15.00 (1.26)
Cd	05	91.54 (1.21)	38.46 (2.83)
	10	91.47 (1.30)	27.58 (3.36)
	20	91.47 (1.80)	28.57 (3.11)
	40	92.55 (1.34)	11.40 (1.80)
	80	91.59 (1.04)	14.95 (1.70)
Fe	05	96.33 (2.34)	78.33 (0.97)
	10	96.49 (4.89)	77.86 (1.50)
	20	95.70 (4.32)	88.67 (1.61)
	40	96.35 (3.48)	88.42 (3.21)
	80	96.45 (2.04)	90.19 (1.19)

Figures in parentheses are standard deviation values.

Table 8. Metal biosorption by Penicillium chrysogenum *biomass in percolation columns.*

	Lead	Zinc
50 g biomass	>90	28
	(Column packing after passing 50 L solution)	
700 g biomass coarse	29	28
	(Column packing after passing 200 L solution)	
700 g carbon-biomass discs	138	46
	(No packing of columns after passing 1000 L solution)	

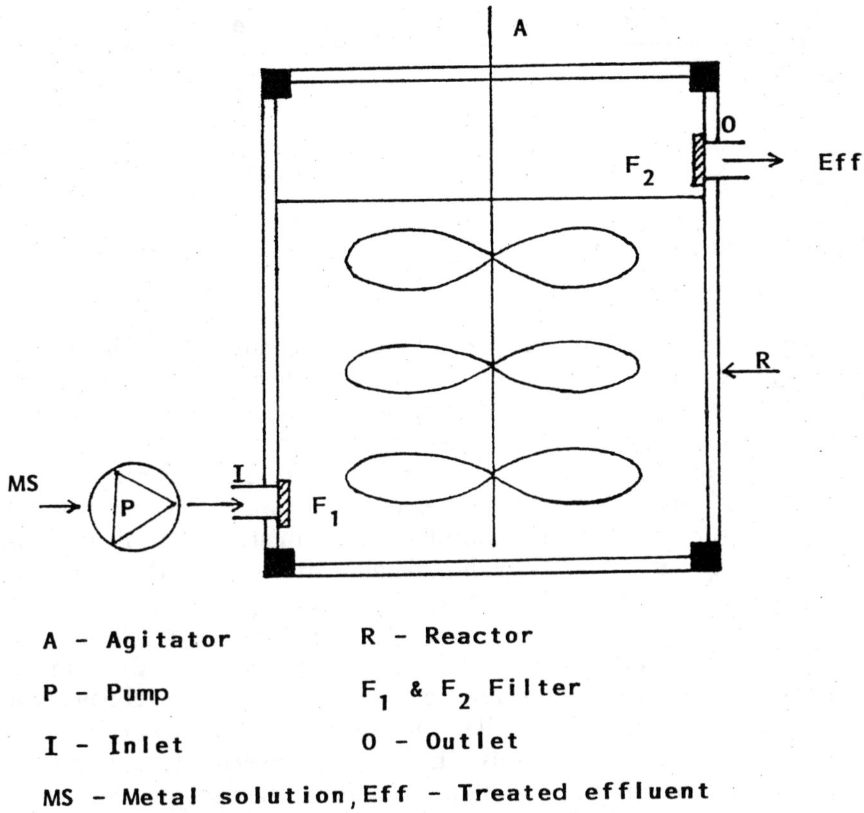

Figure 8. Proposed stirred tank reactor for metal biosorption.

FUTURE WORK PLANNED

The process is largely based on solid–liquid contact and is very similar to that used in the ion exchange or activated carbon processes. Appropriate contact time (residence time) between the solution and biosorbent can be achieved by adopting suitable process equipment, viz., stirred tank reactor, a fixed/packed-bed contactor, pulsating beds, fluidized-bed, etc. Simultaneously, immobilization or engineering of the biomass into a suitable particulate or granular form is essential in order to improve its mechanical properties for the desired level of performance. This will largely depend upon the inputs of physical chemists and polymer/process engineers.

Table 9. Parameters of scaleup experiments.

Reactor				
Capacity (l)	Dimensions (dia. × height)	Biomass (kg)	Flowrate (bed vol. h^{-1})	Retention Time (min)
0.25	3.8 cm × 30 cm	0.05	4.89	12.26
3.0	7.0 cm × 80 cm	0.7	5.14	11.67
2000.0	1.5 m × 3.6 m	500	0.3–1.2	200–500

The presently available research data on the metal biosorbent prepared by us are inadequate to define the necessary process parameters that are essential for designing the scaleup experiments to analyze the process economics. Since the scope of this work is very wide, we need to prioritize and channelize the efforts from several scientific disciplines, viz., physical chemistry (underlining mechanisms of metal sorption, desorption, and selective uptake); biochemistry (identification of cell components responsible for metal sequestration and their genetic modifications for increased metal binding); microbiology (cultivation of a suitable type of microbial biomass by understanding nutritional effects and environmental factors on the morphology and growth of fungal strains); polymer science (formulation of biosorbent material from the active "biomass powder" as pellets or granules having desirable mechanical properties); and finally, chemical engineering (evolving process designs for different types of biosorbent materials and assessment of their economics).

CONCLUSION

The metal biosorbents prepared from microbial biomass can be applied on a large scale for removal/recovery of metals from industrial wastewaters. The potential of natural and cheap microbial biomass waste (which fermentation industries often discard) could be a new useful biotechnological material for cleaning up environmental pollution. However, a concerted effort of several scientific disciplines needs to be mobilized in a coordinated manner to develop different types of economically viable metal biosorption processes.

We would like to give gentle reminders to those who believe that microorganisms are interesting only if they synthesize a valuable molecule that it is not true, but emphasize that microorganisms, whether dead or alive, can be harnessed for the benefit of mankind as exemplified by their use in metal bioremediation processes.

ACKNOWLEDGEMENT

The authors are highly thankful to Mr. A. K. Basu, Managing Director, Hindustan Antibiotics Ltd., Pimpri, Pune, India, for his keen interest in this work.

REFERENCES

Aksu, Z., Sag, Y. and Kutsal, T. 1990 "A comparative study of the adsorption of chromium VI ions to *C. vulgaris* and *Z. ramigera*," *Environ. Technol.* 11: 33–40.

Brierley, J.A. and Brierley, C.L. 1983. "Biological accumulation of some heavy metals—biotechnological applications," In: *Biomineralization and biological metal accumulation* (Eds. P. Westbroek and E.W. De Jong). Reidel, Dordrecht, pp. 449–509.

Darnall, D.W. 1991. "Removal and recovery of heavy metals from waste waters using a new biosorbent ALGASORB," In: *Biological Processes—Innovative Hazardous Waste Treatment Technology Series, Vol. 3* (Eds. H.M. Freeman and P.R. Sferra), Technomic, Lancaster, PA pp. 65–72.

Gadd, G.M. 1986. "Fungal responses towards heavy metals," In: *Microbes in extreme environments* (Eds. R.A. Herbert and G.A. Codd), Academic Press, London, pp. 83–110.

Galun, M., Keller, P. and Malki, D. 1983. "Removal of uranium (VI) from solution by fungal biomass and fungal wall-related biopolymers," *Science,* 219: 285–286.

Holan, Z.R. and Volesky B. and Prasetyo I. 1993. "Biosorption of cadmium by biomass of marine algae," *Biotechnol Bioeng.,* 41: 819–825.

Holan, Z.R. and Volesky B. 1994. "Biosorption of lead and nickel by biomass of marine algae," *Biotechnol. Bioeng.,* 43: 1001–1009.

Kubiak, W.W., Wang, J. and Damall, D. 1989. "Algal columns with anodic stripping voltametric detection," *Anal. Chem.,* 61: 468–471.

Kuyucak N. and Volesky, B. 1988. "Biosorbents for recovery of metals from industrial solutions," *Biotechnol. Lett.,* 10: 137–142.

Muraleedharan, T.R. and Venkobacher C. 1990. "Mechanism of biosorption of copper (II) by *Ganederma lucidum,*" *Biotechnol. Bioeng.,* 35: 320–325.

Paknikar, K.M., Palnitkar, U.S. and Puranik, P.R. 1993. "Biosorption of metals from solution by myclial waste of *Penicillium chrysogenum,*" In: *Biohydrometallurgical Technologies, Volume II,* Fossil Energy Materials Bioremediation, Microbial Physiology (Eds. A.E. Torma, M.L. Apel, C.L. Brierley). The Minerals, Metals and Materials Society, Warrendale, pp. 229–235.

Ramelow, G.J., Yumo, Z. and Liu, L. 1991. "Uptake of metallic ions from aqueous solution by dried lichen biomass," *Microbios,* 66: 95–105.

Schumate, S.E. and Strandberg, G.W. 1985. "Accumulation of metals by microbial cells," In: *Comprehensive biotechnology, Vol. 4* (Eds. M. Moo-Young, C.N. Robinson and J.A. Howell) Pergamon Press, New York, pp. 235–247.

Ting, Y.P., Lawson, F. and Prince, I.G. 1989. "Uptake of cadmium and zinc by the alga *Chlorella vulgaris:* Part I. Individual ion species," *Biotechnol. Bioeng.,* 34: 990–999.

Tsezos, M. and Volesky, B. 1981. "Biosorption of uranium and thorium," *Biotechnol. Bioeng.,* 23: 583–604.

Tsezos, M. 1984. "Recovery of uranium from biological adsorbents—desorption equilibrium," *Biotechnol. Bioeng.,* 26: 973–981.

Volesky, B. 1990. "Biosorption of fungal biomass," In: *Biosorption of heavy metals* (Ed. B. Volesky), CRC Press, Boca Raton, pp. 139–172.

Vymazal, J. 1990. "Toxicity and accumulation of lead with respect to algae and cyanobacteria: A review," *Acta Hydrochim. Hydrobiol.,* 18: 513–535.

Wong, P.K. and Wong, C.K. 1990. "Toxicity of nickel and nickel electroplating water to *Chlorella pyrenoidosa,*" *Bull. Environ. Contam. Toxicol.,* 45: 752–759.

Removal of Heavy Metals and Sulphate from Contaminated Groundwater Using Sulphate-Reducing Bacteria: Development of a Commercial Process

L. J. Barnes
164 Sandyhurst Lane
Ashford, Kent TN25 4NX, UK

INTRODUCTION

Historically, the metal industry, in particular metal refineries and mining operations that process sulphidic minerals, has produced substantial amounts of sludges and solid wastes that contain toxic heavy metals or sulphate or both. These wastes were often stored on production sites or used for landfill. There are probably many examples where leaching of these wastes by percolating rain and surface waters has caused substantial contamination of the underlying groundwater that permeates through the strata, spreading the contamination further as time progresses. With the increased demand for potable water from underlying aquifers and with the advent of tighter environmental controls, it has become essential to at least prevent the contaminated groundwaters from spreading. Authorities are now insisting that the contamination under such sites should be contained or, preferably, be eradicated.

Problem

One example of a site that has this groundwater problem [1] is at Budel (Brabant, the Netherlands) where zinc production has been carried out for about a century. In 1892 the company Kempensche Zinkmaatschappy commenced zinc production on this site with a pyrometallurgical process that used coke to reduce oxidic zinc ores in horizontal retorts. More recently, 50,000 tonnes per annum of zinc were produced from sulphidic ores by this thermal process. Solid residues (so-called "zinc ashes") from this process were used as landfill to level the swampy areas of the Budel site making it suitable for plant extensions. The zinc ashes, which contained up to 15% of heavy metals, were originally considered to be inert; however, we now know that heavy metals are leached by percolating ground- and rainwater.

Budelco B.V., the owners of the site, is a company jointly owned by Billiton (the Netherlands) and Pasminco (Australia). In 1973 the old thermal process was replaced by a modern electrolytic, hydrometallurgical process with increased capacity (210,000 tonnes of zinc per annum) and improved zinc recovery that reduced environmental contamination. This new refining process produces a slurry of ammonium jarosite $[(NH_4)_2SO_4{\cdot}Fe_2(SO_4)_3{\cdot}4Fe(OH)_3]$ as the major by-product. Since this material contains traces of heavy metals, it has to be treated as a chemical waste and therefore has to be stored in watertight, high-density polyethylene-lined ponds on site.

Both the transmission rate and direction of movement of the contaminated groundwater are governed by the strata underlying the Budel site (Figure 1). Geologically, the region forms part of the "Central Slenk" of the Brabant massive, which is intersected by NW–SE faults. Underneath the medium permeability top soil, where rainwater penetration is estimated to be 0.2 $m{\cdot}year^{-1}$, is a permeable sand formation known as the "1st aquifer" where water flows horizontally from SW to NE at about 10 $m{\cdot}year^{-1}$. Below the 1st aquifer is the impermeable Brunssem clay, which is underlain by a sand bed, the "2nd aquifer," from which domestic and industrial water is drawn.

Since 1983 a series of boreholes were constructed within the Budel site boundary to determine the extent of groundwater contamination. These studies showed that the severe contamination of groundwater is confined within the industrial site boundary to a depth of 30 meters. The shaded area in Figure 2 represents zinc and cadmium contamination that exceeds the Dutch "C" levels for contaminated soil; therefore, this area is subject to isolation, monitoring, and control. The major source of the contaminating metals was shown to originate from the "zinc ashes" with a relatively small contribution from a leaking jarosite pond where high concentrations of heavy metals occur locally. Zinc concentrations in groundwater vary considerably across the site with up to 200 $mg{\cdot}L^{-1}$ of zinc being found at depths up to 20 meters beneath areas containing the zinc ashes. Groundwater analysis data, coupled with computer modeling indicate that the contamination will eventually go beyond the site perimeter (Figure 3) and could contaminate the deeper underlying drinking water aquifers in the future.

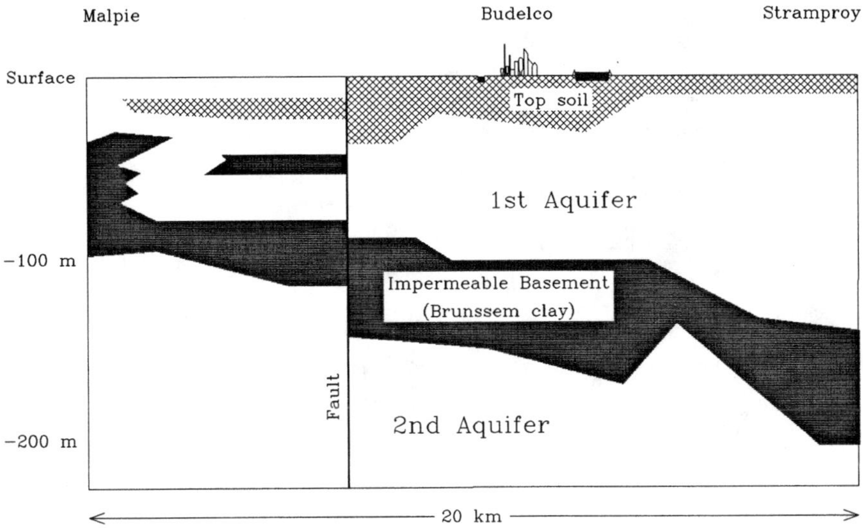

Figure 1. Stratigraphic profile below the Budelco area.

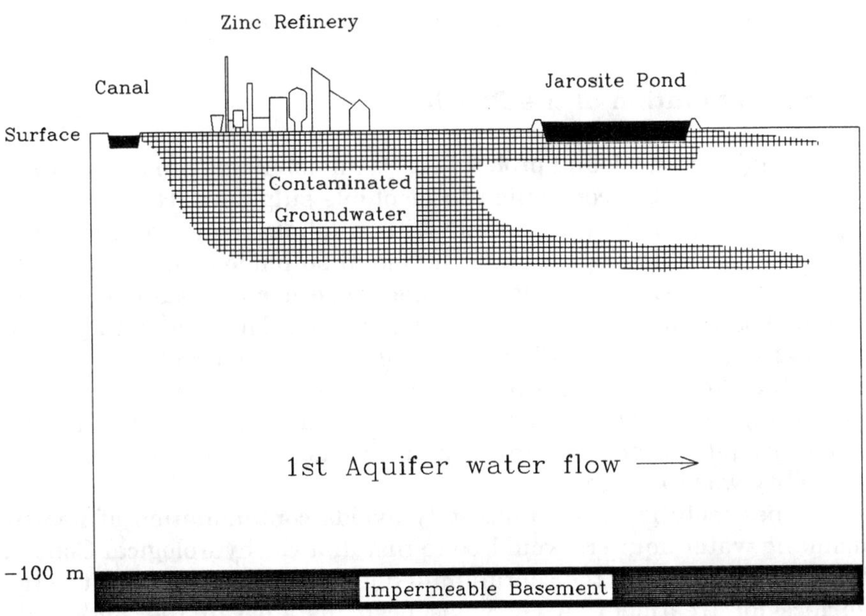

Figure 2. Section through the Budelco site.

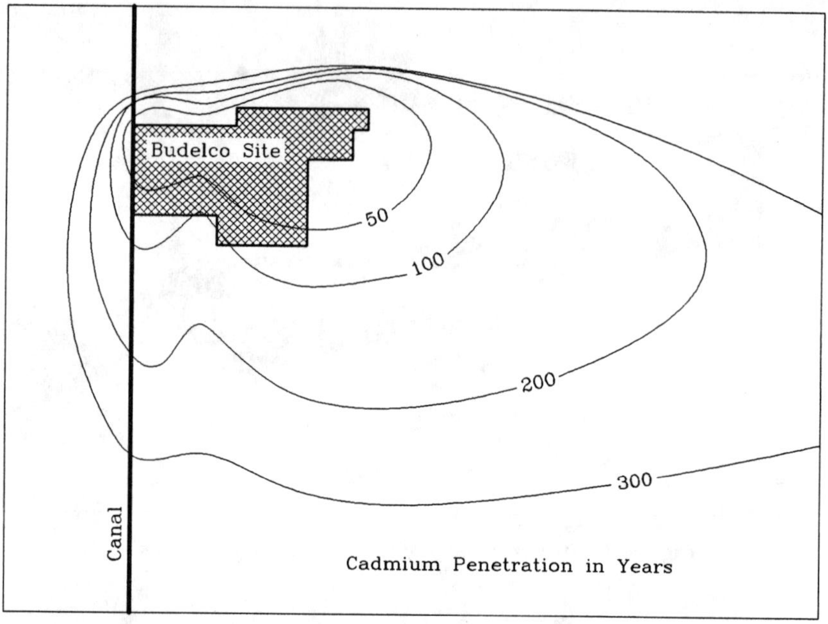

Figure 3. Potential future cadmium contamination under Budelco's site, if untreated.

Potential Solution of the Problem

The major natural, and probably best, immobilizing process for toxic heavy metals is the conversion to insoluble sulphides. However, if *in situ* immobilization were possible with sulphate-reducing organisms, then the water flowing through the metal sulphide containing strata would have to remain anoxic in perpetuity; otherwise, sulphide oxidation will occur and the metals will be mobilized. This oxidation process is most probably responsible for the groundwater problem at Budel and is the likely cause of the problematical "acid mine drainage" created by mining operations worldwide. Hence, stabilization of the metals underground was deemed unrealistic for this site since the water permeability was too high.

The best solution that completely avoids contamination of nearby drinking water aquifers would be to install a Geohydrological Control System (GCS) where the contaminated groundwater is pumped to the surface for treatment. A GCS has been installed at the Budel site and was designed [2] by using a computer-simulated model that was generated from groundwater analysis data, geological profiles of the site, and the requirement to contain the contamination within the site

Table 1. Approximate analysis of groundwater and permitted discharge levels ($mg \cdot L^{-1}$)

Component	Nominal Groundwater Feed	Maximum Discharge Level
Sulphate	1000	200
Zinc	100	0.3
Cadmium	0.6	0.01
Copper	0.3	0.02
Cobalt	0.2	0.01
Nickel	0.3	0.02
Iron	15	1
BOD	—	4
pH	4.5	6.5–8.5

boundary. The final design, which incorporated a combination of 17 strategically placed shallow and deep wells, would be capable of extracting 7000 $m^3 day^{-1}$ of groundwater. Since the water extracted by this system is contaminated with sulphate, zinc, and traces of other heavy metals, it has to be purified before it can be discharged into the public domain or used as process water. Table 1 shows typical groundwater data and the required discharge levels.

Candidate Treatment Processes

In the mid-1980s Shell Research initiated a program to examine suitable candidate processes (Table 2) for decontaminating the abstracted groundwater. Three of the most promising processes that would produce a water effluent acceptable to the authorities, at that time, were selected for research; they were ion exchange, liquid mem-

Table 2. Candidate processes for treatment of groundwater.

Candidate Process	Components Removed
Biological sulphate reduction	Metals, arsenic, and sulphate
Biosorption	Metals
Electrolytic deposition	Metals
Ion exchange	Metals, arsenic, and sulphate
Iron hydroxide flocculation	Metals and arsenic
Lime precipitation	Metals, arsenic, and some sulphate
Liquid membrane extraction	Metals
Reverse osmosis	Metals and sulphate
Solvent extraction	Metals

brane extraction, and anaerobic microbial sulphate reduction. Of these three processes, only the last will remove sulphate economically.

The ion exchange process was investigated at Shell's Billiton Research Laboratory in Arnhem [3], the Netherlands. After limited research on a process based on exchanging heavy metals with hydrogen ions, the work was curtailed and the following conclusions drawn. Heavy metals could be removed from the groundwater to acceptable levels, but the treated groundwater would have to be neutralized before being discharge. In addition, it was not economic to use the ion exchange technology to remove anions such as sulphate and arsenate. Also, the relatively large volume of acid required to regenerate the ion exchange resin will contain low concentrations of heavy metals, making it difficult to incorporate this stream into Budelco's zinc refining process. Although scaleup to a commercial process is well documented, the process operation costs were estimated to be relatively high.

The continuous liquid membrane extraction process, which was developed by the University of Graz (Austria) [4,5], uses the principle of solvent extraction. Metal ions in the abstracted groundwater are exchanged with the hydrogen ions of the extractant, which resides in the spherical nonaqueous liquid membrane. Within the spherical membrane is an acidic stripping liquor that regenerates the extractant by exchanging the metals with hydrogen ions. Data obtained at Budelco in 1988 from a pilot plant operating with groundwater were used to design a 7000 m^3day^{-1} commercial process; the following conclusions were drawn from the test program. The process is more metal ion specific than ion exchange, with some heavy metals being removed to levels only just acceptable to the authorities. Unlike ion exchange, ferric iron affects extraction efficiency and anions cannot be removed. As with ion exchange, the treated groundwater will require neutralization before discharge. The low-volume, high-metal concentration of the acid stripping liquor can be incorporated into Budelco's existing zinc refining process. A combination of the process limitations and rather high capital costs contributed to the demise of this option for treating the abstracted groundwater.

The anaerobic sulphate-reducing bacterial process [6] uses an environmental process that widely occurs in nature to precipitate heavy metals. Although this process is well documented, only a limited number of engineered "reed bed processes" that use this biological system have been used [7]. Since 1987 research was carried out at Shell's Research Center to investigate the feasibility of applying this bioprocess, in reactors, for the removal of heavy metals and sulphate from wastewaters. The key features of the process are the *in situ* generation

Table 3. Solubility of divalent metal ions as sulphides (mg · L^{-1}).

Metal	Solubility
Zinc	10^{-7}
Cadmium	10^{-9}
Copper	10^{-14}
Cobalt	10^{-8}
Iron	10^{-5}
Calcium	10^{+2}

of sulphide by microbes in a reactor and the concomitant precipitation of heavy metals as highly insoluble sulphides (Table 3), therefore removing both sulphate and heavy metals from the aqueous phase (note: soluble calcium sulphide is also shown in Table 3 for comparison). This process was adopted as the treatment option for Budelco's contaminated groundwater and is the subject of this chapter.

THE SULPHATE-REDUCING BACTERIA (SRB) PROCESS

Background

Many natural anaerobic aqueous environments, especially sediments such as those found in oceans, lakes, and rivers, contain microorganisms [8–19] that use organic compounds for growth and as the energy source required to reduce sulphate to sulphide. When heavy metals are present, they combine with the generated sulphide to produce extremely insoluble sulphides. This process plays a major role in the formation of sedimentary metal sulphide deposits [20] as well as producing the iron sulphides found in coal [21–23]. The overall reaction in such an anaerobic environment can be represented by Reaction (1).

$$\text{Metal Sulphate} + \text{Carbon Source} \rightarrow \text{Metal Sulphide} + CO_2 + H_2O + \text{Cells} \quad (1)$$

Sulphate-reducing bacteria normally exist in nature as part of an interdependent, anaerobic consortium of organisms that exists because of the presence of the complex mixture of sedimentary carbonaceous compounds that are found in these natural ecosystems. Because only sulphate reducers utilize a limited range of simple organic substrates, they generally take advantage of the extracellular products generated by primary organic compound degrading organisms present in the consortium. Table 4 gives some examples of sulphate reducers, together with their maximum growth temperature, preferred energy/growth substrates, and extracellular products.

In order to sustain active growth of these bacteria, two important environmental conditions must be satisfied [24]: neutral pH and low Redox potential. In nature, neutral pH is achieved through chemical equilibria involving the buffering effect of carbonates and sulphides [Reactions (2) to (5)], the products of organism growth. Low redox potential is maintained by the presence of excess sulphide ions in the aqueous phase. Any oxygen that is transmitted into this ecosystem rapidly reacts with the sulphide [Reaction (6)], thus maintaining a low redox potential.

$$HCO_3^- + H^+ \rightarrow H_2O + CO_2 \tag{2}$$

$$CO_2 + OH^- \rightarrow HCO_3^- \tag{3}$$

$$HS^- + H^+ \rightarrow H_2S \tag{4}$$

$$H_2S + OH^- \rightarrow HS^- + H_2O \tag{5}$$

$$4HS^- + O_2 \rightarrow 2H_2O + 4S^o \tag{6}$$

These two strict conditions can be attained in an anaerobic bioreactor, therefore, with excess sulphide in solution; then concentrations of heavy metals in the aqueous phase of such a reactor would be expected to be

Table 4. Some sulphate-reducing and methanogenic organisms.*

Organism	°C	Growth Substrate	Product	Reference
Desulfovibrio vulgaris	44	La	Ac	25
Desulfomonas pigra	45	Yeast	Ac + H_2	25
Desulfobulbus propioicus	43	La, Pr, Et	Ac	12
Desulfococcus multivorans	40	La, Ac, Et, Me	CO_2	25
Desulfobacter postgatei	40	Ac, Et	CO_2	17
Desulfosarcina variabilis	38	La, Ac, Et, Me	CO_2	25
Desolfonema magnum	35	Be	CO_2	25
Desolfotomaculum orientis	38	H_2 + CO_2		25
Thermodesulfobacterium commune	70	La	Ac + H_2	26
*Methanosarcina barkeri**	37	Ac, Me	CH_4 + CO_2	27
*Methanobacterium soehngennii**	45	Ac	CH_4 + CO_2	25
*Methanococcus mazei**	37	Ac	CH_4 + CO_2	25
*Methanobacterium thermoautrophicum**	70	H_2 + CO_2	CH_4	25

*Ac = Acetate, Be = Benzoate, Et = Ethanol, La = Lactate, Me = Methanol, Pr = Propionate.

insignificant (Table 3). Also, the bioreactor may be operated under nonsterile conditions because sulphide is known to be a potent inhibitor for many other microorganisms; hence, microbial contamination is prevented, and consequently a stable sulphate-reducing regime can be established.

It should be noted that the quantity of sulphate reduced is proportional to the growth of the sulphate-reducing bacteria because the sulphate acts as the terminal electron acceptor in the organism's energy-producing electron transport chain. Growth is also related to the energy available in the organism's carbonaceous growth substrate, the electron donor; therefore, the quantity of sulphate reduced can be controlled by varying the amount of carbon substrate supplied to the growing culture. Acetate, which is often produced as a by-product by growing sulphate reducers, is only slowly degraded by these organisms. However, this compound is rapidly degraded by methanogenic organisms, such as those given in Table 4, which normally coexist in the naturally occurring bacterial consortium; hence, a low biological oxygen demand process effluent can be obtained, a prerequisite for discharge at the Budelco site.

Development

Development of the SRB process for treating wastewaters was divided into two phases. Initially, laboratory investigations were carried out by Shell Research and involved screening organisms to obtain suitable mixed bacterial cultures, identifying an economically attractive carbon substrate, and studies in continuously operated bench-scale reactors with synthetic groundwater to determine the process operating parameters. The second phase was executed at Budelco using the site's groundwater. This scaleup phase involved using a simple 9 m^3 sludge blanket reactor to establish the feasibility of the process, an integrated pilot plant to obtain plant design data, and finally, construction of a commercial process.

LABORATORY INVESTIGATIONS

Test Feed Compositions

Nonsterile synthetic solutions, which simulated the Budelco groundwater (Table 1) but fortified with additional growth supplements (carbon, nitrogen and phosphorus), were made up with general purpose grade chemicals. These solutions were used as media throughout the laboratory investigations.

Screening for Suitable SRB Organisms

Selected environmental samples, which were expected to contain sulphate-reducing bacteria, were obtained from a variety of locations, such as an oil refinery, sewage works, effluent treatment plant, river estuary, stagnant pools, water channels, Budelco's zinc refinery, and an anaerobic biotreater. These samples were used to produce a suitable mixed culture for the development of the SRB process.

Batch screening of the environmental samples was carried out by inoculating separate 100-ml portions of medium with 5 g of each sample in 150-ml bottles. The contents of each bottle were made anaerobic by adding sodium sulphide (10 mg) and purging with nitrogen and then sealed with a septum. A selection of potentially useful carbonaceous substrates were used in these experiments to establish their suitability. Lactate, the most extensively studied substrate, was used since it is known [9,24,28,29] to be utilized by most sulphate reducers. Both ethanol [8,12,19,29–32] and methanol [11,15] were incorporated into the screen since these are also used by some sulphate reducers and are considerably cheaper than lactate. Resazurin, a redox indicator, was added to the medium to show when an excess of hydrogen sulphide was produced through sulphate-reducing activity. The nonagitated bottles were inspected daily for solution color change and colony growth in the sediment; this initially showed up as small black specks in the grey sediment, which eventually turned completely black. Small samples (5 ml) of the aqueous phase were taken weekly and analyzed for carbon substrate, sulphate, various metals, and acetate. Acetate, a known [8,9,29,32] metabolite of these organisms (see Table 4) when grown on lactate or ethanol, was determined in order to monitor the degree of degradation of the carbon substrate.

The data obtained during the microbial screen show that sulphate-reducing bacteria are very widely distributed in the environment. The majority of the environmental samples contained organisms that were able to reduce sulphate when grown on lactate, and a large number of these cultures were also able to utilize ethanol for this purpose. Although methanol was consumed by some cultures, practically no sulphate was reduced. See Figure 4 for representative time profiles obtained during typical mixed culture growth on methanol and ethanol.

The organisms that made up the mixed bacterial culture used in this work were not specifically identified because, first, they were obtained

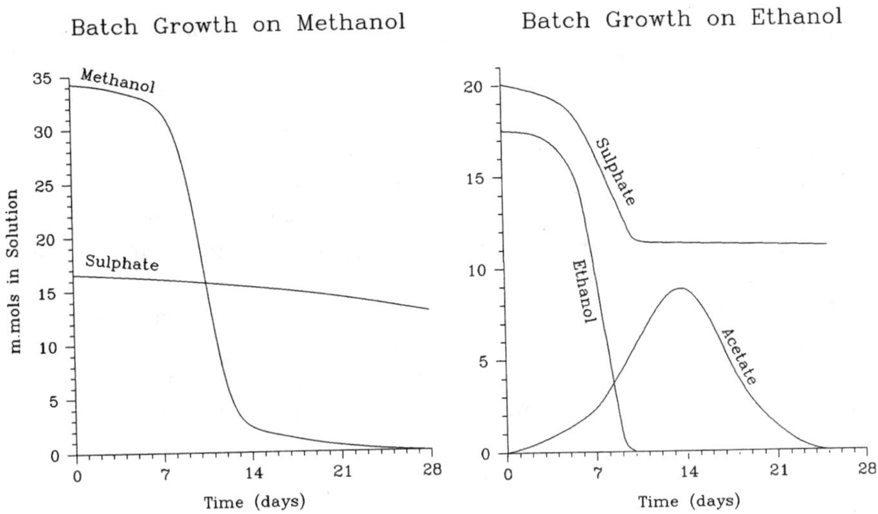

Figure 4. Biological sulphate reduction with ethanol and methanol carbon sources.

from an open natural environment and second, the SRB process would have to operate in a completely sealed system to ensure that *all* hydrogen sulphide produced was safely contained.

ORGANISM SELECTION

In order to obtain the best chance of isolating the most robust mixed culture for the SRB process, subsamples of all the organisms isolated during the screening procedure were combined. This mixed culture was initially grown batch-wise in a large anaerobic vessel before being subjected to continuous culture operation, which selected the most appropriate mixed organism population for the process.

Operating Window for the Microorganisms

Continuous experiments were carried out in the laboratory using three reactor types:

1. Stirred tank reactors are very suitable for studying the effects of changing operating parameters on the microbiology and chemistry of the process. It was possible to reliably operate at constant, predetermined conditions and rapidly attain equilibria for steady-state analysis. Basic data required for the process were obtained from experiments carried out in 1.6-L, stirred (200 rpm) reactors with

pH and temperature control. Figure 5(a) is a schematic diagram of the equipment used.

2. Sludge blanket reactors [Figure 5(b)] have the advantage of operating with a minimal disturbance of the very fine particulate sludge bed; therefore, this type of reactor avoids dislodging the microorganisms from the metal sulphide sludge. This effectively immobilizes the organisms, giving a high biomass concentration in the reactor, which facilitates the use of liquid residence times that are shorter than the organisms' doubling times.
3. Raked sludge blanket reactors [Figure 5(c)] were used to prevent compaction and channelling of the high-density sludge that occurred when long liquid residence times were used in the simple sludge blanket reactor. Successful operation of this type of reactor was demonstrated; consequently, this reactor configuration was used for the process demonstration unit.

CARBON SOURCE

Growing sulphate-reducing organisms use sulphate as an electron acceptor and an energy-rich carbon source as the electron donor. Now reduction of sulphate is energy-intensive; hence, the SRB process uses a considerable quantity of an energy-rich reductant. Consequently, the choice of carbon substrate has an important effect on the economics of the process; therefore, the most cost-effective carbon substrate will be required. As shown in Table 4, these organisms use simple carbonaceous compounds as the energy source and growth substrate during sulphate reduction [hydrogen (an energy source) and carbon dioxide (the growth substrate [33,34]) can be used by some organisms; see Table 4]. Since organism growth is directly linked to sulphate reduction, then the reductant must be readily assimilated by the organism; otherwise, no sulphate reduction takes place.

Cheap complex natural organic wastes have been used by others [35–38] and were considered as potential carbon sources for the SRB process to be installed at Budelco, but such ill-defined materials were rejected on two accounts. First, an additional aerobic biotreater would have been required to degrade any nonassimilated carbon components in order to produce an effluent with an acceptable biological oxygen demand [36–38] before it could be discharged to the environment. Second, reliability and consistency of supply at low cost is questionable as these wastes are likely to become commodities when the wastes are shown to be useful.

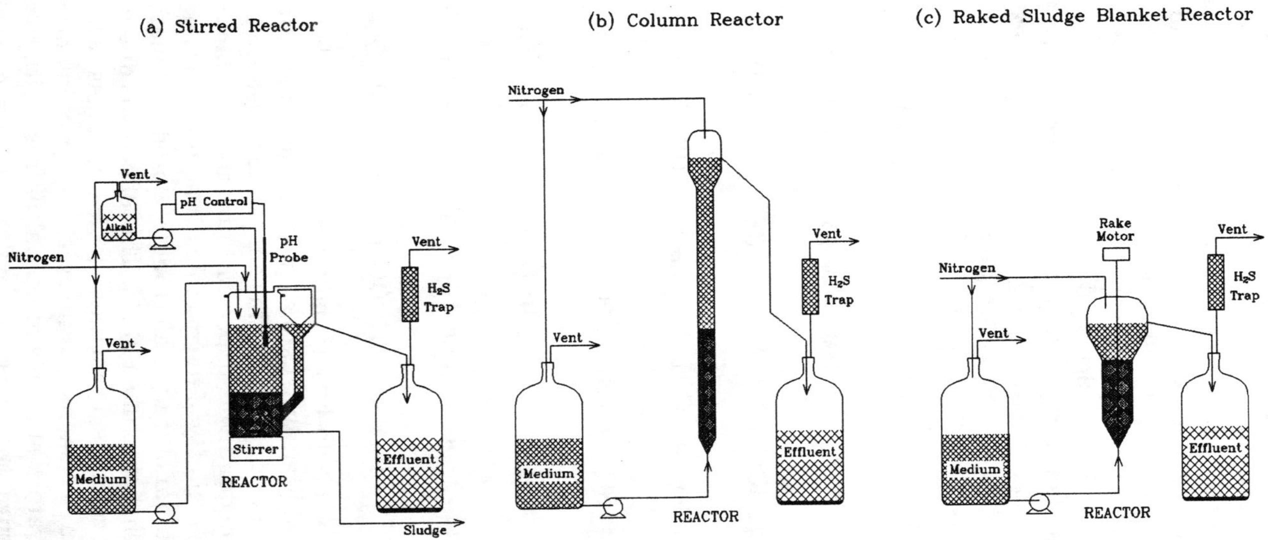

Figure 5. Schematic of laboratory reactor systems.

Although hydrogen would be a relatively inexpensive substrate, this was deemed not to be an acceptable energy source because of engineering and safety requirements at commercial scale. Sulphate-reducing bacteria normally use simple carboxylic acids or alcohols as substrates; therefore, in the first instance, simple chemical reactions [Reactions (7) to (14)] for the complete reaction were used and their thermodynamic data calculated (Table 5). This data indicates that all reactions should proceed since they have negative free energies, large equilibrium constants, and positive entropies (except hydrogen). Estimated substrate costs per kilogram of sulphate reduced were determined from compound costs [39] and are also shown in Table 5.

$$SO_4^{2-} + 4H_2 \rightarrow S^{2-} + 4H_2O \tag{7}$$

$$3SO_4^{2-} + 4CH_3OH \rightarrow 3S^{2-} + 8H_2O + 4CO_2 \tag{8}$$

$$3SO_4^{2-} + 2C_2H_5OH \rightarrow 3S^{2-} + 6H_2O + 4CO_2 \tag{9}$$

$$9SO_4^{2-} + 4C_3H_7OH \rightarrow 9S^{2-} + 16H_2O + 12CO_2 \tag{10}$$

$$SO_4^{2-} + 4HCOOH \rightarrow S^{2-} + 4H_2O + 4CO_2 \tag{11}$$

$$SO_4^{2-} + CH_3COOH \rightarrow S^{2-} + 2H_2O + 2CO_2 \tag{12}$$

$$7SO_4^{2-} + 4C_2H_5COOH \rightarrow 7S^{2-} + 12H_2O + 12CO_2 \tag{13}$$

$$3SO_4^{2-} + 2CH_3CHOHCOOH \rightarrow 3S^{2-} + 6H_2O + 6CO_2 \tag{14}$$

Methanol seems to be the most cost-effective substrate (excluding hydrogen), and literature data show [15] that some organisms, in *pure* culture, are capable of using this substrate to reduce sulphate; therefore, this compound was expected to be an ideal candidate material. Although the majority of the microbially active environmental samples consumed methanol, they unfortunately produced methane in preference to reducing sulphate [10,11]. This effect was due to the presence, in the mixed cultures, of methanogenic organisms, which occur naturally in the environmental samples. Since it would not make economic sense to operate a sterile commercial process with a pure sulphate-reducing culture, then methanol will *not* be a good carbon source for an SRB process.

The screening experiments showed that most cultures grew on lactate and many used ethanol as substrate. Although lactate would be too expensive (see Table 5) for the commercial process, most of our preliminary investigations were carried out using this substrate because this compound could be used by most sulphate-reducing bacteria (see Table 4). This approach gave us the best chance of selecting a large range of organisms while establishing the process operating window. However, all cultures formed acetate, which is only very slowly degraded by the sulphate reducers to carbon dioxide and water. Presence of acetate in the SRB process effluent adversely affects the process economics because an additional biodegradation step would be needed to produce an effluent with a low biological oxygen demand, which is necessary in order to discharge this effluent to the environment.

Although lactate was initially used as the sole carbon source, degradation of the acetate by-product was never achieved (see Table 6). Steady-state data from experiments using stirred tank reactors showed that the ratio of sulphate reduced to carbon substrate consumed was independent of operating conditions, indicating that sulphate reduction is directly coupled to the growth of the organisms. The averaged data from all our experiments showed that, for every mol of lactate consumed, 0.88 mol of acetate was formed and 0.4 mol of sulphate was reduced, irrespective of operating conditions. This can be represented by Reaction (15) where CH_2O represents organism composition.

$$25CH_3CHOHCOOH + 10SO_4^{2-} \rightarrow \qquad (15)$$
$$10S^2 + 22CH_3COOH + 20H_2O + 20CO_2 + 11CH_2O$$

Table 5. Thermodynamic data (per mol of substrate) and carbon substrate cost.

	Energy Substrate	Enthalpy (kcal)	Free Energy (kcal)	Entropy ($cal \cdot °C^{-1}$)	Equilibrium Constant	Sub/SO_4^{2-} (m/m)	$\$ \cdot kg^{-1}$ SO_4^{2-}
7	Hydrogen	−12.4	−7.4	−15.9	$4 \cdot 10^{12}$	0.042	
8	Methanol	−8.4	−16.0	25.6	$1 \cdot 10^{27}$	0.445	0.08
9	Ethanol	2.9	−15.7	62.5	$3 \cdot 10^{26}$	0.318	0.19
10	*n* Propanol	7.0	−20.5	92.5	$4 \cdot 10^{34}$	0.278	0.32
11	Formate	−9.4	−15.8	21.6	$5 \cdot 10^{26}$	1.916	1.75
12	Acetate	6.3	−12.6	63.3	$2 \cdot 10^{21}$	0.625	0.45
13	*n*-Propionate	13.0	−10.5	78.7	$5 \cdot 10^{17}$	0.441	0.40
14	Lactate	−4.0	−27.0	79.5	$6 \cdot 10^{46}$	0.625	1.42

Table 6. Lactate utilization at various conditions (components in mg/L^{-1}).

Temperature		Effluent					
Residence		22°C			31°C		
(h)	Feed	33	31[a]	5	29	28[a]	5
Lactate	1140	<1	20	310	<1	<1	<1
Acetate	0	640	1900	500	640	2080	680
Sulphate	1590	1050	480	1200	970	150	1210
Zinc	115	<0.01	<0.01	<0.01	<0.01	0.13	0.50
Cadmium	1.8	0.01	<0.01	<0.01	<0.01	<0.01	<0.01
Copper	1.4	<0.01	0.02	0.03	<0.01	<0.01	<0.01
Cobalt	1.6	0.05	0.03	0.03	0.03	0.01	0.01
Iron	3.5	0.04	0.04	0.03	0.02	0.03	0.04
Manganese	5.9	0.10	<0.01	0.05	0.03	0.03	0.03

[a]Feed contained 3380 mg · L^{-1} lactate.

ORGANISM ADAPTATION TO ETHANOL

Complete degradation of ethanol by the lactate culture was possible even though acetate is initially produced (Figure 4). Unfortunately, in continuous culture (a fundamental SRB process requirement) the level of acetate in solution was shown to be inversely proportional to the liquid residence time in the reactor. Attempts to adapt this culture, by continuous selection, failed to completely degrade the acetate at short residence time; however, this problem was solved by using methanol as an intermediate substrate. This procedure encouraged organisms (especially methanogens, see Table 4) to grow on acetate [10,27,40,41] such that all of the compound that is produced by the sulphate reducers was degraded (Table 7).

Subsequently, data from our experiments, using well-established continuous cultures, showed that 1.25 mol of sulphate were reduced for each mol of ethanol completely consumed, irrespective of operating conditions. This can be represented by Reaction (16).

$$4C_2H_5OH + 5SO_4^{2-} \rightarrow 5S^{2-} + 10H_2O + 6CO_2 + 2CH_2O \qquad (16)$$

NUTRIENTS

All organisms need nitrogen, phosphorus, and trace elements to sustain growth. Ammonium salts or urea were found to be ideal nitrogen sources for the sulphate-reducing bacteria. Nitrate can be used as a nitrogen source, but it was not considered acceptable because it acts as an electron acceptor in preference to sulphate [42–44], therefore

Table 7. Effect of changing growth substrate.

Feed (m · mol)			Effluent (m · mol)	
Lactate	Methanol	Ethanol	Sulphate Reduced	Acetate Produced
12.6			6.4	10.7
12.7	21.5		6.9	50.0
4.3	17.4	11.8	7.6	3.3
1.3	16.9	12.1	7.1	0.6
	17.8	11.1	7.0	0.4
	5.1	11.6	7.3	<0.1
		12.5	8.6	<0.1

consuming carbon substrate without reducing sulphate. Inorganic phosphates are the most suitable source of phosphorus. All other trace elements were known to be present in Budelco's groundwater at a level that will sustain growth.

Experimental data showed that complete degradation of ethanol to carbon dioxide, methane, and biomass is achieved if the molar ratios of ethanol/nitrogen and ethanol/phosphorus are maximally 20 and 500, respectively (i.e., 18 kg ammonia and 4 kg phosphate per tonne of ethanol). Insufficient addition of any nutrient leads to reduced organism activity and sulphate reduction, which shows initially as an increase in the reactor acetate concentration due to increased redox potential, which is caused by the lack of excess sulphide in solution. If this condition prevails, then pH drops and sulphate reduction ceases. Fortunately, the nongrowing organisms remain viable and the SRB process can be easily returned to full operation by the addition of the depleted nutrients and sodium sulphide, therefore correcting both redox potential and pH.

pH REQUIREMENT

The optimum pH for sulphate reduction and organism growth was found to be between 6.5 and 8. Outside this range organism growth is affected; however, these microorganisms will survive for a short period of time, without growing, at pH's as low as 4 and as high as 9. Budelco's groundwater has a pH of about 4.5; however, it may not be necessary to neutralize the feed before it enters the reactor since it has been shown in both the laboratory and the demonstration sludge blanket reactor that the sludge bed has substantial buffer capacity (see Table 8). The ability of this system to maintain a neutral pH, even with an acid feed, is due to both the presence of carbonates and sulphides, the products of microbial growth and sulphate reduction, and the conver-

Table 8. Effect of feed pH on culture performance, without pH control ($mg \cdot L^{-1}$).

	Feed	Effluent				
Feed pH		7.5	6.4	5.2	3.7	3.2
Reactor pH		7.5	7.8	7.1	7.1	7.1
Ethanol	572	<1	<1	<1	<1	<1
Sulphate	1250	120	19	108	108	27
Zinc	140	0.032	0.030	0.039	0.022	<0.004
Cadmium	3.7	0.010	0.007	<0.004	<0.004	0.005
Copper	2.6	<0.004	<0.004	<0.004	<0.004	<0.004
Cobalt	2.2	0.027	0.027	0.020	<0.004	<0.004

sion of sulphuric acid (pK_a 1.9) to the hydrogen sulphide (pK_a 7.0). However, a neutral pH must be maintained within the reactor; otherwise, sulphate reduction is curtailed.

LIQUID RESIDENCE TIME

The minimum liquid residence time required in a sludge blanket reactor to achieve the desired sulphate reduction and complete degradation of carbon substrate and microbial products (i.e., ethanol and acetate, respectively) will be limited by the concentration of active organisms maintained within the reactor. Since the organisms adhere to the sulphide sludge, then the settling velocity of the sludge bed particles will be strongly influenced by the characteristics of the solid particles. Because the sludge had a very long residence time (weeks), washout of the organisms is very slow; therefore, it is possible to operate the process at a liquid residence time that is less than the organism's doubling time. A minimum liquid residence time for a sludge bed reactor was not determined since this required maintaining true steady-state conditions for long periods (months). However, laboratory data from the column reactor indicated that a liquid residence of less that 4 h was readily attainable even though the organisms require a longer time to double [14,45]. This short residence time was subsequently confirmed during the studies with the process demonstration unit.

EFFECT OF TEMPERATURE

For the mixed culture used, a maximum operating temperature between 40 and 45°C (Figure 6) was demonstrated; above 45°C the organisms died. Experimental data indicates that at temperatures lower than 15°C, degradation of acetate was incomplete at liquid residence times of less than 4 h.

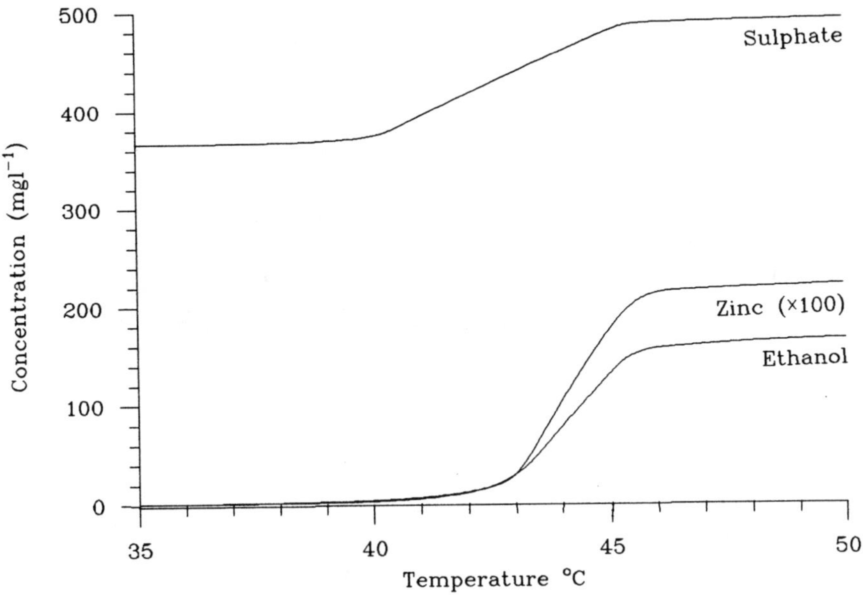

Figure 6. Temperature effect on sulphate reduction.

EFFECT OF PRESSURE

Hydrostatic pressure has little effect on sulphate reducers since they occur in deep ocean basins [46] and are problematic in souring oil wells [18]; therefore, this parameter is not likely to limit reactor design.

EFFECT OF REDOX POTENTIAL

Experimental results show that a low redox potential in the reactor is essential in order to maintain a stable culture, maximize sulphate reduction, and ensure complete degradation of the carbon substrate. At pH 7 a maximum redox potential of –100 mV (with respect to a standard calomel electrode) was found to be sufficient for SRB activity [13,24]; however, the methanogens required a value nearer to –300 mV for optimum activity. Figure 7 shows the equilibrium Eh-pH diagram for the sulphate/sulphide and carbonate/methane systems and the steady-state operating point for typical mass balance data (see later). This low redox potential is maintained by the presence of soluble sulphide [47]; therefore, for this to occur then *all* heavy metals must have be precipitated as their sulphides (Table 3). Too high a redox potential lowers methanogen activity, which results in an increased acetate concentration; consequently, the reactor pH drops and sulphate

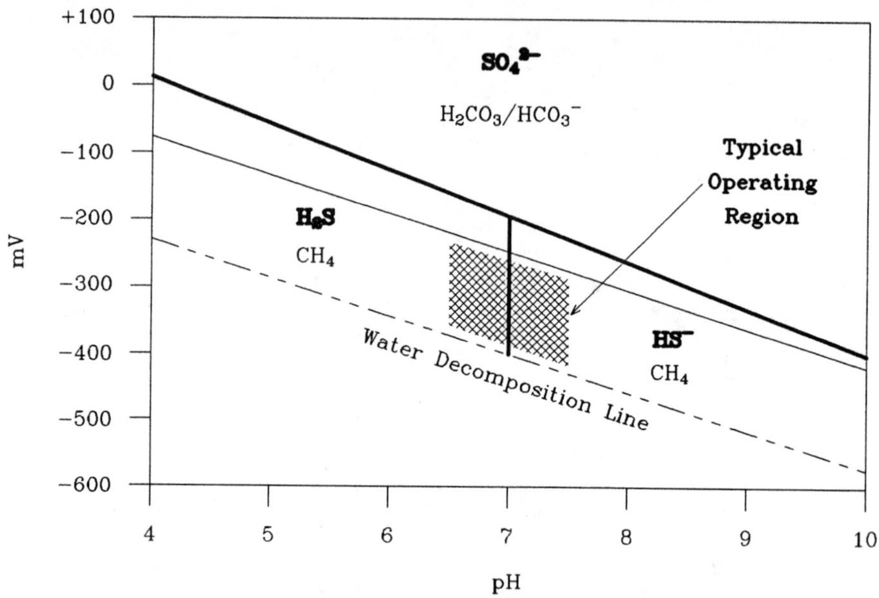

Figure 7. Eh-pH diagram for sulphate/sulphide and carbonate/methane systems.

reduction declines. Short excursions to a high redox potential have no detrimental effect on the organisms [24]; however, recovery times are long unless the redox potential is lowered by addition of a soluble sulphide. In addition, a high redox potential, through lack of sulphide reduction, will increase the metal concentration in solution until the solubility of metal hydroxide (or carbonate) is reached, with the final metal concentration in solution being highly pH dependent (Figure 8).

EFFICIENCY OF HEAVY METAL REMOVAL

Efficient removal of heavy metals is possibly the most important attribute of the SRB process. Providing sufficient sulphate is reduced, and soluble sulphide is present to maintain a low redox potential; then heavy metal concentrations of only parts per billion will remain in the aqueous phase of the reactor's aqueous effluent. Whenever our reactors were operated under conditions that sustained good microbial growth, heavy metal removal was practically complete even for reactor feeds containing in excess of 1 $g \cdot L^{-1}$ of heavy metals, as shown in Table 9. Tables 6 and 8 show additional examples of such data. Although the metal concentrations in solution are extremely low, the practical effi-

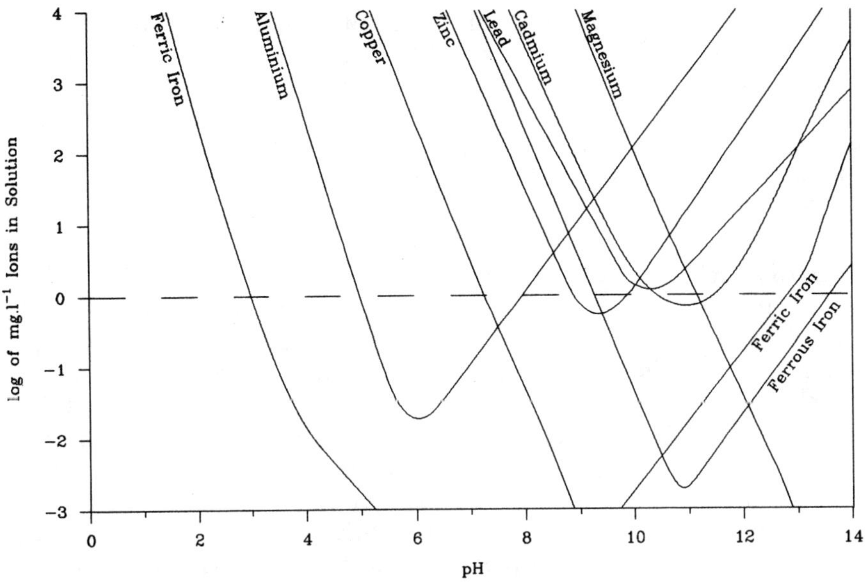

Figure 8. Equilibrium solubilities of metal hydroxides as a function of pH.

ciency of heavy metal removal is determined by separation of solid metal sulphides from the reactor's aqueous phase.

FATE OF CALCIUM AND MAGNESIUM

For our particular application at Budelco, it was necessary to determine the fate of calcium and magnesium in the SRB process. Since the sludge produced in the reactor will be added to the zinc refinery feed, a buildup of these elements in the sludge should be minimized to prevent calcium precipitation in the zinc refining process and magne-

Table 9. Effect of high metals in feed ($mg \cdot L^{-1}$).

Component	Feed	Effluent
Ethanol	1040	<1
Sulphate	3200	1340
Zinc	1070	0.22
Cadmium	18	<0.01
Copper	6.8	<0.02
Cobalt	0.5	<0.02
Iron	67	0.10
Magnesium	140	116
Calcium	380	390

sium buildup in the electrowinning circuit. Experiments show that the majority of both elements end up in the aqueous effluent provided that phosphate, which is added for microbial growth, is kept to a minimum. The effect of pH on phosphate solubility in the presence of calcium is shown in Figure 9, and typical reactor data, which shows that the majority of the calcium and magnesium remains in solution, is given in Table 9.

INHIBITORY EFFECTS OF FEED COMPONENTS

The inhibitory effect of potential feed components on microbial growth and, hence, sulphate reduction must be considered. As expected, alkali and alkaline earth cations were found to have no deleterious effect since the organisms used in this process can be isolated from marine environments [25]. Heavy metal cations appear to be nontoxic, presumably because the presence of excess soluble sulphide maintains a very low metal concentration in solution (see Table 3). Certain anions are, in general, known to be powerful microbial inhibitors. However, the following anions were found not to inhibit microbial growth or sulphate reduction at the concentrations tested ($mg \cdot L^{-1}$): molybdate (10), selenate (3.5), arsenate (3.3), fluoride (48), and sulphide (500).

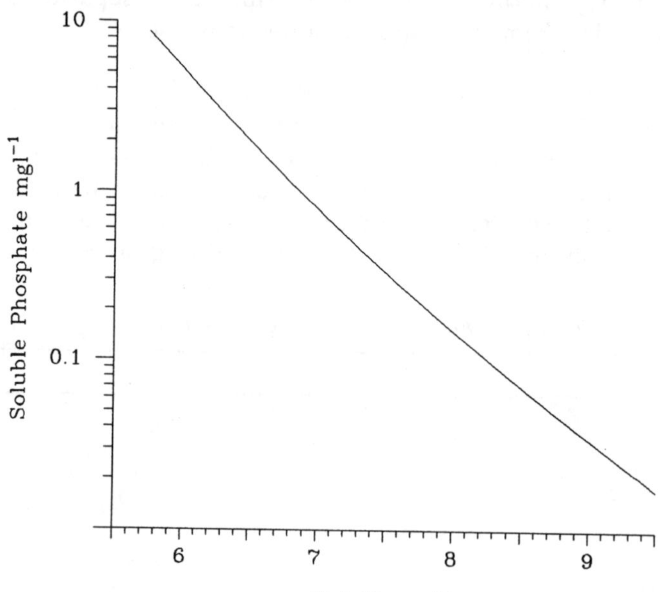

Figure 9. Effect of pH on phosphate solubility in the presence of 300 $mg \cdot L^{-1}$ calcium.

MASS BALANCE OVER THE SRB REACTOR

The buildup of a sludge blanket in the reactor is very slow, and it can only be considered to be in perfect equilibrium with the feed and effluents after a long period of operation (months), providing the feed composition and reactor conditions have remained constant during the whole period. However, a sensible and reasonably accurate overall mass balance can be attained from analysis of the feed, effluent liquid, and gas streams, assuming the remaining elemental balance is associated with the sludge. Figure 10 shows such a mass balance over a reactor that has been operated under pseudo-steady-state conditions for several weeks.

It should be noted that, although there is a multitude of equilibria within the reactor, only the relative concentrations of sulphur and carbon compounds in each reactor effluent stream change with operating conditions. The levels of each compound depend on the

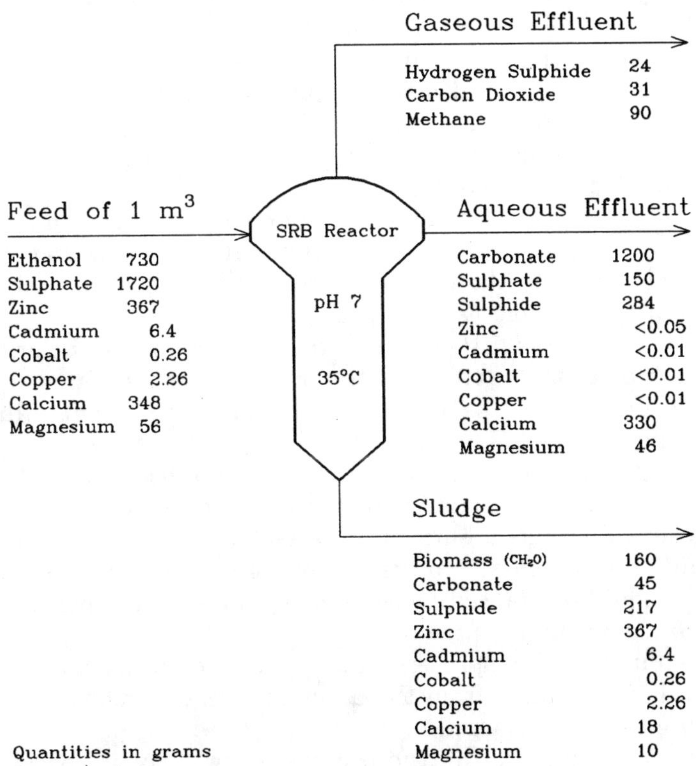

Figure 10. Mass balance over the SRB reactor.

quantity of sulphate reduced, the heavy metal concentration and carbon substrate in the feed, and reactor pH, provided that optimal microbial conditions exist within the reactor.

Conclusions from Laboratory Experiments

The following conclusions were drawn from data obtained during batch and continuous experiments in the laboratory:

1. Sulphate and a wide range of heavy metals can be removed from the aqueous phase to levels acceptable to the authorities.
2. Sulphate-reducing bacteria utilize ethanol as a carbon/energy substrate for reducing sulphate and, subsequently, form acetate as a major product. The acetate can be successfully degraded to carbon dioxide, methane, and biomass by methanogens.
3. The final sulphate concentration in the anaerobic reactor effluent can be controlled by the amount of ethanol fed to the reactor. One mol of ethanol is capable of reducing 1.25 mol of sulphate.
4. Dissolved heavy metal concentrations in the aqueous effluent from the anaerobic reactor are extremely low (ppb), provided that sufficient sulphate is reduced and the redox potential in solution is maintained below −100 mV.
5. Microbial growth is sustained, provided that nitrogen and phosphorus are supplied in the reactor feed at 50 m·mol and 2 m·mol for each mol of ethanol consumed. Excess phosphate should be minimized to avoid calcium precipitation.
6. Temperature has little effect on the performance of the organisms in the range 15°C to 40°C. For the organisms investigated, temperatures higher than 42°C must be avoided to prevent killing the organisms.
7. pH control in the anaerobic reactor may not be required for treatment of Budelco's groundwater if the pH is greater than 4 because the sludge acts as a buffer in this system.
8. Addition of flocculant to the anaerobic reactor feed is essential if short liquid residence times and adequate solid retention are to be obtained within a single reactor.
9. Unwanted sulphide present in the three product streams (i.e., gas, liquid, and sludge) from the anaerobic reactor must be removed before being discharged.
10. Energy consumption in the process is low since it operates near ambient temperature.

PROCESS DEMONSTRATION UNIT

After 2 years of experimental research in the laboratory, sufficient data were obtained to progress the experimental program to a larger scale. Budelco's local authority insisted that the contaminated groundwater problem should be solved as quickly as possible; therefore, it was decided to scale up the laboratory SRB process by 6000 times in a single step. A 9 m^3 raked, sludge blanket reactor process was designed, built, and commissioned within 3 months in order to rapidly prove that the anaerobic sulphate-reducing process concept was a viable option for the treatment of the groundwater in the zinc refining environment. Figure 11 shows a schematic diagram of the experimental process demonstration unit.

The demonstration unit contained the following components:

- 9 m^3 upflow raked, sludge blanket reactor, which was constructed of resin bonded fiberglass lined with nonplasticized polyvinyl chloride
- 15 m^3 resin bonded fiberglass buffer tanks for the groundwater feed
- polypropylene feed tanks for solutions of nutrients (ammonium sulphate and phosphoric acid), ethanol, alkali (sodium hydroxide), flocculant, ferric chloride, and zinc sulphate, plus pumps to transfer these solutions
- in-line heat exchanger to control the temperature of the reactor's contents
- in-line mixer for blending the various chemicals with the groundwater feed
- stripper, which was constructed of polyvinyl chloride lined resin-bonded fiberglass, to remove hydrogen sulphide from the gaseous product stream
- mixing tank to remove soluble sulphide from the aqueous effluent stream
- pH, temperature, and feed flow control system with associated pumps

Process Description

Nutrients and ethanol are mixed with the groundwater; this mixture may be pH or temperature controlled or both before being passed into the base of the anaerobic upflow sludge blanket reactor where the sulphate is reduced to sulphide. A flocculant was added to prevent the solid sludge particles being carried over in the aqueous effluent. The gaseous stream is stripped with zinc sulphate solution to remove hydrogen sulphide, as zinc sulphide, before being flared to destroy the small quantity of methane produced. The reactor's aqueous effluent was treated with ferric chloride solution to precipitate soluble sulphide as iron sulphide before it was discharged. Excess sludge (which contained the biomass) produced in the anaerobic reactor, along with the zinc sulphide and iron sulphide, were disposed of in the existing zinc refining process.

PROCESS OPERATION

Investigations carried out in laboratory bench-scale reactors showed that a suitable inoculum could be obtained by using equal volumes of an environmental sediment from one of Budelco's clarifying lakes (sulphate-reducing bacteria source) and an anaerobic biotreater sludge (source of methanogens). About 1 m^3 of this mixed inoculum was added to the demonstration reactor, which contained neutralized groundwater, nutrients (nitrogen and phosphorus), and the ethanol (energy source). The ratio of nutrients and ethanol to groundwater was determined by the level of sulphate that needed to be reduced in order to precipitate all the heavy metals. Four days after inoculation, the process was switched from batch to continuous operation using a long liquid residence time of 40 h (3 days was required for the same batch operation in the liter-size laboratory reactors). During the early part of these reactor studies, it became apparent that the settling behavior of the sludge particles was different from that observed in the laboratory experiments. Substantial quantities of organism-containing sulphide sludge was washed out of the demonstration reactor when liquid residence times were reduced below 30 h. This problem was solved by the addition of the flocculant "Kemfloc 254," which was shown in laboratory studies not to have any deleterious effect on the mixed culture at a concentration of 15 $mg{\cdot}L^{-1}$, which was five times higher than that required for adequate settling of the sludge particles. Shorter liquid residence times were thereafter achieved without substantial sludge particle carryover in the reactor's aqueous effluent. An example of data produced during 1 year of continuous operation of this demonstration unit is shown in Table 10.

Demonstration Process Conclusions

The process demonstration unit studies showed that

1. The sulphate-reducing bacterial process concept could be readily scaled up without any loss in aqueous effluent quality.

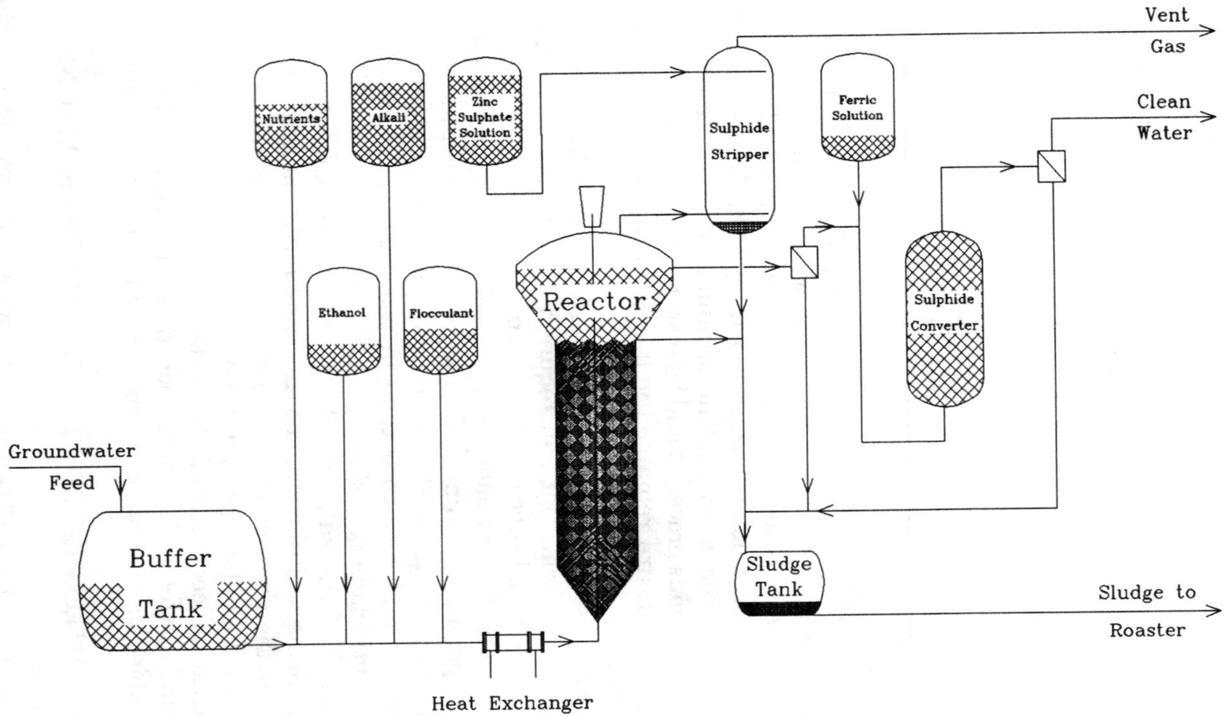

Figure 11. Schematic of the demonstration unit.

Table 10. Typical data from the demonstration unit (mg · L^{-1}).

Component Analyzed	Groundwater Feed	Reactor Effluent
Ethanol	260	<1
Acetate	0	<10
Sulphate	850	335
Zinc	107	0.17
Cadmium	0.87	<0.01
Copper	0.40	<0.02
Cobalt	0.14	<0.02
Iron	49	0.03
Calcium	360	365
Magnesium	18	18

2. The operating window was confirmed, and recovery from operational upsets was rapid. These organisms were shown to be able to withstand conditions outside the operating window, allowing recovery to normal continuous operation within 24 h. Even after a shutdown of several weeks, without maintaining strict anaerobic conditions, the continuous process could be restarted within a few days, showing that the culture remained viable without the addition of growth substrates.
3. This process was shown to be robust; it readily fitted into the metal refining environment and required no specific microbial expertise once the initial process was set up.
4. A shorter liquid residence time (<4 h) than that required (7 h) for a potential commercial process was achieved after addition of flocculant. Addition of this reagent increased the sludge, associated biomass, and residence time in the reactor, as well as reducing sludge thickening costs.
5. Settling of the sulphide sludge, which contains the immobilized organisms, was shown to be the rate-limiting process in the reactor, rather than microbial sulphate reduction, acetate degradation, or chemical sulphide precipitation.
6. Startup procedure could be scaled up by a factor of 6000 without extending the initial batch operation time required for establishing a stable mixed culture capable of continuous operation.

INTEGRATED PROCESS INVESTIGATIONS

The process demonstration unit successfully proved that the SRB process could be scaled up without compromising the effective removal of sulphate and heavy metals. In addition, provisional estimates of operating costs indicated that the process would be economically viable; therefore, Budelco initiated the development of a commercial-scale plant capable of treating 7000 m^3day^{-1} of groundwater. Paques, a Dutch company experienced in anaerobic wastewater treatment systems, was contracted by Budelco to produce the basic design of a commercial process. Although the microbial sulphate reduction process was adequately demonstrated, the subsequent treatment of the reactor effluent streams to remove excess hydrogen sulphide and solids still had to be developed and proven. In addition, further design data were required; therefore, an integrated pilot plant that contained all basic components needed in the commercial process was installed at the Budelco site. This pilot plant, which is shown schematically in Figure 12, was operated for 8 months to demonstrate the efficiency of an integrated system containing all the process operations needed, as well as to fix and finalize the design parameters.

Main process units tested during this period were

- an upflow anaerobic sludge blanket (UASB) reactor for the SRB process
- a gas treatment system
- a submerged fixed film (SFF) reactor for the aerobic removal of sulphide from the aqueous UASB effluent
- a tilted-plate settler (TPS) for solids removal
- a continuous sand bed filter (DynaSand) to polish the aqueous solution before discharge

Upflow Anaerobic Sludge Blanket (UASB) Reactor

The Biopaq UASB reactor (Figure 13), which was initially developed at the University of Wageningen [48] for wastewater treatment, has been used in a number of commercial anaerobic processes [49] for the degradation of carbonaceous materials in wastestreams. The reactor has an influent system designed to spread the incoming groundwater evenly below the active biological sludge bed. In order to maintain the maximum reaction time within the reactor, channelling (which is usually caused by compaction of the sludge) must be avoided. Best results are given by a consistent gently mixed sludge bed; this is achieved by recycling some of the reactor's aqueous effluent to ensure that a consistent vertical liquid velocity is maintained, when groundwater feed rates are lowered, and there is gas bubble formation within the sludge bed. A three-phase separator, which is incorporated in the top of the reactor, consists of three layers of inverted V-shaped channels.

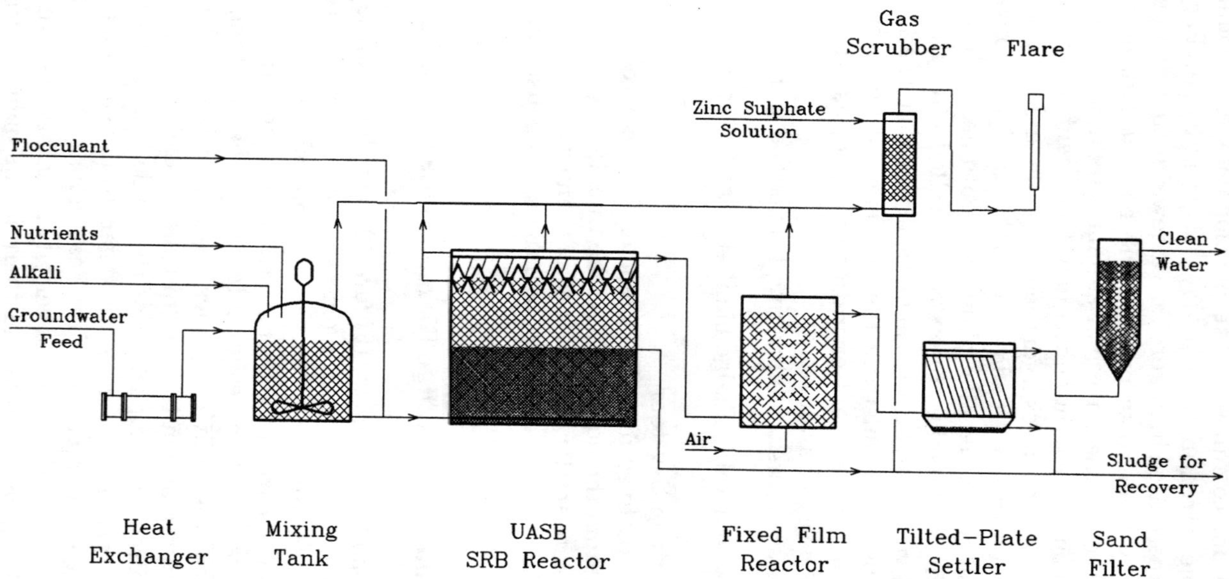

Figure 12. Schematic of integrated pilot plant.

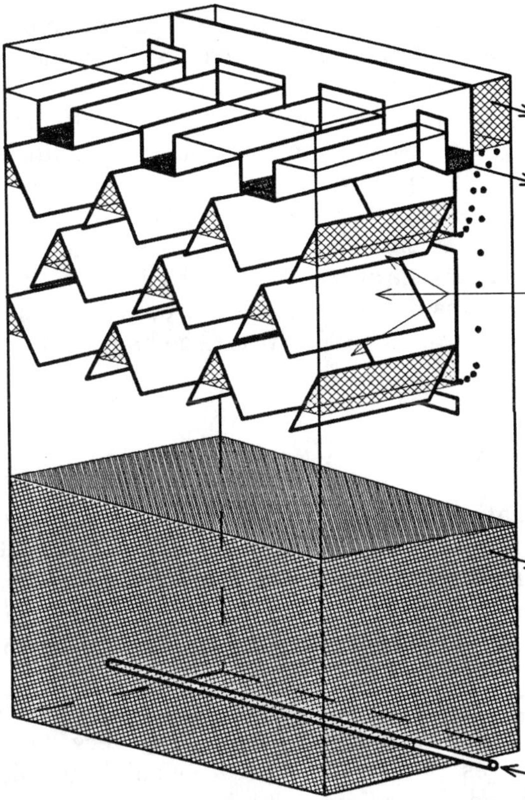

Figure 13. Schematic view of UASB reactor.

Because the gas bubbles produced are caught in these "hoods" below the water surface, the uppermost water layer is relatively undisturbed; hence, the solids settle on the sloping sides of the hoods and subsequently sink to the sludge bed.

UASB OPERATING CONDITIONS AND RESULTS

The 12 m^3 UASB reactor was operated between pH 6 and 8, at 20 to 38°C, and with a liquid residence time of less than 8 h, which was determined by the settling characteristics of the sludge. The groundwater used originated from a borehole under a jarosite pond and often contained more than 200 $mg \cdot L^{-1}$ zinc and 1000 $mg \cdot L^{-1}$ sulphate. Biosludge from the demonstration reactor was used as inoculum for the UASB reactor, and the flocculant "Synthofloc" was dosed at 1 $mg \cdot L^{-1}$ to prevent solids being flushed out in the aqueous effluent. Higher concentrations of flocculant led to frothing with flotation of the solids, which could potentially block the aqueous overflow and reduce the biomass in the reactor. Hydraulic flows of between 1.5 to 3 m^3h^{-1} (equivalent to residence times of 8 and 4 h, respectively) were employed. At flow velocities greater than 3 m^3h^{-1}, solids were flushed out of the reactor, whereas lower flow rates (<1.5 m^3h^{-1}) encouraged bed compaction and channelling; this greatly reduced the exposure of the groundwater to the organisms and reduced the efficiency of sulphate reduction. Table 11 shows the averaged data obtained using the reactor operating at a liquid residence time of 4 h, culture temperature of 20°C, pH 7, redox potential of −390 mV, and a sludge concentration of 100 $g{\cdot}L^{-1}$.

Gaseous Stream Treatment

Anaerobic degradation of organic materials in wastewater treatment systems produces methane and carbon dioxide. In the SRB process small quantities of methane are produced along with hydrogen sulphide and carbon dioxide. During the operation of this UASB reactor, 7 m^3day^{-1} of gas was generated when groundwater, with the composition given in Table 11, was used at a feed rate of 50 m^3day^{-1}. The gas contained about 85% methane, 8% carbon dioxide, and 7% hydrogen sulphide with the absolute concentrations being very dependent on reactor pH (*cf.*, Figures 14 and 15). Since zinc sulphate solution was readily available at the zinc refining site, it was expedient to strip the

Table 11. Average UASB inlet and outlet concentrations in the aqueous phase.

Component	Inlet $mg \cdot L^{-1}$	Outlet $mg \cdot L^{-1}$	% Precipitated
Ethanol	510	<1	
Sulphate	1400	120	91
Zinc	250	<0.05	>99.9
Cadmium	1.1	<0.001	>99.9
Iron	33	0.03	99.9
Lead	13	<0.02	>99
Copper	2.2	<0.02	>99
Cobalt	0.2	<0.02	>90
Nickel	0.3	<0.01	>96
Manganese	35	20	43
Magnesium	40	35	12
Calcium	300	290	3
Arsenic	0.01	<0.005	>50

hydrogen sulphide from the gas stream with this solution prior to the flaring, thus preventing sulphur dioxide emissions.

Aqueous Stream Treatment

For optimal microbial sulphate reduction and degradation of acetate, a low redox potential must be maintained within the reactor. This is best achieved by ensuring that excess sulphate, over heavy metals, is reduced so that some sulphide remains in solution. In addition, it may be necessary to reduce excess sulphate in order to meet the sulphate discharge levels. This aqueous reactor effluent could not be discharged directly to the environment because of the soluble sulphide's toxicity and odor; therefore, it had to be treated. Various treatment regimes were considered, the most promising approaches being stripping, sulphide precipitation, and oxidation.

Stripping the sulphide from solution with an inert gas [38] (e.g., nitrogen) would be best achieved if the UASB effluent pH is lowered to less than 5 (Figure 15); unfortunately, this would require the addition of acid, which introduces an anion such as chloride (or sulphate). When the sulphide has been stripped, the pH may need to be increased to at least 6.5, the allowable discharge minimum. Both acidification and neutralization will increase operating costs and add components to the

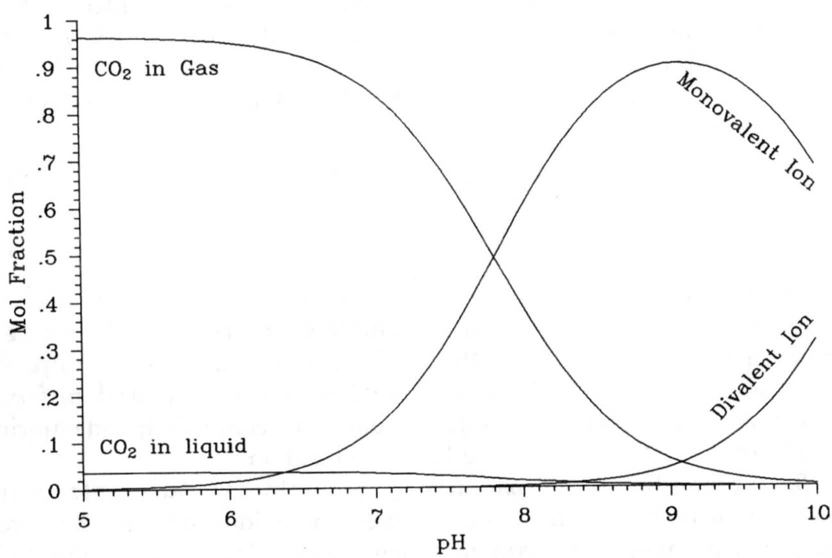

Figure 14. Equilibrium concentration of carbon dioxide species as a function of pH.

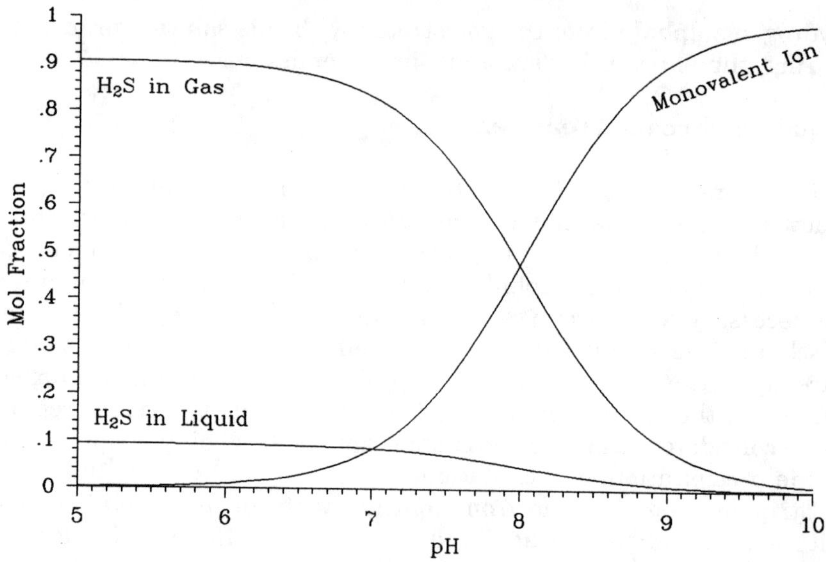

Figure 15. Equilibrium concentration of hydrogen sulphide species as a function of pH.

effluent. If stripping is carried out at the UASB effluent pH, then excessive quantities of inert gas will be required since stripping of both hydrogen sulphide and carbon dioxide causes the solution pH to increase [Reactions (17) and (18)], making it difficult to strip the last traces of sulphide (see Figure 15). In addition, acid will need to be added to decrease the pH to <8.5, the allowable discharge maximum. This option was therefore rejected.

$$HS^- + H_2O \rightarrow OH^- + H_2S \tag{17}$$

$$HCO_3^- \rightarrow OH^- + CO_2 \tag{18}$$

Chemical precipitation with iron salts [50] would remove sulphide from solution; however, the iron's soluble counter-ion would create a problem for disposal of the effluent from the Budelco site. Sulphate and chloride, the most suitable counter-ions, were considered; unfortunately, sulphate levels in the effluent had to be controlled, and chloride could not be discharge into the local environment.

Chemical and biological oxidation of the sulphide to elemental sulphur [Reaction (6)] gave the best option. Chemical oxidation was considered expensive because of the oxidizing agent's cost. Two biological sulphide oxidation processes capable of operating at neutral pH are reported in

Table 12. Influence of air supply on sulphide conversion in SFF.

	Soluble Sulphide		
Air Flow m^3h^{-1}	Inlet $mg \cdot L^{-1}$	Outlet $mg \cdot L^{-1}$	Percent Removed
30	156	<1	>99
28	136	8	94
26	156	13	92

the literature. One uses photosynthetic bacteria [51], which was deemed not to be a viable option for our application. A second option uses controlled bacterial oxidation [52,53] in a submerged fixed film (SFF) reactor, which consists of a Pall ring packed reaction chamber where air and the sulphidic water flow co-currently upwards. The *Thiobacillus* bacteria present in the reactor use the oxygen in the air to convert soluble sulphide to elemental sulphur. The bacteria and sulphur initially adhere to the surfaces within the reactor; however, agitation by the air slowly dislodges both, thus preventing clogging of the reactor.

SFF OPERATING CONDITIONS AND RESULTS

Since these aerobic organisms are autotrophs, they use the energy of sulphide oxidation to fix carbon dioxide (their sole carbon source) that is present in the UASB effluent since it is a product of sulphate reduction [Reaction (1)]. In addition, the UASB effluent contains all the other nutrients at concentrations sufficient for growth; therefore, no further additions are required. Control of the redox potential within this reactor is necessary because the elemental sulphur can be oxidized by these organisms to sulphate at high redox, which is achieved by minimizing the oxygen supply to the reactor. However, as shown in Table 12, insufficient air will prevent complete removal of sulphide. A liquid residence time of about 30 minutes is required for this process; therefore, the SFF reactor need only be about one-tenth of the size of the UASB reactor.

Table 13. Solids removal in TPS and DynaSand filter for a typical run.

	Tilted-Plate Settler			DynaSand Filter			
Component	Inlet ($mg \cdot L^{-1}$)	Outlet ($mg \cdot L^{-1}$)	Percent Removed	Inlet ($mg \cdot L^{-1}$)	Outlet ($mg \cdot L^{-1}$)	Percent Removed	Combined Percent Removed
Total solid	2400[a]	385	84	385	37	90	98
Zinc	114	19	83	19	1.3	93	99
Sulphur[b]	336	15	96	15	<1	>93	>99

[a]High solids due to inadequate flocculant dosing in UASB reactor.
[b]Elemental sulphur plus heavy metal sulphides.

A 1.5-m^3 SFF reactor was used in the pilot plant to obtain the design data for this process operation. The conversion of sulphide to elemental sulphur is highly efficient in the SFF reactor; for example, 300 mg·L^{-1} influent sulphide can be treated in the SFF reactor to give a reactor effluent that contains less than 1 mg·L^{-1} of sulphide. Sulphate concentration may continue to decrease in the SFF reactor because sulphate-reducing bacteria that are carried over from the UASB reactor may remain active [54], especially if insufficient oxygen is used in this process, as shown in Table 14.

Effluent Polishing

Solids, such as sulphur from the SFF reactor, metal sulphides that have carried over from the UASB reactor, and microorganisms, must be removed before the aqueous phase can be discharged. This is necessary because, otherwise, the metal sulphides and sulphur can be oxidized by microorganisms to metal sulphate and sulphuric acid, respectively, thus recontaminating the environment. Acceptable removal of the solids was achieved in the pilot plant with a two-stage process. A 12-m^3 tilted plate settler was used as the first stage to remove the majority of the solids, and a continuous 0.3-m^3 sand-bed filter (DynaSand) was employed in the second stage to polish the aqueous effluent.

Several flocculants were tested to improve the efficiency of the tilted plate settler. Although all experiments showed that more than 80% of the solids could be recovered, none of the flocculants outperformed the others. The DynaSand filter consistently removed more than 90% of the solids from the tilted-plate settler's aqueous effluent, giving an overall solids removal for both steps in excess of 98%. Solids removal data are shown in Table 13.

Table 14. Sulphur mass balance ($g \cdot h^{-1}$) over the UASB and SFF reactors.

Component	UASB Reactor ($g \cdot h^{-1}$)	Inlet	Outlet	SFF Reactor ($g \cdot h^{-1}$) Outlet
Sulphate	795	330	330	225
Soluble sulphide	0	368	368	8
Elemental sulphur	0	0	0	435
Gaseous hydrogen sulphide	0	34	0	60
Metal sulphides	0	75	0	0
Total	795	807	698	728

Integrated Results from Pilot Plant Test

Although pilot plant trials were affected by variations in feed concentration, all the test objectives were achieved. Stable operations were achieved for 3 days, a period equivalent to 18 reactor volumes of groundwater passing through the UASB reactor. Zinc and cadmium data for integrated plant effluent obtained during this test are presented in Figure 16, which also shows the mean and standard deviation. Sulphur mass balances over the UASB and SFF reactors, using a groundwater feed of 1.5 m^3h^{-1} to the UASB and an air flow of 20 m^3h^{-1} to the SFF, are presented in Table 14.

Pilot Plant Conclusions

Pilot plant studies showed that

1. A fully integrated process was economically viable.
2. The soluble sulphide present in the UASB's aqueous effluent could be completely removed in the fixed-film reactor, converting it to elemental sulphur.

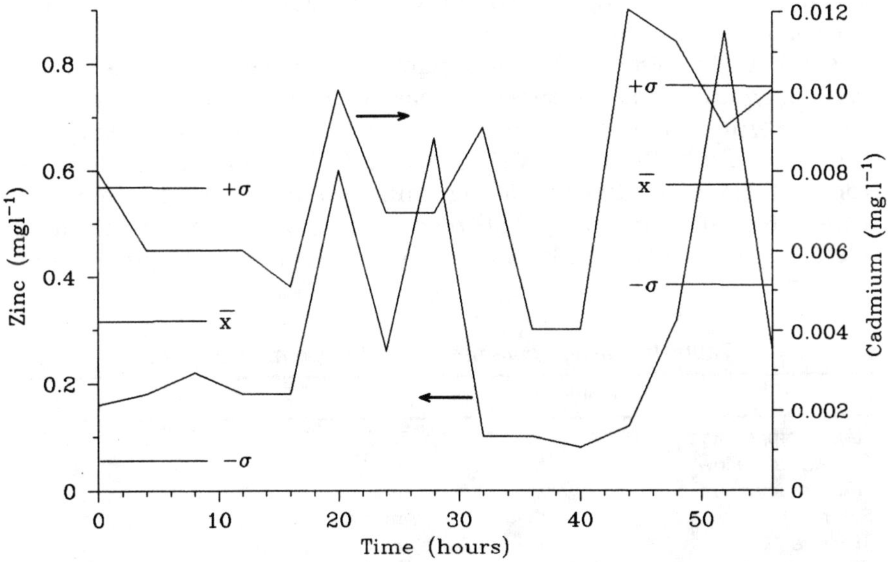

Figure 16. Variance test of SRB process for zinc and cadmium.

3. A tilted-plate settler, in combination with a continuous sand filter, could remove more than 99% of suspended solids, zinc, and insoluble sulphur.
4. A completely satisfactory gas handling system was demonstrated.
5. A "Biopaq" UASB reactor design was very suitable for the SRB process, giving the same quality effluent obtained previously in the laboratory.
6. Total startup, giving an acetate-free effluent, of the process can take up to 3 weeks, although sulphate reduction and metal removal reached acceptable levels after 3 days.

COMMERCIAL SRB PROCESS

Successful operation of the integrated pilot plant demonstrated that it should be possible to scale up this unit to a commercial process. Therefore, a commercial SRB process was designed to treat 7000 m^3day^{-1} of groundwater extracted by a geohydrological control system from beneath Budelco's zinc refining site. The design criteria used for groundwater composition and authorized discharge limits are given in Table 1. Since this was the first commercial SRB process, then all design parameters were set conservatively; these are given in Table 15.

The design of the commercial plant, shown in Figure 17, was based on the pilot plant configuration that was evaluated in the experimental program.

The UASB reactors in both the pilot and commercial plants were of similar depths, so a 150 times increase of surface area was required; therefore, no problems were anticipated in scaleup. This full-scale plant consisted of a buffer tank (V1) to even out variations in groundwater concentrations and flow; blending tank (V2) for mixing groundwater, substrate, and nutrients; UASB reactor (V3); SFF reactor (V4); tilted-plate settler (V5); DynaSand filter (V6); and a gas handling system

Table 15. Design parameters for the commercial plant.

Parameter		Value
UASB Temperature	°C	25
Groundwater Flow	m^3h^{-1}	300
UASB reactor size	m^3	1800
SFF reactor size	m^3	165
TPS size	m^3	120
Biofilter surface area	m^2	30

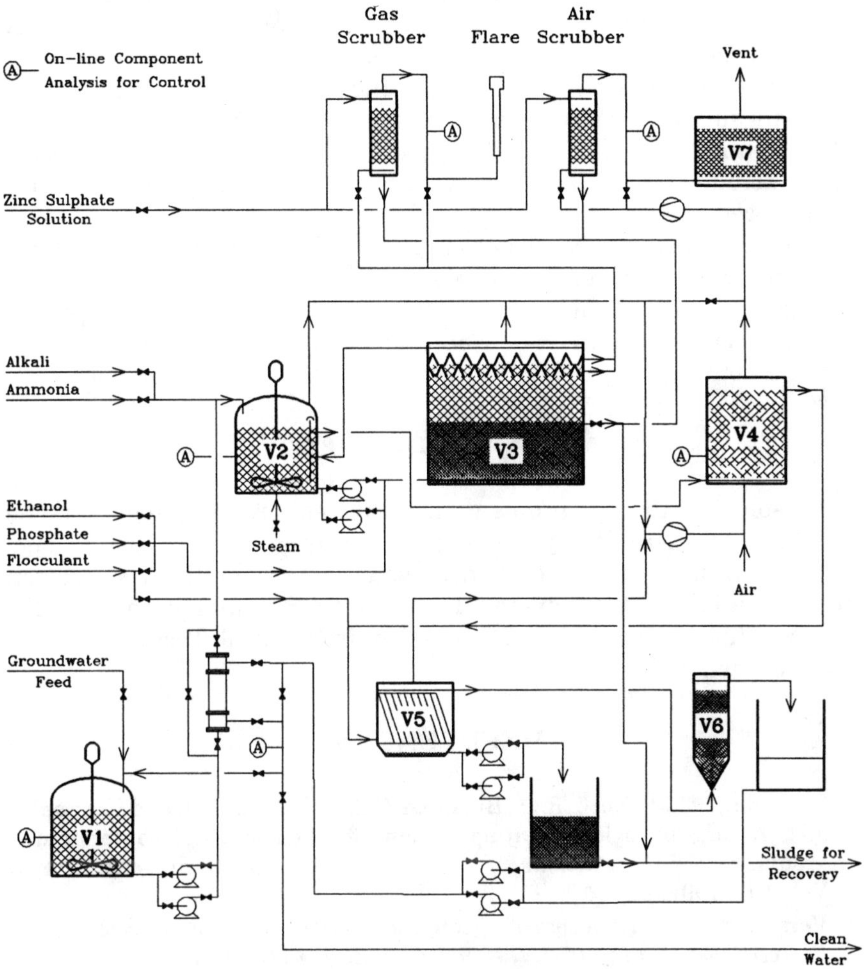

Figure 17. Schematic of commercial plant.

and a biofilter (V7). Because the process produces hydrogen sulphide, a very toxic material, and methane, an explosive material, special attention was paid to safety and environmental protection. Leakage of these gases from the process potentially causes odor and flammability problems; therefore, all relevant plant units operate under slightly reduced pressure. The head gases from each vessel are combined and then scrubbed with concentrated zinc sulphate solution to remove hydrogen sulphide before being treated in a biofilter.

The commercial plant was designed for fully automated operation, which included on-line analysis with feedback control and safe shutdown procedures.

The commercial process was commissioned in October 1992 and has performed very successfully to date.

In conclusion, although the SRB process has been developed to solve a specific problem, it can be used to clean wastewater streams that are contaminated with heavy metals or sulphate or both.

ACKNOWLEDGEMENT

The solution to this problem necessitated bringing together the disciplines of microbiology, chemistry, and process engineering. Success of this project was only achieved because of the cooperation and dedication of the personnel at Shell's Research Centre (UK), Budelco Zinc Refinery, and Paques Environmental Technology (the Netherlands).

REFERENCES

1. Scheeren, P.J.M., Koch, R.O., Buisman, C.J.N., Barnes, L.J. and Versteegh, J.H. A new biological treatment plant for heavy metal contaminated groundwater. *Transactions of the Institute of Mining and Metallurgy,* Vol. 101, 1992, pp. C190–C199.
2. Versteegh, J.H. The design of a geohydrological control system. *Symposium Proceedings: Dealing with the environmental problem in the mining and petroleum industry.* 1990, pp. 9–15.
3. Konings, A.J.A. Private communication. 1987.
4. Draxler, J. and Marr, R. Emulsion liquid membranes, Part I. *Chemical Engineering Proceedings.* Vol. 20, 1986, pp. 319–329.
5. Lorbach, D. and Marr, R. Emulsion liquid membranes, Part II. *Chemical Engineering Proceedings.* Vol. 21, 1987, pp. 83–93.
6. Barnes, L.J., Janssen, F.J., Scheeren, P.J.H., Versteegh, J.H. and Koch, R.O. Simultaneous microbial removal of sulphate and heavy metals from

waste water. *Transactions of the Institute of Mining and Metallurgy,* Vol. 101, 1992, pp. C183–C189.

7. Wildeman, T.R., Updegraff, D.M., Reynolds, J.S. and Bolis, J.L. Passive bioremediation of metals from water using reactors or constructed wetlands. *Proceedings of the 2nd International Symposium on In-Situ and On-Site Bioreclaimation.* San Diego, USA. 1993.
8. Laanbroek, H.J., Geerligs, H.J., Sijtsma, L. and Veldkamp, H. Competition for sulphate and ethanol among *Desulfobacter, Desulfobulbus* and *Desulfovibrio* species isolated from intertidal sediments. *Applied and Environmental Microbiology.* Vol. 47, 1984, pp. 329–334.
9. Laanbroek, H.J. and Pfennig, N. Oxidation of short-chain fatty acids by sulphate-reducing bacteria in freshwater and in marine sediments. *Archives of Microbiology,* Vol. 128, 1981, pp. 330–335.
10. Oremland, R.S. and Polcin, S. Methanogenesis and sulphate reduction: competitive and noncompetitive substrates in estuarine sediments. *Applied and Environmental Microbiology,* Vol. 44, 1982, pp. 1270–1276.
11. Oremland, R.S., Marsh, L.M. and Polcin, S. Methane production and simultaneous sulphate reduction in anoxic, salt marsh sediments. *Nature.* Vol. 296, 1982, pp. 143–145.
12. Widdel, F. and Pfennig, N. Studies on dissimilatory sulphate-reducing bacteria that decompose fatty acids II. Incomplete oxidation of propionate by *Desulfobulbus propionicus* gen. nov., sp. nov. *Archives of Microbiology.* Vol. 131, 1982, pp. 360–365.
13. Battersby, N.S., Stewart, D.J. and Sharma, A.P. Microbiological problems in the offshore oil and gas industries. *Journal of Applied Bacteriology Symposium Supplement.* 1985, pp. 227S–235S.
14. Widdel, F. and Pfennig, N. A new anaerobic, sporing, acetate-oxidising, sulphate-reducing bacterium, *Desulfotomaculum* (emend.) *acetoxidans. Archives of Microbiology.* Vol. 112, 1977, pp. 119–122.
15. Klemps, R., Cypionka, H., Widdel, F. and Pfennig, N. Growth with hydrogen, and further physiological characteristics of *Desulfotomaculum* species. *Archives of Microbiology.* Vol. 143, 1985, pp. 203–208.
16. Sorensen, J., Christensen, D. and Jorgensen, B.B. Volatile fatty acids and hydrogen as substrate for sulphate-reducing bacteria in anaerobic marine sediment. *Applied and Environmental Microbiology.* Vol. 42, 1981, pp. 5–11.
17. Widdel, F. and Pfennig, N. Studies on dissimilatory sulphate-reducing bacteria that decompose fatty acids I. Isolation of new sulphate-reducing bacteria enriched with acetate from saline environments. Description of *Desulfobacter postgatei* gen. nov., sp. nov. *Archives of Microbiology.* Vol. 129, 1981, pp. 395–400.
18. Stott, J.F.D. and Herbert, B.N. The effect of pressure and temperature on sulphate-reducing bacteria and the action of biocides in oilfield water injection systems. *Journal of Applied Bacteriology.* Vol. 60, 1986, pp. 57–66.

19. Ollivier, B., Caumette, P., Garcia, J. and Mah, R.A. Anaerobic Bacteria from Hypersaline Environments. *Microbiological Reviews.* Vol. 58, 1994, pp. 27–38.
20. Ivarson, K.C., Hallberg, R.O. and Wadsten, T. The pyritization of basic ferric sulphates in acid sulphate soils: a microbiological interpretation. *Canadian Journal of Soil Science.* Vol. 56, 1976, pp. 393–406.
21. *Biogeochemical Cycling of Mineral-Forming Elements.* Trudinger, P.A., Swaine, D.J. Published by Elsevier. 1979, pp. 401–430.
22. *Microbial Biogeochemistry.* James E. Zajic. Published by Academic Press, 1969, pp. 87–95.
23. Howarth, R.W. Pyrite: its rapid formation in a salt marsh and its importance in ecosystem metabolism. *Science.* Vol. 203, 1979, pp. 49–51.
24. Brown, D.E., Groves, G.R. and Miller, J.D.A. pH and Eh control of cultures of sulphate-reducing bacteria. *Journal of Applied Chemical Biotechnology.* Vol. 23, 1973, pp. 141–149.
25. *Bergey's Manual of Systematic Bacteriology. Volume 1.* Krieg, N.R. and Holt, J.G. Published by Williams & Wilkins. 1984, pp. 663–679.
26. Zeikus, J.G., Dawson, M.A., Thompson, T.E., Ingvorsen, K. and Hatchikian, E.C. Microbial ecology of volcanic sulphidogenesis: isolation and characterization of *Thermodesulfobacterium commune* gen. nov. and sp. nov. *Journal of General Microbiology.* Vol. 129, 1983, pp. 1159–1169.
27. Weimer, P.J. and Zeikus, J.G. Acetate metabolism in *Methanosarcina barkeri. Archives of Microbiology.* Vol. 119, 1978, pp. 175–182.
28. Le Gall, J. and Postgate, J.R. The physiology of sulphate-reducing bacteria. *Advances in Microbiological Physiology.* Vol. 10, 1973, pp. 81–133.
29. Bryant, M.P., Campbell, L.L., Reddy, C.A. and Crabill, M.R. Growth of desulfovibrio in lactate or ethanol media low in sulphate in association with H_2-utilising methanogenic bacteria. *Applied and Environmental Microbiology.* Vol. 33, 1977, pp. 1162–1169.
30. Nanninga, H.J. and Gottschal, J.C. Properties of *Desulfovibrio carbinolicus* sp. nov. and other sulphate-reducing bacteria isolated from an anaerobic-purification plant. *Applied and Environmental Microbiology.* Vol. 53, 1987, pp. 802–809.
31. Isa, Z., Grusenmeyer, S. and Verstraete, W. Sulphate reduction relative to methane production in high-rate anaerobic digestion: technical aspects. *Applied and Environmental Microbiology.* Vol. 51, 1986, pp. 572–579.
32. Laanbroek, H.J., Abee, T. and Voogd, I.L. Alcohol conversions by *Desulfobulbus propionicus Lindhorst* in the presence and absence of sulphate and hydrogen. *Archives of Microbiology.* Vol. 33, 1982, pp. 178–184.
33. Brandis, A. and Thauler, R.K. Growth of *Desulfovibrio* species on hydrogen and sulphate as sole energy source. *Journal of General Microbiology.* Vol. 126, 1981, pp. 249–252.

34. Badziong, W., Thauer, R.K. and Zeikus, J.G. Isolation and characterization of *Desulfovibrio* growing on hydrogen plus sulphate as the sole energy source. *Archives of Microbiology.* Vol. 116, 1978, pp. 41–49.
35. Jenkins, R.L. Scheybeler, B.J., Smith, M.L., Baird, R., Lo, M.P. and Haug, R.T. Metals removal and recovery from municipal sludge. *Journal of Water Pollution Control Federation.* Vol. 53, 1981, pp. 25–32.
36. Maree, J.P. and Strybom, W.F. Biological sulphate removal from industrial effluent in an upflow packed bed reactor. *Water Research.* Vol. 21, 1987, pp. 141–146.
37. Maree, J.P. and Strybom, W.F. Biological sulphate removal in an upflow packed bed reactor. *Water Research.* Vol. 9, 1985, pp. 1101–1106.
38. Maree, J.P., Geber, A. and Hill, E. An integrated process for biological treatment of sulphate-containing industrial effluents. *Journal of Water Pollution Control Federation.* Vol. 59, 1987, pp. 1069–1074.
39. *Chemical Marketing Reporter.* June 1994.
40. Bull, M.A., Sterritt, R.M. and Lester, J.N. An evaluation of four start-up regimes for anaerobic fluidized bed reactors. *Biotechnology Letters.* Vol. 5, 1983, pp. 333–338.
41. Smith, M.R. and Mah, R.A. Growth and methanogenesis by *Methanosarcina* Strain 227 on acetate and methanol. *Applied and Environmental Microbiology.* Vol. 36, 1978, pp. 870–879.
42. Sorensen, J. Reduction of ferric iron in anaerobic, marine sediment and interaction with reduction of nitrate and sulphate. *Applied and Environmental Microbiology.* Vol. 43, 1982, pp. 319–324.
43. van der Hoek, J.P., Latour, P.J.M. and Klapwijk, A. Effect of hydrolic residence time on microbial sulphide production in an upflow sludge blanket denitrification reactor fed with methanol. *Applied Microbiology and Technology.* Vol. 28, 1988, pp. 493–499.
44. Seitx, H. and Cypionka, H. Chemolithotrophic growth of *Desulfovibrio desulfuricans* with hydrogen coupled to ammonification of nitrate or nitrite. *Archives of Microbiology.* Vol. 146, 1986, pp. 63–67.
45. Ingvorsen, K. and Jorgensen, B.B. Kinetics of sulphate uptake by freshwater and marine species of *Desulfovibrio. Archives of Microbiology.* Vol. 139, 1984, pp. 61–66.
46. Jorgensen, B.B. Mineralization of organic matter in the sea bed—the role of sulphate reduction. *Nature.* Vol. 296, 1982, pp. 643–645.
47. *Atlas of Electrochemical Equilibria in Aqueous Solution.* Marcel Pourbaix. Published by National Association of Corrosion Engineers, Houston, USA.
48. Lettinga, G., van Velsen, A.F.M., Hobma, S.W., de Zeeuw, W. and Klapwijk, A. Use of the upflow sludge blanket (USB) reactor concept for biological waste water treatment, especially for anaerobic treatment. *Biotechnology and Bioengineering.* Vol. 22, 1987, pp. 699–734.

49. Maat, D.Z. and Habets, L.H.A. The upflow anaerobic sludge blanket wastewater treatment system: a technological review. *Pulp and Paper Canada 88*. Vol. 11, 1987, pp. T410–T414.
50. Maree, J.P. and Hill, E. Biological removal of sulphate from industrial effluents and concomitant production of sulphur. *Water Science Technology*. Vol. 21, 1989, pp. 265–276.
51. Cork, D.J. Microbial conversion of sulphate to sulphur—an alternative to gypsum synthesis. *Advances in Biotechnological Processes*. Vol. 4, 1985, pp. 183–209.
52. Buisman, C.J.N., Post, R., Ijspeert, P., Geraats, S. and Lettinga, G. Biotechnological process for sulphide removal with sulphur reclamation. *Acta Biotechnology*. Vol. 9, 1989, pp. 271–283.
53. Buisman, C.J.N., Geraats, B.G., Ijspeert, P. and Lettinga, G. Optimization of sulphur production in a biotechnological sulphide-removing reactor. *Biotechnology and Bioengineering*. Vol. 35, 1990, pp. 50–56.
54. Buisman, C.J.N., Stams, A.J.M., Meijer, H. and Lettinga. G. Sulphur and sulphate reduction with acetate and propionate in an aerobic process for sulphide removal. *Applied Microbiology and Biotechnology*. Vol. 32, 1989, pp. 363–372.

Biotreatment of Aqueous Wastes from Pesticide Manufacture Using Sequencing Batch Reactors (SBR) with PAC

ARTHUR BINDER, LLOYD C. EDMUNDS AND DONALD R. WILCOX
Zeneca Inc. Western Research Center
P.O. Box 4023
1200 S. 47th Street
Richmond, CA 94804-0023

KARL A. WILKS
Alfa Tech Consulting Engineers
120 Montgomery Street, Suite 715
San Francisco, CA 94104

INTRODUCTION

Treatability studies were undertaken to determine the feasibility and subsequent design of a wastewater treatment process for pesticide manufacturing wastes. The wastewater treatment process consisted of pretreatment of the individual wastestreams, primary treatment of the combined streams, and secondary treatment of those wastewaters. This chapter presents a portion of the laboratory treatability results and the performance of the full-scale plant.

At the plant site, 15 technical herbicides and insecticides are produced in nine production units. The production units are a combination of continuous and batch operations with the products produced in seasonal campaigns. Consequently, the production site aqueous wastes are highly variable with respect to flow, contaminants, and contaminant concentration. The aqueous waste, which is a commingling of the wastes from the individual production units, is a brine of 12 to 18% salinity, with fluctuations as low as 8 to 10%. In order to facilitate biological treatment, it is necessary to reduce the salt concentration to about 3%. The principal sources of water for salinity reduction are stormwater from process and nonprocess areas, noncontact process cooling water, boiler blowdown, laundry, and treated sewage. Supplemental water is supplied from on-site wells, as necessary.

Process development studies were conducted to determine if biological treatment was a cost-effective and technically feasible means of removing or reducing conventional pollutants (BOD, TSS, and TOC), specific compounds (pesticides, cyanide, benzene, and toluene), and effluent toxicity. Initially, treatability studies, which demonstrated feasibility, were also used to compare the performance of continuous versus batch biological treatment. Other aspects of the design, such as the need for PAC addition, primary treatment unit operations, and the specification of the operating parameters, were then determined.

The final overall treatment scheme consisted of pretreatments for the individual wastestreams, primary treatment of the combined streams, and secondary (biological) treatment. There are several unit operations contained in the primary and secondary treatment areas. Primary treatment consists of waste brine collection, pH adjustment, organics and solids removal, equalization, and salinity adjustment. Secondary treatment consists of biological treatment with powdered activated carbon (PAC) in sequencing batch reactors (SBR), pH adjustment, sand filtration, and discharge to the river through an effluent diffuser. The waste sludge is concentrated and dewatered for off-site disposal. Figure 1 shows a block flow diagram of the wastewater treatment plant (WWTP) unit operations.

LABORATORY TREATABILITY STUDIES

Process Selection

Several options were considered for the treatment of the production site waste brines, e.g., brine electrolysis, incineration, biological treatment, and off-site disposal. Evaluation of these processes showed that biological treatment was a cost-effective and technically feasible means of removing or reducing conventional pollutants (BOD, TSS, and COD), specific compounds (pesticides, cyanide, benzene, and toluene), and effluent toxicity.

During this screening phase, process development studies were conducted to determine the effectiveness of different primary and secondary treatments of the waste brines. From these studies, a portion of which are not presented here, it became apparent that primary treatment to remove separate phase organics and solids was essential to consistently achieve the necessary reductions in toxicity. A primary treatment step was then included in all further treatability studies.

Treatability studies were carried out to determine the effect of salt concentration on the treatability of these wastes. There was no inhibition

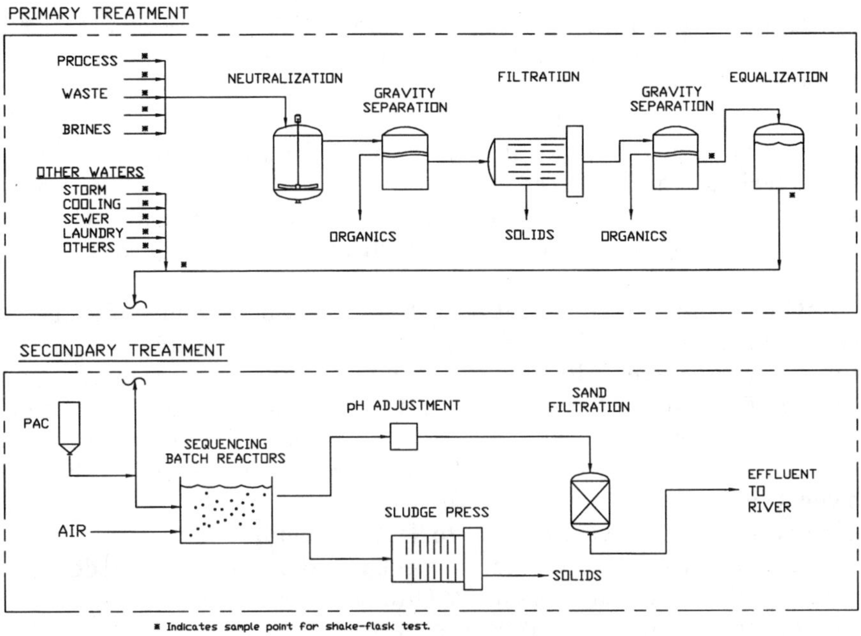

Figure 1. Wastewater treatment plant.

to TOC reduction for salt concentration below 4%. A concentration of 3% salt was chosen for the remaining studies. A 3% salt concentration provides a safety margin while minimizing the volume of the treatment equipment. See Section 6.1 for the treatability data with regard to salt inhibition.

Screening studies were also used to compare the performance of continuous versus batch biological treatment. Both processes showed similar removal efficiencies for conventional pollutants, specific compounds, and toxicity. SBRs were chosen for the biological treatment step primarily due to the need for flexibility, since there are wide fluctuations in flow and load of the wastewater. It was also felt that SBRs had a slight advantage in this situation due to their better sludge settling characteristics under low F/M (food to biomass) conditions (i.e., $F/M \leq 0.1$ mg BOD/mg biomass/day) and to possible kinetic advantages (Irvine et al., 1989). The process variables were optimized in subsequent studies.

Selection of PAC

A number of performance advantages have been reported in the literature for the PACT® process (Flynn, 1975; Schultz et al., 1984; Thibault et al., 1977). The PACT® process combines biological treatment with powdered activated carbon (PAC) treatment. The reported benefits include improved removal of organics, improved resistance to organic shock loads, lower effluent toxicity, improved sludge settling characteristics, and reduced emissions of volatile compounds. In some instances (Sublette et al., 1982), an apparent synergistic effect between PAC and biological treatment was observed with regard to TOC reduction.

PAC treatment in combination with biotreatment was evaluated to determine the potential benefits of this process for treating the production site wastewaters. Results of these studies are in agreement with several of the above listed benefits. Treatability studies showed that the combination of biological and carbon treatment resulted in an effluent with a lower toxicity, little or no color, low or no odor, and a slightly lower TOC. In some tests on individual wastestreams, a synergistic effect was observed with respect to TOC reduction.

In the first study, an evaluation was carried out in side-by-side laboratory SBRs with and without PAC to compare its effect on effluent toxicity, color, odor, and TOC removal. The biotreaters were fed a composite waste mixture A from several individual production units. During the initial 55 days, the PAC biotreater was dosed with 0.4 g of PAC per liter of feed. Thereafter, the dose was increased to 0.8 g per liter of feed.

The study showed that the presence of PAC significantly reduced the toxicity. Figure 2 shows the effluent toxicity from identical biotreaters with and without PAC. The effluent of the PAC biotreater had a toxicity of 13 toxic units (TUs), compared to 26 TUs in the biotreater without PAC. TUs are defined as 100%/EC_{50}. The EC_{50} (Effect Concentration) is the percentage of the effluent sample in the standardized test sample that kills or "effects" 50% of the test organisms. Higher TUs indicate that the sample is more toxic. The increase of the PAC dose from 0.4 g/L to 0.8 g/L did not result in further toxicity reduction, indicating that all the carbon adsorbable toxic compounds had been removed at the lower PAC dose. The presence of residual toxicity indicated that the effluent contained other nonadsorbable toxic compounds.

The effluent color was significantly reduced by the PAC treatment. The composite waste mixture A exhibited a yellow color. The effluent of the biotreater without PAC was still yellow, indicating that the compounds causing the color were not removed. In the biotreater with PAC, the effluent was essentially colorless, presumably because the compounds causing color were adsorbed on the carbon (Figure 3).

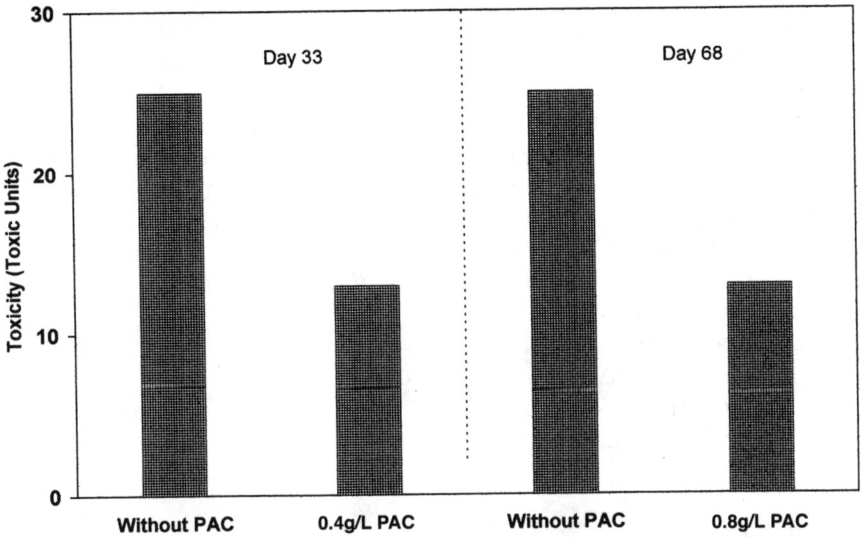

Figure 2. Toxicity of SRB effluent to mysid shrimp. Feed was waste composite "A." PAC dose was 0.4 mg/L of feed for days 1–55 and 0.8 mg/L after day 55.

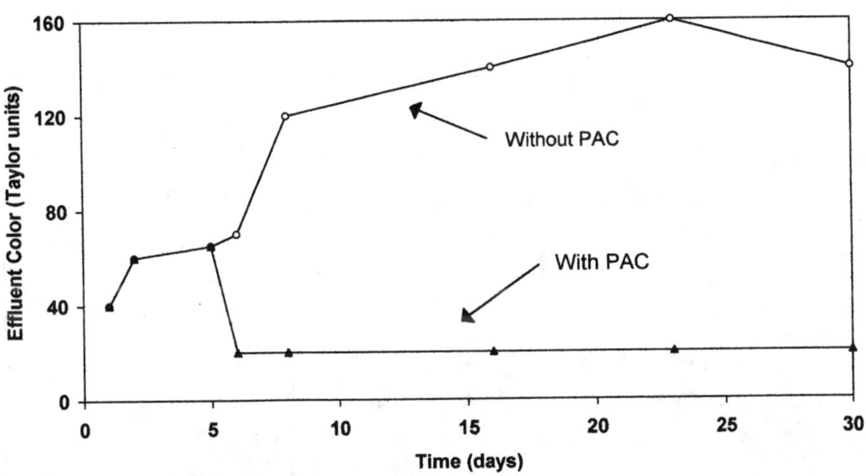

Figure 3. Color of SRB effluent with and without PAC. Feed was waste composite "A." PAC dose was 0.4 mg/L of feed starting on day 1.

The odor from the PAC and non-PAC biotreaters was compared. The pesticide brines have a strong chemical odor, which was evident in the biotreater without PAC. However, the biotreater with PAC had an acceptable odor similar to PAC sludge treatment of nonodorous feeds.

Both the PAC and non-PAC reactors effectively reduced the TOC. However, the PAC biotreater appeared to exhibit a slightly lower effluent TOC compared to the biotreater without PAC. Figure 4 shows the feed and effluent TOC profiles for this 85-day test. The PAC dose increase from 0.4 g/L to 0.8 g/L at Day 55 had no effect on the overall TOC reduction.

The overall rate of TOC removal was comparable in the biotreaters with and without PAC. Figure 5 shows React cycle TOC profiles for these biotreaters on days 28 and 57. The data were fitted to a first order degradation rate equation:

$$\frac{dc}{dt} = -kmc$$

where c is the biodegradable TOC in milligrams per liter, t is the reaction time in hours, m is the biomass concentration in grams per liter, and k is the TOC degradation rate constant in liters per gram per hour. The solid lines were generated by fitting the data to the above

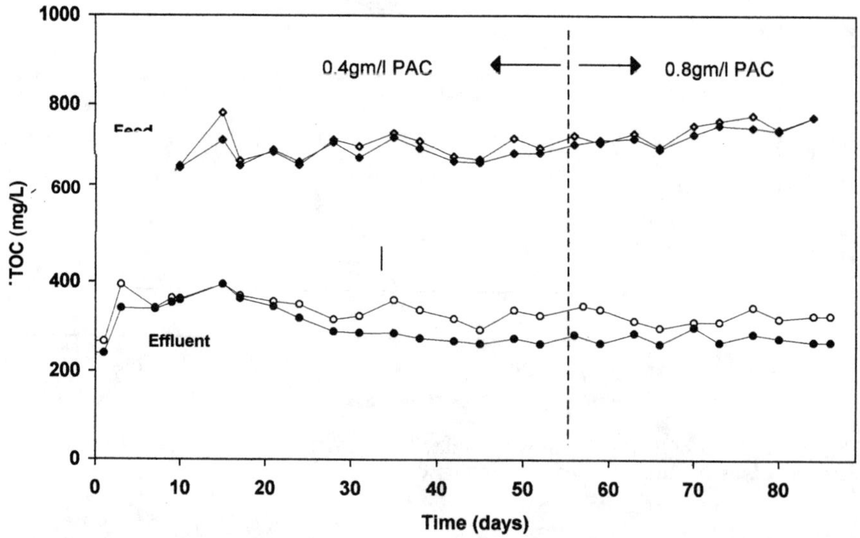

Figure 4. TOC profiles for SBR biotreaters with and without PAC. PAC biotreater is closed symbols. Feed was waste Composite "A." PAC dose was in gm/L of feed.

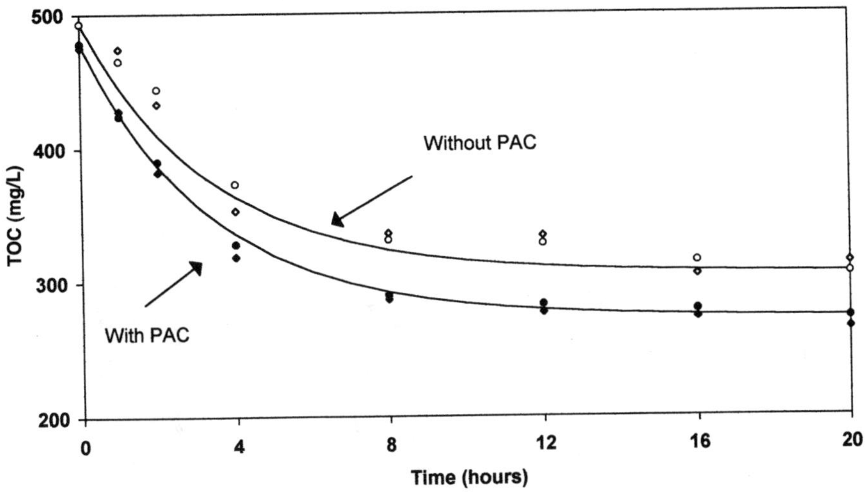

Figure 5. TOC profiles during react cycle. Feed was waste Composite "A." PAC dose was 0.4 mg/L of feed for days 1–55 and 0.8 mg/L after day 55. TOC profiles taken at days 28 (open circle), 29 (closed circle), 58 (closed diamond), 63 (open diamond).

rate equation using linear regression analysis. The estimated overall TOC degradation rate constant was 0.11 L/g/h for both biotreaters.

In another treatability study, it was shown that for some wastestreams, the combination of biotreatment and carbon treatment had a synergistic effect on TOC reduction. This study was done in side-by-side SBR biotreaters with and without PAC. The average TOC reductions were 57% and 24% for the biotreaters with and without PAC, respectively (see Figure 6). Assuming that the effects of biotreatment and carbon treatment are additive, a theoretical TOC reduction was calculated. For this calculation, it was assumed that the effluent from the biotreater without PAC was treated in a separate reactor with the same dose of PAC that was added to the SBR with PAC. The TOC removal accomplished by the carbon alone was estimated using the isotherm in Figure 7. The theoretical TOC reduction for the additive effects of PAC and biotreatment was 33%. This is significantly lower than the actually observed TOC reduction of 57%. Therefore, a synergistic effect was obtained when carbon treatment was combined with biotreatment.

Process Design Studies

Laboratory treatability studies were conducted to generate the design information for the full-scale treatment plant. Daily grab samples were

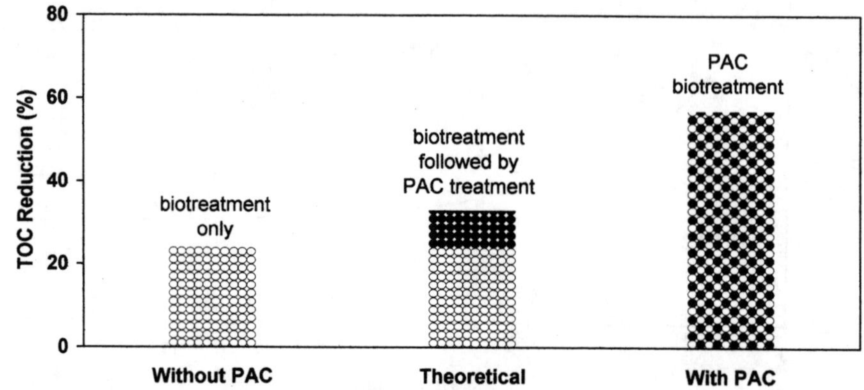

Figure 6. Comparison of TOC reduction. Feed was waste Composite "B." Theoretical TOC reduction calculation was based on the isotherm in Figure 7, a PAC dose of 0.28 g/L of feed, and an effluent TOC of 45 mg/L for a biotreater without PAC.

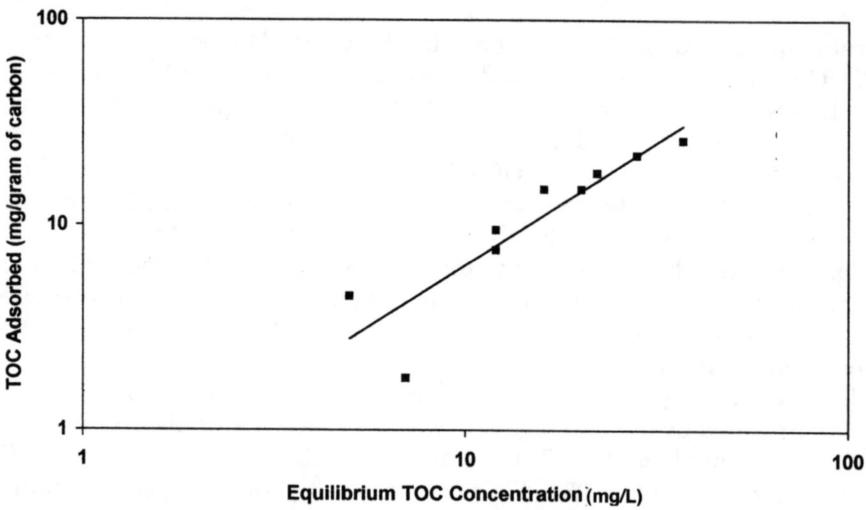

Figure 7. Carbon adsorption isotherm for effluent from biotreatment of waste composite "B."

taken at the production site from tanks containing the newly commingled wastes from the individual production units. Samples were taken prior to and after a primary treatment consisting of an organic phase separation and a solids filtration followed by a second organic phase separation. The samples were transferred in the laboratory into carboys that simulated equalization tanks with a 4-day hydraulic retention time (HRT). The equalized waste was introduced to biotreaters that operated at 1-, 2-, or 4-day hydraulic retention time. The biotreaters were operated at oxygen transfer rates (OTR) of 50 to 100 mg/L/h, which were equivalent to those in conventional commercial equipment. This was done by adjusting the aeration and agitation rates such that the OTR capacity was in the correct range. With this approach, the HRT required to achieve the treatment objectives for a given organic loading was the same in the laboratory and in the full-scale treatment plant.

Early in the treatability studies, it was determined that primary treatment of the waste brines was necessary to achieve the required effluent quality. Samples of the combined waste brines collected prior to primary treatment contained varying amounts of dispersed organics. Biotreatment tests were conducted with both untreated and primary treated waste. Samples receiving a primary treatment generally exhibited lower pesticide levels before and after biotreatment (see Table 1) and were consistently less toxic (see Figure 8). It was concluded that

Table 1. Effect of HRT on pesticide removal.

	Primary Treated Waste Brine			Untreated Waste Brine
	1-day HRT	2-day HRT	4-day HRT	4-day HRT
Herb. C, mg/L				
Feed	8.5	8.5	8.5	35.3
Effluent	0.5	0.4	0.2	1.3
Herb. D, mg/L				
Feed	33	33	33	47
Effluent	20	17	10	17
Insect. D, μg/L				
Feed	4070	4070	4070	26,200
Effluent	10	21	24	1200
Efflu. susp. solids (mg/L)	47	73	93	65

a primary treatment to remove separable organics and solids was necessary to achieve a high-quality effluent.

Biotreatment tests were conducted with the primary treated samples to evaluate the effect of HRT and *F/M* ratio on the effluent quality. Figure 9 shows the results of biotreaters operated at 1-, 2-, and 4-day HRTs with *F/M* ratios ranging from less than 0.1 to 1.0. These tests showed that, for a wide range of waste mixtures, a 4-day HRT and an *F/M* loading of less than 0.2 mg BOD per milligram biomass per day resulted in consistent BOD removal. The SBR biotreaters showed excellent sludge settling under the range of operating conditions used in the study.

The effect of HRT on pesticide removal was studied. Table 1 presents sample data of biotreater feed and effluent concentrations for two herbicides and an insecticide. In general, the concentration of residual pesticides in the effluent decreased with increasing HRT in the range studied. However, the insecticide did not follow this general trend. It is believed that the concentration of the insecticide in the biotreater effluent depended upon the effluent suspended solids (ESS) concentration, since this insecticide tended to adhere to solids. The average ESS from the biotreater with a 4-day HRT was 93 mg/L, compared to 47 mg/L for the biotreater with a 1-day HRT. This may explain the higher insecticide concentration in the effluent of the biotreater with the 4-day HRT. Other specific compounds, such as toluene, were removed, presumably by biodegradation and partially by air stripping.

Treatment at the 4-day HRT reduced the effluent toxicity. An example of typical toxicity data is shown in Figure 8. The effluent from the biotreater with a 4-day HRT is much less toxic than the effluents from the biotreaters with a 1- or 2-day HRT. A possible explanation for this observation is that some of the toxic compounds are slowly biodegradable. The longer reaction time is needed to reduce the concentration of these compounds below the toxicity threshold.

Acclimation and "Washout" Tests

Two additional laboratory studies were performed to support the operation of the full-scale plant. In one study, the acclimation of the seed sludge to salt (NaCl) was studied in order to maintain optimum treatment performance while achieving the fastest acclimation time. This "fast acclimation" strategy using a seed of domestic sludge had two uses: (1) startup of the full-scale treatment plant and (2) restart of a reactor after discarding sludge that was contaminated with separate phase organics. In the second study, the effect of liquid turnover or "washing" of the sludge was studied to observe the effect on toxicity

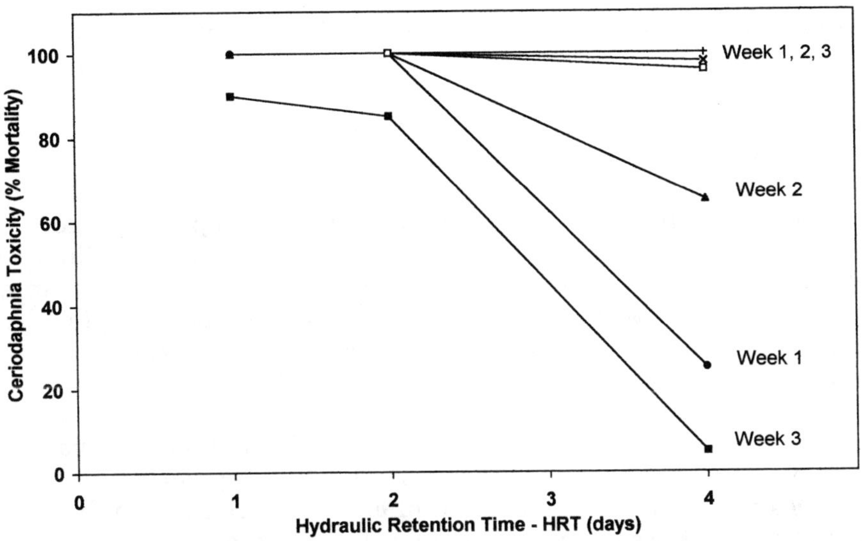

Figure 8. Effect of HRT and primary treatment on toxicity. Reactor feeds were daily grab samples from the site, which were either untreated (open symbols) or primary treated (closed symbols).

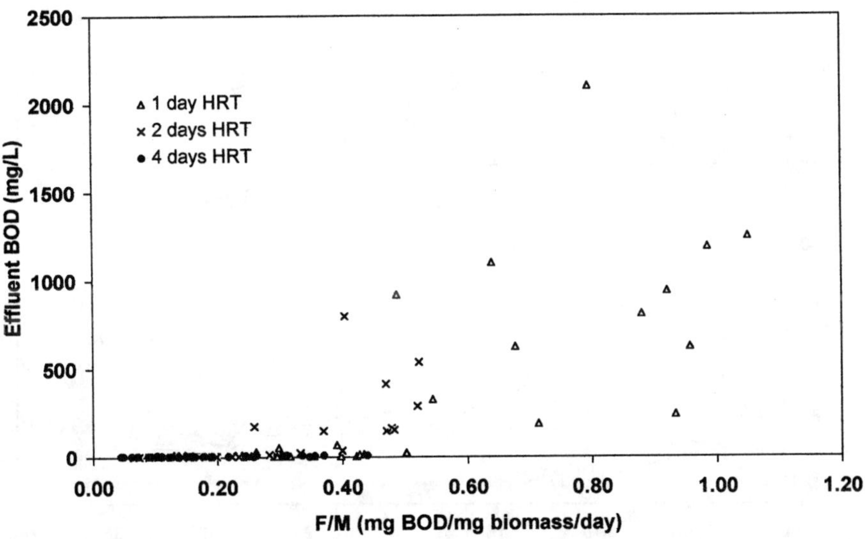

Figure 9. Effect of organic loading on BOD removal. Reactor feeds were daily grab samples from the site.

and on subsequent treatment performance. This "washout" test became an assessment tool to judge if a biotreater with a toxic effluent could be quickly recovered by continuing operation with a nontoxic feed. Alternately, if the sludge had to be discarded, then the reactor could be restarted using the "fast acclimation" strategy.

The effect of different salt acclimation rates was studied to find the fastest acclimation rate while maintaining treatment performance. The salt acclimation rates in the biotreaters were 2, 4, and 8 g per liter per day for a 2-day HRT. With the most aggressive scenario, the biotreaters treated the full hydraulic load on the third day of operation. Full hydraulic loading was achieved before the salt concentration reached 3%, the steady-state salt concentration (see Table 2).

Treatment performance was judged based on TOC reduction, i.e., the viability of the biomass. The effluent TOC profiles for the various acclimation strategies are shown in Figure 10. Surprisingly, the TOC profiles were very similar for all the different scenarios, and the biomass acclimated to the salt environment much faster than was expected. In all of the tests conducted without PAC, the effluent TOC during the first 10 days exceeded the TOC of the fully acclimated biotreater. In contrast, the biotreater that was spiked initially with 1000 mg/L of PAC showed TOC reductions equivalent to the fully acclimated biotreater. It

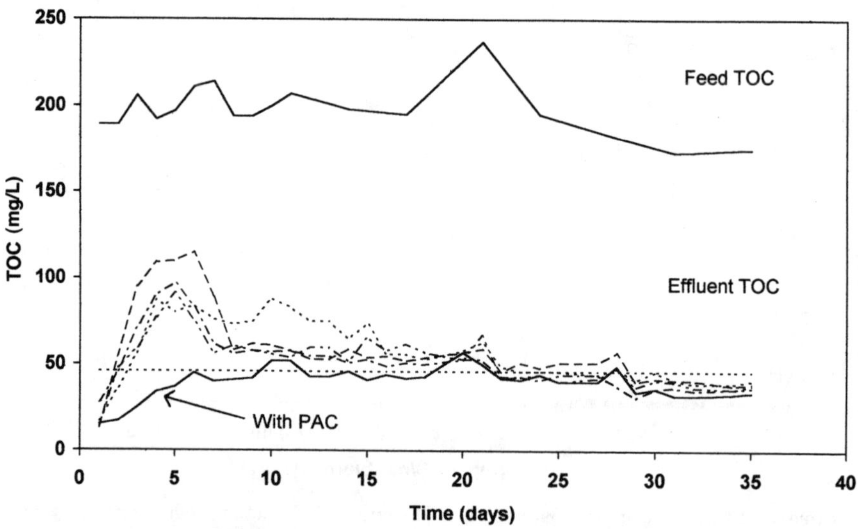

Figure 10. Effect of salt acclimation rate on effluent TOC. (– – – – 2 g/L/d; — — — 4 gm/L/d; ——— 4 gm/L/d with 1000 mg/L PAC; – – – 8 gm/L/d with one cycle/d; — — — 8 gm/L/d with two cycles/d; horizontal dashed line is fully acclimated).

Table 2. Effect of salt acclimation rate on time to full loading.

Salt Acclimation Rate (g/L/day)	Time to Full Hydraulic Loading (Days)	Time to Reach 3% Salt (Days)
2	14	16
4	7	11
8	3	10

appeared that the PAC in this biotreater acted as a sink for TOC that was not degraded during the first few days of the test. In addition, the TOC reduction in the PAC reactor appeared to be slightly better than those reactors without PAC.

In the second study, a simple test was developed to determine whether the contaminated sludge could be washed to quickly recover its normal treatment performance with respect to TOC, BOD, and toxicity reduction. Toxicity was of particular interest since these wastes generally did not show inhibition effects from the wastes. If toxicity were observed in a biotreater, this simple test could determine if the toxicity was from a "dissolved" species that could be "washed" from the biotreater or from contaminants adhering to the biomass, such as an organic phase. If the toxicity were dissolved in the aqueous phase, the test could be used to estimate the recovery time.

For this test, the biomass in a biotreater exhibiting toxicity was subjected to a series of quick feed, mix, settle, and decant cycles in the laboratory. The feed was a 3% salt solution which was added all at once. The volume was proportional to the actual biotreater feed volume to maintain the same HRT. The mix and settling times were 1 h each. In this case the sludge was not being fed during the washing cycles, which took less than a day. In parallel, the same sludge was subjected to regular biotreatment cycles with a nontoxic feed. These sludges needed a food source since the study took several weeks. Decant samples from both sets of tests were then tested for toxicity.

Figure 11 summarizes the results of these tests. The triangles show the toxicity of the decant samples from the "salt wash" test. The circles show the toxicity profile in the effluent from the biotreater with regular SBR cycles and the nontoxic feed. The dashed line shows the theoretical toxicity based on washing out the toxicant, assuming that the toxic compound is completely dissolved in the liquid phase. There was good agreement between the three sets of data. Since the theoretical, salt wash, and the actual toxicity profiles were very similar, it was concluded that the toxic compounds were completely dissolved in the aqueous phase and not held up in the sludge.

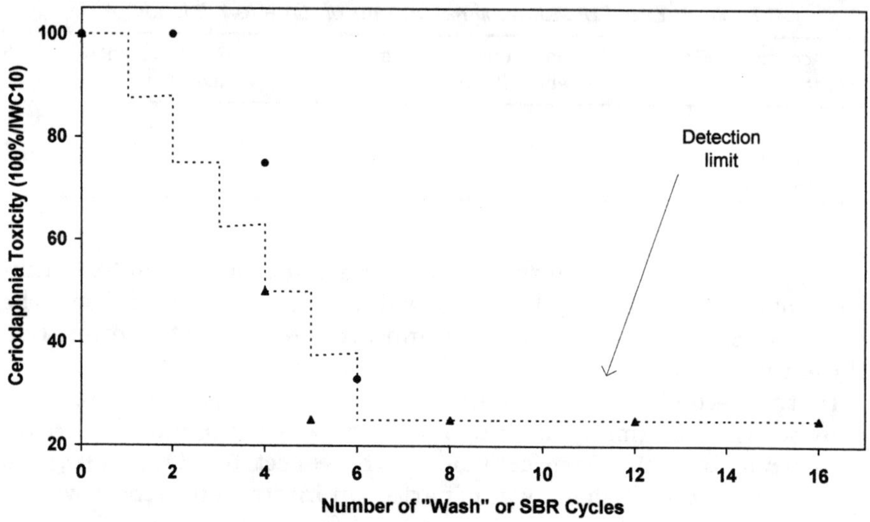

Figure 11. Effluent toxicity as a function of treatment cycles [3% "salt wash" cycles (triangle); biotreatnent cycles (circle); theoretical profile (dashed line) is based on toxicant being fully mixed in the liquid phase].

It was concluded that if toxicity were observed in the full-scale reactors, a "wash test" in the laboratory could be used to estimate the recovery time of the sludge in a full-scale plant. For example, in the laboratory, seven "wash" cycles can be conducted in 1 day. This simulates 1 week of actual biotreatment time. Two days would then be required for ceriodaphnia testing. Therefore, in less than 1 week in the laboratory, it can be determined whether the best recovery strategy for the full-scale reactors is to restart with fresh sludge or to quickly cycle the biotreaters with nontoxic feed.

This type of scenario was encountered during the operation of a 500-gal pilot plant biotreater. A separate organic phase, which contained a toxic compound with a low solubility in water, entered the biotreater as a result of an upset condition. In this case, the organic phase absorbed on the biomass. Therefore, the organic phase residence time in the biotreater was the same as that of the biomass. As a result, on a daily basis, the toxicant leached from the organic phase on the sludge into the bulk liquid phase. Effluent toxicity continued for several months. Use of a "washout" test at that time would have changed the strategy from sludge recovery to discarding the contaminated sludge and restarting with fresh seed.

PRETREATMENT IN THE PRODUCTION UNITS

Some of the waste brines from the individual production units had high volumes, were amenable to biotreatment, and yielded a nontoxic effluent. Other streams were not easily biotreatable and showed toxic effluents when treated in the laboratory. Laboratory and pilot plant studies demonstrated that the most critical parameter for successful operation of the WWTP was not BOD_5 removal, but toxicity reduction. The operational philosophy focused upon reducing toxicity at the source by removing nonbiodegradable and nonadsorbable toxic compounds in the waste brines. Those waste brines were pretreated in the individual production units before being transferred to the WWTP. In a few cases where pretreatments were too expensive or ineffective, waste brines were segregated for off-site disposal.

The waste brine from each production unit was generally a composite of as many as eight individual wastes from different reaction steps. If a waste brine from a production unit exhibited toxicity in the laboratory tests, the waste from each individual was characterized and tested to determine if that step was the source of toxicity. Pretreatment techniques were then investigated to reduce or eliminate the toxic compound(s). For example, biotreatment of the insecticide C waste brine resulted in a toxic effluent, but further investigation revealed that only one of the eight individual steps was responsible for the toxicity. That stream was isolated from the others for off-site disposal. A cost-effective pretreatment is being developed to remove or significantly reduce the toxic compounds.

Pretreatment in the production units had benefits other than avoiding off-site disposal costs. Pretreatment employing recovery and recycling of products, raw materials, intermediates, and solvents improved product yields. Another cost-reducing strategy was to introduce process improvements that minimized the amount of wastes being generated that were sent off-site. A key benefit of pretreatment was that improved waste brine "quality" resulted in higher process reliability in the treatment plant.

Previously, these wastestreams were viewed as "wastes for disposal." With biological treatment, they were then considered "products to be delivered" to the WWTP. As such, it was imperative that the effectiveness of pretreatment in the production units was measurable. Therefore, specifications were established for each of the waste brines originating in the individual production units. All waste brines were sampled twice per day. The brines were not to contain free organics or visible suspended solids. Certain brines had to satisfy pH ranges, density requirements, and/or meet maximum levels of specific toxic compounds. Table 3 shows the waste brine specifications for the individual production units. In all cases, it was critical that the pretreated waste brine met its specifications in the production unit before being transferred to the WWTP.

General Pretreatment Techniques

While pretreatments must be effective, they cannot be prohibitively expensive in terms of capital costs for new equipment or added labor. In most cases, the existing equipment in each production unit was utilized with minor additions and modifications.

Nearly all production processes (Table 3) used phase (gravity) separation techniques to recover products and/or raw materials. In many of the units, instrumentation was upgraded to improve control of phase separations and prevent phase inversions. Floating skimmers and organic detectors were installed to prevent the transfer of large quantities of organics to the WWTP during potential upset conditions.

One process under development for future production at that site has an acidic waste filtrate rich in dissolved triethylamine. Adjusting the filtrate to pH 12 allows the formation and separation of a triethylamine layer, which will be recycled back to the production process. As a result of this simple pretreatment step, the BOD load in the aqueous

Table 3. Production unit waste brine specifications.

Production Unit	Product	Specific Gravity	pH	Other
10	Insecticide A Herbicide A, B	>1.06	10–13	Insecticide A <25 ppb
30/230	Herbicides C, D, E, F, G, H	>1.06	4–10	
40	Herbicides I, J Intermediate B	>1.06	4–10	
50/150	Intermediate B	None	<3	Iodine <1000 ppm Iodide <1000 ppm
100	Insecticide B	>1.06	<3	Insecticide B <10 ppm[a]
160	Insecticide C Herbicide K	None	>7	Intermediate C <0.5 ppm
240	Insecticide D Intermediate C	>1.06	>12	Insecticide D <75 ppb

[a]This is a current guideline being proposed as a specification.
Note: Specifications of "No Free Organics" and "No Visible Suspended Solids" are in effect for all wastestreams.

phase is significantly reduced. Additionally, when an organic layer forms, it has the ability to extract organic compounds that may contribute to toxicity.

Specific Pretreatment Techniques

Solvent extraction of the waste brine to remove toxic compounds was an important pretreatment technique. Solvents were typically recycled back to the process for recovery and reuse. Solvent extraction was implemented in two production units. The insecticide A and insecticide D pretreatment processes were based on a toluene solvent extraction. These insecticides are extremely toxic to aquatic organisms such as *Ceriodaphnia dubia;* EC_{10} = 5–10 parts per trillion (Meier, 1993). It was necessary to pretreat these waste brines in the production units to satisfy the effluent toxicity requirement. The large volume of waste brine produced by these processes precluded off-site disposal as a cost-effective option.

The pretreatment scheme developed for these insecticides was a double extraction of the waste brine with the process solvent. This reduced the insecticide concentration by five orders of magnitude. The extraction solvent was then returned to the process. Additional equipment was required in the production unit necessitating a capital investment. A 1,500 gallon extraction vessel, a bag filter for the recycled solvent, and a 15,000 gallon waste brine check tank were installed. The recycled solvent extracts, which contained product, resulted in yield improvements of 4% for insecticide D and approximately 1% for insecticide A. The costs for the capital improvements were recovered in the first year of operation. In addition to removal in pretreatment, approximately 95% of the remaining insecticides were removed in primary treatment and overall removal including biotreatment was better than 99.8%.

Another pretreatment process recovered iodine, a potent toxicant, from the waste brine. In the manufacture of Intermediate A, an acidic waste stream was generated which contained about 30,000 mg/L of iodide. The iodide was oxidized to iodine with hydrogen peroxide, and the solid iodine was filtered and returned to the process. Toxicity studies demonstrated that both iodide and iodine were toxic to *Ceriodaphnia dubia* at relatively low concentrations. In fact, iodine (EC_{50} < 0.1 mg/L) was at least an order of magnitude more toxic than iodide (EC_{50} = 2.5 mg/L).

In the production plant, the filtrate from the iodine recovery process generally contained acceptable levels of iodide and iodine under normal operating conditions. The efficiency of the oxidation and recovery depended upon the pH, stoichiometry, and the concentration of the oxidant. To avoid upset conditions that would release iodine or iodide, analytical procedures were added to insure the operators hit the correct endpoint. A check tank was then used to insure that the iodine and iodide levels were within specification before release to the waste treatment area.

The fate of iodine and iodide in the waste treatment process was also studied. In the laboratory, iodine was removed during biotreatment by carbon adsorption and/or stripping during aeration. However, iodide passed through the biotreaters unaffected, due to its nonvolatile, ionic character. Figures 12 and 13 show the concentrations of iodine and iodide, respectively, in the laboratory biotreater feed and effluent.

The iodine/iodide analytical procedure (utilizing an ion-specific electrode) was sensitive enough to detect toxic levels of iodide in the biotreater effluent. However, that was not the case for iodine because it exhibited toxicity at levels below the detection limit. To overcome this limitation, specifications for iodine were set at the production unit where they could be measured prior to any significant commingling with the other waste brines.

A maximum level of 1000 mg/L was set for iodide and iodine in the waste brine leaving the production unit. This specification was based on the expected reduction in concentration from mixing with the other wastes as well as the removal efficiencies observed in the laboratory.

Wet air oxidation (WAO) is a very aggressive technique that can be used to pretreat waste brines that are not amenable to biotreatment. In this process, the waste brines are subjected to high pressures (800–2000 psig) and temperatures (450–550°F). At these conditions refractory organics are oxidized to less toxic and more readily biodegradable compounds. Laboratory and pilot plant tests confirmed that WAO was effective in reducing the TOC and the toxicity of several of the individual waste brines. WAO, followed by biotreatment, resulted in TOC reductions from 50–80% for several previously nonbiotreatable waste brines. The toxicity of these effluents was reduced to nontoxic levels similar to other biotreatable waste brines. The combination of WAO and biotreatment of the most toxic waste brines resulted in a nontoxic effluent suitable for discharge to surface waters. The major drawback of the WAO technology is cost. Currently, this pretreatment technique is not being utilized at the production site because it is more cost-effective to pretreat or segregate the highly toxic or nonbiotreatable wastes. Current research is focused upon the development of other more cost-effective techniques.

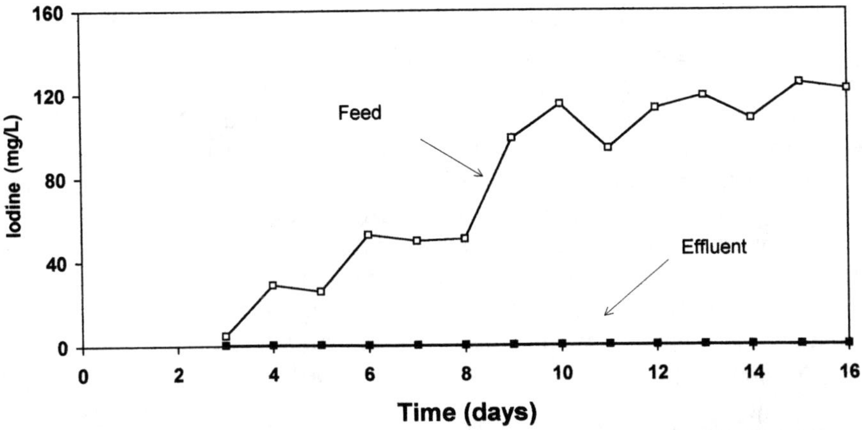

Figure 12. Concentration of iodine in biotreater feed and effluent. Iodine in feed was stepwise increased from 5 to 125 mg/L.

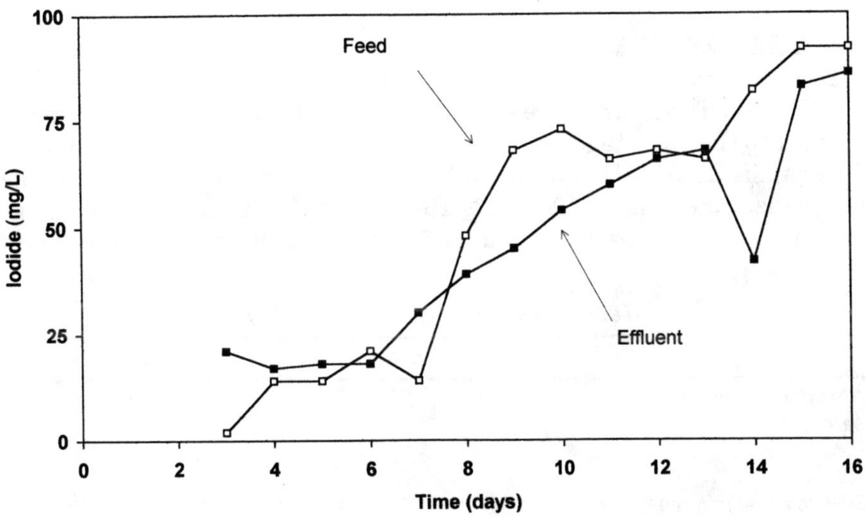

Figure 13. Concentration of iodide in biotreater feed and effluent.

PROCESS DESIGN

The process was designed using PAC biological treatment with sequencing batch reactors (SBRs). Treatability studies were carried out to define the process conditions that would maximize the removal of conventional pollutants, specific constituents, and effluent toxicity. PAC treatment, as part of biological treatment, was essential for reducing toxicity. Primary treatment of the waste brine to remove organics and solids was also essential to achieving reductions in toxicity.

The process can be more fully understood by studying the process block flow diagram (Figure 1) in conjunction with the description of the individual steps. The block flow diagram shows the layout of the primary and secondary treatment areas. Table 4 contains the overall design basis for the plant. A description of the individual process steps is as follows.

The wastewater, prior to treatment, is RCRA hazardous due to the presence of benzene and toluene. The WWTP was designed to RCRA standards for secondary containment and leak detection.

Primary Treatment

NEUTRALIZATION

Each production unit has an individual waste brine transfer line to the WWTP. These lines are combined in the feed header just prior to entry into the two-stage neutralization system. Provision to add either 50% caustic or 33% HCl for pH adjustment is provided. The first stage neutralization tank is 1900 gal (16-min residence time at 120 gpm), and the second stage is 900 gal (7.5-min residence time at 120 gpm).

Table 4. Design basis for the WWTP.

	Average	Design or Maximum
Waste brine flow (gpm)	71	120
Salinity (%)		
Waste brine	16	23
Wastewater	3	3
BOD removal (lb/day)	1,570	17,300
MLSS (mg/L)	6,000	20,000
PAC dose (mg/L)	75	1,000
PAC to MLSS ratio	1:1	2:1
Sludge yield (lb solids/1b BOD removed)	0.35	0.35

Each tank has an agitator. The waste brine pH can be adjusted to between 6.0 to 8.5 standard units. Typically, it is run at 7.0 to 7.5 at ±0.1 of the set point.

PRIMARY GRAVITY SEPARATION

The waste brine overflows the second stage neutralization tank into an organics separation tank, which holds approximately 30,000 gallons. The light organics are removed using floating skimmers. The heavy organics are gravity separated and removed. Both organic phases are sent off-site for disposal.

FILTRATION

The waste brine then enters one of two precoat pressure leaf filters. Each filter has approximately 105 sq ft of filtration area. The precoat filters are followed by bag filters to prevent any diatomaceous or precoat solids from being carried forward. The filter aid is diatomaceous earth. The spent filter cake is vibrated off of the screens, through a drop chute, and into an enclosed dumpster for off-site disposal. Vapors from the filter and dumpster are vented to an existing incinerator.

SECONDARY GRAVITY SEPARATION

The waste brine then undergoes a second gravity separation in a duplicate 30,000-gal tank from which the light and heavy organics separate and are similarly removed.

EQUALIZATION

Two waste brine equalization tanks are provided. The tanks are sized to provide 4 days of equalization for the site waste brines at the design flow rate of 120 gpm. They are operated at a level that provides 2 days of surge capacity based on 140% of design flow. The waste feed pumps are used to provide continuous back mixing of the contents of both equalization tanks. The tank contents are turned over three times per day. The net volume of the two tanks is 840,000 gal. The tanks are operated with 691,000 gal of equalization capacity and 147,000 gal of surge capacity. The tanks can be operated in series or parallel.

PROCESS AND OTHER WATERS

Stormwater, process water, cooling water, boiler blowdown, laundry water, treated sewage, well water (as needed), and other miscellaneous waters are collected in a lined basin. The process and other waters are combined with the waste brine in a mixing tee to produce wastewater with 3% salinity for biotreatment. The average water requirement is

355 gpm and the maximum (design) is 680 gpm. Typically, well water accounts for one-third of the water.

The dissolved solids level is continuously monitored downstream of the mixing tee. This signal is used in a feedback control loop to adjust the water flow to maintain the desired salinity going into biological treatment.

Secondary Treatment

PAC SYSTEM

Carbon arrives on-site in bulk trucks. A carbon storage silo is provided, which holds two truck loads of carbon (80,000 lb). The carbon is pneumatically transferred into the silo by blowers on the carbon trucks. A dust collection system is provided on the silo to control PAC emissions during transfer. The carbon from the silo is fed by gravity into an eductor in the wastewater line feeding the SBR biotreaters. The PAC system can deliver a maximum of 1000 mg/L of PAC to the wastewater feed.

SEQUENCING BATCH REACTORS (SBRs)

There are four SBRs, each with a 1.152 million-gal capacity. Each SBR can operate a minimum of one complete SBR cycle per day at the design flow rate. Currently, each cycle consists of a 6-h fill, 15.5-h react, 1-h settle, 1-h decant, and 0.5-h sludge draw-off. The total cycle time is 24 h.

The design flow is for a 4-day hydraulic retention time (HRT) using one cycle per day with a PAC dose of 50 to 300 mg/L of feed. Therefore, the design flow rate is 1.152 MGD of wastewater using four SBR basins. This provides a reasonable degree of flexibility in operation for reduced flows and fluctuating waste loads. At the design flow rate, one of the four SBRs is always in the 6-h fill cycle.

SAND FILTERS

Three multi-media filters are provided to remove suspended solids (biomass and PAC) from the effluent. These units are downflow in filter service and upflow in the backwash mode. The design is based upon a conservative filter loading of 2 lb of suspended solids per square foot of filter area and a 3–7 gpm/sq ft of filtration rate. Each filter is 10 ft in diameter. Typically, two filters will be on-line and the third will be a stand-by. The design sand filter performance is 30 ppm TSS in the filtered effluent.

The filtration media is 9–12″ of 0.4–0.5 mm silica sand and 24″ of 1.0–1.2 mm anthracite coal. A backwash rate of 20 gpm/sq ft is used, which provides for 100% bed expansion. Operation of the filter is at constant flow rate. Under this constraint, the pressure drop through the filter increases as the bed loading increases. When the pressure drop reaches a maximum of 6 psig, the filter automatically is taken off-line and backwashed. The filter is backwashed for 8 min, with the backwash directed to an on-line SBR.

pH ADJUSTMENT

The wastewater pH changes in the SBR during biological treatment. Effluent pH adjustment is made prior to sand filtration. The effluent pH is adjusted with hydrochloric acid or sodium hydroxide, as appropriate, to maintain the effluent pH at 7.

EFFLUENT DIFFUSER

The treated effluent is pumped to the river through a 1.75-mile pipeline for discharge through a diffuser. A two-arm multiport diffuser is installed in the river to disperse the final treated effluent. Each diffuser arm contains 30 3″ diameter nozzles on 1-m centers angled 60 degrees upstream. The arms are parallel with a 15-m separation. The maximum (design) flow is 800 gpm. The average flow is 426 gpm. The diffuser provides 1 : 68 mixing ratio of the effluent in the river at the "7Q10" low-flow conditions.

SLUDGE FILTER PRESS

At design flow and BOD load, approximately 75,200 gal/day of 3% waste sludge is produced. Mixing and conditioning of the sludge is carried out in a 150,000-gal waste sludge tank. The sludge is conditioned with ferric chloride. The optimum pH for dewatering the sludge is pH 6.3. Two plate and frame filter presses dewater the waste sludge, producing a filter cake. The waste sludge production at the maximum design conditions is 28,000 lb/day at 50% moisture content. Each press contains 72 30″ by 30″ plates. A complete press cycle is approximately 4 h. The filter presses are in an elevated structure to facilitate dropping the pressed cake directly into a roll-off dump box for off-site disposal.

OPERATING PERFORMANCE OF FULL-SCALE SBR BIOTREATERS

The WWTP was commissioned with seed biomass from a municipal wastewater treatment plant. The initial feed to the biotreaters consisted of several "low-risk" streams. "Low-risk" streams were those wastes that were known to be easily biotreatable and that were not likely to present a toxicity problem. To acclimate the biomass, the salt concentration in the biotreaters was increased at a rate of 2 g per liter per day. With this strategy, about 2 weeks were required before the biotreaters reached 3% salt and were able to treat the full wastewater flow (hydraulic load). After 1 month, additional streams were added one by one at about 2-week intervals. This approach was chosen to make troubleshooting of potential problems easier.

The treatment plant performance met the design objectives for removal of conventional pollutants, specific compounds, and toxicity. The BOD reduction was better than 90%, except for months 5–7, 11–13, and 17 when several abnormal conditions were encountered (Figure 14). Elevated BOD in month 5 was caused by a high concentration of suspended solids in the effluent. In general, ESS is less than 25 mg/L in the discharged effluent. The elevated BOD in months 6, 7, and 17 was caused by a deficiency of nitrogen. The laboratory treatability studies had indicated that there could be time periods when nitrogen would have to be supplemented to the feed. However, this nitrogen deficiency was much larger than anticipated. The elevated BOD continued because an insufficient quantity of nitrogen was supplemented. This led to the implementation of a respiration test based on measuring the oxygen uptake rate (OUR) at the end of the react cycle. The respiration test enabled the operators to determine whether any unreacted BOD was in the reactor before the effluent was decanted.

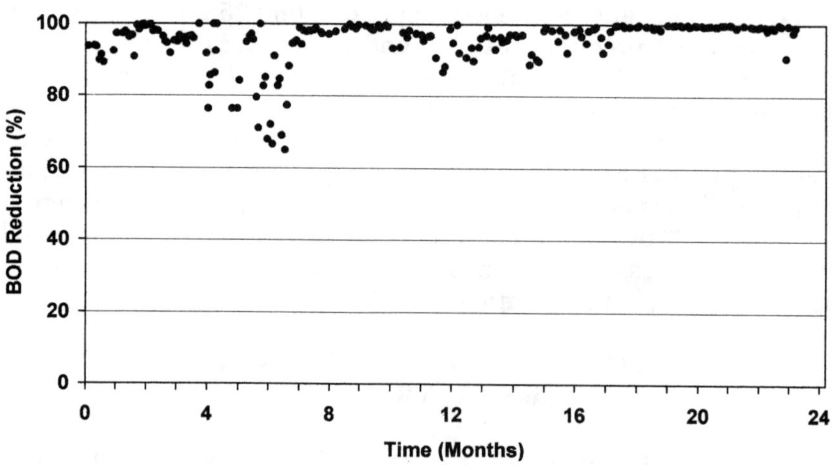

Figure 14. BOD reduction in full-scale biotreaters.

A different type of BOD anomaly was experienced during months 11–13. Whenever the biotreater feed contained a certain wastestream, the BOD of the effluent started to increase. Extending the reaction time did not correct the problem. Nutrient deficiencies were ruled out. Eventually, additional data suggested that the problem was related to solids entering the biotreater from this wastestream. The solids did not dissolve or fully degrade in the biotreater. However, in the more dilute environment and during the longer reaction time of the BOD_5 test, they slowly degraded, thereby contributing to the measured BOD of the effluent samples.

TOC removal in the full-scale biotreaters varied from approximately 60% to 90% (see Figure 15). The increasing and decreasing trends in TOC removal were due to several of the treatability problems that were encountered in the startup phase and to the changing waste mixture, which is typical of this multiproduct plant. Unfortunately, Study #1 (Figure 4) is not directly comparable to the full-scale since there was not a comparable waste mixture. There were some stretches of time when the waste mixture was fairly consistent (see months 18–23). Acclimation effects are difficult to detect since the waste mixture changes frequently, but it is suspected that acclimation can be seen at months 8–10 and 13–15.

Toxicity reduction in the full-scale biotreaters was quite effective, except for the treatment of one wastestream. Figure 16 shows the effluent toxicity to ceriodaphnia for the first 2 years of operation. The three episodes of elevated toxicity were due to the admission of the insecticide B wastestream, which had been treated successfully in the laboratory treatability tests. When this stream was first admitted in month 2 and then again after minor improvements in month 7, the elevated toxicity was immediately evident, and the process was shut down immediately. A more detailed inspection of the process revealed inadequate pretreatment equipment and procedures.

Substantial pretreatment improvements (i.e., solvent extraction, phase separation tank, and improved controls) were added to prevent phase inversions and reduce dispersed organics. Another excursion was seen at month 17 due to startup problems with the new equipment. These problems were quickly resolved. No toxicity was observed in the effluent when this insecticide waste pretreatment process was successfully operated during months 19–21 and later in months 27 and 28.

The removal of pesticides in the full-scale biotreaters was much improved when compared to the laboratory treatability tests, which are summarized in Table 1. Herbicide C was generally not detected after the first 6 months of operation. If it was detected, the concentration was

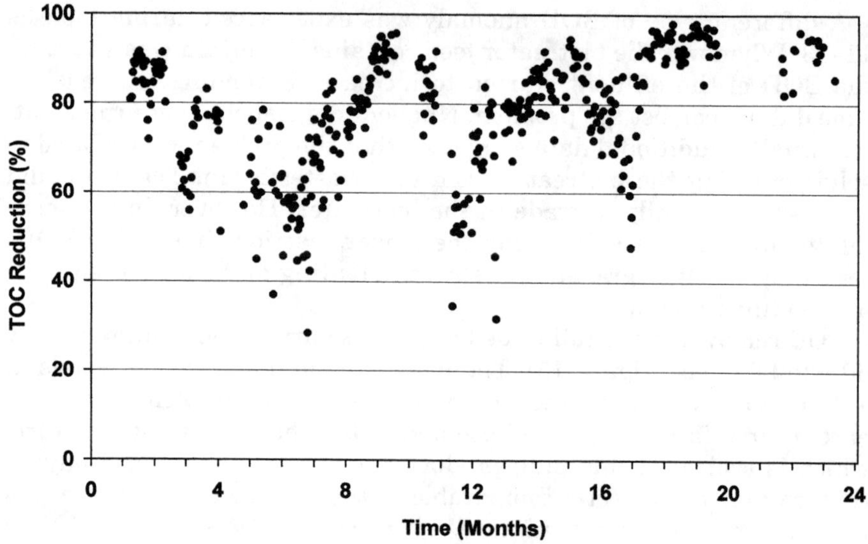

Figure 15. TOC reduction in full-scale biotreaters.

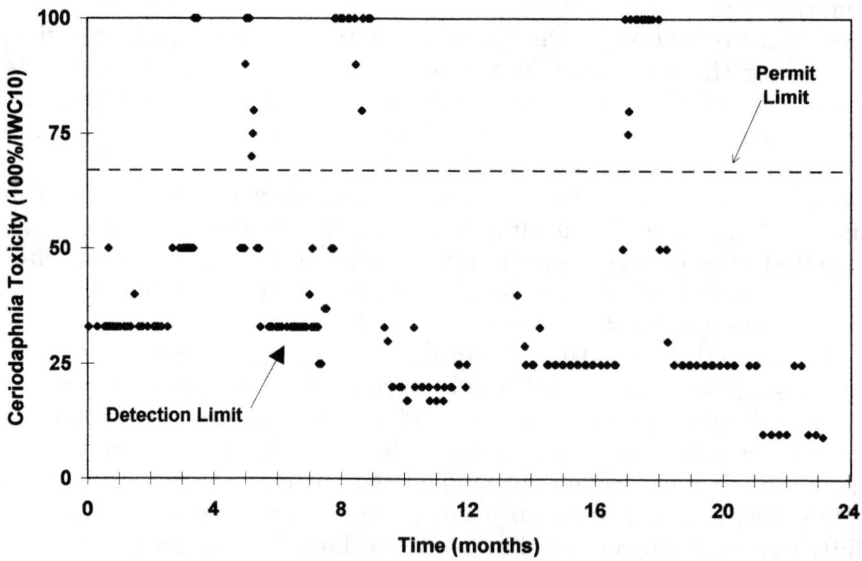

Figure 16. Effluent toxicity to *Ceriodaphnia* in full-scale bioeaters.

near the detection limit of 0.1 mg/L. Herbicide D was present in the effluent during the first 9 months at a concentration of a few milligrams per liter. Since then, it has generally not been detected, even though the concentration in the feed was actually higher than during the first 9 months (due to a higher production rate). It is believed that the improved performance was due to biological adaptation. Insecticide D was generally not detected in the plant effluent (detection limit is 0.2 ppb). The main reason for the improved performance with the insecticides was the implementation of the solvent extraction process in the production units (see Section 3.2). Other specific compounds, such as toluene, were also removed during biotreatment. Sludge-settling characteristics in the full-scale SBRs were as good or better than experienced in the laboratory. The sludge solids settled to a relatively dense concentration of approximately 2 wt%. The sludge volume index was typically about 50 ml/g.

TOXICITY REQUIREMENTS AND MANAGEMENT PROGRAM

Toxicity compliance is one of the main driving forces with respect to (1) the treatment plant design, (2) the management (and monitoring) program, and (3) the development of future products. Since permit toxicity requirements are site-specific and are based on the "use designation" of the receiving waters, the toxicity requirements of the permit will be described first. These requirements are the driving forces for the toxicity monitoring and management programs. The monitoring program measures the performance of the wastewater treatment plant with respect to the permit requirements. The toxicity management program provides a structure for overseeing the daily operation, future planning, and troubleshooting of the plant. The philosophy for managing toxicity is also discussed.

Toxicity Requirements

The production site operates under an NPDES permit with effluent discharge to the local river, a freshwater receiving body with a "Use Designation" of fishable and swimmable. In order to achieve a discharge that is consistent with the "Use Designation" and protective of the environment, it is necessary to biologically treat the effluent for the removal of conventional and priority pollutants and to meet the toxicity standard. For determining compliance with the toxicity standard, the regulating body used a mixing zone approach with the Cormix Diffuser Model.

Table 5. Effect of salt concentration on TOC reduction.

	3% Salt			6% Salt		
Feed Mixture	Feed TOC (mg/L)	Effluent TOC (mg/L)	TOC Reduction (%)	Feed TOC (mg/L)	Effluent TOC (mg/L)	TOC Reduction (%)
"C"	135	30	78	250	50	80
"D"	18	7	61	32	9	72
"E"	790	200	75	1460	1460	0

As previously discussed, the process wastes are concentrated salt (sodium chloride) brines in the 12 to 18% range. For biological treatment to remove conventional and priority pollutants, treatment at a lower salt concentration was necessary to avoid the inhibitory effects of the salt on the biomass. Treatability studies with these wastes showed that inhibitory effects were not seen at a 3% salt concentration. Inhibition was seen with some feeds at 6%. In one study, parallel biotreaters were sequentially fed three different waste mixtures (see Table 5). The TOC reduction in both SBRs was similar for composite wastes C and D. When a step change to composite E was made, the biotreater with 6% salt showed inhibition. In another study, the TOC degradation rate was measured at salt concentrations between 3% and 8%. Figure 17

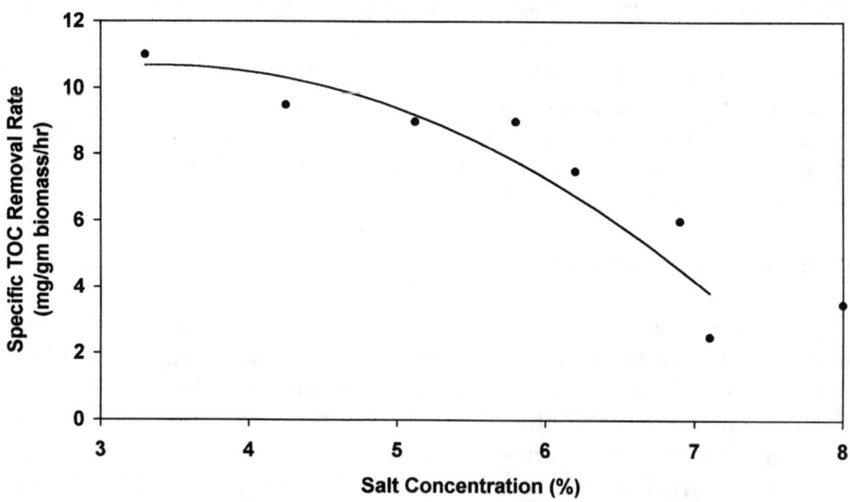

Figure 17. Effect of salt concentration on specific TOC removal rate. After a 6-day acclimation at each salt concentration, the specific TOC removal rate was the average TOC removal rate during the first 3 h of the react cycle.

shows that the specific TOC removal rate reached its maximum at a salt concentration of 3%. Based on these results, the biotreaters are operated at a 3% salt concentration.

The toxicity requirement is for ≥90% survival of the test organisms after a 48-h exposure at the instream waste concentration (IWC) occurring at the edge of the zone of initial dilution (ZID). Diffuser modeling studies determined that a 68 to 1 mixing ratio could be achieved in the river during the "7Q10" low-flow conditions. This resulted in a permit requirement that a 1.5% IWC be used for testing purposes. The toxicity test is conducted with 1.5% wastewater in a "standardized" dilution water matrix. The NPDES permit requires monthly bioassays on two freshwater species, the invertebrate ceriodaphnia (*Ceriodaphnia dubia*) and the fathead minnow (*Pimephales promelas*). Compliance bioassays are performed by a contract aquatic laboratory according to USEPA methods (Weber, 1993). At a 1.5% IWC, the salt concentration is below the no observable effect concentration (NOEC) for the test species (Meier 1993).

Toxicity Monitoring

Monitoring the toxicity of the laboratory-treated effluent was a concern during the development studies. Before permit conditions were set, two saltwater species, mysid shrimp (*Mysidopsis bahia*) and inland silverside (*Menidia beryllina*), were used to measure effluent toxicity from the laboratory biotreaters. Once the compliance organisms (i.e., *Ceriodaphnia* and fathead minnow) were specified, considerable experience was gained using these freshwater organisms to test effluent toxicity. Testing of laboratory-treated effluent showed that the ceriodaphnia was a more sensitive organism than the fathead minnow to these treated wastewaters. As a result, most toxicity screening was done with the ceriodaphnia.

To monitor toxicity in the laboratory treatability studies, and eventually for permit compliance at the production site, the capability for "in-house" testing using the ceriodaphnia was developed. This test had some limitations. It was time-consuming (i.e., 2 days) and could not be used to screen toxicity at an IWC greater than 4% due to the toxicity from the salt in the effluent. In order to overcome these limitations, an attempt was made to develop a faster, alternative bioassay with a more salt-tolerant species that would correlate with the ceriodaphnia test. Tests were conducted using a wide variety of toxicants with an invertebrate *Daphnia magna* (Daphnia IQ Test™), and saltwater bacteria (Microtox™). Both tests use luminescence as an endpoint, rather than lethality. However, both fast bioassays correlated poorly with the ceriodaphnia bioassay, although some success was seen by others with the *Daphnia magna* bioassay (Janssen and Persoone, 1993; Hayes, 1992). The only bioassay that could be used with complete confidence was the ceriodaphnia compliance test.

Currently, on-site toxicity monitoring is conducted using the ceriodaphnia test. Typically, monitoring is carried out at several effluent concentrations up to 4% IWC. Above that, organism "effects" can be observed due to the salt. Under normal operating conditions, samples from the individual biotreaters and the final effluent are tested for toxicity twice a week. Monitoring for toxicity at IWCs above the permit limit provides some advanced warning that toxicity may be developing in the effluent.

Philosophy for Managing Toxicity

The overall philosophy is to eliminate toxicity at the source, i.e., at the individual production units. In the development program, it was necessary to (1) identify the production units that had toxic wastestreams that were unaffected by biotreatment, (2) segregate the nontoxic from the toxic discharges within the process, and (3) add a pretreatment step in the production unit to remove toxic compounds. Table 6 shows that, for some pesticide wastestreams, toxicity was readily removed during biotreatment. An important aspect of choosing biological treatment was that biotreatment removed toxicity from a large portion of the waste from the production site without the need for a specific pretreatment. For the organophosphate wastes, toxicity was only slightly removed or reduced during biotreatment. Those streams required pretreatments to remove toxicity before entering the WWTP.

Experience showed that the biomass and PAC had a somewhat limited capacity to remove certain dissolved toxic compounds. For the

Table 6. Effect of biotreatment on pesticide wastewater toxicity.

Wastestream	*Ceriodaphnia* Toxicity before Biotreatment (TU)	*Ceriodaphnia* Toxicity after Biotreatment (TU)
Herbicide C brine	80, >200	<33
Herbicide D brine	80	<33
Herbicide E brine	<33	<33

TU: Toxicity Units = 100%/EC_{50}.

biomass and PAC to be effective in reducing toxicity, the toxic compounds either had to be readily degradable by the biomass or be adsorbable on the PAC. Chemical compounds such as some organophosphates and metals passed through biotreatment and were relatively unaffected by the presence of PAC. These compounds need to be removed by a pretreatment in the production units.

The biomass and PAC were also ineffective in removing BOD or toxic compounds when contaminated with an organic phase. When organics were present as a separate, dispersed phase, they tended to adhere to or coat the biomass. This dispersed phase was not easily removed, depending upon its volatility. Pilot plant studies showed that some organic compounds persisted at low levels and contributed to toxicity over several months when a separate organic phase was adhering to the sludge (e.g., Herbicide E and Herbicide J). In this case, the toxic compounds were being continuously released to the noncontaminated feed.

Some toxic compounds in the dissolved state were not effectively removed by biological treatment (given their slow degradation rates and the quantity of biomass present). When toxicants were present in the dissolved state, "washout" tests showed that these compounds were easily purged when nontoxic feeds were introduced (see Figure 11).

The treatability studies demonstrated that the biomass appeared to be robust for the pesticide wastes from this production site. Inhibitory effects from the pesticide compounds, separate phase organics, or suspended solids were not observed. The biomass generally continued to remove BOD, TOC, and TSS even when toxic compounds were present.

Toxicity Management Program

There are four aspects that comprise the toxicity management program: (1) pretreatment at the production units, (2) primary treatment of the combined wastes, (3) analytical monitoring, and (4) management for toxicity prevention and correction.

The pretreatments in the production units (Section 3) reduced toxic compounds before the waste brines were commingled in the primary treatment area. Specifications were developed for the various individual wastestreams such that they are treated as a product that is then delivered to the waste treatment area.

Primary treatment of the combined waste brines was necessary to remove organics and solids. These "insolubles" were not readily biodegradable and contributed to toxicity. The performance of the primary treatment area was monitored for removal of organics and solids by turbidity measurements and direct observations.

Regular analytical monitoring comprised the third major aspect of the toxicity management program. As previously discussed, compounds in the individual wastestreams were routinely monitored (see Section 3). The water for salinity adjustment was monitored for elevated levels of pesticides, which could come from stormwater run-off in the production units. The equalization tanks, biotreater decants, and effluent discharges were monitored for pesticides and other compounds.

The biotreater decants and the final effluent were analyzed for toxicity with the ceriodaphnia test. Toxicity was screened at higher IWCs than specified by the permit. The final effluent was also analyzed for a variety of other compounds, as required by the NPDES permit. The Quality Assurance Laboratory employed a full-time chemist to support the biotreatment operation. Additional tests were also performed by the operators in the wastewater treatment plant and at the individual production units.

The final aspect of the toxicity management program was the "managed" response to a toxicity excursion. The objectives of the "response" were to (1) quickly identify the source of the toxicity and isolate the stream, (2) rectify the problem, and (3) minimize the effects of the toxicity by manipulating process variables.

In order to minimize the extent of the toxic event, the source must be quickly identified. A decision tree, called the toxicity response plan, was developed. Figure 18 graphically illustrates the steps taken when responding to observed toxicity. The three main "tools" used to identify the source of the toxicity are direct physical observation of the wastestreams, analytical monitoring, and a shake flask screening program. These three investigative techniques are used simultaneously to identify the responsible stream as quickly as possible. This approach can identify the source of the toxicity but not necessarily the responsible chemical compound. If the source stream is not easily identified, then additional studies may be necessary, i.e., a TIE and/or TRE.

Inspection of the wastestreams at various points in the process can provide qualitative information concerning the quality of the waste and the performance of the waste pretreatment and primary treatment systems. Very often, a simple visual inspection can reveal which wastestream does not look normal and is experiencing an upset. This inspection step is highly valuable but can be easily overlooked.

Routine analytical monitoring for pesticides and their intermediates can often confirm the presence of known toxicants. The baseline levels for the pesticides and intermediates are usually well below the NOEC.

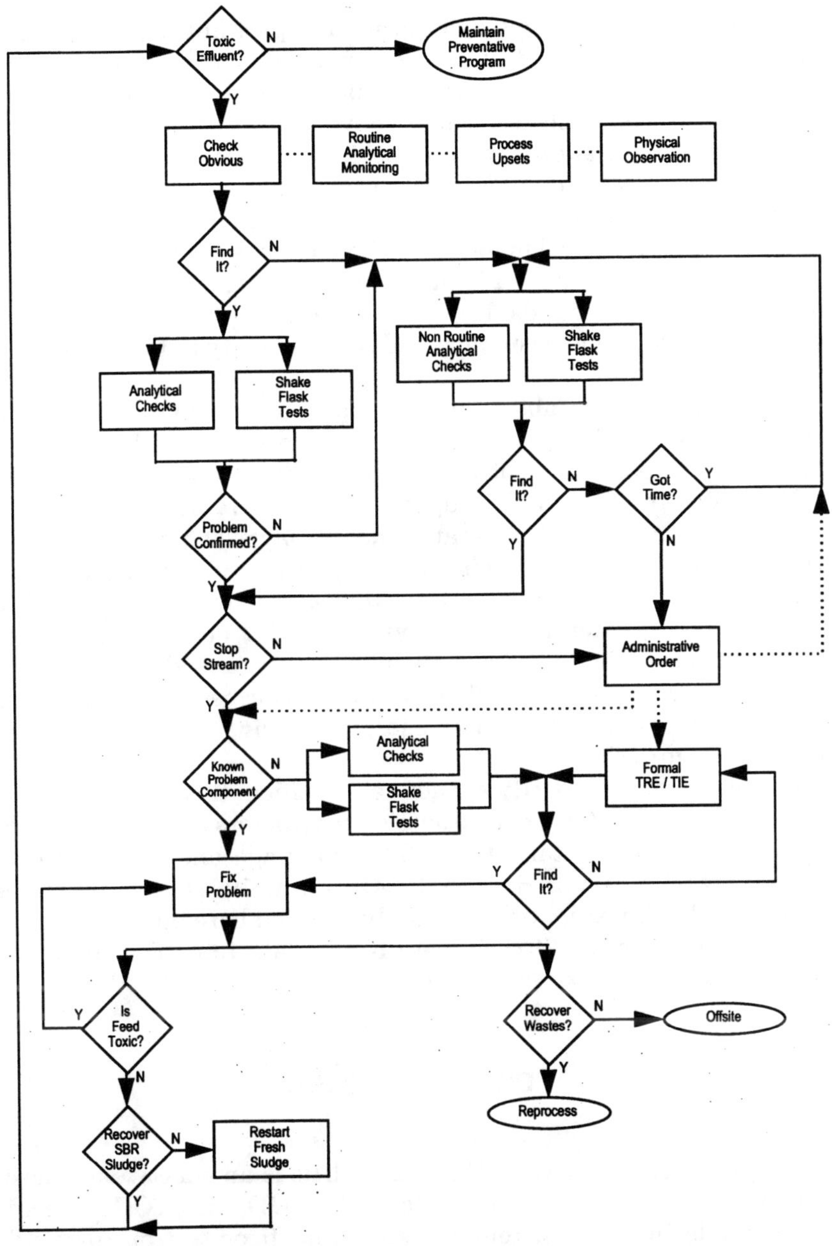

Figure 18. Toxcicity response plan.

Levels above that baseline may indicate an upset in the production or pretreatment unit. An additivity toxicity model was developed to predict toxicity based on the analytical results for the levels of the known toxicants. This approach "checks the obvious."

When necessary, additional nonroutine analyses are initiated for heavy metals, new unidentified chemical compounds, toxic solids, organics on the sludge, etc. Since there are many other possible contributors to toxicity, analytical monitoring is not always successful in identifying specific toxic compounds or potential antagonistic effects from combinations of compounds. However, when these techniques are successful, they may avoid a complex and costly toxicity reduction evaluation (TRE).

Several process variables can be manipulated to minimize the effects of observed toxicity: hydraulic retention time (HRT), sludge age, and PAC dose. The biotreaters were usually operated with a 4-day HRT. Treatability studies suggested that there was no additional toxicity reduction for HRTs greater than 4 days. The sludge age can be reduced by increasing sludge wasting to purge toxicity adsorbed on the biomass and PAC. Typically, the biotreaters are operated at sludge ages of greater than 15 days. The PAC dose can be raised to provide more carbon to adsorb the toxic compounds. Before increasing the PAC dose, shake flask tests are performed to determine if PAC reduces toxicity. In some cases, increasing the PAC had no effect (see Figure 2).

If the source of the toxicity cannot be identified, then a TRE may be necessary. It can be costly to hire outside contractors to perform a TRE. The evaluation could start in the $30,000 range and go up to over $100,000 (Martin, 1992; Hanson 1994; Smith, 1994). Details of the nature of a TRE investigation will not be covered here but are summarized elsewhere (Ford, 1992; Norberg-King et al., 1991; Durhan et al., 1993; Mount, 1989).

CONCLUSIONS

1. Biological treatment with PAC using SBRs is an effective means of removing conventional pollutants (TSS, BOD, and TOC), specific compounds (pesticides, toluene, etc.), and toxicity from pesticide manufacturing wastewaters. However, biological treatment is not a stand-alone process but requires pretreatment of some individual streams and primary treatment of the combined wastes.

2. Primary treatment of the combined waste brines is necessary for the removal of solids and dispersed organics in order to meet toxicity and BOD requirements. Dispersed organics and solids were shown to directly contribute to toxicity. Certain solids were found to elevate BOD_5 when carrying through to and slowly degrading during the BOD_5 measurement.
3. Pretreatment at the individual productions units are a cost-effective means of removing toxicants that are not biotreatable or carbon adsorbable. Pretreatment can also remove compounds that are recycled to the process, thereby increasing product yield. These yield advantages were demonstrated in several cases.
4. A toxicity management program is essential to staying in compliance. One key aspect of the program is to identify toxicants and their source before toxicity manifests itself in the effluent. This is accomplished with the aid of on-site toxicity testing.

FUTURE WORK

Future efforts will focus in three major areas. There will be strides to improve reliability in the pretreatment and primary treatment process areas to insure effective biological treatment. There will also be an effort to reduce individual pesticides at the production units through pretreatment improvements and through optimization of the biotreatment processing variables. Finally, treatability studies will be undertaken for the introduction of new wastes from new products.

ACRONYMS

BOD biological oxygen demand, mg/L

BOD_5 5-day biological oxygen demand, a standardized test, mg/L

EC_{10} effect concentration; concentration of test sample in the dilution water matrix where 10% of the test species are killed or "effected" %

EC_{50} effect concentration; concentration of test sample in the dilution water matrix where 50% of the test species are killed or "effected" %

ESS effluent suspended solids, mg solids/liter

F/M food to mass ratio, mg BOD/mg biomass/day

HRT hydraulic retention time, days

IWC10 instream waste concentration; concentration of test sample in the dilution water matrix where 10% of the test species are killed or "effected," %
MGD million gallons per day
NOEC no effect concentration
NPDES national pollutant discharge elimination system
OTR oxygen transfer rate, mg/L/hour
OUR oxygen uptake rate, mg/L/hour
PAC powdered activated carbon
RCRA Resource Conservation and Recovery Act
SBR sequencing batch reactor
TIE toxicity identification evaluation
TOC total organic carbon, mg/L
TRE toxicity reduction evaluation
TSS total suspended solids, mg solids/L
TU_{10} toxicity unit defined as 100%/EC_{10}
TU_{50} toxicity unit defined as 100%/EC_{50}
USEPA United States Environmental Protection Agency
WAO wet air oxidation
WWTP wastewater treatment plant
ZID zone of initial dilution

ACKNOWLEDGEMENTS

The authors wish to gratefully acknowledge the advice and assistance of: Prof. Robert L. Irvine for consultation services, Dr. Jonathan P. Earhart for his technical assistance in pretreatment, Markus Meier for his expertise and advice in the area of toxicology, and Steve W. Duym and his team who operate the production site WWTP.

REFERENCES

Durhan, T.J., Norberg-King, T.J. and Burkhard, L.P. eds. 1993. *Methods for Aquatic Toxicity Identification Evaluations: Phase II Toxicity Identification Procedures for Samples Exhibiting Acute and Chronic Toxicity.* Duluth, MN: USEPA EPA/600/R-92/080, pp. 1–71.

Flynn, B.P. 1975. *Proceedings of the 30th Industrial Waste Conference,* May 6–8, 1975, Purdue University, Lafayette, Indiana. pp. 233–252.

Ford, D.L., ed. 1992. *Toxicity Reduction Evaluation and Control.* Lancaster, PA: Technomic Publishing Co., Inc.

Hanson, S.R., and Assoc., personal communication, 1994.

Hayes, K.R. 1992. "Fluorescent Aquatic Bioassay and Procedure," U.S. Patent No. 5,094,944; March 10, 1992.

Irvine, R.L. and Ketchum. 1989. "Sequencing Batch Reactors for Biological Wastewater Treatment," *CRC Critical Reviews in Environmental Control,* 18(4):255–294.

Janssen, C.R. and Persoone, G. 1993. "Rapid Toxicity Screening Tests for Aquatic Biota. I. Methodology and Experiments with *Daphnia magna.*" *Environmental Toxicology and Chemistry,* Vol. 12:711–717.

Martin, A.M. 1992. "Use Biomonitoring Data to Reduce Effluent Toxicity," *Chemical Engineering Progress,* September, 43–49.

Meier, M. 1993. Personal Communication, Company confidential.

Mount, D.I., ed. 1989. *Methods for Aquatic Toxicity Identification Evaluations: Phase III Toxicity Confirmation Procedures.* Duluth, MN: USEPA EPA/600/3-88/036, pp. 1–26.

Norberg-King, T.J., Mount, D., Durhan, E., Ankley, G.T., Burkhard, L., Amato, J., Lukasewycz, M., and Anderson-Carnahan, L. eds. 1991. *Methods for Aquatic Toxicity Identification Evaluations: Phase I Toxicity Characterization Procedures* (Second Ed.). Duluth, MN: USEPA EPA/600/6-91/003, pp. 1–84.

Schultz, J.R. and Keinath, T.M. 1984. "Powdered Activated Carbon Treatment Process Mechanism," *Journal of the Water Pollution Control Federation,* 56(2): 143–151.

Smith, D.S. 1994. Personal Communication.

Sublette, K.L., Snider, E.H. and Sylvester, N.D. 1982. "A Review of the Mechanism of Powdered Activated Carbon Enhancement of Activated Sludge Treatment," *Water Res.,* 16: 1075–1082.

Thibault, G.T. Tracy, K.D., and Wilkinson, J.B. 1977. "PACT Performance Evaluated," *Hydrocarbon Processing,* 56 (5): 143–146.

Weber, C.I., ed. 1993. *Methods for Measuring the Acute Toxicity of Effluents and Receiving Waters to Freshwater and Marine Organisms.* Cincinnati, OH: USEPA EPA/600/4-90/027F, pp. 57–62.

Index